Control of
Color
Imaging
Systems

Analysis and Design

Control of
Color
Imaging
Systems

Analysis and Design

Lalit K. Mestha

Sohail A. Dianat

CRC Press
Taylor & Francis Group
Boca Raton   London   New York

CRC Press is an imprint of the
Taylor & Francis Group, an **informa** business

CRC Press
Taylor & Francis Group
6000 Broken Sound Parkway NW, Suite 300
Boca Raton, FL 33487-2742

First issued in paperback 2017

ISBN 13: 978-1-138-11227-8 (pbk)
ISBN 13: 978-0-8493-3746-8 (hbk)

Library of Congress Cataloging-in-Publication Data

Mestha, L. K.
    Control of color imaging systems : analysis and design / authors, L.K. Mestha and Sohail A. Dianat.
        p. cm.
    "A CRC title."
    Includes bibliographical references and index.
    ISBN 978-0-8493-3746-8 (alk. paper)
    1. Imaging systems--Automatic control. 2. Color display systems--Automatic control. 3. Digital printing--Automatic control. 4. Image processing--Digital techniques. 5. Color printing. I. Dianat, Sohail A. II. Title.

TK8315.M47 2009
681'.62--dc22   2008054552

Visit the Taylor & Francis Web site at
http://www.taylorandfrancis.com

and the CRC Press Web site at
http://www.crcpress.com

Dedication

This book is dedicated to
our families, Suhan, Savan, and Veena Mestha,
Ahrash Dianat, and Mitra Nikaein

Contents

Preface

Digital color printing technology offers many new avenues and opportunities for rendering color pages on demand at a lower run cost as compared with conventional printing. Many digital printers are based on electrophotographic technology, which is used in the process of laser printing. In this process, the image content controls the amount of light that selectively discharges a uniformly charged photoreceptor material with a laser or light emitting diodes to form an image. The electrostatic image is then developed with a thermoplastic powder containing charged pigment that is transferred and fused to paper under heat and pressure. The quality and productivity issues of these devices are addressed using a variety of new technologies including optical sensing, imaging, and closed loop feedback controls. Since the process tends to vary more over time, print quality defects and subtle variations in output are more noticeable in color printing than in monochrome printing. Also, color is rendered on different imaging devices (variety of printing and display devices) with varying color capabilities. As a result, the ability to produce accurate and pleasing color across numerous output devices is extremely complex and, to some extent, impossible. This is further made difficult due to variation in the workflow requirements, stocks and environment.

Practical feedback control systems used in digital production printers touch on a range of interconnected subsystems. Although there are numerous digital printers serving today's market, many new challenges must be overcome to improve the output quality and enable further growth and opportunity. At a system level, a good basic understanding of the end-to-end color printing process may be helpful and at a subsystem level, theoretical knowledge of the physical printing process, device technology, and the principles associated with electronic imaging. Someone new to the field, who could be a researcher, a practitioner, or a student, should have a good foundation in the major disciplines used for managing & controlling color in digital production printers. Unlike off-set printers, digital printers offer many new actuators that give control engineers access to numerous process steps (e.g., laser intensity, toner mass, CMYK primaries etc.) for developing real-time, closed-loop algorithms and architectures that can be self-tuned using various sensors along the print path and self-corrected for process variations and uncertainties.

Although an extensive number of patents and conference papers have been written on the control of digital color printing, to the best of our knowledge, no design books have been written on this subject matter. This book brings together the numerous complex disciplines associated with digital color printing and presents a technical story with proper mathematical rigor and design examples. The objective of this book is to provide a good understanding of fundamental techniques and push the frontier of this field. We have attempted to keep the mathematics at a moderate level. So, it can also be used as a textbook for an introductory course in printing, digital control, digital imaging systems, color management and control, color printing etc., at a senior level or at a first-year graduate level. In our opinion, it is also suitable for a graduate level course in imaging and computer applications. Additionally, this book can serve as a reference for instructors or a self-learning book for practicing engineers.

The material covered in this book is based on our 27 years of combined experience in applying control algorithms to modern digital printing systems, where precision color controls and sensing are needed to achieve offset quality color images. This book also captures some of our combined experience of over 30 years in teaching fundamental courses in engineering, especially in the areas of digital signal/image processing, modern control theory, color, and applied linear algebra. This book largely discusses the imaging systems developed over recent years, which have appeared as products for automating color consistency in digital presses. They use algorithms and capabilities far beyond standard printing and copying machines.

It is assumed that the readers have a basic understanding of color, linear system theory (both continuous and discrete), including topics like convolution and Fourier analysis. The contents of the book are outlined as follows: Chapter 1 provides an overview of digital printing systems, workflows of data, digital front-end system, and the major elements in an end-to-end production print path. Chapter 2 covers some fundamental topics, which form the cornerstone of digital image processing, such as digital image formation, sampling and quantization, image coding, image transform, optical and modulation transfer function, and image de-noising. Chapter 3 covers the mathematical tools needed for the analysis and design of feedback control systems. Topics such as differential and difference equations (DE), the numerical solution of DE, the z-transform and its properties, the relationship between the z-transform and the Laplace transform, and the pulse transfer function are described. Chapter 4 deals with state variable techniques used to analyze continuous and discrete linear systems. State variable representation, state transition matrix, the solution of state equations, controllability, observability, and stability of linear systems are major topics discussed in this chapter. Techniques for closed-loop linear system analysis and design are covered in Chapter 5. Design techniques such as pole placement for single-input single-output (SISO) as well as multi-input multi-output systems (MIMO) and linear quadratic regulator design are among the topics that are covered in this chapter. Chapter 6 explains different techniques used for interpolation of multidimensional functions. Techniques such as trilinear interpolation, tetrahedral interpolation, sequential linear interpolation, Shepard, moving matrix, iterative clustered interpolation (ICI) algorithm, dynamic optimization and 1-D, 2-D, 3-D smoothing algorithms are described. Chapter 7 covers the 3-D control of color management systems with International Color Consortium (ICC) profiles generated using gray-component replacement (GCR) constraints, control-based inversion and control-based gamut mapping approaches. In this chapter, various empirical and first principle based Neugebauer models are covered in detail. In Chapter 8, the basics of 1-D and 2-D tone control of color management systems are discussed. State variable representation of imaging systems, control-based 1-D and 2-D tone management, and spot color control methods are presented. In Chapter 9, a detailed modeling and analysis of internal closed-loop process controls, closed-loop controller design for charging and development systems are presented. Current best practices used in toner concentration system modeling with time delay & controller design for time delay systems are covered. Optical sensing methods used for measuring various process outputs (e.g., toner density measurements on photoreceptors), including measurement of spectral functions of color are considered beyond the scope of this book. Components of a digital laser printing system, charging, exposure, development, transfer, and fusing subsystems are modeled in Chapter 10.

The control theory and methods presented in this book are state-of-the art for color printing systems. We deliberately limit the control theory to MIMO pole placement, state feedback and linear quadratic regulator design so that the mathematics is reasonably simple and can be taught at a senior undergraduate or a first year graduate level class in color, imaging, control or computer engineering disciplines. Formulations and illustrations presented emphasize simplicity so that the readers can easily understand the concepts and use them in their systems in the same way we did for high-end printers. Emphasis is on essential theoretical design principles and algorithms needed to build high quality, accurate, and offset-like output at lower cost through automation. While constructing end-to-end system models with process and subsystem parameters, we tried to capture the meaningful & essential behavior of subsystems in terms of parameters accessible for designing control systems. System models are presented in Chapter 10 to provide a nonlinear simulation platform for students, faculty and practicing engineers to explore more advanced approaches for designing future imaging systems that compare and contrast with experimental data when they become available. Tuning of model parameters is a rich area for applying modern system identification methods. Our approach shown in Chapter 10, although elementary, is considered useful to bring some reality to simulations since most control systems designed today are first simulated before a real test is done. We hope this book will bridge the gap between current and future theoreticians and practitioners, as well as generate new ideas, algorithms and methods in imaging systems to achieve full autonomy.

Lalit K. Mestha
Sohail A. Dianat

Acknowledgments

I would like to thank Graham Rees of Rutherford Appleton Laboratory, Professor Richard Talman of Cornell University, Bob Webber of Fermi National Accelerator Laboratory, and Professor Kai Yeung of the University of Texas at Arlington for injecting the seeds of physics and controls into my mind before I moved on to Xerox Research and realized the need for sophisticated learning when designing large scale physical systems.

It was Dr. Charles B. Duke who brought me to Xerox Corporation while I was looking for a new job in Texas and pondering my next big challenging control application. Thanks to the support of his staff, he exposed me to the complex world of digital color printing. Therefore, before I go any further, I would like to express my most sincere gratitude and appreciation to Dr. Charles B. Duke. If it wasn't for him, my research would have been different and the book would have been written on a different topic.

I would like to acknowledge and express special gratitude to my Senior Management, Sophie Vandebroek, Steve Hoover and Steve Bolte, who always gave me the support I needed for carrying out the research and producing the text book, from approval to completion, while trying to meet their business objectives. Several Industry-University collaborations were initiated since I started writing the book in 1996: a National Science Foundation's Grant Opportunity for Academic LIason (GOALI), New York State's Center for Electronic Imaging Systems (CEIS), one of 15 NYSTAR sponsored Centers for Advanced Technology (CATs) in the area of control theory and electronic imaging. These academic relationships led to many important research results. Industry-University interactions and teaching as an adjunct professor at the University of Texas at Arlington and the Rochester Institute of Technology (RIT) greatly improved my ability to internalize and leverage existing methods and algorithms for real world system applications. Multi-dimensional smoothing, dynamic optimization, linear MIMO state feedback, state variable methods, MIMO pole-placement design, input-output experimental processes, principal component analysis, singular value decomposition, optimal linear estimators, design approaches for time delay systems, anti-windup compensators for saturation, and system identification are just a few of the methods that are now routinely used in our research regarding next generation systems. Applying all of these methods and algorithms is truly a great success for me, and I would like to acknowledge the support of all the faculty members and students who worked with me for all of those years.

I would like to recognize my managers, Tracy E. Thieret, Michael R. Furst, Debbie Wickham, Kenneth J. Mihalyov, William J. Hannaway, Norm W. Zeck, and Lisa Purvis, as well as Peter A. Crean, a senior research fellow, who have supported me at various stages in my research career while developing my knowledge in color and xerographic systems, and working with product development groups so that our research led to the creation of value for our customers. I would like to especially thank Peter S. Fisher and Paul A. Kaufman of the Xerox Special Information Systems group for supporting the early injection of our recent research solutions into products.

Lalit K. Mestha

We would like to thank CEIS and Xerox Corporation for supporting our research in the areas of control for imaging and printing applications over the past ten years. Special thanks goes to Prudhvi Krishna Gurram, a PhD student at RIT, who not only got his MSEE degree while working on the CAT funded research project, but also contributed to the tuning of printing system models and read and edited the entire manuscript. We would also like to thank Bruce Brewington, our student intern at RIT, who helped us in developing a tensor based multidimensional smoothing algorithm that is covered in this book. Nikolaos M. Freris and Kunal Srivatsava of University of Illinois at Urbana-Champaign also worked in the early stages with us while developing the Printing System models. Very special thanks to Palghat Ramesh, a Principal Scientist at Xerox Corporation, without his efforts and guidance, we would not have created the material for Chapter 10. Finally, it is Jack G. Elliot, our colleague at Xerox Research, who carefully edited the proof line by line. Without his focused effort and numerous suggestions, many errors may have gone to press unnoticed.

Writing a book of this magnitude is a major undertaking and producing the material for writing it is even harder. A special thanks to all the administrative staff, scientists, engineers, technicians, and developers at Xerox Corporation who contributed to the growth of the technology with us over the years and made it useful to our customers. Special thanks goes to Martin S. Maltz, a Principal Scientist at Xerox Research, whose experience and intuition helped us to understand color, color transforms and ICC profiles particularly well.

We thank Barbara Zimmerli, who greatly helped with the administrative support while completing the final manuscript.

Lastly, we would like to thank our families, Suhan, Savan, and Veena Mestha, Ahrash Dianat, and Mitra Nikaein, for their support and understanding when dedicating our personal time on this project.

Lalit K. Mestha
Sohail A. Dianat

1 An Overview of Digital Printing Systems

1.1 INTRODUCTION

The printing industry includes a number of segments [1] including commercial printing and publishing. The integration of computing, imaging, and controls technology has enabled advanced digital color printing and publishing systems for the office and production.

In this chapter, we describe some of the key elements of an end-to-end digital printing and publishing system used for the production of high print volumes and the management of complex print jobs. This overview will help the reader to better understand the functional and processing system-level components involved when designing optimal printing systems.

1.2 PRINTING AND PUBLISHING SYSTEM

Printing and publishing is a large industry composed of many shops, that vary in size. These shops use equipments based on a variety of printing methods. Lithography, letterpress, flexography, gravure, and screen-printing use plates or some other form of image carrier, and digital or electronic printing such as electrostatic or ink-jet is plateless. Lithography, often called "offset printing," is the dominant printing process in the industry. Flexography produces vibrant colors with little rub-off qualities, valued for newspapers, directories, and books. Gravure's high-quality reproduction, flexible pagination and formats, and consistent print quality are used in packaging and printing of periodicals.

In offset printing presses, the press control system controls and monitors the ink, water, and print registration systems, and often robots are used to move parts in and out of the presses in print shops [2]. Unlike digital printers, the traditional offset press does not allow the changing of pixels on page boundaries while printing. This limitation brings new challenges and opportunities to the digital printing and publishing value chain.

Some publishing processes became digital for several reasons: (1) variable information electronic documents containing fragments of text, graphics, and images from either the electronic or the scanned input stream can be merged, edited, and assembled into laid-out pages forming a complete job; (2) the printing technology can handle digital stream of data; and (3) the benefits of digital data brought additional value, for example, 'print-on-demand'. This led to variable data printing with a lower cost short run, personalization, and versioning, that is difficult to create with offset printing. Understanding some of the key steps involved in the printing and publishing workflow can shed some light on the complexity of the system.

Workflows (various steps required all the way from receiving the orders in a print shop to the production of a job in finished form) are generally unique to each print shop. The typical workflows used in both areas of printing (digital and offset) can be divided into three main components: (1) business management, (2) output production, and (3) process management/supporting functions (see Figure 1.1 for a block diagram view of the workflow process). It includes not just the actual production steps, but all the necessary supporting tasks like billing, archival, etc.

1.2.1 BUSINESS MANAGEMENT

Business management tasks (the top portion of Figure 1.1) involve taking orders from customers, and assigning job-tracking numbers for monitoring purposes. Order information may include artwork, text, illustrations, design/layout, variable information rules, demographic data, etc. Pricing is estimated for the order and compared with the actual cost of running the job. Aspects related to handling the shipment of the finished job, archiving, and billing are an integral part of business management functions. Customers and sales/service representatives are involved at the order-taking stage. A project manager handles the job tracking and billing issues.

1.2.2 OUTPUT PRODUCTION

After the order taking step, customer data flow into the output production stage (the middle portion of Figure 1.1). This category can be further divided into creative, prepress, print (press), finish, and fulfill stages. For high-volume, single-shipment, sheet-oriented documents, offset presses are used. Customer data is directly sent to the prepress area. For low-volume, multishipment, and variable information documents, the customer data is processed in the creative stage before sending it to prepress. In the creative stage, concepts and drawings are developed; documents are designed by assembling the content using various layout tools. Image capture from scanner or other document input devices are done in this stage.

Electronic documents are then sent to the prepress area for further processing and assembly. Since documents could be of various formats and color spaces (RGB/CMYK), proper design choices are required before converting the documents to the language required by the raster image processor (RIP), which converts a document's strings of character codes to pixels. Generally speaking, the prepress stage (see Figure 1.2) encompasses all the steps involved in creating a digital electronic master. In a typical prepress system, multiple workstations are networked together to serve as the publishing desktop for generating, editing, managing, manipulating, and integrating multimedia content. Scanners are usually connected to the workstations to convert hardcopy documents (photos on film, paper handouts, etc.) to electronic form by utilizing a variety of image scanning software packages. Similarly, digital still and/or video cameras offer the user the ability to capture single snapshots and/or video imagery. Once captured, the images—stored electronically in

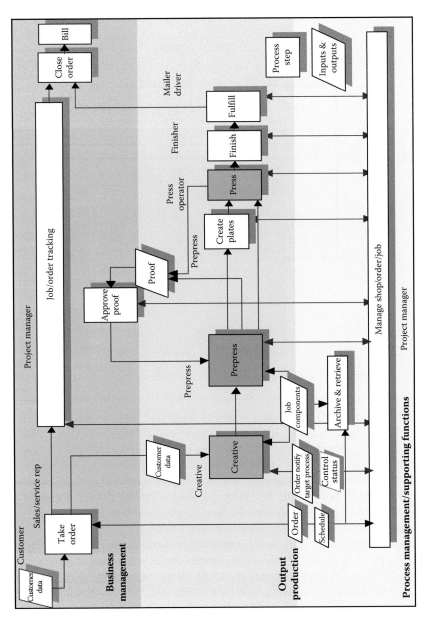

FIGURE 1.1 An illustration of the typical print shop workflow.

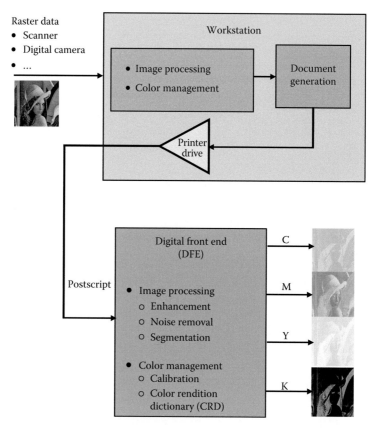

FIGURE 1.2 High level steps in the prepress area and DFE for digital printing.

a raster format (JPEG, TIF, etc.)—can be integrated into documents using a variety of desktop publishing software packages such as Microsoft Office (Word, PowerPoint, etc.), Quark XPress, Freehand, and so on. Defining spot/process colors, checking for fonts, setting the right screens (halftone dots), inserting/editing multiple pages are done in the prepress area. Supported stocks are entered to meet the customer order. Trapping is done to the documents to compensate for the small amount of misregistration in the printing system. Without trapping, unnecessary white gaps may appear between two colors that are supposed to be touching.

The assembled job is then sent for proofing without actually printing it on the press. The proofing is sometimes done on another digital printer and often on a well-calibrated monitor. If problem pages are found within the proofing stage, the job is further checked to handle corrections in the prepress area. Most documents contain a mixture of vector and raster type data. Vector data includes line drawings, arrows, boxes, line art, etc., as created by applications like Microsoft Word. Raster data, on the other hand, are usually generated by a scanner or digital camera (still or video). Once the documents have been designed on the desktop, they are assembled into jobs, evaluated for color quality, and previewed/proofed on a monitor or on a workgroup digital printer. After the proofs are approved for content, color, format, and quality the job is submitted to the print queue. For high-volume offset jobs, digital documents are imaged on the plate. For digital printing, the jobs are submitted to the press—in page description language (PDL) format such as postscript (PS) or PDF—through the digital front end (DFE) for further processing and printing. In the finishing (post-press) stage, the printed documents take on their final form through cutting, folding, collating, bundling, stapling, and/or packaging operations prior to shipment to customers. The finishing work has a major impact on the final product's quality, which may be a book, folded/collated sheets, booklets nicely cut and bound, brochures, etc.

1.2.3 PROCESS MANAGEMENT

Process management (the lower portion of Figure 1.1) tasks include scheduling production, performing process-engineering functions, and organizing file folders and servers related to all jobs components and job archivals. Job notification in the event a new job arrives for production, job intervention, and customer communication handling is another key task executed in this step.

1.3 DIGITAL FRONT END

The printing stage normally contains: (a) a DFE (see Figure 1.2) and (b) a print engine (see Figure 1.3). Unlike the workstation, where processing by the user may be independent of the print engine, a DFE or a network of DFEs from multiple vendors are used to convert the electronic "master" documents or job (through a series of image processing applications such as trapping, segmentation, rasterization, color management, image resolution enhancement, and antialiasing) to a form cyan (C), magenta (M), yellow (Y), and black (K) that is specifically designed and optimized for a particular digital printing system [3]. Multidimensional, industry standard source profiles are used to transform RGB images to a device-independent form like $L^*a^*b^*$ or standard Web offset printing (SWOP) CMYK files to device-independent form. In some cases, the transformation may be direct between device-specific forms to printer-specific form. To this effect, the input document is transformed from its PDL format such as PS or PDF, tiff, etc., to CMYK color separations to be printed by the engine. For postscript images, this is done by first utilizing an interpreter (e.g., PS interpreter) to identify the commands found in the PDL. An imaging module then generates a rasterized format of the PDL document at the correct resolution (e.g., 600 dpi). The above is usually referred to as RIP. During the RIP, color profiles (e.g., International Color Consortium [ICC]) comprising of multidimensional lookup tables (LUTs) are applied that transform the color from RGB to CMYK separations. Some DFEs employ object-oriented rendering (OOR) algorithms intended to enhance the color reproduction by utilizing custom profiles for specific image objects such as a "skin" profile for fleshtone or a "sky" profile for a blue sky background. For OOR to be effective, segmentation algorithms must be utilized to identify the objects of interest. Some DFEs also use trapping to mask registration errors; it is a

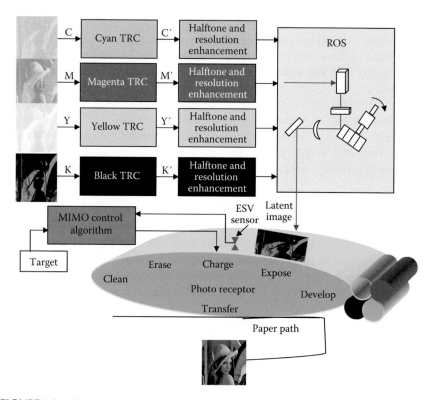

FIGURE 1.3 Key processing elements in the DFE and the electrophotographic (EP) process.

part of the rendering process. Trapping parameters are specific to a device and its colorant set.

Once the RIP is complete, the input job is transformed from a PDL format to CMYK separations ready for engine consumption. The separations are usually generated at the engine resolution (e.g., 600 dpi) for a given paper size (e.g., 8.5 × 11 in.), where each separation is made up of 8 b/pixel. Dependent on the option selected by the user, DFEs may RIP a given input document to a higher resolution (e.g., 1200 dpi) than what the printer is designed to handle (e.g., 600 dpi) and then subsample it to the appropriate printer resolution using standard techniques. This sequence, of course, results in a slight loss of sharpness but reduces aliasing effects.

1.4 DIGITAL PRINT ENGINE (ELECTROPHOTOGRAPHIC)

The print engine—sometimes referred to as the "marking engine"—is designed to convert the electronic CMYK separations provided by the DFE into hardcopy color prints. Figure 1.3 illustrates a typical digital press or printing system based on the principles of electrophotography (EP) invented by Chester Carlson in 1938 [4]. Unlike offset presses, digital print engine technologies are still evolving to improve image quality, productivity, and substrate latitude. Although ink-jet printing is another digital technology that is architecturally simple, it has its own limitations regarding volume and speed. Kodak stream ink-jet technology is a continuous system that reportedly enables offset caliber reliability, productivity, cost, and quality with the full benefits of digital printing for high-volume commercial applications. On the other hand, the EP process is utilized today as a key technology for high-volume full digital color printing in which four primary colors (cyan, magenta, yellow, and black) are developed by architecting the engine in six basic xerographic process steps: charge, expose, develop, transfer, fuse, and clean.

The toner-based digital printing process involves a circulating photoreceptor (PR) in the form of a belt or a drum. The PR is light sensitive; it is insulating in the absence of light and conductive when light is present. The first step in the EP process is "charging," where a high voltage wire deposits electrons or ions on the PR in the dark causing a uniform charge buildup. The CMYK separations, provided by the DFE, are then utilized to selectively expose—through the use of raster output scanners (ROS)—the charged PR drum or belt according to the binary halftoned image pattern. The resulting spatial charge distribution, called the latent image, corresponds to the desired image to be printed. It is then "developed" by depositing oppositely charged toner particles exclusively in the charged regions thus forming a toned image on the PR. The toned image is "transferred" to paper by electrostatic forces and made permanent by "fusing," where heat and pressure are applied to melt the toner particles and adhere them to the paper. Finally, the PR is mechanically and electrostatically "cleaned" of residual toner and then recirculated to the charging system for the next image.

1.4.1 IMAGE-ON-IMAGE AND TANDEM PRINT ENGINES

Several print engine architectures have been developed to produce full-color prints image-on-image (IOI), tandem, etc. [5–7]. The fundamental difference between tandem architecture and IOI architecture is where the four-color CMYK image is constructed. In the IOI architecture, used in the iGen3 and iGen4 production presses, the four-color image is constructed one on top of the other on the PR belt in one complete revolution. Figure 1.4 shows the "skeleton" view of the system, with outlines of the feeders, marking paper handling, and finishing systems. Once the magenta layer is developed, yellow, cyan, and black are developed on top of the prior toner layer. This four-color image is then transferred in a single step to the substrate. The basic steps of the IOI xerographic process used in iGen3 and iGen4 print engines are precharge, charge/recharge, expose, develop, pretransfer, transfer, detack/stripping, and cleaning. Three of these steps (charge/recharge, expose, and develop) are performed up to four times to achieve the single pass, four color process.

The advantages of single step transfer are that it eliminates at least four opportunities for image misregistration or disturbance or transfer efficiency loss. In the tandem architecture (Figure 1.5), used in the DC8000 printer, each primary is first developed on an individual PR and then separately transferred to the paper directly (or through an intermediate transfer drum or belt), thus making it more difficult to maintain high accuracy in the registration of each primary because of the four separate transfer steps.

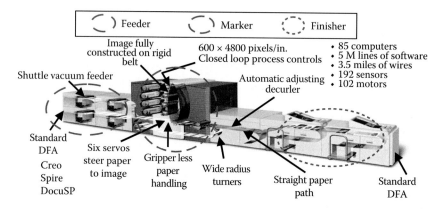

FIGURE 1.4   Physical layout of the iGen3 print engine (IOI architecture). (From Mestha, L.K. et al., Control elements in production printing and publishing systems: DocuColor iGen3, *42nd IEEE Conference on Decision and Control*, Vol. 4, pp. 4096–4108, Dec. 9–11, 2003. With permission.)

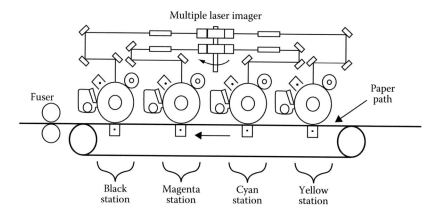

FIGURE 1.5   Physical layout of the DC8000 print engine (tandem architecture).

1.4.2   PARALLEL PRINTING SYSTEMS

The tightly integrated parallel digital printer (TIPP) architecture developed by Xerox is designed to deliver the power of two or more printers with the simplicity of one. The TIPP architecture includes the software and hardware that enables multiple print engines to work together seamlessly as one printing device. This architecture also makes the printer smart enough, depending upon the circumstances, to allow each individual engine to work independently of the other.

A side view of a parallel printing system is shown in the schematic of Figure 1.6, which includes two printers stacked vertically, a media transporting system, and a network of flexible paper paths, which delivers media to and from each of the printers. At any time, both printers can be simultaneously printing so that they

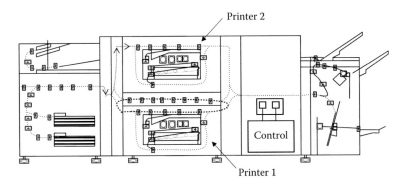

FIGURE 1.6 Schematic side view of a parallel printing system. (US Patent Application 20060197966, Gray balance for a printing system of multiple marking engines, Sept. 7, 2006.)

act as one. As such, multiple printers can be employed in the printing of a single print job. More than one print job can be in the course of printing at any single time instant.

The printers can be fed with paper from a variety of feeder modules loaded with a variety of paper types. A finisher with different finishing capabilities may receive printed sheets from any of the printers. Job output trays may include one or more special trays for multiple job collections. The finisher may also include a purge tray for diverting sheets in order to maintain the integrity of the job.

The media handling system may include highways that extend from the feeder module to the finisher and pathways to transport the print media between the downstream media highways and selected ones of the printers. For example, when printing a two-page document, page one may be printed by printer one and page two by printer two, where pages one and two may be formed on opposite sides of the same sheet (duplex) or on separate sheets (simplex). Thereafter, these sheets are sent to finisher in the correct order. The highways and/or pathways may include inverters, reverters, interposers, bypass pathways, and the like, as known in the existing art, to direct the print substrate between the highway and a selected printer or between two printers.

One of the printers may be designated as a reference or master printer with the remaining printer considered a slave. The master is linked by a network of paper pathways to sensors in the printing system. Printers may include electrophotographic printers, ink-jet printers, including solid ink printers, thermal head printers that are used in conjunction with heat-sensitive paper, and other devices capable of marking an image on a substrate. Thus many different printing technologies can be incorporated within a TIPP system with a complex network of media paths so that all engines work as one, thus providing the user with the productivity and speed of a multi-engine printer. A tandem Xerox Nuvera 288 digital perfecting system is a twin-engine monochrome press that uses a technology that keeps one engine running at full speed, even if the other stops.

At the Drupa 2008 trade show, Xerox demonstrated the Xerox ConceptColor 220 (Figure 1.7) with its tightly integrated serial printing (TISP) architecture. It takes

FIGURE 1.7   Schematic side view of a TISP parallel printing system.

the iGen3 press and doubles the effective print speed to 220 images per minute to produce 110 cut-sheet duplex pages per minute. Print engine 1 prints side 1 and print engine 2 prints side 2, both at regular iGen3 speeds. For simplex jobs, the speed will be 110 pages per minute, but with duplex jobs 110 pages per minute. With the ConceptColor 220, users can achieve greater printing economics by getting twice the speed and twice the productivity with a single operator, which saves time and labor. By integrating two iGen3 engines in-line, it can approach a monthly print volume of up to 7 million color pages. The ConceptColor 220 builds on the tandem architecture technology of the Xerox Nuvera 288 Digital Perfecting System.

As increasing numbers of press systems within the print shop become parallel, as in TIPP, TISP, or cluster printing, with similar print engines or with loosely connected heterogeneous print engines, there will be enhanced need for distributed optimization of interconnected workflows and outputs to ensure print quality consistency. Realization of enterprise-wide optimization will require substantial progress in a number of key technical areas. They are (1) automation of the entire publishing, production, and decision processes via feedback control using real-time functional press models, real-time scheduling policy decisions based on sampling the current state of the press; (2) optimization of workflow layouts; (3) the use of sensors to measure, control, and export color to proofing, prepress, and creation stages and hence standardization of the color interfaces between heterogeneous press modules; and (4) the management of sensor to sensor variabilities throughout the press, which becomes significant when distributed color optimization is required. A single book is not enough to cover all of the science and technology that have gone into developing these systems. Hence, in this book, we concentrate on the most important imaging and control technology. These are the technologies that show promise for making the printed images appear like offset and consistent in a single job, job to job, and between multiple machines whether intended for use in an office or an enterprise-wise production printing and publishing systems.

1.5   EVOLUTION OF CONTROLS TECHNOLOGY FOR DIGITAL PRINTERS—COLOR CONTROLS VIEW

Xerographic printing process used for copying and printing has evolved over many years starting from Xerox' 914 era. A brief review of the history of the evolution of controls in EP products can shed some light on why controls technology is considered important in this process. This can later help us to understand the complexity of advances involved in designing digital production color printers.

In the EP process, as the electrostatic image is developed with a charged pigmented thermoplastic powder that is transferred and fused to paper under heat and pressure, mass of the toner particles on the paper can vary. The control functions should maintain the mass by adjusting the electrostatic charge, development field, and transfer currents. Although dependency of various parameters to output quality was reasonably well understood by developers, no closed-loop control was applied to the early copiers and printers (the Xerox' 914, 813, and 2400 copiers) to adjust these control actuators since the sensors were not reliable. In 1970, automatic density control (ADC) sensors were introduced with Xerox 4000 duplicators and subsequently used in various forms in the 5600 and 9200 families [8]. Using the ADC sensors, somewhat frequent manual adjustments were made to the toner control system. Since selenium alloy PRs were used, the system was fairly stable. No separate charge control was required in those printers.

The use of different PRs and the demand for improved copy quality led to the need for better controls in the 1980s. At that time, copy quality tune up was required every 50k prints and process quality drifted due to environmental conditions. A two-patch control system was developed in the 1980s that controlled the toner concentration (TC), electrostatic charge, and hence the developability on the PR. Low- and high-density patches were created as surrogates to customer images on the PR. Low-density patches were measured by the sensor, and the data were used in a single-input single-output (SISO) closed-loop configuration to control the electrostatic charge. Similarly, a high-density patch was used to adjust the TC in the developer housing at a lower rate, which subsequently improved developability. Although high-density patch measurements are sensitive to electrostatic charge and TC, the coupling was removed by running the electrostatic charge control loop at a much faster rate.

The presence of a two-patch control scheme in the Xerox 1075, 1090, and 4050 copiers helped in reducing service calls for background and density variation by more than a factor of 10 and resulted in lower subsystem costs. Due to architectural constraints and unstable charge on the PR, Xerox developed a low cost charge-measuring sensor called an electrostatic voltmeter (ESV). A reduced-cost version of the infra red density (IRD) sensor was also developed. The charge control was done using a separate ESV sensor followed by the developability control with the low cost IRD sensor, which appeared in the 1065 marathon copiers in 1987. This control strategy was subsequently adopted in various forms in the 5090, the DocuTech 135 (1990), 4135 (1991), 5390 (1993), and the DocuPrint 4635 (1994). The same sensors were also used with more advanced software in the 5100 (1991). In addition to runtime controls, these sensors helped to accurately set up the process and perform good diagnostics to reduce cost. With runtime controls, the perceptible page-to-page differences were minimized. In many accounts, automated setups changed PR replacements from a 45 min service call to a 10 min customer operation, which contributed to increased customer satisfaction rating.

For color printers, toner mass has to be tightly regulated so that the printer maps the desired tone to the actual output. This is achieved by creating and implementing inverse maps in a one-dimensional (1-D) coordinate space, called a tone reproduction curve (TRC), at various stages in the printer path. At the least, a midpoint tone adjustment is required during runtime. New density sensors were developed for measuring various tone densities for monochrome and color toners. A hierarchical and multilevel control loop architecture was simultaneously developed to address these problems, which required the use of modern control theory and methods [9–12].

For production quality color EP printers, the control challenges are several orders of magnitude higher since they compete in the traditional offset market that has a reputation for high quality. Materials affect the print quality stability and stability of color balance in prints. The control loops should not only maintain process stability for individual color separations, but also adjust the color for varying media conditions and wide array of media stocks (e.g., coated, uncoated, textured, smooth, and specialty) in order to compensate for overlay colors, sheet-to-sheet differences, temperature, humidity, PR aging and wear in drives, etc. They should maintain much tighter control on image registration between separations (for simplex and duplex functions) and paper motion at various regions in the paper path. To make them competitive with offset printers in terms of operational cost, many of the press makeready costs should be eliminated using automated setups. Thus, due to many new challenges, to deliver quality and high productivity at a low run cost, technological advances are required in different areas including sensing, algorithms, and processes. To compete, color EP printers also need color accuracy improvements against industry standards (e.g., General Requirements for Applications in Commercial Offset Lithography [13] [GRACoL], International Organization for Standards [ISO]). In addition to these, they should also closely match offset printing standards in time, that is, not require too much time on the press to adjust the color manually while following the international color consortium (ICC) [14] workflow for producing higher performance photo quality prints. Neutral grays and highlights should be reproduced at offset quality; photo smoothness in faces should be retained while maintaining sharp background and shadow details. Color EP printers should also retain highlight, midtone color balance without injecting any contours or blocking and offer high definition image quality at high speed.

Because measurement devices have different response rates, accurate color control methods require accurate and repeatable color measurements referenced to some "golden measurement standard" device or sensor in the master printer. Each color standard may specify a particular device, such as an X-Rite iSis or DTP70 Autoscan spectrophotometer, for generating the reference target (aim) measurements. For example, in the Pantone® matching system, the device-independent targets supplied by Pantone are measured by their standard instrument, an iSis spectrophotometer. A correction that adjusts the sensor output to correlate between color sensors in different printers may be required to maintain instrument-to-instrument variability to be within tightly specified limits.

To achieve paper/media-based contactless, noninvasive measurements at high-speed, in-line embedded spectrophotometers are used in the paper path to measure "just-fused" toner patches. There are many sensor-related issues to consider when the embedded spectrophotometers are located inside the print engine. In-line spectrophotometers sensors have been found to give different measurements on just-fused sheets as compared to those when the printed sheets have been cooled. A change in temperature can cause a chromatic shift in color pigments. Similarly, the

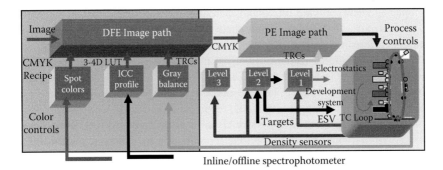

FIGURE 1.8 Hierarchical time-based color and process control functions used for improving consistency in a color EP print engine system.

cooling of glossy images can cause a shift in the lightness component, that is, L^*, of the color. These are some of the factors creating error in color measurements. Such temperature related shifts have to be compensated for, so that the measurements are closer to those of a golden measurement standard instrument. Sensor readings are very dependent upon the displacement of the media from the focal point of the sense head. As the medium moves through the point where the sensors are mounted in the paper path, the effect of displacement of the medium on sensor output is significant and should be compensated for, either through tight control of the media in the path, which may be counterproductive for paper path control system, or through special displacement insensitive (DI) optics [15]. Achieving accurate and repeatable readings from the sensor also depends on the stability of the electronics, the light source and its wavelength band, and the number of photons the system can integrate while the color samples are present underneath the sensor on a passing medium.

As pointed out before (Figure 1.1), the processing of images can occur at various levels inside and outside the printing and publishing system hierarchy. Many of the processing techniques required for imaging and control functions occur at multiple levels. A time-based separation is adopted with higher level functions occurring at a slow timescale near the prepress and faster real-time control functions typically occur in the print engine. The processing that goes on in the DFE will be at slower timescale than in the print engine, but at a much faster timescale than the prepress. There is also a time-based hierarchy being adopted local to the print engine. In Figure 1.8, an example multilevel hierarchical structure is shown, which becomes the foundation for navigating through the material of this book.

1.6 PREPRESS-BASED PROCESSING

Prepress refers to the preparation of digital files for printing that begins after the design decisions are made in the creation stage and ends when the document hits the press. A whole segment of the printing industry is devoted to prepress, expanding on the traditional prepress and removing limitations of printing methods. Digital prepress work is largely aided by modern computing hardware, software tools, and imaging instruments. Image capture and manipulation software, such as desktop publishing, color management software, multimedia handling software (speech to text conversion) are used heavily. Preflight is a final checklist to ensure that the files are ready for printing. Failure to prepare all digital files can cause delays and cost overruns.

Scanners and digital cameras usually capture color images in RGB type format, where each channel is quantized to 8 b/pixel. They can also act as sensors to perform color analysis of the image. They are defined in an 8 bit three-dimensional (3-D) color space whose components are red, green, and blue (RGB). One typical function of the color management module would be to correct/compensate for scanner or camera artifacts with respect to the tone reproduction by calibrating and character-izing the devices and further compensating for device differences. Generally speak-ing, two different scanners imaging the same spot in a hardcopy will generate different RGB values. As a result, the color management module needs to transform the color from a device-dependent space (a specific scanner RGB) to a device-independent space ($L^*a^*b^*$ or independent RGB) to ensure quality color reproduc-tion. The transformation usually is in the form of a multidimensional LUT, which is generated by measuring known color targets for sample of colors [16–20]. Any color that falls between is interpolated using a standard technique like trilinear or tetrahedral interpolation. Image processing techniques like denoising, deblurring, up/down sampling, cropping, color manipulations, and histogram modification, are applied to the image or video frame as dictated by the user prior to its inclusion in a given document [3].

In an ideal preferred print workflow, the prepress environment would incorporate an accurate color model of the production environment within the design tool, which in turn helps to view the color images on a calibrated monitor (soft proofing) or a proofer (hard proofing). The design inspection involves looking for image quality defects (e.g., the loss of shadow or chromatic details, color balance, contours, smoothness, etc.) prior to running production jobs. This process requires accurate characterization and calibration of the monitors prior to viewing images so that the soft-proofed images match the actual prints. A good test image is useful for evaluating monitor's quality and calibration as well as the match between the monitor and printer. For viewing images, the multidimensional LUTs are used to transform the file first to a device-independent color space and then on to the monitor color space. These transformations have to be accurate and should not induce unnecessary image artifacts.

Thus, in this stage, the control functions are discretely handled at a lower timescale based on capturing the model of the imaging system and processing images with a myriad of algorithms. In recent years, image processing has become more sophisticated and more prevalent in the digital print production industry. Chapter 2 contains relevant theoretical fundamentals of important digital image processing topics such as image formation, image sampling, quantization, filtering, transforms, denoising, resizing, etc. To help the reader understand how to extract spatial frequency-based models, we also introduce the optical transfer function and modulation transfer function of imaging systems. These models can be incorp-orated in the printing workflow of the production environment. The spatial models could help to perform diagnostics and design inspection for production anomalies. Material covered in Chapter 2 is also helpful for processing images in the DFEs.

1.7   DFE-BASED PROCESSING

The role of the DFE is to convert specifications of the user's intent into print engine data and control information. The image data and control information will cause the print engine to produce the best possible rendering of the user's intent. As described in Section 1.3, color jobs are separated into C, M, Y, and K images and rasterized. These images are then compressed to optimize for different types of input and stored on the image disk of the DFE. When the print engine requests the job files, it is decompressed in real-time off of the image disk. This data is then transferred over high-speed image data lines to the marker module of the print engine. The data are then further processed for anti-aliasing before the appropriate halftone is applied. The image data are then sent to the analog controls of the ROS, which writes the digital bits of color separated image to the PR belt, and the rest of the xerographic process follows (in the marker module) to create the printed jobs.

Normally, DFEs will provide the user with rendering control. In the DFE image path, tone adjustment and the multidimensional color transforms are critical control points where feedback from the internal and external paper-based spectral or color measurements are used to develop 1-D (single-channel linearization or gray balance), two-dimensional (2-D) or 3-D transforms [21]. Advanced color profiling solutions offered by Xerox [22,23] with an in-line spectrophotometer automatically generate multidimensional profiles for each halftone screen and media so that the colors match offset printing standards and the rendering is of photo quality. Various rendering intents (or user preferences), gray component replacement (GCR)/under color removal (UCR), gamut mapping, black point compensation are included in this control function. These profiles are updated at user's request. Underlying image and signal processing and control algorithms of these control functions are described in Chapters 7 and 8.

Often, multidimensional profiles do not accurately render process spot colors. Customers printing applications such as marketing collaterals and direct mail can be very sensitive to spot color consistency and, for many, repeatability is as critical as or even more important than accuracy. Accurate and consistent spot colors are also important for catalogs, business cards, and design documents. Automatic spot color editing (or control) (ASCE) function corrects with respect to device-independent $L^*a^*b^*$ reference values (e.g., Pantone matching system). ASCE automatically reads print engine $L^*a^*b^*$ values using in-line sensors, compares them to the reference values, and modifies the CMYK recipe for each spot color to minimize the difference. Chapter 8 contains same basic ASCE algorithms. In addition, that chapter shows how the spot color control approach can be applied to create a 1-D gray balance TRC and 2-D transforms. Furthermore, users can adjust tone curves manually in the DFE to modify image appearance to match their intent.

Modern DFEs also perform an automated color check using in-line spectral sensors to tell the press operator if the press is ready to go into full production or if other activities are needed. This eliminates unneeded adjustments and ensures the press is put into service as quickly as possible.

We now discuss the press control functions by first describing the physical print station, where the jobs are actually printed on the media. All of the key press control loops are resident inside the marker module.

1.8 PRINT ENGINE-BASED PROCESSING

The print engine and the DFE are responsible for most of the constraints regarding image quality, color balance, and color stability. These constraints limit both nominal device performance and the ability to achieve that performance repeatedly. Nominal performance is a function of engineering trade-offs in the design process. Repeatability is a function of the process control system used in the print engine.

The control algorithms employed to control the process are often customized for the underlying system architecture to achieve optimum stability results. A generic implementation is described in Refs. [10–12], which contains time-based hierarchy which is architecturally named as levels 1, 2, and 3 controls [11] and not associated with any particular print job. In real-time, controls of this nature run at a much faster rate and include charge, density (or dot gain), background, and developed tone reproduction control functions for each of the separations and provide information for online remote interactive diagnostics. Surrogate patches, placed in the interdocument zone (IDZ) in between images, are utilized to provide the appropriate feedback for process control. For example, PR voltages are read using an ESV sensor. Charge control loop adjusts the PR charge and the intensity of the laser in a level 1 subsystem loop as indicated in Figure 1.7 so that the voltages on the PRs maintain or track the desired values within a small tolerance (generally less that 1%) to prevent the appearance of unwanted variations in prints. The amount of toner mass deposited on the PR is measured at different tone levels using calibrated optical sensors. This information is then used to control the dot gain and development reproduction curves by actuating the charging and development system actuators. These are called level 2 control loops. TC, which is the ratio of toner mass to carrier plus toner mass, is maintained to some desired set point for each of the color station. This is accomplished with digital controllers in the developer housing using TC sensors and actuating the dispense rate. This is perhaps one of the most complex digital SISO control loops to analyze which comprises of unstable time varying plant with time delays, actuator saturation, sensor noise, disturbance due to demand for toner usage (coming from each page of the job), and a variable actuation cycle, as in a typical inventory management system. In addition to the real-time process adjustments, to control primary color mixtures for optimum color quality, the number of actuators required is more than those currently available in levels 1 and 2. One obvious place to look for more actuators is in the image, since electronically produced color documents contain pixels that are described in a 3-D RGB space. These color pixels are transformed to corresponding digital CMYK values to a printable form, before being sent to the printer. So, by linearizing the tone levels to each of the input tone values of the primaries, we can achieve improved controls for all separations individually [21]. This type of tone adjustment is called level 3 controls and can be performed on the PR belt or on the paper, depending on the sensing method. For levels 1, 2, and 3, only individual separations are controlled. Chapter 9 describes relevant theory and practical controller design based on state variable methods with pole-placement and regular linear quadratic design. We also show how observers can be used to compensate for time delays in the TC control system.

In recapitulation, this book includes a collection of theoretical techniques, algorithms, and methods required for developing and optimizing a digital printing system and producing state-of-the-art color quality. In addition to the fundamental knowledge in image processing and image transforms, mathematical tools for the analysis and design of open- and closed-loop control systems are required to optimize an internal press control system. Therefore, Chapters 3 through 5 provide a strong mathematical foundation to design modern multivariable discrete control systems, which are required for improving the press stability using process actuators. Chapter 6 describes the multidimensional interpolation and smoothing/filtering techniques useful for generating inverse maps. Chapters 7 and 8 describe the use of image actuators (e.g., CMYK separations), largely in the DFEs and to some extent in the prepress, to provide an accurate system inverse. These two chapters also contain necessary information for characterizing printers, mapping out-of-gamut colors to the surface, efficient use of multidimensional interpolation algorithms, and methods to generate good color transformations. Chapter 10 presents actual physical models of the EP process with access to principal image, process, and marking subsystem parameters that simulate the development of fused prints. Color gamuts for different settings of printer parameters can be generated using these physical models. In addition to the opportunities provided by these models to create robust control and imaging system, they can be extended to control the color by rendering spatially color corrected pixels. Spatial color correction is required in EP printing process [24] due to the streaks and bands that are inherent to the print process. A dynamic 1-, 2-, or 3-D spatial model structure for the development of component primaries and then to mixed colors on the paper are some of the new developments that can enable spatial image quality compensation [25]. They are briefly mentioned in this book.

REFERENCES

1. US Department of Labor, Bureau of Labor Statistics, *Career Guide to Industries: Printing, Publishing*, Mar. 2006, www.bls.gov
2. BF Kuvin, Modular press control, *Metalforming*, pp. 37–39, Nov 2002.
3. E Saber, S Dianat, LK Mestha, and PY Li, DSP utilization in digital color printing, *IEEE Signal Processing Magazine*, Jul. 2005.
4. DA Hays and KR Ossman, Electrophotographic copying and printing (xerography), in *The Optics Encyclopedia*, Wiley-VCH, Berlin, 2003.
5. R Lux and H-J Yuh, Is image-on-image color printing a privileged printing architecture for production digital printing applications? *NIP20: Proceedings of the IS&T's International Conference on Digital Printing Technologies*, Salt Lake City, UT, pp. 323–327, Oct. 31–Nov. 5, 2004.
6. JJ Folkins, Five cycle image on image printing architecture, US Patent 5,576,824, Nov. 19, 1996.
7. JJ Folkins, Transfer, cleaning and imaging stations spaced within an interdocument, US Patent 5,576,824, May 5, 1998.
8. LK Mestha et al., Control elements in production printing and publishing systems: DocuColor iGen3, in *42nd IEEE Conference on Decision and Control*, Vol. 4, pp. 4096–4108, Dec. 9–12, 2003.

9. LK Mestha, Control advances in production printing and publishing systems, *NIP20: Proceedings of the IS&T's The International Conference on Digital Printing Technologies*, Salt Lake City, UT, Oct. 31–Nov. 5, 2004.

10. ES Hamby et al., A control-oriented survey of xerographic systems: Basic concepts to new frontiers, in *Proceedings of the American Control Conference*, Boston, MA, Jun. 30–Jul. 2, 2004.

11. LK Mestha et al., A multilevel modular control architecture for image reproduction, in *Proceedings of the IEEE International Conference on Control Applications*, Trieste, Italy, Sep. 1–4, 1998.

12. CB Duke et al., *Color System Integration*, 1997 (contributions by R.E. Grace).

13. A full GRACoL Technical Specification document, Calibrating, printing and proofing to the G7 method, V4, Mar. 2006.

14. International Color Consortium Specification, ICC. 1:2004-10 (Profile version 4.2.0.0), Image technology colour management—Architecture, profile format, and data structure.

15. FF Hubble III and JA Kubby, Spectrophotometer for color printer color control with displacement insensitive optics, US Patent 6,384,918, May 7, 2002. Other US Patents 6,603,551; 6,633,382; 6,809,855; 7,259,853; 7,271,910; 6,721,692.

16. CS Chan, Method and system for providing closed loop color control between a scanned color image and the output of a color printer, US Patent 5,107,332, Apr. 21, 1992.

17. KD Vincent, Colorimeter and calibration system, US Patent 5,272,518, Dec. 21, 1993.

18. M Stokes, Method and system for analytic generation of multidimensional color lookup tables, US Patent 5,612,902, Sep. 13, 1994.

19. G Bestmann, Method and apparatus for calibration of color values, US Patent 5,481,380, Jan 2, 1996.

20. LK Mestha, R Bala, and LK Mestha, Use of spectral sensors for automatic media identification and improved scanner correction, US Patent 6,750,442, Jun. 15, 2004.

21. PK Gurram, SA Dianat, LK Mestha, and R Bala, Comparison of 1-D, 2-D and 3-D printer calibration algorithms with printer drift, *NIP21: Proceedings of the IS&T's 21st International Conference on Digital Printing Technologies*, Baltimore, MD, Sep. 18–22, 2005.

22. Xerox News Release, New Xerox color press delivers breakthrough image quality: Drives more profitability from digital printing, Dusseldorf, Germany, May 29, 2008.

23. Xerox News Release, The best gets better: Xerox elevates high performance and image quality of iGen3 digital press, Rochester, NY, May 22, 2008.

24. HA Mizes, Systems and methods for compensating for streaks in images, US Patent 7,347,525, Mar. 25, 2008.

25. LK Mestha and ER Viturro, Method for spatial color calibration using hybrid sensing systems, US Patent Application, 20080037069, Feb. 14, 2008.

2 Fundamentals of Digital Image Processing

2.1 INTRODUCTION

This chapter is devoted to providing an overview of fundamentals of digital imaging, including topics such as digital image formation, imaging systems, image sampling, quantization, filtering, and image transformation. Section 2.2 briefly covers image formation and systems. Section 2.3 covers optical and modulation transfer functions. Section 2.4 discusses image sampling and quantization. Section 2.5 deals with image transforms, and image filtering is covered in Section 2.6. Issues such as image resizing and its practical implementation are addressed in Section 2.7. Image enhancement is covered in Section 2.8. Image degradation and restoration are briefly discussed in Section 2.9. Finally, basic image halftoning techniques are described in Section 2.10.

2.2 DIGITAL IMAGE FORMATION AND SYSTEMS

Linear system theory provides a powerful tool for the modeling and analysis of various imaging systems [1–3]. A linear system is characterized as a system that obeys the superposition principle, that is, if the input I_1 to a system results in the output O_1, and the input I_2 to the system results in the output O_2, then the input $aI_1 + bI_2$ results in the output $aO_1 + bO_2$ for any I_1, I_2 signal and scale factors a and b. A linear system provides a convenient model for an imaging system. Unfortunately, none of the imaging systems encountered in the real world are completely linear. However, such systems are almost always approximated by linear systems to make their analysis mathematically tractable. Conventional and digital cameras, scanners, printers, and the human visual system (HVS) are among many examples of imaging systems that are modeled and analyzed by using linear system theory. A two-dimensional (2-D) linear imaging system is characterized by a function $h(x, y; \lambda_1, \lambda_2)$, referred to as the point spread function (PSF) of the imaging system that specifies the output of the system when the input is a point (impulse) at location (λ_1, λ_2) in the input image plane as shown in Figure 2.1. To find the output of an imaging system $g(x, y)$ to a given input $f(x, y)$, first the input is broken up into sum of weighted impulses (points)

$$f(x, y) = \int_{-\infty}^{\infty} \int_{-\infty}^{\infty} f(\lambda_1, \lambda_2) \delta(x - \lambda_1, y - \lambda_2) \, d\lambda_1 d\lambda_2 \qquad (2.1)$$

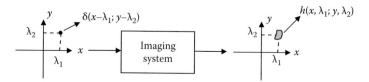

FIGURE 2.1   PSF of an imaging system.

Next, the output for each weighted impulse is determined from the impulse response

$$f(\lambda_1, \lambda_2)\delta(x - \lambda_1, y - \lambda_2) \rightarrow f(\lambda_1, \lambda_2)h(x, \lambda_1; y, \lambda_2) \qquad (2.2)$$

Finally, using the linearity principle, the output image is determined as the sum (integral) of the outputs of all the impulses that make up the input.

$$g(x, y) = \int_{-\infty}^{\infty} \int_{-\infty}^{\infty} f(\lambda_1, \lambda_2)h(x, \lambda_1; y, \lambda_2)d\lambda_1 d\lambda_2 \qquad (2.3)$$

This is a general input–output relationship for a linear imaging system. The integral equation given by Equation 2.3 is known as Fredholm equation of first kind. It assumes that the shape of the PSF depends on its location in the input image plane. Such a system is referred to as a linear space-variant system. Examples for such system are the subject motion when a picture is being captured with a stationary camera, or capturing a picture of a scene with objects at various planes. In many instances, the shape of the PSF is independent of its location and is thus the same everywhere, that is,

$$h(x, \lambda_1; y, \lambda_2) = h(x - \lambda_1, y - \lambda_2) \qquad (2.4)$$

As a result of this, the PSF can be characterized as a function with two arguments x and y, or simply $h(x, y)$. Such a system is known as a linear space-invariant (LSI) system. The input–output relationship simplifies for such a linear system and is called convolution integral.

$$g(x, y) = \int_{-\infty}^{\infty} \int_{-\infty}^{\infty} f(\lambda_1, \lambda_2)h(x - \lambda_1, y - \lambda_2)d\lambda_1 d\lambda_2 \qquad (2.5)$$

A major advantage of LSI systems is that they can be analyzed using Fourier transform theory. For that purpose, many imaging systems are approximated by LSI systems. Now, we give simple examples of LSI imaging systems.

2.2.1   POINT SPREAD FUNCTION OF A DEFOCUSED LENS

A simple example of an LSI imaging system is a defocused lens. An exact method of finding the PSF of such a system is based on physical optics and has been described in detail in a paper by Lee [4]. An alternative approach based on geometrical optics results in a reasonable approximation of the 2-D PSF as a circle of unit area with a radius R. The PSF is given by

$$h(x, y) = \begin{cases} \frac{1}{\pi R^2} & x^2 + y^2 \leq R^2 \\ 0 & \text{otherwise} \end{cases} \tag{2.6}$$

In this equation, R is the blur radius and is given by

$$R = \frac{f^2}{2F} \frac{|L - L_0|}{L L_0} \tag{2.7}$$

where
  f is the focal length of the lens
  F is the f number of the lens that is the focal length divided by diameter of the
    lens aperture
  L_0 is the distance at which the lens has been focused
  L is the distance from the object to the lens

2.2.2 POINT SPREAD FUNCTION OF MOTION BLUR

Another example of a linear imaging system is the motion blur resulting from camera motion or object motion. Consider an object with uniform motion in the horizontal direction characterized by a constant speed v. During the exposure time T, a point on the object moves by a distance $x_0 = vT$. Thus, the 2-D PSF is a line of length x_0, that is,

$$h(x, y) = \begin{cases} \frac{1}{x_0} \delta(y) & 0 \leq x \leq x_0 \\ 0 & \text{otherwise} \end{cases} \tag{2.8}$$

Consequently, the image $g(x, y)$ of an object $f(x, y)$, which is moving with a speed of v in the horizontal direction, is found by the following equation:

$$g(x, y) = f(x, y) * h(x, y) = \frac{1}{x_0} \int_0^{x_0} f(x - \lambda, y) d\lambda \tag{2.9}$$

where $*$ stands for convolution. Using a change of variable $\lambda = vt$, we can rewrite the above integral as

$$g(x, y) = \frac{1}{T} \int_0^T f(x - vt, y) dt \tag{2.10}$$

The above equation can be generalized for any type of motion as

$$g(x, y) = \frac{1}{T} \int_0^T f(x - x_0(t), y - y_0(t)) dt \tag{2.11}$$

where $x_0(t)$ and $y_0(t)$ are the equations for motion in the x- and y-directions, respectively. For example, for uniform motion in both x- and y-directions $x_0(t) = v_x t$ and $y_0(t) = v_y t$. If there is acceleration in the x-direction, then $x_0(t) = 0.5at^2 + v_x t$, where a is acceleration in the x-direction.

2.2.3 POINT SPREAD FUNCTION OF HUMAN VISUAL SYSTEM

The third example of an LSI imaging system is the HVS. The optics of the eye, that is, the cornea, the pupil, and the lens, constitute a blur type (low-pass) PSF, while the rod and cone detectors' inhibitory response constitute a sharpening type (high-pass) linear system. The actual PSF of the eye is the cascade of these two PSFs [3,5].

Finally, the PSF of a printing system consisting of a scanner, printer, and display device is the convolution of the PSFs of the three imaging subsystems. This will be described in detail in Chapter 3.

2.3 OPTICAL AND MODULATION TRANSFER FUNCTIONS

The optical transfer function (OTF) of an imaging system is defined as the 2-D Fourier transform of its 2-D PSF and is given by the following integral

$$H(\omega_x, \omega_y) = \int_{-\infty}^{\infty} \int_{-\infty}^{\infty} h(x, y) e^{-j(\omega_x x + \omega_y y)} dx dy \tag{2.12}$$

Either the OTF or the PSF completely characterizes an LSI imaging system. The OTF of an imaging system is a complex quantity and consists of magnitude and phase. The modulus of the OTF normalized to a value of 1 at zero frequency is referred to as the modulation transfer function (MTF), that is,

$$M(\omega_x, \omega_y) = \left| \frac{H(\omega_x, \omega_y)}{H(0,0)} \right| \tag{2.13}$$

As an example, consider uniform motion in the x-direction; the OTF is one-dimensional (1-D) since there is no motion in the y-direction. This 1-D OTF is given by

$$OTF(\omega_x) = \frac{1}{x_0} \int_0^{x_0} e^{j\omega_x x} dx = \frac{e^{j\omega_x x_0} - 1}{j\omega_x x_0} = \frac{\sin\left(\frac{\omega_x x_0}{2}\right)}{\frac{\omega_x x_0}{2}} e^{-j\frac{\omega_x x_0}{2}} \tag{2.14}$$

The corresponding MTF is

$$MTF(\omega_x) = \left| \frac{OTF(\omega_x)}{OTF(0)} \right| = \left| \frac{\sin\left(\frac{\omega_x x_0}{2}\right)}{\frac{\omega_x x_0}{2}} \right| \tag{2.15}$$

Figure 2.2 shows MTF of uniform motion in the x-direction.

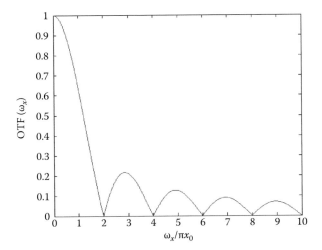

FIGURE 2.2 MTF of uniform motion in x-direction.

The previous discussion is fairly general in the sense that it applies to imaging systems with nonsymmetrical PSF too. However, in many practical systems, such as the lens and the optics of a digital camera, printer, or scanner, the PSF is rotationally invariant or circularly symmetric. In such cases, the previous analysis can be simplified. Now, we consider a circularly symmetric imaging system. The PSF of such a system is 1-D and is given by

$$h(r) = h\left(\sqrt{x^2 + y^2}\right) \tag{2.16}$$

where r is the radial distance. The OTF of this system like its PSF is circularly symmetric. To show this, consider

$$H(\omega_x, \omega_y) = \int_{-\infty}^{\infty} \int_{-\infty}^{\infty} h(x, y)e^{-j(\omega_x x + \omega_y y)}dxdy$$

$$= \int_{-\infty}^{\infty} \int_{-\infty}^{\infty} h(\sqrt{x^2 + y^2})e^{-j(\omega_x x + \omega_y y)}dxdy \tag{2.17}$$

Using polar coordinates, Equation 2.17 can be rewritten as

$$H(\omega_x, \omega_y) = \int_0^{2\pi} \int_0^{\infty} h(r)e^{-j(\omega_x r \cos\theta + \omega_y r \sin\theta)}rdrd\theta \tag{2.18}$$

Using the trigonometric identity

$$\omega_x \cos\theta + \omega_y \sin\theta = \omega_\rho \cos(\theta - \varphi) \tag{2.19}$$

where

$$\omega_\rho = \sqrt{\omega_x^2 + \omega_y^2}$$

and

$$\varphi = \tan^{-1}\frac{\omega_y}{\omega_x}$$

we have

$$H(\omega_x, \omega_y) = \int_0^{2\pi}\int_0^\infty h(r)e^{-jr\omega_\rho\cos(\theta-\varphi)}r\,dr\,d\theta = \int_0^\infty rh(r)\int_0^{2\pi} e^{-jr\omega_\rho\cos(\theta-\phi)}d\theta\,dr \tag{2.20}$$

Since function $\cos(\theta)$ is periodic with period of 2π and integration is over one period, the inner integral is independent of ϕ. Therefore, we set $\phi = 0$.

$$H(\omega_x, \omega_y) = \int_0^\infty rh(r)\int_0^{2\pi} e^{-jr\omega_\rho\cos\theta}d\theta\,dr \tag{2.21}$$

The inner integral is given by

$$\int_0^{2\pi} e^{-jr\omega_\rho\cos\theta}d\theta = 2\pi J_0(r\omega_\rho) \tag{2.22}$$

where J_0 is the Bessel function of first kind and order 0. Hence,

$$H(\omega_\rho) = 2\pi \int_0^\infty rh(r)J_0(r\omega_\rho)dr \tag{2.23}$$

This is known as Hankel transform. Thus, the OTF of a rotationally symmetric PSF is also rotationally symmetric and is related to its 1-D PSF through 1-D Hankel transform. The inverse Hankel transform can be used to express the PSF in terms of OTF. This is given by

$$h(r) = \frac{1}{2\pi} \int_0^\infty \omega_\rho H(\omega_\rho)J_0(r\omega_\rho)d\omega_\rho \tag{2.24}$$

As an example of a circularly symmetric OTF, consider a defocused lens. The OTF is given by Hankel transform as

$$H(\omega_p) = 2\pi \int_0^\infty rh(r)J_0(r\omega_p)dr = 2\pi \int_0^R r\frac{1}{\pi R^2}J_0(r\omega_p)dr = \frac{2}{\omega_p^2 R^2}\int_0^{R\omega_p} xJ_0(x)dx \quad (2.25)$$

Using the known relationship $xJ_0(x) = \frac{d[xJ_1(x)]}{dx}$, we have

$$H(\omega_p) = \frac{2}{\omega_p^2 R^2}\int_0^{R\omega_p}\frac{dxJ_1(x)}{dx}dx = \frac{2}{\omega_p^2 R^2}xJ_1(x)\Big|_0^{R\omega_p} = 2\frac{J_1(R\omega_p)}{R\omega_p} \quad (2.26)$$

The MTF is given by

$$\text{MTF}(\omega_p) = \left|\frac{H(\omega_p)}{H(0)}\right| = 2\left|\frac{J_1(R\omega_p)}{R\omega_p}\right| \quad (2.27)$$

where $J_1(x)$ is a Bessel function of the first kind of order 1 and its Taylor series expansion is

$$J_1(x) = \sum_{i=0}^\infty (-1)^i \frac{x^{2i+1}}{2^{2i+1}i!(i+1)!} \quad (2.28)$$

Figure 2.3 shows the MTF of a defocused lens.

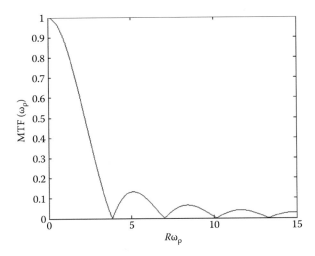

FIGURE 2.3   MTF of defocused lens.

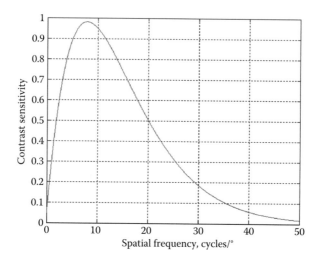

FIGURE 2.4 MTF of the HVS.

As a final example, we consider the MTF of the HVS. The measurement on human subjects has confirmed that the MTF of the HVS can be approximated by the empirical function given by [3,5],

$$\mathrm{MTF}(\omega_x, \omega_y) = \mathrm{MTF}\left(\sqrt{\omega_x^2 + \omega_y^2}\right) = \mathrm{MTF}(\omega_\rho) = A\left[\alpha + \frac{\omega_\rho}{\omega_0}\right]e^{-\left(\frac{\omega_\rho}{\omega_0}\right)^\beta} \quad (2.29)$$

In image processing applications, the following parameter values have been used: $A = 2.6$, $\alpha = 0.0192$, $\omega_0 = 8.772$ cycles/degree and $\beta = 1.1$. The plot of MTF is shown in Figure 2.4.

The MTF of LSI discrete systems (such as a digital printer) can be measured using DFT concepts. In a discrete system, a sampled input (an image described as pixels) results in an output (printed image) that is the convolution of the input with the system PSF (in this case resulting from the inefficiencies of the development process). Using linear system theory concepts, the MTF of the system can be measured by applying a sampled sinusoid with a certain frequency to the system and measuring its output. The ratio of the output to input modulation gives the MTF at that particular frequency and orientation. The main problem with this approach is that even for a given orientation, many such measurements are needed before a good approximation to the MTF curve can be constructed. An approach that is based on square-wave analysis is more efficient. Since the harmonics of a square wave contain significant energy at high frequencies, a single measurement can determine several MTF points. To find the MTF of such a linear system, an input level is chosen and a square wave with two amplitudes near that input level is constructed. The number of pixels contained in each cycle of the square wave (which defines the fundamental frequency or the first harmonic of the wave) would depend on the resolution needed for the MTF measurement. In practice, several square waves with various fundamental frequencies are applied and the results of the measurements are fitted to a curve. For example, a square wave with 50 pixels/cycle will have a fundamental frequency of 0.02 cycles/pixel. When the target is printed with a printer, the output frequency will be converted to cycles/mm, for example, for a printer at 300 pixels/in. (12 pixels/mm), the fundamental frequency is 0.24 cycles/mm. In summary, to measure the MTF of an imaging system (e.g., a digital printer), the following steps are followed:

1. Square-wave target with a given fundamental frequency is applied to the system (note that the DFT of a square wave does not contain any even harmonics). The output image is scanned (e.g., 300 pixels/in. for a printer with that resolution) and its DFT is measured.
2. MTF at each discrete frequency is measured as the magnitude of the DFT of the output divided by the magnitude of the DFT of the input at that frequency. The MTF is normalized to 1 at zero frequency.
3. To make the measurement more robust, this process is repeated for several square waves with different fundamental frequencies and the results are fitted to a curve. Figure 2.5 shows square-wave target patches for MTF measurement.

FIGURE 2.5 Square-wave target patches.

2.4   IMAGE SAMPLING AND QUANTIZATION

To process an image digitally (using a computer), it must be available in digital form. The digitized image is obtained by sampling the continuous tone image on a discrete grid and quantizing the sample values using a finite number of bits. To understand the process of image sampling, we need to revisit the 2-D sampling theorem [3,5]. Assume that we have a band-limited 2-D signal (such as an image). Real-world images are not truly band limited; however, they can be approximated by band-limited functions. A 2-D function $f(x, y)$ is said to be band limited if its Fourier transform $F(\omega_x, \omega_y)$ is zero outside a bounded region in the frequency plane (ω_x, ω_y), as shown in Figure 2.6.

The ideal rectangular image sampler is a 2-D array of discrete delta functions placed on a rectangular grid (Figure 2.7) with spacing of Δx and Δy, that is,

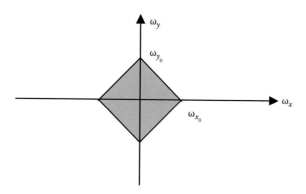

FIGURE 2.6   Region defining a band-limited signal.

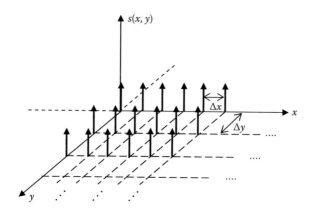

FIGURE 2.7   Sampling function $s(x, y)$.

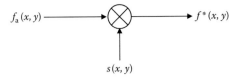

FIGURE 2.8 Sampling process.

$$s(x, y) = \sum_{n=-\infty}^{\infty} \sum_{m=-\infty}^{\infty} \delta(x - n\Delta x, y - m\Delta y) \qquad (2.30)$$

The sampling frequencies in the x- and y-directions are $f_{sx} = \frac{1}{\Delta_x}$ and $f_{sy} = \frac{1}{\Delta_y}$, respectively.

The sampling process is basically the multiplication of $f_a(x, y)$ and $s(x, y)$, as shown in Figure 2.8.

$$f^*(x, y) = \sum_{n=-\infty}^{\infty} \sum_{m=-\infty}^{\infty} f_a(n\Delta_x, m\Delta_y)\delta(x - n\Delta_x, y - m\Delta_y) \qquad (2.31)$$

The spectrum of the sampled signal is obtained by taking the Fourier transform of $f^*(x, y)$. That is,

$$F^*(\omega_x, \omega_y) = \int_{-\infty}^{\infty} \int_{-\infty}^{\infty} f^*(x, y)e^{-j(\omega_x x + \omega_y y)} dx dy$$

$$= \sum_{n=-\infty}^{\infty} \sum_{m=-\infty}^{\infty} f_a(n\Delta_x, m\Delta_y) \int_{-\infty}^{\infty} \int_{-\infty}^{\infty} \delta(x - n\Delta_x, y - m\Delta_y)e^{-j(\omega_x x + \omega_y y)} dx dy$$

$$= \sum_{n=-\infty}^{\infty} \sum_{m=-\infty}^{\infty} f(n, m)e^{-j(\omega_x n\Delta_x + \omega_y m\Delta_y)} \qquad (2.32)$$

where $f(n, m) = f_a(n\Delta_x, m\Delta_y)$. Since $f_a(x, y)$ can be obtained by taking the inverse Fourier transform of $F(\omega_x, \omega_y)$, then we have

$$f(n, m) = f_a(x, y)|_{x=n\Delta_x, y=m\Delta_y} = \frac{1}{4\pi^2} \int_{-\infty}^{\infty} \int_{-\infty}^{\infty} F(\omega_x, \omega_y)e^{-j(\omega_x x + \omega_y y)} d\omega_x d\omega_y|_{x=n\Delta_x, y=m\Delta_y}$$

$$= \frac{1}{4\pi^2} \int_{-\infty}^{\infty} \int_{-\infty}^{\infty} F(\omega_x, \omega_y)e^{-j(\omega_x n\Delta_x + \omega_y m\Delta_y)} d\omega_x d\omega_y \qquad (2.33)$$

Substituting Equation 2.33 into Equation 2.32 yields

$$F^*(\omega_x, \omega_y) = \sum_{n=-\infty}^{\infty} \sum_{m=-\infty}^{\infty} \frac{1}{4\pi^2} \left[\int_{-\infty}^{\infty} \int_{-\infty}^{\infty} F(u, v) e^{-j(un\Delta_x + vm\Delta_y)} du dv \right] e^{-j(\omega_x n\Delta_x + \omega_y m\Delta_y)}$$

$$= \frac{1}{4\pi^2} \int_{-\infty}^{\infty} \int_{-\infty}^{\infty} F(u, v) \left[\sum_{n=-\infty}^{\infty} \sum_{m=-\infty}^{\infty} e^{-j(u-\omega_x)n\Delta_x - j(v-\omega_y)m\Delta_y} \right] du dv$$

$$= \frac{1}{4\pi^2} \int_{-\infty}^{\infty} \int_{-\infty}^{\infty} F(u, v) \sum_{m=-\infty}^{\infty} e^{-j(v-\omega_y)m\Delta_y} \sum_{n=-\infty}^{\infty} e^{-j(u-\omega_x)n\Delta_x} du dv \qquad (2.34)$$

We now use the well-known identity

$$\sum_{k=-\infty}^{\infty} e^{-j2\pi xk} = \sum_{k=-\infty}^{\infty} \delta(x - k) \qquad (2.35)$$

Then, Equation 2.34 can be reduced to

$$F^*(\omega_x, \omega_y) = \frac{1}{4\pi^2} \int_{-\infty}^{\infty} \int_{-\infty}^{\infty} F(u, v) \sum_{n=-\infty}^{\infty} \sum_{m=-\infty}^{\infty} \delta \left[\frac{(u - \omega_x)\Delta_x}{2\pi} - n \right]$$

$$\times \delta \left[\frac{(v - \omega_y)\Delta_y}{2\pi} - m \right] du dv \qquad (2.36)$$

Let $u' = \frac{(u-\omega_x)\Delta_x}{2\pi}$ and $v' = \frac{(v-\omega_y)\Delta_y}{2\pi}$, then

$$F^*(\omega_x, \omega_y) = \frac{1}{4\pi^2} \int_{-\infty}^{\infty} \int_{-\infty}^{\infty} F\left(\omega_x + \frac{2\pi}{\Delta_x} u', \omega_y + \frac{2\pi}{\Delta_y} v' \right) \sum_{n=-\infty}^{\infty} \sum_{m=-\infty}^{\infty} \delta[u' - n]\delta[v' - m] \frac{2\pi}{\Delta_x} \frac{2\pi}{\Delta_y} du' dv'$$

$$= \frac{1}{\Delta_x} \frac{1}{\Delta_y} \sum_{n=-\infty}^{\infty} \sum_{m=-\infty}^{\infty} \int_{-\infty}^{\infty} \int_{-\infty}^{\infty} F\left(\omega_x + \frac{2\pi}{\Delta_x} u', \omega_y + \frac{2\pi}{\Delta_y} v' \right) \delta[u' - n]\delta[v' - m] du' dv'$$

$$= f_{sx} f_{sy} \sum_{n=-\infty}^{\infty} \sum_{m=-\infty}^{\infty} F(\omega_x + n2\pi f_{sx}, \omega_y + m2\pi f_{sy}) \qquad (2.37)$$

Therefore, the Fourier transform of the sampled signal is the periodic extension of the Fourier transform of the analog signal with period given by $(\omega_{sx}, \omega_{sy})$ as shown in Figure 2.9.

As a result of this, the original signal can be reconstructed by low-pass filtering the sampled signal, provided that there is no overlap between the spectrum components in frequency domain or in other words there is no aliasing. This is possible only if the sampling frequencies satisfy.

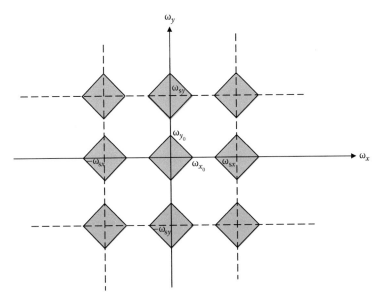

FIGURE 2.9 Fourier transform of the sampled signal.

$$\begin{aligned} \omega_{sx} - \omega_{x0} &\geq \omega_{x0} \\ \omega_{sy} - \omega_{y0} &\geq \omega_{y0} \end{aligned} \rightarrow \begin{aligned} \omega_{sx} &\geq 2\omega_{x0} \\ \omega_{sy} &\geq 2\omega_{y0} \end{aligned} \rightarrow \begin{aligned} f_{sx} &\geq 2f_{x0} \\ f_{sy} &\geq 2f_{y0} \end{aligned} \tag{2.38}$$

Thus we obtain the Nyquist theorem for sampling,

$$\begin{aligned} \Delta_x &\leq \frac{1}{2f_{x0}} \\ \\ \Delta_y &\leq \frac{1}{2f_{y0}} \end{aligned} \tag{2.39}$$

2.4.1 Two-Dimensional Sampling Theorem

Suppose a 2-D function $f_a(x, y)$ is band limited to $(\omega_{x0}, \omega_{y0})$, that is, the Fourier transform of the function is zero for all frequency pairs outside the support shown in Figure 2.6. Then the function $f_a(x, y)$ is completely determined by samples of the function spaced $(\Delta x, \Delta y) = \left(\frac{\pi}{\omega_{x0}}, \frac{\pi}{\omega_{y0}}\right)$ units apart. This means that the signal has to be sampled at the minimum rate of $(2f_{x0}, 2f_{y0})$ samples per unit length in the x- and y-directions. This minimum required rate is called the Nyquist rate or the Nyquist frequency. Consequently, if a certain sinusoid contained in the original signal needs to be accurately reproduced after sampling, it has to be sampled at least twice at each cycle. Thus, any sampling rate will capture all the information contained in frequencies below one-half of the sampling rate. If a signal is undersampled (i.e., sampled at a rate less than the Nyquist rate), aliasing will occur. Aliasing causes the high frequencies to appear as low frequencies in the sampled signal. To avoid aliasing,

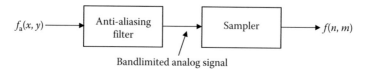

FIGURE 2.10   Anti-aliasing filter.

the original signal should be low-pass filtered prior to sampling. In particular, the low-pass filter should ideally eliminate all the frequencies that are higher than half the sampling rate. The low-pass filter used to reduce or eliminate aliasing is referred to as the anti-aliasing filter. It should be noted that the anti-aliasing filtering has to be applied prior to the sampling. Once the image has been inadequately sampled and the high frequencies have been aliased into the low frequencies, there is no way to eliminate the aliasing effect. The block diagram of a sampler with an anti-aliasing filter is shown in Figure 2.10.

Aliasing is a common problem in digital cameras that use CCD sensor arrays. This is usually due to the fact that the camera lens is of high quality and is capable of accurately reproducing scenes containing frequencies higher than half the Nyquist frequency (determined by the sensor resolution). The anti-aliasing filter should be placed somewhere between the lens and the sensor.

Example 2.1

In the first example, we consider the 1-D analog signal

$$f_a(x) = 2\cos(400\pi x)$$

The signal has one frequency component $f_0 = 200$ cycles per unit length. The Nyquist rate for this signal is $f_{Nyquist} = 2f_0 = 2 \times 200 = 400$. If we undersample this signal at the rate of $f_s = 300$ samples per unit length, we obtain

$$f(n) = 2\cos\left(\frac{400\pi}{300}n\right) = 2\cos\left(\frac{4\pi}{3}n\right) = 2\cos\left(2\pi n - \frac{4\pi}{3}n\right) = 2\cos\left(\frac{2\pi}{3}n\right)$$

Reconstructing the signal from these samples will result in a signal of lower frequency. This lower frequency is given by

$$\omega_1 = f_s\frac{2\pi}{3} = 200\pi$$

Figure 2.11 shows the original analog signal and the reconstructed signal. As is obvious from this figure, the signal is aliased due to undersampling.

Example 2.2

Consider the analog signal

$$f_a(x, y) = 2\cos(400\pi x + 600\pi y)$$

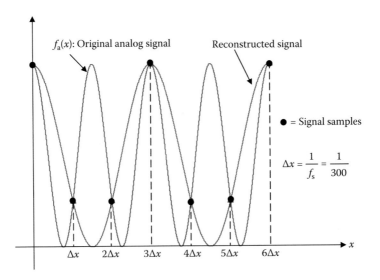

FIGURE 2.11   One-dimensional example of aliasing.

The signal has frequency components of 200 and 300 cycles per unit length in the x- and y-directions, respectively. Thus the Nyquist rate for the signal is $f_{Nyquist} = 2 \times (200, 300) = (400, 600)$ cycles per unit length. To avoid aliasing, the sampling rate must exceed the Nyquist rate. Now suppose we sample the signal at a rate below the Nyquist frequency, say 300 and 400 samples per unit length in the x- and y-directions, respectively. Then the sampled signal is

$$f(n, m) = 2\cos\left(\frac{400\pi}{300}n + \frac{600\pi}{400}m\right) = 2\cos\left(\frac{4\pi}{3}n + \frac{3\pi}{2}m\right)$$

$$= 2\cos\left(\frac{4\pi}{3}n - 2\pi n + \frac{3\pi}{2}m - 2\pi m\right) = 2\cos\left(-\frac{2\pi}{3}n - \frac{\pi}{2}m\right)$$

$$= 2\cos\left(\frac{2\pi}{3}n + \frac{\pi}{2}m\right)$$

Now, if we send the sampled signal through an ideal digital to analog converter (D/A), the resulting continuous signal would be

$$g(x, y) = 2\cos\left(\frac{2\pi}{3} \times 300x + \frac{\pi}{2} \times 400y\right) = 2\cos(200\pi x + 200\pi y)$$

This means that the reconstructed signal appears as a low-frequency signal.

Example 2.3

To show aliasing in an image, we first downsample the monochromic LENA image shown in Figure 2.12a by a factor of 4 and then upsample it to the original size without using any anti-aliasing filter. The resulting image is aliased as shown in Figure 2.12b.

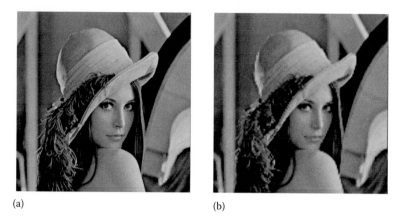

(a) (b)

FIGURE 2.12 (a) Original LENA Image (b) aliased LENA Image.

2.4.2 IMAGE QUANTIZATION

Once the signal has been sampled (discretized), the amplitude of the sampled values is quantized (digitized). A quantizer is a many-to-one mapping that maps a range of input values into a single output value. There are two ways to quantize a signal: scalar quantization (SQ), where each signal sample is individually quantized, and vector quantization (VQ), where a block of signal samples is jointly quantized. The concept of SQ for 1-D signals is shown in Figure 2.13. A typical B-bit scalar quantizer for a signal with a dynamic range of $x_{\min}$ to $x_{\max}$ is defined as

$$Q(x) = r_i \quad \text{if } d_{i-1} < x < d_i \quad i = 1, 2, \ldots, L \quad \text{where } L = 2^B \qquad (2.40)$$

d_i's are called decision levels and r_i's are reconstruction levels.

The function $Q(x)$ is shown in Figure 2.14.

2.4.2.1 Uniform Quantization

In uniform SQ, the quantization decision levels are of equal length Δ, referred to as the quantization step size. The quantization step size is related to the signal dynamic range $[x_{\min} \ x_{\max}]$ and number of quantization bits B by

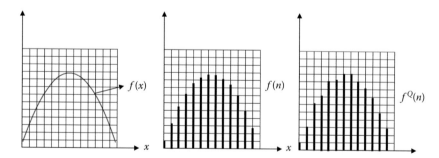

FIGURE 2.13 SQ for 1-D signals.

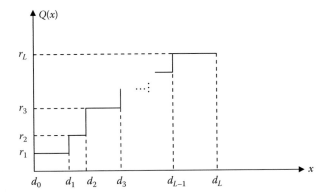

FIGURE 2.14 Quantization function $Q(x)$.

$$\Delta = \frac{x_{max} - x_{min}}{L} = \frac{x_{max} - x_{min}}{2^B} \qquad (2.41)$$

The decision boundaries are related to the quantization step size Δ by

$$d_i - d_{i-1} = \Delta \quad 1 \le i \le L \qquad (2.42)$$

The reconstruction levels are chosen to be halfway between the decision boundaries in order to minimize the mean-squared error (MSE).

$$r_i = \frac{d_i + d_{i-1}}{2} \quad 1 \le i \le L \qquad (2.43)$$

Example 2.4

Design a uniform 2 bit (4-level) quantizer for a signal with dynamic range between 0 and 1.

SOLUTION

The decision boundaries and reconstruction levels are

$$d_0 = 0, \quad d_1 = 0.25, \quad d_2 = 0.5, \quad d_3 = 0.75, \quad d_4 = 1$$
$$r_1 = 0.125, \quad r_2 = 0.375, \quad r_3 = 0.625, \quad r_4 = 0.875$$

2.4.2.2 Signal-to-Quantization Noise Ratio (SQNR)

Let Δ denote the quantization step size of a uniform quantizer and let's assume that the signal distribution (histogram) over a given quantizer decision interval can be approximated by a uniform probability density function (PDF) as shown in

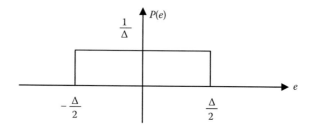

FIGURE 2.15 PDF of quantization noise.

Figure 2.15. In such a case, the quantization error for any signal sample would be uniformly distributed between $-\frac{\Delta}{2}$ and $\frac{\Delta}{2}$.

The quantization noise power is the variance of this uniform distribution, that is,

$$\sigma_e^2 = \int_{-\infty}^{\infty} e^2 P(e) de = \int_{-\frac{\Delta}{2}}^{\frac{\Delta}{2}} e^2 \frac{1}{\Delta} de = \frac{\Delta^2}{12} \qquad (2.44)$$

A B-bit quantizer for a signal in the range of $[x_{min}\ x_{max}]$ has a step size of

$$\Delta = \frac{x_{max} - x_{min}}{2^B} \qquad (2.45)$$

The signal to quantization noise power ratio (SQNR) is given by

$$SQNR = \frac{\sigma_s^2}{\sigma_n^2} = \frac{12\sigma_s^2}{\Delta^2} = \frac{12\sigma_s^2 2^{2B}}{(x_{max} - x_{min})^2} \qquad (2.46)$$

The SQNR in decibels (dB) is

$$(SQNR)_{dB} = 10\log\left(\frac{\sigma_s^2}{\sigma_n^2}\right) = 6B + 10\log_{10}\left(12\sigma_s^2\right) - 20\log_{10}(x_{max} - x_{min}) \quad (2.47)$$

Therefore, the SQNR increases by 6 dB for each additional quantization bit.

2.4.2.3   Optimum Minimum Mean-Square Error Quantizer

Assume that signal X has a PDF $p(x)$. If $p(x)$ is uniformly distributed between x_{min} and x_{max}, then the optimum quantizer in the minimum mean-square error (MMSE) sense is a uniform quantizer. However, if the PDF of X is nonuniform, the optimum quantizer will also be nonuniformly spaced. When the PDF of the input signal is not available, the smoothed data histogram can be used as an estimate of the PDF. In general, one would expect that in order to minimize the average error the quantizer structure should be finer around the peaks of the signal distribution and coarser in regions where the signal occurs infrequently. Mathematically, the reconstruction and decision levels are determined by minimizing the average distortion (MSE) given by the equation

$$D = \sum_{i=1}^{L} \int_{d_{i-1}}^{d_i} (x - r_i)^2 p(x) dx \tag{2.48}$$

To minimize the average distortion D, we set the derivatives of D with respect to r_k and d_k equal to zero. This yields

$$\frac{\partial D}{\partial d_k} = 2(d_k - r_k)p(d_k) - 2(d_k - r_{k+1})p(d_k) = 0$$

$$\frac{\partial D}{\partial r_k} = - \int_{d_{k-1}}^{d_k} 2(x - r_k)p(x) dx = 0 \tag{2.49}$$

Solving these equations results in

$$d_i = \frac{r_i + r_{i+1}}{2} \quad i = 1, 2, \ldots, L-1 \quad \text{with} \quad d_0 = x_{\min} \quad d_L = x_{\max} \tag{2.50}$$

and

$$r_i = \frac{\int_{d_{i-1}}^{d_i} x p(x) dx}{\int_{d_{i-1}}^{d_i} p(x) dx} \quad i = 1, 2, \ldots, L \tag{2.51}$$

The above MMSE solution, also known as the Lloyd–Max quantizer [6], has the following properties: Decision levels are halfway between two adjacent reconstruction levels and reconstruction levels are given by the centroids of the signal probability density that are enclosed between two adjacent decision levels. The Lloyd–Max quantizer results in the minimum-mean-square quantization error for a given number of reconstruction levels. The MMSE quantizer decision and reconstruction levels are solutions to a set of integral equations. In rare occasions (such as the Laplacian distribution), a closed form solution exists. However, in most cases, a numerical solution needs to be determined. The following popular iterative technique due to Lloyd can be used to design an MMSE quantizer with L reconstruction levels:

Step 1: Divide the range of the signal values into L uniform reconstruction levels.

Step 2: For the current set of reconstruction levels, compute the optimum decision levels using Equation 2.50.

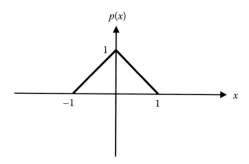

FIGURE 2.16   PDF of Example 2.5.

Step 3: For the current set of decision levels, find the optimum reconstruction levels by using Equation 2.51. If the changes in the reconstruction levels with respect to the previous values are less than a threshold, stop. Otherwise, go to Step 2. Now the process is illustrated using a simple example.

Example 2.5

The PDF of random variable X is shown in Figure 2.16. For this random variable,

   a. Design the 2 bit (4 levels) uniform quantizer and calculate the MSE.
   b. Design the 2 bit (4 levels) Lloyd–Max quantizer and calculate the MSE.

SOLUTION

The step size for uniform quantizer is $\Delta x = \frac{x_{max}-x_{min}}{4} = \frac{2}{4} = 0.5$. The reconstruction and decision levels are

$$d_0 = -1 \quad d_1 = -0.5 \quad d_2 = 0 \quad d_3 = 0.5 \quad d_4 = 1$$
$$r_1 = -0.75 \quad r_2 = -0.25 \quad r_3 = 0.25 \quad r_4 = 0.75$$

The MSE is computed as

$$D = \sum_{i=1}^{L} \int_{d_{i-1}}^{d_i} (x - r_i)^2 p(x)dx = \int_{-1}^{-0.5} (x + 0.75)^2 p(x)dx + \int_{-0.5}^{0} (x + 0.25)^2 p(x)dx$$
$$+ \int_{0}^{0.5} (x - 0.25)^2 p(x)dx + \int_{0.5}^{1} (x - 0.75)^2 p(x)dx$$

or

$$D = 2 \int_{0}^{0.5} (x - 0.25)^2 (1 - x)dx + 2 \int_{0.5}^{1} (x - 0.75)^2 (1 - x)dx = 0.0208$$

Since the PDF is symmetric, the optimum quantizer would be symmetric, that is,

$$d_0 = -1 \quad d_1 = -d \quad d_2 = 0 \quad d_3 = d \quad d_4 = 1$$

and

$$d_1 = \frac{r_1 + r_2}{2} = -d, \quad d_2 = \frac{r_2 + r_3}{2} = 0 \quad \text{and} \quad d_3 = \frac{r_3 + r_4}{2} = d$$

Therefore the reconstruction levels are

$$r_1 = -2d - r_2$$
$$r_3 = -r_2$$
$$r_4 = 2d + r_2$$

where r_2 and r_1 are given by

$$r_2 = \frac{\int_{-d}^{0} xp(x)dx}{\int_{-d}^{0} p(x)dx} = \frac{\int_{-d}^{0} x(1+x)dx}{\int_{-d}^{0} (1+x)dx} = \frac{2d^2 - 3d}{6 - 3d}$$

and

$$r_1 = \frac{\int_{-1}^{-d} xp(x)dx}{\int_{-1}^{-d_i} p(x)dx} = \frac{\int_{-1}^{-d} x(1+x)dx}{\int_{-1}^{-d_i} (1+x)dx} = \frac{3d^2 - 2d^3 - 1}{3d^2 - 6d + 3}$$

The above equations can be reduced to

$$\begin{cases} r_2 = \frac{2d^2 - 3d}{6 - 3d} \\ -2d - r_2 = \frac{3d^2 - 2d^3 - 1}{3d^2 - 6d + 3} \end{cases}$$

Eliminating r_2, we have

$$d^4 - 5d^3 + 8d^2 - 5d + 1 = 0$$

Factoring this polynomial yields

$$(d - 1)(d - 1)(d - 0.382)(d - 2.618) = 0$$

The only acceptable solution is $d = 0.382$, which results in $r_2 = -0.176$. Therefore, the decision boundaries and the reconstruction levels are

$$d_0 = -1 \quad d_1 = -0.382 \quad d_2 = 0 \quad d_3 = 0.382 \quad d_4 = 1$$
$$r_1 = -0.588 \quad r_2 = -0.176 \quad r_3 = 0.176 \quad r_4 = 0.588$$

The MMSE is

$$D = \sum_{i=1}^{L} \int_{d_{i-1}}^{d_i} (x - r_i)^2 p(x)dx = \int_{-1}^{-0.382} (x + 0.588)^2 p(x)dx + \int_{-0.382}^{0} (x + 0.176)^2 p(x)dx$$

$$+ \int_{0}^{0.382} (x - 0.176)^2 p(x)dx + \int_{0.382}^{1} (x - 0.588)^2 p(x)dx$$

This can be simplified as

$$D = 2 \int_{0}^{0.382} (x - 0.176)^2 (1 - x)dx + 2 \int_{0.382}^{1} (x - 0.588)^2 (1 - x)dx = 0.01548$$

Notice that the resulting MSE from Lloyd–Max quantizer is smaller than the MSE obtained in part (a) from uniform quantizer.

Example 2.6

Design a 3 bit optimum quantizer for a signal with unit variance Laplacian distribution. The unit variance Laplacian distribution is given by the following PDF.

$$P(x) = \frac{\sqrt{2}}{2} e^{-\sqrt{2}|x|} \quad -\infty < x < \infty$$

SOLUTION

Using the Lloyd–Max iterative algorithm, the following values are obtained for a 3 bit optimum quantizer

$$d_0 = -\infty \quad d_1 = -2.3654 \quad d_2 = -1.2474 \quad d_3 = -0.5313 \quad d_4 = 0$$
$$d_5 = 0.5313 \quad d_6 = 1.2474 \quad d_7 = 2.3654 \quad d_8 = \infty$$
$$r_1 = -3.0658 \quad r_2 = -1.664 \quad r_3 = -0.83 \quad r_4 = -0.2327$$
$$r_5 = 0.2327 \quad r_6 = 0.83 \quad r_7 = 1.664 \quad r_8 = 3.0658$$

The PDF of the Laplacian distribution and its 3 bit quantization levels are shown in Figure 2.17.

2.4.2.4 Perceptual Quantization

Digital images are obtained either through the process of scanning a raw continuous tone image (e.g., facsimile) or direct acquisition (e.g., digital cameras). In digital cameras, the sampling is accomplished by a CCD sensor array. At each pixel, a photodiode converts the incident photons into electrons that are subsequently counted as the pixel intensity. In many imaging systems, it is customary to quantize the pixel value to 8 bits (256 levels), since it is argued that the human eye cannot distinguish more than 100–200 brightness levels. Quantization to 12 or even 14 bits is not

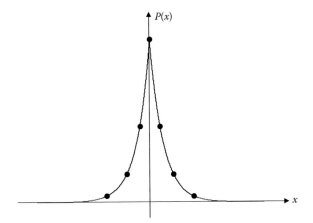

FIGURE 2.17   Laplacian distribution with a 3 bit optimum quantizer.

uncommon in medical imaging systems. It is well known that uniform quantization of the intensity values (photon counts) usually creates visual artifacts. This is because equal changes in the number of photons do not correspond to equal changes in the perceived brightness. For example, a small change in the photon count at low intensities would be visible, while a large change in the photon count at high intensities may not be visible. Consequently, to take full advantage of the available quantization levels, in many imaging systems the quantization is performed in a perceptual domain as opposed to an intensity domain. It has been argued that the HVS responds almost to the percentage changes in the intensity. This suggests an approximate logarithmic relationship between the light intensity and the perceived brightness. There are two methods for implementing the perceptual quantization in imaging systems: In the first technique, the intensity of the pixel value is first quantized to a large number of levels (e.g., 10–14 bits) during the image digitization stage. Subsequently, the pixel value is requantized to 8 bits through a nonlinear digital lookup table (LUT) that simulates a logarithmic or a cube-root transformation. For example, a high-resolution scanner measures the transmittance of each pixel in roughly 13 bits and uses a cube-root LUT to convert them into 8 bit values. Similarly, some digital cameras initially digitize the pixel intensity to 10 bits and subsequently requantize it to 8 bits using LUTs based on piecewise linear curves simulating a cube-root nonlinearity. In the second approach, the analog signal is first converted to a perceptual domain through an analog nonlinearity (such as a log amplifier) and the output of the amplifier is quantized to 8 bits. This approach, although less costly, is also less accurate.

2.4.2.5   Vector Quantization

In VQ, the concept of a reconstruction level is generalized to a reconstruction vector and the decision levels are generalized to decision regions. In general, VQ introduces less quantization error than SQ for the same number of bits [5,7,8]. As an example, let's assume that each signal value is scalar quantized to 4 bits (16 levels). To achieve the same number of bits, the joint vector should be quantized to 8 bits, which provides a choice of 256 reconstruction vectors. To show that VQ outperforms SQ, consider two random variables X_1 and X_2 with a joint PDF $p_{X_1, X_2}(x_1, x_2)$ given by

$$p_{X_1, X_2}(x_1, x_2) = \begin{cases} \frac{1}{2} & (x_1, x_2) \in \text{dashed area} \\ 0 & \text{otherwise} \end{cases} \tag{2.52}$$

The marginal densities of random variables X_1 and X_2 are

$$p_{X_1}(x_1) = \int_{-\infty}^{\infty} p_{X_1, X_2}(x_1, x_2) dx_2 = \begin{cases} \frac{1}{2} & -1 < x_1 < 1 \\ 0 & \text{otherwise} \end{cases} \tag{2.53}$$

$$p_{X_2}(x_2) = \int_{-\infty}^{\infty} p_{X_1, X_2}(x_1, x_2) dx_1 = \begin{cases} \frac{1}{2} & -1 < x_2 < 1 \\ 0 & \text{otherwise} \end{cases} \tag{2.54}$$

Therefore, X_1 and X_2 are uniformly distributed between -1 and $+1$. They are not independent since $p_{X_1, X_2}(x_1, x_2) \neq p_{X_1}(x_1) p_{X_2}(x_2)$. The correlation coefficient between x_1 and x_2 is

$$\rho = \frac{E(X_1 X_2) - E(X_1)E(X_2)}{\sigma_{X_1} \sigma_{X_2}} = \frac{\frac{1}{4} - 0}{\sqrt{\frac{4}{12}}\sqrt{\frac{4}{12}}} = 0.75 \tag{2.55}$$

If we quantize X_1 and X_2 separately, each to 1 bit, the optimum quantizer will be a uniform quantizer with decision boundaries $\{d_1 \quad d_2 \quad d_3\} = \{-1 \quad 0 \quad 1\}$ and quantization levels of -0.5 and 0.5. The total average distortion is

$$D = \int_{-1}^{0} (x_1 + 0.5)^2 p(x_1) dx_1 + \int_{0}^{1} (x_1 - 0.5)^2 p(x_1) dx_1$$

$$+ \int_{-1}^{0} (x_2 + 0.5)^2 p(x_2) dx_2 + \int_{0}^{1} (x_2 - 0.5)^2 p(x_2) dx_2 = \frac{1}{6} \tag{2.56}$$

The quantization levels for X_1 and X_2 are shown in Figure 2.18. In this case, 2 bits are used and the average distortion is $\frac{1}{6}$. The same average distortion can be achieved using 1 bit VQ. The reconstruction vectors would be $r_1 = [0.5 \quad 0.5]^T$ and $r_2 = [-0.5 \quad -0.5]^T$ as shown in Figure 2.18. As can be seen from Figure 2.18, the two quantization vectors c and d are wasted if we use SQ.

The performance of the vector quantizer improves by increasing vector dimension. The design of the MMSE vector quantizer is conceptually similar to the scalar quantizer but is computationally more complex. The generalized iterative Lloyd algorithm (also known as the LBG (Linde–Buzo–Gray) or K-means algorithm) is used to design the optimum decision regions and reconstruction vectors.

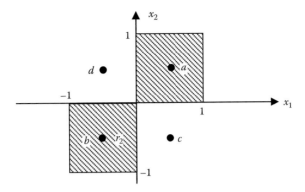

FIGURE 2.18 VQ and SQ.

Definition: VQ. Let $X = [x_1 \quad x_2 \quad \ldots \quad x_N]^T$ be an N-dimensional vector consisting of N scalars x_i, $1 \le i \le N$. One approach to quantize X is to quantize each component of vector X separately (SQ). An alternative approach is joint quantization of scalars x_i's. This is called VQ.

Quantization Rule:

$$X_Q = r_i \quad \text{if } X \text{ is in cell } D_i \quad i = 1, 2, \ldots, L \tag{2.57}$$

where $r_i, i = 1, 2, \ldots, L$ are N-dimensional reconstruction vectors and L is the number of levels. The concept of VQ for $N = 2$ and $L = 4$ is shown in Figure 2.19. The reconstruction vectors r_i and the cells D_i are determined by minimizing the total distortion defined by mean square quantization error, that is,

$$D = E\left[(X - X_Q)^T (X - X_Q)\right] \tag{2.58}$$

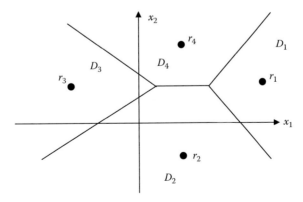

FIGURE 2.19 VQ concept when the number of scalars in the vector is 2 and the number of reconstruction vectors is 4.

K-means Algorithm: Given M training vectors $\{X_1 \quad X_2 \quad \dots \quad X_M\}$, we need to find the L reconstruction vectors $\{r_1 \quad r_2 \quad \dots \quad r_L\}$ and their corresponding regions $\{D_1 \quad D_2 \quad \dots \quad D_L\}$. This is a classification problem that can be solved iteratively by K-means algorithm. Following is a summary of the K-means algorithm:

1. *Initialization*: Start with an initial estimate of reconstruction vectors r_i for $i = 1, 2, \dots, L$. They can be picked randomly or we can pick L vectors from the training data set.
2. *Classification*: Classify the M training vectors into L groups. Vector X belongs to D_i, if and only if $\|X - r_i\| < \|X - r_j\|$, or all $1 \le j \le L$, where $\|\ \|$ stands for Euclidean norm.
3. *Total distortion*: Compute the overall average distortion $D = \frac{1}{M} \sum_{i=1}^{M} \|X_i - X_i^Q\|$ where $X_i^Q = r_k$ if X_i is in cell D_k.
4. If the average distortion is small (less than a prescribed threshold), stop the iteration. Otherwise, continue to the next step.
5. *Centroid*: Find a new estimate of the reconstruction vector r_i by computing the centroid of the ith cell D_i and go to step 2. The centroid of D_i is computed by $r_i = \frac{1}{K_i} \sum (\text{all vectors in } D_i)$, where K_i is the number of vectors in cell D_i.

Example 2.7

Given the five training vectors:

$$X_1 = \begin{bmatrix} -1 \\ -2 \\ 2 \end{bmatrix}, \quad X_2 = \begin{bmatrix} -1 \\ 2 \\ -2 \end{bmatrix}, \quad X_3 = \begin{bmatrix} 0 \\ 2 \\ 2 \end{bmatrix}, \quad X_4 = \begin{bmatrix} -1 \\ 0 \\ 2 \end{bmatrix} \quad \text{and} \quad X_5 = \begin{bmatrix} 2 \\ -2 \\ 0 \end{bmatrix}$$

design a 1 bit vector quantizer using K-means algorithm.

SOLUTION

For a 1 bit quantizer $L = 2$ and there are two reconstruction vectors. Let the initial vectors be $r_1 = \begin{bmatrix} -1 \\ -2 \\ 2 \end{bmatrix}$ and $r_2 = \begin{bmatrix} -1 \\ 2 \\ -2 \end{bmatrix}$.

$$\|X_1 - r_1\| = 0, \qquad \|X_1 - r_2\| = 5.656 \rightarrow X_1 \in D_1$$

$$\|X_2 - r_1\| = 5.656, \qquad \|X_2 - r_2\| = 0 \rightarrow X_2 \in D_2$$

$$\|X_3 - r_1\| = 4.123, \qquad \|X_3 - r_2\| = 4.123 \rightarrow X_3 \in D_1$$

$$\|X_4 - r_1\| = 2, \qquad \|X_4 - r_2\| = 4.472 \rightarrow X_4 \in D_1$$

$$\|X_5 - r_1\| = 3.605, \qquad \|X_5 - r_2\| = 5.385 \rightarrow X_5 \in D_1$$

Therefore,

$$D_1 = \{ X_1 \quad X_3 \quad X_4 \quad X_5 \} \quad D_2 = \{X_2\}$$

The total average distortion is

$$D = \frac{0 + 0 + 4.123 + 2 + 3.605}{5} = 1.9456$$

The new reconstruction vectors are computed as

$$r_1 = \frac{X_1 + X_3 + X_4 + X_5}{4} = \begin{bmatrix} 0 \\ -0.5 \\ 1.5 \end{bmatrix} \quad r_2 = X_2 = \begin{bmatrix} -1 \\ 2 \\ -2 \end{bmatrix}$$

Classifying the training vectors with new centroid results in

$$\|X_1 - r_1\| = 1.8708, \quad \|X_1 - r_2\| = 5.656 \rightarrow X_1 \in D_1$$

$$\|X_2 - r_1\| = 4.415, \quad \|X_2 - r_2\| = 0 \rightarrow X_2 \in D_2$$

$$\|X_3 - r_1\| = 2.549, \quad \|X_3 - r_2\| = 4.123 \rightarrow X_3 \in D_1$$

$$\|X_4 - r_1\| = 1.2247, \quad \|X_4 - r_2\| = 4.472 \rightarrow X_4 \in D_1$$

$$\|X_5 - r_1\| = 2.915, \quad \|X_5 - r_2\| = 5.385 \rightarrow X_5 \in D_1$$

Therefore,

$$D_1 = \{ X_1 \quad X_3 \quad X_4 \quad X_5 \} \quad D_2 = \{X_2\}$$

The total average distortion is

$$D = \frac{1.8708 + 0 + 2.549 + 1.2247 + 2.915}{5} = 1.7119$$

The new reconstruction vectors are

$$r_1 = \frac{X_1 + X_3 + X_4 + X_5}{4} = \begin{bmatrix} 0 \\ -0.5 \\ 1.5 \end{bmatrix} \quad r_2 = X_2 = \begin{bmatrix} -1 \\ 2 \\ -2 \end{bmatrix}$$

Carrying out one more iteration does not change the overall average distortion.

The K-means algorithm is locally optimal and there is no guarantee that it will converge to the global optimal solution. It is also very slow because for every iteration, all the vectors in the data base are compared with each codeword vector. There are other algorithms, such as pairwise nearest neighbor and simulated annealing for VQ design, that are faster than K-means algorithm.

An important application of VQ is image data compression. Image data compression using VQ is a lossy compression technique. Once the codebook has been designed, the uncompressed image is divided into blocks and each block is converted

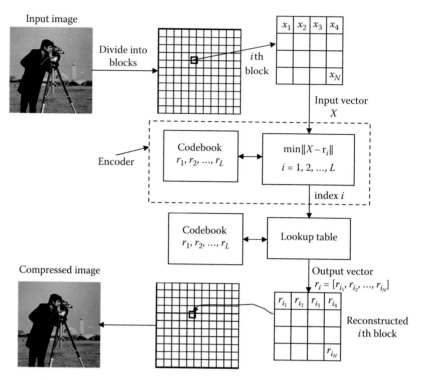

FIGURE 2.20　Encoder and decoder for VQ-based image compression technique.

into a vector. These vectors are input to the encoder. The encoder will compare each input vector with every codevector in the codebook and output an index corresponding to the minimum distortion codevector from the codebook. The decoder will use this index to pick the corresponding codeword from the codebook and generate the output vector. These output vectors are used to reconstruct the compressed image. The encoder and decoder for a VQ image data compression system are shown in Figure 2.20.

2.5　IMAGE TRANSFORM

There are different transforms that can be used in image processing. The most common transforms are discrete Fourier transform (DFT), discrete cosine transform (DCT), discrete Hadamard transform, and Karhunen–Loeve (KL) transform [9–12].

2.5.1　TWO-DIMENSIONAL DISCRETE FOURIER TRANSFORM

The 2-D DFT provides an expansion of $N \times M$ samples of a discrete 2-D sequence in terms of discrete sine and cosine signals, in much the same way as a continuous signal can be expanded into sine and cosine using the continuous Fourier transform. The DFT of an $N \times M$ discrete signal $f(n, m)$ is defined as

$$F(k, l) = \sum_{m=0}^{M-1} \sum_{n=0}^{N-1} f(n, m) e^{-j\frac{2\pi}{N}nk} e^{-j\frac{2\pi}{M}ml} \qquad (2.59)$$

For $k = 0, 1, 2, \ldots, N - 1$, and $l = 0, 1, 2, \ldots, M - 1$. The DFT coefficients are samples of $F(\omega_x, \omega_y)$ uniformly spaced at $\omega_x = \frac{2\pi}{N}k$ and $\omega_y = \frac{2\pi}{M}l$. They are generally complex valued and can be used to expand $f(n, m)$ in terms of 2-D complex exponential signals (basis functions). This expansion is called the inverse DFT and is given by

$$f(n, m) = \frac{1}{MN} \sum_{l=0}^{M-1} \sum_{k=0}^{N-1} F(k, l) e^{j\frac{2\pi}{N}nk} e^{j\frac{2\pi}{M}ml} \qquad (2.60)$$

The discrete frequency indices k and l correspond to the analog frequencies $\frac{k}{N}f_x^s$ and $\frac{k}{N}f_y^s$, where f_x^s and f_y^s are the sampling frequencies in the x- and y-directions, respectively. The complexity of DFT computation is measured in terms of the number of multiplications required in implementing Equation 2.59. Assuming that $N = M$, the number of complex multiplications needed to compute one sample of $F(k, l)$ is N^2. Since there are N^2 DFT samples to be computed, the total number of multiplications is N^4. We refer to this as direct computation. We can compute 2-D DFT more efficiently by using the fact that the basis functions involved in DFT expansion are separable. As a result of this, we can compute the 2-D DFT by taking the 1-D DFT of each row of the image and then taking the 1-D DFT of the columns of the intermediate image $G(n, l)$, as shown in Figure 2.21.

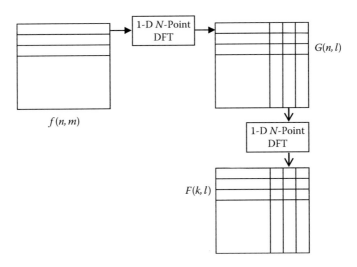

FIGURE 2.21 Row–column decomposition.

TABLE 2.1

Direct 2-D DFT/FFT Comparison

Technique	No. of Complex Multiplications	%
Direct computation	$N^4 = 7.871947 \times 10^{10}$	100
Row–column decomposition (1-D DFT)	$2N^3 = 2.73154048 \times 10^8$	0.4
Row–column decomposition (1-D FFT)	$N^2 \log_2 (N) = 2.359296 \times 10^6$	0.0035

This is called the row–column decomposition technique. With row–column decomposition, we need to compute $2N$ 1-D DFTs and each 1-D DFT requires N^2 multiplications. Therefore, the total number of complex multiplications needed is $2N \times N^2 = 2N^3$. Now, if we use row–column decomposition with fast Fourier transform (FFT) algorithm, the total number of multiplications will be reduced to $2N \times \frac{N}{2} \log_2 (N) = N^2 \log_2 (N)$. This is due to the fact that the total number of multiplications necessary to compute N-point FFT is $\frac{N}{2} \log_2 (N)$. To see the computational advantage of using FFT for 2-D DFT computation, consider calculating the 2-D DFT of a 512×512 image using direct computation, row–column decomposition and row–column decomposition with the FFT algorithm replacing direct DFT computation. The comparison is shown in Table 2.1.

Properties of 2-D DFT

The properties of 2-D DFT are listed below. The proofs are left as an exercise.

a. *Linearity property*: If $f_1(n, m) \overset{\text{2-D DFT}}{\longleftrightarrow} F_1(k, l)$ and $f_2(n, m) \overset{\text{2-D DFT}}{\longleftrightarrow} F_2(k, l)$, then

$$af_1(n, m) + bf_2(n, m) \overset{\text{2-D DFT}}{\longleftrightarrow} aF_1(k, l) + bF_2(k, l) \qquad (2.61)$$

b. *Conjugation property*: If $f(n, m)$ is a real sequence, then its DFT $F(k, l)$ satisfies the following conjugation property

$$F(k, l) = F^*(N - k, M - l) \qquad (2.62)$$

This means that half of the DFT data is redundant data.

c. *Rotation property*: If $f(n, m) \overset{\text{2-D DFT}}{\longleftrightarrow} F(k, l)$ or $f(r, \theta) \overset{\text{2-D DFT}}{\longrightarrow} F(\omega, \varphi)$ where

$$n = r \cos \theta, \quad m = r \sin \theta, \quad k = \omega \cos \varphi, \quad l = \omega \sin \varphi$$

then,

$$f(r, \theta + \theta_0) \xrightarrow{\text{2-D DFT}} F(\omega, \varphi + \theta_0) \tag{2.63}$$

This implies that rotating $f(n, m)$ by an angle θ_0 rotates its transform $F(k, l)$ by the same angle.

d. *Convolution property of DFT*: The convolution theorem for continuous signals states that convolution in space domain is equivalent to multiplication in the Fourier domain. This theorem can be extended to the discrete domain; however, it should be noted that ordinary convolution is replaced by circular convolution, that is,

$$f(n, m) \otimes h(n, m) \xrightarrow{\text{DFT}} F(k, l)H(k, l) \tag{2.64}$$

The circular convolution is defined as

$$f(n, m) \otimes h(n, m) = \sum_{v}\sum_{u} f(u, v)h(n - u, m - v) \tag{2.65}$$

where the shift $(n - u$ or $m - v)$ is circular shift. The linear discrete convolution is the type of convolution used for filtering signals and images and also for finding the output of imaging systems modeled in linear shift-invariant form. Circular convolution is a side effect of DFT.

Example 2.8

If $x(n, m) = \begin{bmatrix} 2 & 1 \\ -1 & 3 \end{bmatrix}$ and $h(n, m) = \begin{bmatrix} 1 & 1 \\ 1 & 1 \end{bmatrix}$, find

a. $y(n, m) = x(n, m) \otimes h(n, m)$, circular convolution of $x(n, m)$ and $h(n, m)$.
b. $g(n, m) = x(n, m)*h(n, m)$, linear convolution of $x(n, m)$ and $h(n, m)$.

SOLUTION

a. Circular convolution

$$y(n, m) = x(n, m) \otimes h(n, m) = \sum_{v=0}^{1}\sum_{u=0}^{1} h(u, v)x(n - u \bmod 2, y - v \bmod 2)$$

Expanding the above equation yields

$$\begin{aligned} y(n, m) &= h(0, 0)x(n, m) + h(0, 1)x(n, m - 1) + h(1, 0)x(n - 1, m) \\ &\quad + h(1, 1)x(n - 1, m - 1) \\ &= x(n, m) + x(n, m - 1) + x(n - 1, m) + x(n - 1, m - 1) \end{aligned}$$

If $n = 0$ and $m = 0$, then

$$y(0, 0) = x(0, 0) + x(0, -1) + x(-1, 0) + x(-1, -1)$$
$$= x(0, 0) + x(0, 1) + x(1, 0) + x(1, 1) = 2 + 1 - 1 + 3 = 5$$

If $n = 0$ and $m = 1$, then

$$y(0, 1) = x(0, 1) + x(0, 0) + x(-1, 1) + x(-1, 0)$$
$$= x(0, 1) + x(0, 0) + x(1, 1) + x(1, 0) = 1 + 2 + 3 - 1 = 5$$

If $n = 1$ and $m = 0$, then

$$y(1, 0) = x(1, 0) + x(1, -1) + x(0, 0) + x(0, -1)$$
$$= x(1, 0) + x(1, 1) + x(0, 0) + x(0, 1) = -1 + 3 + 2 + 1 = 5$$

Finally, if $n = 1$ and $m = 1$, then

$$y(1, 1) = x(1, 1) + x(0, 1) + x(0, 1) + x(0, 0) = 3 + 1 - 1 + 2 = 5$$

Therefore,

$$y(n, m) = x(n, m) \otimes h(n, m) = \begin{bmatrix} 5 & 5 \\ 5 & 5 \end{bmatrix}$$

b. Linear convolution. In this case, we have

$$y(n, m) = x(n, m) * h(n, m)$$
$$= \sum_{v=0}^{1} \sum_{u=0}^{1} h(u, v)x(n - u, y - v) \quad n = 0, 1, 2 \quad m = 0, 1, 2$$

Expanding the above sum results in

$$y(n, m) = h(0, 0)x(n, m) + h(0, 1)x(n, m - 1) + h(1, 0)x(n - 1, m)$$
$$+ h(1, 1)x(n - 1, m - 1)$$
$$= x(n, m) + x(n, m - 1) + x(n - 1, m) + x(n - 1, m - 1)$$

Computing $y(n, m)$ for $n = 0, 1, 2$ and $m = 0, 1, 2$, we would have

$$y(0, 0) = x(0, 0) + x(0, -1) + x(-1, 0) + x(-1, -1) = 2 + 0 + 0 + 0 = 2$$
$$y(0, 1) = x(0, 1) + x(0, 0) + x(-1, 1) + x(-1, 0) = 1 + 2 + 0 + 0 = 3$$
$$y(0, 2) = x(0, 2) + x(0, 1) + x(-1, 2) + x(-1, 1) = 0 + 1 + 0 + 0 = 1$$
$$y(1, 0) = x(1, 0) + x(1, -1) + x(0, 0) + x(0, -1) = -1 + 0 + 2 + 0 = 1$$
$$y(1, 1) = x(1, 1) + x(0, 1) + x(0, 1) + x(0, 0) = 3 + 1 - 1 + 2 = 5$$

$$y(1, 2) = x(1, 2) + x(1, 1) + x(0, 2) + x(0, 1) = 0 + 3 + 0 + 1 = 4$$

$$y(2, 0) = x(2, 0) + x(2, -1) + x(1, 0) + x(1, -1) = 0 + 0 - 1 + 0 = -1$$

$$y(2, 1) = x(2, 1) + x(2, 0) + x(1, 1) + x(1, 0) = 0 + 0 + 3 - 1 = 2$$

$$y(2, 2) = x(2, 2) + x(2, 1) + x(1, 2) + x(1, 1) = 0 + 0 + 0 + 3 = 3$$

Hence

$$y(n, m) = x(n, m) * h(n, m) = \begin{bmatrix} 2 & 3 & 1 \\ 1 & 5 & 4 \\ -1 & 2 & 3 \end{bmatrix}$$

So, in general, to perform fast convolution, one would be interested in computing linear convolution via DFT. As will be shown, the FFT can be employed for this purpose provided that the data is properly formatted to avoid circular convolution. Suppose that signal $f(n, m)$ is $M \times M$ and the filter $h(n, m)$ is $L \times L$ and we would like to compute $g(n, m)$, the linear convolution of $f(n, m)$ and $h(n, m)$ using DFT.

a. Choose N the DFT size to be $N \geq M + L - 1$
b. Define $N \times N$ arrays

$$\hat{f}(n, m) = \begin{cases} f(n, m) & 0 \leq n, m \leq M - 1 \\ 0 & \text{otherwise} \end{cases} \tag{2.66}$$

$$\hat{h}(n, m) = \begin{cases} h(n, m) & 0 \leq n, m \leq L - 1 \\ 0 & \text{otherwise} \end{cases} \tag{2.67}$$

c. Compute

$$\hat{g}(n, m) = \text{IDFT}\{\text{DFT}[\hat{f}(n, m)] \text{DFT}[\hat{h}(n, m)]\} \tag{2.68}$$

d. Find

$$g(n, m) = \hat{g}(n, m) \quad 0 \leq n, m \leq M + L - 1 \tag{2.69}$$

Example 2.9

Convolve the following 2-D sequences using DFT,

$$f(n, m) = \begin{bmatrix} 1 & 2 & 4 \\ 2 & 3 & 2 \\ 3 & 1 & 4 \end{bmatrix} \quad \text{and} \quad h(n, m) = \begin{bmatrix} 1 & 3 \\ 2 & 1 \end{bmatrix}$$

SOLUTION

a. Choose DFT size to be $N \geq 3 + 2 - 1 = 4$ and let $N = 4$.
b. Zero-pad the signal $f(n, m)$ and compute its DFT

$$\hat{f}(n,m) = \begin{bmatrix} 1 & 2 & 4 & 0 \\ 1 & 3 & 2 & 0 \\ 3 & 1 & 4 & 0 \\ 0 & 0 & 0 & 0 \end{bmatrix} \xrightarrow{\text{2-D DFT}} \hat{F}(k,l) = \begin{bmatrix} 21 & -5-j6 & 9 & -5+j6 \\ -1-j6 & -5 & -3 & 1+j2 \\ 9 & -3 & 9 & -3 \\ -1+j6 & 1-j2 & -3 & -5 \end{bmatrix}$$

c. Zero-pad the signal $h(n,m)$ and compute its DFT

$$\hat{h}(n,m) = \begin{bmatrix} 1 & 3 & 0 & 0 \\ 2 & 1 & 0 & 0 \\ 0 & 0 & 0 & 0 \\ 0 & 0 & 0 & 0 \end{bmatrix} \xrightarrow{\text{2-D DFT}} \hat{H}(k,l) = \begin{bmatrix} 7 & 3-j4 & -1 & 3+j4 \\ 4-j3 & -j5 & -2-j & 2+j \\ 1 & -1-j2 & -3 & -1+j2 \\ 4+j3 & 2-j & -2+j & j5 \end{bmatrix}$$

d. Multiply the 2 DFTs

$$\hat{G}(k,l) = \hat{F}(k,l)\hat{H}(k,l) = \begin{bmatrix} 147 & -39+j2 & 9 & -39-j2 \\ -22-j21 & j25 & 6+j3 & j5 \\ 9 & 3+j6 & -27 & 3-j6 \\ -22+j21 & -j5 & 6-j3 & -j25 \end{bmatrix}$$

e. Take the inverse DFT of $\hat{G}(k,l)$

$$g(n,m) = \text{IDFT}[G(k,l)] = \begin{bmatrix} 1 & 5 & 10 & 12 \\ 3 & 11 & 21 & 10 \\ 5 & 17 & 14 & 14 \\ 6 & 5 & 9 & 4 \end{bmatrix}$$

In practice, it may not be practical to compute DFT if the DFT size is large. In this case, block convolution can be employed. As an example, consider designing a convolver for filtering an image f of size $M \times M$ by an finite impulse response (FIR) filter of size $L \times L$, using FFT hardware capable of performing 1-D N-point FFT. We first divide the image data into blocks of size $B \times B$, where $B = N - L + 1$, as shown in Figure 2.22. There are approximately $N_B = (\frac{M}{B})^2$ of such image blocks.

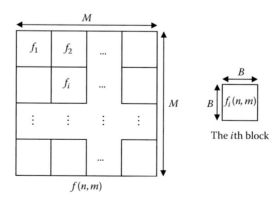

FIGURE 2.22 Dividing the image into subimages.

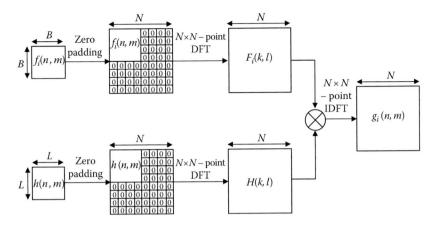

FIGURE 2.23   Filtering an image block using DFT.

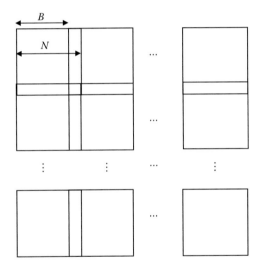

FIGURE 2.24   Overlap and add technique.

We then convolve each block with $h(n, m)$ using the DFT approach as outlined in Section 2.5.1 and shown in Figure 2.23. That is, for $i = 1, 2, \ldots, N_B$ we compute,

$$g_i(n, m) = f_i(n, m) * h(n, m) \tag{2.70}$$

Note that the 2-D DFT computation is performed using row–column decomposition and a 1-D FFT algorithm.

The output image is formed by overlapping and adding these N_B blocks as illustrated in Figure 2.24.

Example 2.10

Compute $g(n, m) = f(n, m) * h(n, m)$ using a 4-point FFT, where

$$
f(n, m) = \begin{bmatrix}
19 & 16 & 19 & 3 & 9 & 17 & 6 & 8 \\
5 & 9 & 18 & 4 & 19 & 0 & 4 & 17 \\
12 & 12 & 8 & 4 & 9 & 14 & 4 & 17 \\
10 & 16 & 18 & 12 & 8 & 8 & 14 & 12 \\
18 & 18 & 1 & 5 & 17 & 17 & 6 & 10 \\
15 & 15 & 7 & 4 & 11 & 10 & 11 & 18 \\
9 & 4 & 16 & 0 & 4 & 14 & 3 & 16 \\
0 & 8 & 0 & 15 & 13 & 9 & 14 & 13
\end{bmatrix}
\quad \text{and} \quad
h(n, m) = \begin{bmatrix} 1 & 1 \\ 1 & 1 \end{bmatrix}
$$

SOLUTION

First we find the block size B

$$
B = N - L + 1 = 4 - 2 + 1 = 3
$$

Now, divide the image into $B^2 = 9$ blocks

$$
f_1(n, m) = \begin{bmatrix} 19 & 16 & 19 \\ 5 & 9 & 18 \\ 12 & 12 & 8 \end{bmatrix}, \quad
f_2(n, m) = \begin{bmatrix} 3 & 9 & 17 \\ 4 & 19 & 0 \\ 4 & 9 & 14 \end{bmatrix}, \quad
f_3(n, m) = \begin{bmatrix} 6 & 8 & 0 \\ 4 & 17 & 0 \\ 4 & 17 & 0 \end{bmatrix}
$$

$$
f_4(n, m) = \begin{bmatrix} 10 & 16 & 18 \\ 18 & 18 & 1 \\ 15 & 15 & 7 \end{bmatrix}, \quad
f_5(n, m) = \begin{bmatrix} 12 & 8 & 8 \\ 5 & 17 & 17 \\ 4 & 11 & 10 \end{bmatrix}, \quad
f_6(n, m) = \begin{bmatrix} 14 & 12 & 0 \\ 6 & 10 & 0 \\ 11 & 18 & 0 \end{bmatrix}
$$

$$
f_7(n, m) = \begin{bmatrix} 9 & 4 & 16 \\ 0 & 8 & 0 \\ 0 & 0 & 0 \end{bmatrix}, \quad
f_8(n, m) = \begin{bmatrix} 0 & 4 & 14 \\ 15 & 13 & 9 \\ 0 & 0 & 0 \end{bmatrix}, \quad
f_9(n, m) = \begin{bmatrix} 3 & 16 & 0 \\ 14 & 13 & 0 \\ 0 & 0 & 0 \end{bmatrix}
$$

Next, we convolve each block with $h(n, m)$ using FFT and inverse FFT, that is, for $i = 1, 2, \ldots, 9$, we compute

$$
g_i(n, m) = IFFT\{FFT[\tilde{f}_i(n, m)]FFT[\tilde{h}(n, m)]\}
$$

where:

$$
\tilde{f}_i(n, m) = \begin{bmatrix} & & & 0 \\ & f_i(n, m) & & 0 \\ & & & 0 \\ 0 & 0 & 0 & 0 \end{bmatrix}
\quad \text{and} \quad
\tilde{h}_i(n, m) = \begin{bmatrix} 1 & 1 & 0 & 0 \\ 1 & 1 & 0 & 0 \\ 0 & 0 & 0 & 0 \\ 0 & 0 & 0 & 0 \end{bmatrix}
$$

All FFTs and IFFTs are 4×4 and are computed using row–column decomposition. The results are

$$g_1 = \begin{bmatrix} 19 & 35 & 35 & 19 \\ 24 & 49 & 62 & 37 \\ 17 & 38 & 47 & 26 \\ 12 & 24 & 20 & 8 \end{bmatrix}, \quad g_2 = \begin{bmatrix} 3 & 12 & 26 & 17 \\ 7 & 35 & 45 & 17 \\ 8 & 36 & 42 & 14 \\ 4 & 13 & 23 & 14 \end{bmatrix}, \quad g_3 = \begin{bmatrix} 6 & 14 & 8 & 0 \\ 10 & 35 & 25 & 0 \\ 8 & 42 & 34 & 0 \\ 4 & 21 & 17 & 0 \end{bmatrix}$$

$$g_4 = \begin{bmatrix} 10 & 26 & 34 & 18 \\ 28 & 62 & 53 & 19 \\ 33 & 66 & 41 & 8 \\ 15 & 30 & 22 & 7 \end{bmatrix}, \quad g_5 = \begin{bmatrix} 12 & 20 & 16 & 8 \\ 17 & 42 & 50 & 25 \\ 9 & 37 & 55 & 27 \\ 4 & 15 & 21 & 10 \end{bmatrix}, \quad g_6 = \begin{bmatrix} 14 & 26 & 12 & 0 \\ 20 & 42 & 22 & 0 \\ 17 & 45 & 28 & 0 \\ 11 & 29 & 18 & 0 \end{bmatrix}$$

$$g_7 = \begin{bmatrix} 9 & 13 & 20 & 16 \\ 9 & 21 & 28 & 16 \\ 0 & 8 & 8 & 0 \\ 0 & 0 & 0 & 0 \end{bmatrix}, \quad g_8 = \begin{bmatrix} 0 & 4 & 18 & 14 \\ 15 & 32 & 40 & 23 \\ 15 & 28 & 22 & 9 \\ 0 & 0 & 0 & 0 \end{bmatrix}, \quad g_9 = \begin{bmatrix} 3 & 19 & 16 & 0 \\ 17 & 46 & 29 & 0 \\ 14 & 27 & 13 & 0 \\ 0 & 0 & 0 & 0 \end{bmatrix}$$

Overlapping and adding blocks g_1 through g_9 yields

$$g = \begin{bmatrix}
19 & 35 & 35 & 19+3 & 12 & 26 & 17+6 & 14 & 8 \\
24 & 49 & 62 & 37+7 & 35 & 45 & 17+10 & 35 & 25 \\
17 & 38 & 47 & 26+8 & 36 & 42 & 14+8 & 42 & 34 \\
12+10 & 24+26 & 20+34 & 8+4+18+12 & 13+20 & 23+16 & 14+4+8+14 & 21+26 & 17+12 \\
28 & 62 & 53 & 19+17 & 42 & 50 & 25+20 & 42 & 22 \\
33 & 66 & 41 & 8+9 & 37 & 55 & 27+17 & 45 & 28 \\
15+9 & 30+13 & 22+20 & 7+4+16+0 & 15+4 & 21+18 & 10+11+14+3 & 29+19 & 18+16 \\
9 & 21 & 28 & 16+15 & 32 & 40 & 23+17 & 46 & 29 \\
0 & 8 & 8 & 0+15 & 28 & 22 & 9+14 & 27 & 13
\end{bmatrix}$$

Hence,

$$g(n,m) = \begin{bmatrix}
19 & 35 & 35 & 22 & 12 & 26 & 23 & 14 & 8 \\
24 & 49 & 62 & 44 & 35 & 45 & 27 & 35 & 25 \\
17 & 38 & 47 & 34 & 36 & 42 & 22 & 42 & 34 \\
22 & 50 & 54 & 42 & 33 & 39 & 40 & 47 & 29 \\
28 & 62 & 53 & 36 & 42 & 50 & 45 & 42 & 22 \\
33 & 66 & 41 & 17 & 37 & 55 & 44 & 45 & 28 \\
24 & 43 & 42 & 27 & 19 & 39 & 38 & 48 & 34 \\
9 & 21 & 28 & 31 & 32 & 40 & 40 & 46 & 29 \\
0 & 8 & 8 & 15 & 28 & 22 & 23 & 27 & 13
\end{bmatrix}$$

Example 2.11

In this example, we compare the processing speed of convolution in the spatial domain to that of the same in the frequency domain.

Consider a gray scale image of size 512×512 pixels. We wish to sharpen the image by processing it with a FIR high-pass filter of size 11×11. For simplicity, let's ignore any symmetrical properties of the filter as it can be equally taken advantage of in both direct convolution and FFT-based convolution. Determine which approach will be faster, block convolution using overlap–add method with 64-point FFT or direct convolution in spatial domain?

SOLUTION

The speed of any digital signal processing algorithm depends on the number of multiplications and additions performed. Since multiplication takes more time than addition, we ignore addition and compute the number of multiplications. Generally speaking, algorithm A will be faster than algorithm B if the number of multiplications in A is less than B.

Let N_1 be the total number of real multiplications when the convolution is performed in spatial domain and N_2 be the total number of real multiplications when the overlap–add method is used with a 64-point FFT and inverse FFT. Assume that the 2-D FFT is computed using row–column decomposition. Then,

$$N_1 = 512^2 \text{ pixels} \times 11^2 \text{ multiplication/pixel} = 31{,}719{,}424 = 31.719424 \times 10^6$$

Each 2-D FFT requires $\frac{64^2}{2} \log_2 (64^2) = 24{,}576$ complex multiplications. The size of each block is $(64 - 11 + 1) \times (64 - 11 + 1) = 54 \times 54$ and the number of blocks is $\left(\frac{512}{54}\right)^2 \approx 100$. Therefore, 100 FFTs are needed to transform the blocks, 1 FFT to transform the filter, $100 \times 64 \times 64 = 409{,}600$ complex multiplications to multiply the transform coefficients, and 100 FFTs to come back to the pixel domain. This is equal to a total of 201 FFTs and 409,600 complex multiplications. Therefore,

$$N_2 = 201 \times 24{,}576 + 409{,}600 = 5{,}349{,}376 \text{ complex multiplications}$$

Now each complex multiplication is equal to three real multiplications using the following algorithm

$$ac - bd = \underset{a(c-d)}{\underbrace{\text{one multiplication}}} + \underset{d(a-b)}{\underbrace{\text{one multiplication}}} \quad \text{Two multiplications}$$

$$ad + bc = d(a-b) + \underset{b(d+c)}{\underbrace{\text{one multiplication}}} \quad \text{One additional multiplication}$$

Note that the term $d(a - b)$ is computed once.
Then,

$$(a + jb)(c + jd) = ac - bd + j(ad + bc)$$

Therefore,

$$\frac{N_2}{N_1} = \frac{3 \times 5{,}349{,}376}{31{,}719{,}424} = 0.507$$

This implies that, in this case, the frequency domain convolution is almost twice faster than the spatial domain convolution.

2.5.2 TWO-DIMENSIONAL DISCRETE COSINE TRANSFORM

The 2-D DCT is used in still image and video data compression [12,13]. The 2-D DCT of 2-D sequence $f(n, m)$ is defined as

$$C(k, l) = a_N(k)a_M(l) \sum_{m=0}^{M-1} \sum_{n=0}^{N-1} f(n, m) \cos\left[\frac{(2n+1)k\pi}{2N}\right] \cos\left[\frac{(2m+1)l\pi}{2M}\right] \quad (2.71)$$

for $k = 0, 1, \ldots, N - 1$ and $l = 0, 1, \ldots, M - 1$

The inverse DCT is defined as

$$f(n, m) = \sum_{l=0}^{M-1} \sum_{k=0}^{N-1} a_N(k)a_M(l)C(k, l) \cos\left[\frac{(2n+1)k\pi}{2N}\right] \cos\left[\frac{(2m+1)l\pi}{2M}\right] \quad (2.72)$$

for $n = 0, 1, \ldots, N - 1$ and $m = 0, 1, \ldots, M - 1$

The function $\alpha_N(.)$ is defined as

$$\alpha_N(k) = \begin{cases} \sqrt{\frac{1}{N}} & k = 0 \\ \sqrt{\frac{2}{N}} & k \neq 0 \end{cases} \quad (2.73)$$

The DCT basis functions for $N = 4$ are shown in Figure 2.25.

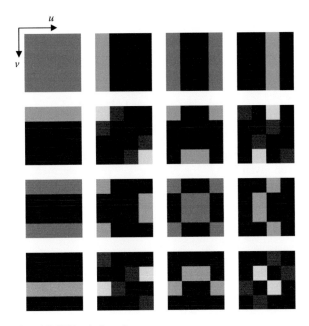

FIGURE 2.25 4×4 DCT basis functions.

Example 2.12

Consider the 4×4 subimage

$$
f(n, m) = \begin{bmatrix} 15 & 15 & 16 & 14 \\ 16 & 17 & 15 & 13 \\ 14 & 11 & 12 & 15 \\ 13 & 12 & 15 & 13 \end{bmatrix}
$$

The DCT of this image is

$$
C(k, l) = \begin{bmatrix} 56.5 & 0.574 & 0 & 1.3858 \\ 3.5042 & 1.7071 & -1.5443 & -0.9142 \\ 0 & -1.0031 & -1.5 & 1.4979 \\ -1.9927 & -1.9142 & 2.8045 & 0.2929 \end{bmatrix}
$$

2.5.3 TWO-DIMENSIONAL HADAMARD TRANSFORM

The 2-D Hadamard transform (HT) is a real transform where the components of the basis functions take values from the binary set $\{1 \ -1\}$. The $N \times N$ HT of an $N \times N$ image f is defined as

$$
F = AfA \tag{2.74}
$$

Here, it is assumed that $N = 2^m$. The $N \times N$ matrix A is called Hadamard transformation matrix and is related to the $N \times N$ Hadamard matrix H by

$$
A = \frac{1}{\sqrt{N}} H_N \tag{2.75}
$$

The 2×2 Hadamard matrix is

$$
H_2 = \begin{bmatrix} 1 & 1 \\ 1 & -1 \end{bmatrix} \tag{2.76}
$$

The $2N \times 2N$ Hadamard matrix is related to the $N \times N$ Hadamard matrix through the recursive equation given by

$$
H_{2N} = \begin{bmatrix} H_N & H_N \\ H_N & -H_N \end{bmatrix} \tag{2.77}
$$

For example,

$$
H_4 = \begin{bmatrix} H_2 & H_2 \\ H_2 & -H_2 \end{bmatrix} = \begin{bmatrix} 1 & 1 & 1 & 1 \\ 1 & -1 & 1 & -1 \\ 1 & 1 & -1 & -1 \\ 1 & -1 & -1 & 1 \end{bmatrix} \tag{2.78}
$$

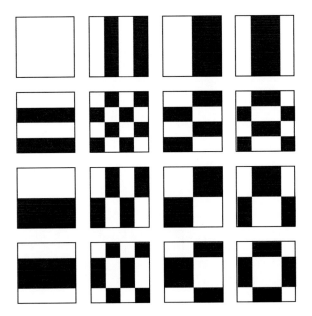

FIGURE 2.26 4 × 4 HT basis functions.

and

$$H_8 = \begin{bmatrix} H_4 & H_4 \\ H_4 & -H_4 \end{bmatrix} = \begin{bmatrix} 1 & 1 & 1 & 1 & 1 & 1 & 1 & 1 \\ 1 & -1 & 1 & -1 & 1 & -1 & -1 & 1 \\ 1 & 1 & -1 & -1 & 1 & 1 & -1 & -1 \\ 1 & -1 & -1 & 1 & 1 & -1 & -1 & 1 \\ 1 & 1 & 1 & 1 & -1 & -1 & -1 & -1 \\ 1 & -1 & 1 & -1 & -1 & 1 & -1 & 1 \\ 1 & 1 & -1 & -1 & -1 & -1 & 1 & 1 \\ 1 & -1 & -1 & 1 & -1 & 1 & 1 & -1 \end{bmatrix}$$ (2.79)

The basis functions for the 4 × 4 HT are shown in Figure 2.26.

2.5.3.1 Inverse Hadamard Transform

The columns of matrix A are forming a set of orthogonal vectors of unit norm. This implies that A is unitary, which implies that $A^{-1} = A$. Now to find the inverse HT, pre-multiply and post-multiply Equation 2.74 by A^{-1}

$$A^{-1}FA^{-1} = f$$ (2.80)

$$f = AFA$$ (2.81)

Example 2.13

Find the HT of the 4×4 image $f(n, m)$ given as

$$f(n, m) = \begin{bmatrix} 15 & 15 & 16 & 14 \\ 16 & 17 & 15 & 13 \\ 14 & 11 & 12 & 15 \\ 13 & 12 & 15 & 13 \end{bmatrix}$$

SOLUTION

$$F(k, l) = Af(n, m)A$$

$$= \frac{1}{2}\begin{bmatrix} 1 & 1 & 1 & 1 \\ 1 & -1 & 1 & -1 \\ 1 & 1 & -1 & -1 \\ 1 & -1 & -1 & 1 \end{bmatrix}\begin{bmatrix} 15 & 15 & 16 & 14 \\ 16 & 17 & 15 & 13 \\ 14 & 11 & 12 & 15 \\ 13 & 12 & 15 & 13 \end{bmatrix}\frac{1}{2}\begin{bmatrix} 1 & 1 & 1 & 1 \\ 1 & -1 & 1 & -1 \\ 1 & 1 & -1 & -1 \\ 1 & -1 & -1 & 1 \end{bmatrix}$$

$$F(k, l) = \begin{bmatrix} 56.5 & 1.5 & 0 & 0 \\ -0.5 & -0.5 & -1 & 2 \\ 4 & 0 & 2.5 & -2.5 \\ 0 & 1 & -1.5 & -1.5 \end{bmatrix}$$

The HT is a real and fast transform. Since the basis vectors contain only ± 1 values, no multiplications are needed in the computation of the HT.

2.6 IMAGE FILTERING

Filtering images has applications in image smoothing (noise removal), image sharpening, edge enhancement, and image deblurring. Filters used in image processing are generally FIR filters of relatively small size. An FIR filter is also referred to as a mask, a kernel, or a window [14]. The filter components are referred to as filter taps, weights, or elements. Some examples of popular small size FIR filters used in image processing are

a. Low-pass filters (LPF)

$$h(n, m) = \frac{1}{9}\begin{bmatrix} 1 & 1 & 1 \\ 1 & 1 & 1 \\ 1 & 1 & 1 \end{bmatrix}, \quad h(n, m) = \frac{1}{16}\begin{bmatrix} 1 & 2 & 1 \\ 2 & 4 & 2 \\ 1 & 2 & 1 \end{bmatrix},$$

$$h(n, m) = \frac{1}{10}\begin{bmatrix} 1 & 1 & 1 \\ 1 & 2 & 1 \\ 1 & 1 & 1 \end{bmatrix} \tag{2.82}$$

b. High-pass filters (HPF)

$$h(n, m) = \begin{bmatrix} 0 & -1 & 0 \\ -1 & 5 & -1 \\ 0 & -1 & 0 \end{bmatrix}, \quad h(n, m) = \begin{bmatrix} 1 & -2 & 1 \\ -2 & 5 & -2 \\ 1 & -2 & 1 \end{bmatrix},$$

$$h(n, m) = \frac{1}{7} \begin{bmatrix} -1 & -2 & -1 \\ -2 & 19 & -2 \\ -1 & -2 & -1 \end{bmatrix}, \quad h(n, m) = \begin{bmatrix} -1 & -1 & -1 \\ -1 & 9 & -1 \\ -1 & -1 & -1 \end{bmatrix} \quad (2.83)$$

The FIR filtering of an image is performed one pixel at a time. The center filter weight is placed on the pixel to be filtered. Each filter tap is multiplied by its corresponding pixel value, and the results are summed together. As an example, consider filtering the pixel $f(5, 2)$ in the following 8×8 block of the LENA image

$$f(n, m) = \begin{bmatrix} 139 & 144 & 149 & 153 & 155 & 155 & 155 & 155 \\ 144 & 151 & 153 & 156 & 159 & 156 & 156 & 156 \\ 150 & 155 & 160 & 163 & 158 & 156 & 156 & 156 \\ 159 & 161 & 162 & 160 & 159 & 159 & 158 & 159 \\ 159 & 160 & 161 & 162 & 162 & 155 & 155 & 155 \\ 161 & 161 & 161 & 161 & 160 & 157 & 157 & 157 \\ 162 & 162 & 161 & 163 & 162 & 157 & 157 & 157 \\ 162 & 162 & 161 & 161 & 163 & 158 & 158 & 158 \end{bmatrix} \quad (2.84)$$

Using the 3×3 FIR filter

$$h(n, m) = \frac{1}{16} \begin{bmatrix} 1 & 2 & 1 \\ 2 & 4 & 2 \\ 1 & 2 & 1 \end{bmatrix} \quad (2.85)$$

the output pixel $g(5, 2)$ is computed as

$$g(5, 2) = \frac{1}{16}(4 \times 161 + 2 \times 161 + 2 \times 161 + 2 \times 161 + 2 \times 161 + 1 \times 162$$
$$+ 1 \times 162 + 1 \times 160 + 1 \times 163) = 161 \quad (2.86)$$

FIR filters are used in many image processing tasks such as smoothing, sharpening, noise removal, and edge detection [15–17]. Examples of filters are

LPF: If the filter weights are all positive, the filtering operation creates a smoother (softer) image that consists mostly of low frequencies. These types of filters are referred to as LPF. LPF are used as anti-aliasing filters or prior to subsampling in image decimation. The larger the filter, and the flatter the weights, the smoother the resulting image.

HPF: The weight at the center is positive, and some or all of the filter weights around the center weight are negative. This kind of filter often creates a sharper image by accentuating the differences among pixels. The larger the relative contribution of the negative taps of the filter, the sharper is the filtered image. These filters are referred to as HPF because they generally boost the high-frequency components of the image. Usually, the filter taps in both LPF and HPF sum up to 1, so as to maintain the average brightness of the output image at the same level as the input image. Following filters are examples of common FIR filters used in image processing:

a. LPF

$$
h(n,m) = \frac{1}{60}
\begin{bmatrix}
1 & 2 & 2 & 2 & 1 \\
2 & 2 & 4 & 2 & 2 \\
2 & 4 & 8 & 4 & 2 \\
2 & 2 & 4 & 2 & 2 \\
1 & 2 & 2 & 2 & 1
\end{bmatrix},
\quad
h(n,m) = \frac{1}{25}
\begin{bmatrix}
1 & 1 & 1 & 1 & 1 \\
1 & 1 & 1 & 1 & 1 \\
1 & 1 & 1 & 1 & 1 \\
1 & 1 & 1 & 1 & 1 \\
1 & 1 & 1 & 1 & 1
\end{bmatrix},
$$

$$
h(n,m) = \frac{1}{45}
\begin{bmatrix}
1 & 1 & 2 & 1 & 1 \\
1 & 2 & 3 & 2 & 1 \\
2 & 3 & 5 & 3 & 2 \\
1 & 2 & 3 & 2 & 1 \\
1 & 1 & 2 & 1 & 1
\end{bmatrix}
$$

b. HPF

$$
h(n,m) =
\begin{bmatrix}
+1 & -2 & +4 & -2 & +1 \\
-2 & +4 & -8 & +4 & -2 \\
+4 & -8 & 13 & -8 & +4 \\
-2 & +4 & -8 & +4 & -2 \\
+1 & -2 & +4 & -2 & +1
\end{bmatrix}
$$

$$
h(n,m) = \frac{1}{4}
\begin{bmatrix}
-1 & -1 & -1 & -1 & -1 \\
-1 & -1 & -1 & -1 & -1 \\
-1 & -1 & 28 & -1 & -1 \\
-1 & -1 & -1 & -1 & -1 \\
-1 & -1 & -1 & -1 & -1
\end{bmatrix}
$$

c. Filter to detect horizontal line

$$h(n, m) = \begin{bmatrix} -1 & -1 & -1 \\ 2 & 2 & 2 \\ -1 & -1 & -1 \end{bmatrix} \tag{2.87}$$

d. Filter to detect vertical line

$$h(n, m) = \begin{bmatrix} -1 & 2 & -1 \\ -1 & 2 & -1 \\ -1 & 2 & -1 \end{bmatrix} \tag{2.88}$$

e. Filter to detect lines oriented at $45°$

$$h(n, m) = \begin{bmatrix} -1 & -1 & 2 \\ -1 & 2 & -1 \\ 2 & -1 & -1 \end{bmatrix} \tag{2.89}$$

f. Filter to detect lines oriented at $-45°$

$$h(n, m) = \begin{bmatrix} 2 & -1 & -1 \\ -1 & 2 & -1 \\ -1 & -1 & 2 \end{bmatrix} \tag{2.90}$$

g. Filter for detecting horizontal edges

$$h_h(n, m) = \begin{bmatrix} -1 & -2 & -1 \\ 0 & 0 & 0 \\ 1 & 2 & 1 \end{bmatrix} \tag{2.91}$$

h. Filter for detecting vertical edges

$$h_v(n, m) = \begin{bmatrix} -1 & 0 & 1 \\ -2 & 0 & 2 \\ -1 & 0 & 1 \end{bmatrix} \tag{2.92}$$

As an example of FIR filtering, consider edge detection based on the two filters specified in g and h above. These filters are known as Sobel operators. To create an edge map of image f, we first compute the gradient of f in the x- and y-directions by filtering the image using filters $h_h(n, m)$ and $h_v(n, m)$, respectively. That is,

$$G_x(n, m) = f(n, m) * h_h(n, m) \tag{2.93}$$

$$G_y(n, m) = f(n, m) * h_v(n, m) \tag{2.94}$$

To find the edge map of the image, quantize the magnitude of the gradient image estimated by

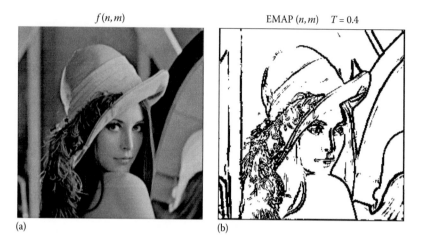

$f(n, m)$ EMAP (n, m) $T = 0.4$

(a)   (b)

FIGURE 2.27 (a) Original LENA image and (b) edge map of LENA image.

$$G(n, m) = |G_x(n, m)| + |G_y(n, m)| \tag{2.95}$$

to 1 bit using the threshold T

$$\text{EMAP}(n, m) = \begin{cases} 1 & G(n, m) \leq T \\ 0 & G(n, m) > T \end{cases} \tag{2.96}$$

The result of applying the above edge detection algorithm to the LENA image is shown in Figure 2.27b.

2.6.1 Design of 2-D FIR Filters

Two-dimensional FIR filters can be designed to have zero-phase response. Zero-phase filters introduce no phase distortion. The necessary and sufficient condition for 2-D FIR filters to have zero-phase response is that the filter impulse response be symmetric with respect to the origin (center point), that is,

$$h(n_1, n_2) = h(-n_1, -n_2) \tag{2.97}$$

The first step in designing 2-D FIR filters using the window method is to compute the desired filter impulse response by inverting the desired frequency response using the inverse Fourier transform, that is,

$$h_d(n_1, n_2) = \frac{1}{4\pi^2} \int_{-\pi}^{\pi} \int_{-\pi}^{\pi} H_d(j\omega_1, j\omega_2) e^{j(\omega_1 n + \omega_2 n_2)} d\omega_1 d\omega_2 \tag{2.98}$$

Closed form solutions exist for the 2-D impulse response sequences of LP, HP, and band-pass filters for both rotationally symmetrical and nonsymmetrical filters [5].

TABLE 2.2
Ideal Impulse Response of Symmetric and Nonsymmetric Filters

Filter Type	$h_d(n_1, n_2)$	Support of $H_d(j\omega_1, j\omega_2)$
Low pass	$\dfrac{\sin(\omega_{cx}\pi n_1)}{\pi n_1}\dfrac{\sin(\omega_{cy}\pi n_2)}{\pi n_2}$	
High pass	$\delta(n_1, n_2) - \dfrac{\sin(\omega_{cx}\pi n_1)}{\pi n_1}\dfrac{\sin(\omega_{cy}\pi n_2)}{\pi n_2}$	
Band pass	$\dfrac{\sin(\omega_{cx_2}\pi n_1)}{\pi n_1}\dfrac{\sin(\omega_{cy_2}\pi n_2)}{\pi n_2}$ $-\dfrac{\sin(\omega_{cx_1}\pi n_1)}{\pi n_1}\dfrac{\sin(\omega_{cy_1}\pi n_2)}{\pi n_2}$	
Low pass (circularly symmetric)	$\dfrac{\omega_c}{2\pi}\dfrac{J_1\left(\omega_c\sqrt{n_1^2+n_2^2}\right)}{\sqrt{n_1^2+n_2^2}}$	
High pass (circularly symmetric)	$\delta(n_1, n_2) - \dfrac{\omega_c}{2\pi}\dfrac{J_1\left(\omega_c\sqrt{n_1^2+n_2^2}\right)}{\sqrt{n_1^2+n_2^2}}$	
Band pass (circularly symmetric)	$\dfrac{\omega_{c_2}}{2\pi}\dfrac{J_1\left(\omega_{c_2}\sqrt{n_1^2+n_2^2}\right)}{\sqrt{n_1^2+n_2^2}} - \dfrac{\omega_{c_1}}{2\pi}\dfrac{J_1\left(\omega_{c_1}\sqrt{n_1^2+n_2^2}\right)}{\sqrt{n_1^2+n_2^2}}$	

They are tabulated in Table 2.2. The support for the impulse response $h_d(n_1, n_2)$ is the entire (n_1, n_2) plane. Therefore, it has to be truncated for a finite support. The truncation is done using a 2-D window function, which is

$$h(n_1, n_2) = h_d(n_1, n_2)w(n_1, n_2) \qquad (2.99)$$

The 2-D window $w(n_1, n_2)$ is generally designed using 1-D windows. For example, for symmetrical filters the 2-D window is

$$w(n_1, n_2) = w_1\left(\sqrt{n_1^2 + n_2^2}\right) \qquad (2.100)$$

TABLE 2.3

Spectral Property of Different Windows

Window Type	$w(n)$, $n = -M, -M+1, \ldots, M$	Distance between the Peak of the Mainlobe and the Peak of the First Sidelobe (dB)
Uniform	1	13
Hamming	$0.54 + 0.46 \cos\left(\dfrac{n\pi}{M}\right)$	43
Hanning	$0.5 + 0.5 \cos\left(\dfrac{n\pi}{M}\right)$	46
Blackman	$0.42 + 0.5 \cos\left(\dfrac{n\pi}{M}\right) + 0.08 \cos\left(\dfrac{2n\pi}{M}\right)$	50

For separable filters,

$$w(n_1, n_2) = w_1(n_1)w_2(n_2) \tag{2.101}$$

The functions $w_1(n_1)$ and $w_2(n_2)$ are 1-D windows. The frequency response of the designed filter can be obtained by taking the Fourier transform from both sides of Equation 2.99. Using the convolution property of the Fourier transform, we get

$$H(\omega_1, \omega_2) = H_d(\omega_1, \omega_2) * W(\omega_1, \omega_2) \tag{2.102}$$

This means that the frequency response of the designed filter is the convolution of the desired frequency response and the frequency response of the truncated window. If the truncated window is a uniform window, the designed filter will have ripples in its passband. This is due to the side lobes in the spectrum of the uniform window. To minimize the effect of these side lobes and reduce the ripples in the passband, we can use other windows such as Hamming, Hanning, or Blackman windows. The purpose of a window is to smooth the frequency response. The window must have a small mainlobe width so that the transition width of $H(\omega_1, \omega_2)$ is small. We also need to have a window with the smallest sidelobe amplitude to make the ripple in the passband and stop-band regions as small as possible. The most frequently used 1-D windows and their spectral properties are listed in Table 2.3.

The 1-D uniform, Hamming, Hanning, and Blackman windows of size 32 in. spatial and frequency domains are shown in Figures 2.28 and 2.29, respectively.

Example 2.14

Design a circularly symmetric 2-D low-pass FIR filter for an image of size 57×57 with a cutoff frequency of $\omega_c = \frac{\pi}{4}$.

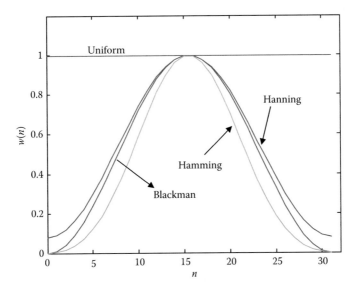

FIGURE 2.28 Windows in spatial domain.

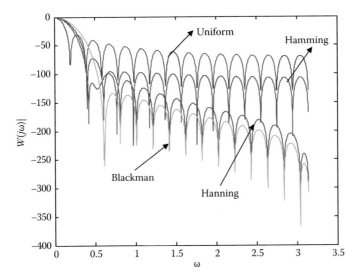

FIGURE 2.29 Windows in frequency domain.

SOLUTION

The desired impulse response is given by

$$h_d(n_1, n_2) = \frac{\omega_c}{2\pi} \frac{J_1\left(\omega_c \sqrt{n_1^2 + n_2^2}\right)}{\sqrt{n_1^2 + n_2^2}} = \frac{1}{8} \frac{J_1\left(\frac{\pi}{4}\sqrt{n_1^2 + n_2^2}\right)}{\sqrt{n_1^2 + n_2^2}}$$

where $J_1(\cdot)$ is Bessel function of the first kind and first order. The designed filter using a hamming window would be

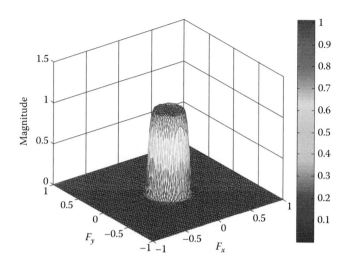

FIGURE 2.30 Frequency response of the design filter.

$$h(n_1, n_2) = h_d(n_1, n_2)w(n_1, n_2) = \frac{1}{8} \frac{J_1\left(\frac{\pi}{4}\sqrt{n_1^2 + n_2^2}\right)}{\sqrt{n_1^2 + n_2^2}} \left[0.54 + \cos\left(\frac{\sqrt{n_1^2 + n_2^2}}{28}\pi\right)\right]$$

$$n_1, n_2 = -28, -27, \ldots, 0, 1, \ldots, 28$$

The frequency responses of the designed filter using uniform and hamming windows are shown in Figure 2.30.

2.7 IMAGE RESIZING

In many imaging applications, it is sometimes necessary to create a low-resolution image from a high-resolution image and vice versa for purposes of viewing, transmitting, or printing. An example for such a situation is when an image is initially scanned at a high resolution (for example: 3072 × 2048 pixels). This image is suitable for high quality printing. However, for the purpose of viewing this image on a high definition television (HDTV) monitor, a lower resolution image is required. Similarly, an even lower resolution image is needed for national television system committe (NTSC) television viewing. Image resizing is basically performed by the process of sampling rate conversion. Sampling rate conversion decouples the spatial resolution of the image source from the spatial resolution requirement of the display or printing device. Thus, the same image source can be viewed on different display systems with different resolutions [18,19].

2.7.1 DEFINITION OF SAMPLING RATE CONVERSION

Given an image sampled at a rate R_1, create a new image at a spatial rate R_2, where the two rates are related by a factor of α, that is,

$$R_2 = \alpha R_1 \tag{2.103}$$

There exist two approaches to sampling rate conversion. The first technique is continuous domain processing. In this approach, the continuous image is first reconstructed and then it is resampled at the new desired rate. This approach is theoretically possible but is not practical. The second approach is digital processing, which implies that the image is processed in the digital domain without conversion to analog or continuous domain. In discrete domain, the sampling rate can be changed by a factor of α only if it is a rational number (i.e., ratio of two integers), that is

$$\alpha = \frac{P}{Q} \tag{2.104}$$

In this case, we first upsample (interpolate) the image by a factor of P and then downsample (decimate) by a factor of Q. We first consider the upsampling process. Here, this process will be presented for 1-D signals. Extension to 2-D signals is straightforward.

2.7.2 UPSAMPLING BY FACTOR OF P

To upsample a 1-D signal by a factor of P, we insert $P - 1$ zeros between the signal samples and LPF the resulting signal. The block diagram of an upsampler is shown in Figure 2.31.

The output signal before low-pass filtering is related to the input by

$$y(n) = \begin{cases} f\left(\frac{n}{P}\right) & n = 0, \pm P, \pm 2P, \ldots \\ 0 & \text{otherwise} \end{cases} \tag{2.105}$$

The DTFT of $y(n)$ is given by

$$Y(\omega) = \sum_{n=-\infty}^{\infty} y(n)e^{-jn\omega} = \sum_{n=-\infty}^{\infty} f\left(\frac{n}{P}\right)e^{-jn\omega} = \sum_{-\infty}^{\infty} f(k)e^{-jkP\omega} = F(P\omega) \tag{2.106}$$

Assume that $f(n)$ is obtained by sampling a truly band-limited analog signal $f_a(x)$ at the Nyquist rate of $f_s = \frac{1}{\Delta x}$, that is

$$F(j\omega) = \frac{1}{\Delta x} \sum_{k=-\infty}^{\infty} F_a\left(j\frac{\omega - 2\pi k}{\Delta x}\right) \tag{2.107}$$

If we now sample $f_a(x)$ at the rate of $Pf_s = \frac{P}{\Delta x}$, to obtain $g(n)$, then the discrete time Fourier transform of $g(n)$ will be

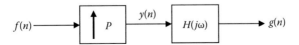

FIGURE 2.31   Upsampling by a factor of P.

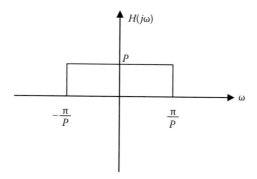

FIGURE 2.32 Frequency response of ideal interpolation filter.

$$G(j\omega) = \frac{P}{\Delta x} \sum_{k=-\infty}^{\infty} F_a\left(j\frac{\omega - 2\pi k}{\frac{\Delta x}{P}}\right) \tag{2.108}$$

Therefore, if the signal $y(n)$ is passed through an ideal LPF with a gain of P and cutoff frequency of $\frac{\pi}{P}$, the output would be the desired upsampled signal $g(n)$. This means that the ideal interpolation LPF has the frequency response given by

$$H(j\omega) = \begin{cases} P & |\omega| < \frac{\pi}{P} \\ 0 & \text{otherwise} \end{cases} \tag{2.109}$$

This frequency response is shown in Figure 2.32.

2.7.3 Downsampling by Factor of Q

To downsample a 1-D signal by a factor of Q, we retain every Qth sample. The block diagram of a downsampler is shown in Figure 2.33.

The output signal is related to the input by

$$y(n) = f(Qn) \tag{2.110}$$

The DTFT of $y(n)$ is given by

$$Y(\omega) = \sum_{n=-\infty}^{\infty} y(n)e^{-jn\omega} = \sum_{n=-\infty}^{\infty} f(Qn)e^{-jn\omega} = \sum_{k=-\infty}^{\infty} f(k)w(k)e^{-j\frac{k}{Q}\omega} \tag{2.111}$$

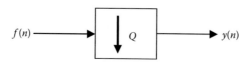

FIGURE 2.33 Downsampling by a factor of Q.

where $w(k)$ is defined as

$$w(k) = \begin{cases} 1 & k = 0, \quad \pm Q, \pm 2Q, \cdots \\ 0 & \text{otherwise} \end{cases} = \sum_{n=-\infty}^{\infty} \delta(k - nQ) \qquad (2.112)$$

Since $w(k)$ is periodic with a period of Q, it can be expanded using discrete Fourier series (DFS) as

$$w(k) = \frac{1}{Q} \sum_{n=0}^{Q-1} e^{j\frac{2\pi}{Q}nk} \qquad (2.113)$$

Substituting Equation 2.113 into 2.111 yields

$$Y(\omega) = \sum_{k=-\infty}^{\infty} f(k)w(k)e^{-j\frac{k}{Q}\omega} = \sum_{k=-\infty}^{\infty} f(k)\frac{1}{Q}\sum_{n=0}^{Q-1} e^{j\frac{2\pi nk}{Q}}e^{-j\frac{k}{Q}\omega} = \frac{1}{Q}\sum_{n=0}^{Q-1}\sum_{k=-\infty}^{\infty} f(k)e^{-j\frac{\omega-2\pi n}{Q}k}$$

$$= \frac{1}{Q}\sum_{n=0}^{Q-1} F\left(\frac{\omega - 2\pi n}{Q}\right) \qquad (2.114)$$

As evident from Equation 2.114, there are Q terms in the expansion of $Y(\omega)$. This expansion shows that there will be aliasing if the signal bandwidth is more than $\frac{\pi}{Q}$. To avoid aliasing, we need to LPF the signal before downsampling. The frequency response of the ideal LPF is

$$H(j\omega) = \begin{cases} 1 & |\omega| < \frac{\pi}{Q} \\ 0 & \text{otherwise} \end{cases} \qquad (2.115)$$

The frequency response of the ideal anti-aliasing LPF is shown in Figure 2.34. The overall block diagram of a downsampler is shown in Figure 2.35.

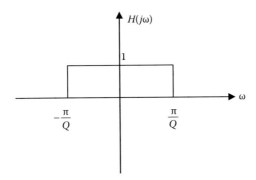

FIGURE 2.34   Frequency response of anti-aliasing filter.

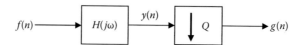

FIGURE 2.35 Downsampling by a factor of Q.

2.7.4 SAMPLING RATE CONVERSION BY A FACTOR OF $\frac{P}{Q}$

To change the sampling rate by a factor of $\frac{P}{Q}$, we first upsample the signal by a factor of P and then downsample it by a factor of Q, as shown in Figure 2.36.

The two LPFs can be combined into one filter with a passband gain of $P \times 1 = P$ and a bandwidth given by the minimum of $\frac{\pi}{P}$ and $\frac{\pi}{Q}$. The overall system is shown in Figure 2.37. In this figure, the LPF bandwidth is given by

$$BW = \min\left\{\frac{\pi}{P}, \frac{\pi}{Q}\right\} \tag{2.116}$$

2.7.5 EXAMPLES OF LOW-PASS FILTERS USED FOR SAMPLING RATE CONVERSION

An ideal LPF is not realizable, so in practice it is approximated by a simple realizable FIR filter. For example, the following simple filters can be used for upsampling. These filters are derived using a linear interpolation algorithm. For example, for upsampling by a factor of P, the LPF would be an FIR filter of size $2P - 1$, given by

$$h(n) = \left[\frac{1}{P} \quad \frac{2}{P} \quad \cdots \quad \frac{P-1}{P} \quad 1 \quad \frac{P-1}{P} \quad \cdots \quad \frac{2}{P} \quad \frac{1}{P}\right] \tag{2.117}$$

The DC gain of this filter is

$$DC\ gain = H(j\omega)|_{\omega=0} = \sum_n h(n)\exp(-j\omega)|_{\omega=0} = \sum_n h(n) = P \tag{2.118}$$

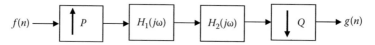

FIGURE 2.36 Sampling rate conversion by a factor of $\frac{P}{Q}$.

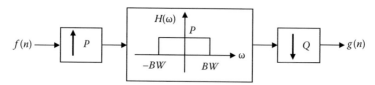

FIGURE 2.37 Sampling rate conversion by factor of a $\frac{P}{Q}$.

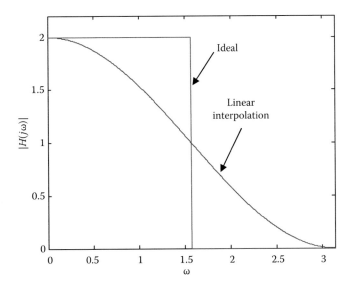

FIGURE 2.38 Frequency response of ideal and linear interpolation filters for $P = 2$.

For $P = 2$ and $P = 3$, the filters are

$$h(n) = \begin{bmatrix} \frac{1}{2} & 1 & \frac{1}{2} \end{bmatrix} \quad \text{and} \quad h(n) = \begin{bmatrix} \frac{1}{3} & \frac{2}{3} & 1 & \frac{2}{3} & \frac{1}{3} \end{bmatrix} \tag{2.119}$$

The frequency response of the interpolation filter for $P = 2$ is shown in Figure 2.38. For downsampling by a factor of Q, the same filter can be used except that the DC gain of the filter should be one. Therefore,

$$h(n) = \frac{1}{Q} \begin{bmatrix} \frac{1}{Q} & \frac{2}{Q} & \cdots & \frac{Q-1}{Q} & 1 & \frac{Q-1}{Q} & \cdots & \frac{2}{Q} & \frac{1}{Q} \end{bmatrix} \tag{2.120}$$

Example 2.15

Given a 4×4 sequence

$$f(n, m) = \begin{bmatrix} 2 & 3 & 1 & 2 \\ 4 & 1 & 3 & 5 \\ 2 & 4 & 1 & 2 \\ 2 & 1 & 3 & 2 \end{bmatrix}$$

upsample the sequence $f(n, m)$ by a factor of 2 using linear interpolation.

SOLUTION

We first insert zeros between pixels and increase the size to 8×8.

$$g(n, m) = \begin{bmatrix} 2 & 0 & 3 & 0 & 1 & 0 & 2 & 0 \\ 0 & 0 & 0 & 0 & 0 & 0 & 0 & 0 \\ 4 & 0 & 1 & 0 & 3 & 0 & 5 & 0 \\ 0 & 0 & 0 & 0 & 0 & 0 & 0 & 0 \\ 2 & 0 & 4 & 0 & 1 & 0 & 2 & 0 \\ 0 & 0 & 0 & 0 & 0 & 0 & 0 & 0 \\ 2 & 0 & 1 & 0 & 3 & 0 & 2 & 0 \\ 0 & 0 & 0 & 0 & 0 & 0 & 0 & 0 \end{bmatrix}$$

Now we filter the signal using a separable 2-D filter. The row and column filters are identical and each is given by $h = \begin{bmatrix} \frac{1}{2} & 1 & \frac{1}{2} \end{bmatrix}$. We first filter each row using h. The resulting image is

$$g_2(n, m) = g_1(n, m)*h(m) = \begin{bmatrix} 2 & 2.5 & 3 & 2 & 1 & 1.50 & 2 & 1 \\ 0 & 0 & 0 & 0 & 0 & 0 & 0 & 0 \\ 4 & 2.50 & 1 & 2 & 3 & 4 & 5 & 2.5 \\ 0 & 0 & 0 & 0 & 0 & 0 & 0 & 0 \\ 2 & 3 & 4 & 2.5 & 1 & 1.5 & 2 & 1 \\ 0 & 0 & 0 & 0 & 0 & 0 & 0 & 0 \\ 2 & 1.5 & 1 & 2 & 3 & 2.5 & 2 & 1 \\ 0 & 0 & 0 & 0 & 0 & 0 & 0 & 0 \end{bmatrix}$$

Apply the same filter h across each column of image $g_2(n, m)$ and this process results in

$$g(n, m) = g_2(n, m) * h(n) = \begin{bmatrix} 2 & 2.5 & 3 & 2 & 1 & 1.50 & 2 & 1 \\ 3 & 2.5 & 2 & 2 & 2 & 2.75 & 3.5 & 1.75 \\ 4 & 2.50 & 1 & 2 & 3 & 4 & 5 & 2.5 \\ 3 & 2.75 & 2.5 & 2.25 & 2 & 2.75 & 3.5 & 1.75 \\ 2 & 3 & 4 & 2.5 & 1 & 1.5 & 2 & 1 \\ 2 & 2.25 & 2.5 & 2.25 & 2 & 2 & 2 & 1 \\ 2 & 1.5 & 1 & 2 & 3 & 2.5 & 2 & 1 \\ 1 & 0.75 & 0.5 & 1 & 1.5 & 1.25 & 1 & 0.5 \end{bmatrix}$$

Example 2.16

Resize the following 6×6 image $f(n, m)$ to a 4×4 image.

$$f(n, m) = \begin{bmatrix} 1 & 2 & 4 & 2 & 1 & 2 \\ 3 & 0 & 2 & 4 & 3 & 2 \\ 2 & 3 & 1 & 0 & 6 & 4 \\ 5 & 3 & 2 & 1 & 0 & 3 \\ 2 & 0 & 0 & 5 & 6 & 2 \\ 4 & 3 & 2 & 7 & 3 & 2 \end{bmatrix}$$

SOLUTION

We need to change the sampling rate by a factor of $\frac{2}{3}$. For this, we first upsample the image by a factor of 2 and then downsample it by a factor of 3. In both cases, we use the linear interpolation filter given by Equation 2.117. The result of upsampling by a factor of 2 is

$$g_1(n,m) = \begin{bmatrix}
1 & 1.5 & 2 & 3 & 4 & 3 & 2 & 1.5 & 1 & 1.5 & 2 & 1 \\
2 & 1.5 & 1 & 2 & 3 & 3 & 3 & 2.5 & 2 & 2 & 2 & 1 \\
3 & 1.5 & 0 & 1 & 2 & 3 & 4 & 3.5 & 3 & 2.5 & 2 & 1 \\
2.5 & 2 & 1.5 & 1.5 & 1.5 & 1.75 & 2 & 3.25 & 4.5 & 3.75 & 3 & 1.5 \\
2 & 2.5 & 3 & 2 & 1 & 0.5 & 0 & 3 & 6 & 5 & 4 & 2 \\
3.5 & 3.25 & 3 & 2.25 & 1.5 & 1 & 0.5 & 1.75 & 3 & 3.25 & 3.5 & 1.75 \\
5 & 4 & 3 & 2.5 & 2 & 1.5 & 1 & 0.5 & 0 & 1.5 & 3 & 1.5 \\
3.5 & 2.5 & 1.5 & 1.25 & 1 & 2 & 3 & 3 & 3 & 2.75 & 2.5 & 1.25 \\
2 & 1 & 0 & 0 & 0 & 2.5 & 5 & 5.5 & 6 & 4 & 2 & 1 \\
3 & 2.25 & 1.5 & 1.25 & 1 & 3.5 & 6 & 5.25 & 4.5 & 3.25 & 2 & 1 \\
4 & 3.5 & 3 & 2.5 & 2 & 4.5 & 7 & 5 & 3 & 2.5 & 2 & 1 \\
2 & 1.75 & 1.5 & 1.25 & 1 & 2.25 & 3.5 & 2.5 & 1.5 & 1.25 & 1 & 0.5
\end{bmatrix}$$

Now before downsampling $g_1(n,m)$ by a factor of 3, we LPF the signal using a separable filter constructed from the 1-D filter $h(n) = \begin{bmatrix} \frac{1}{9} & \frac{2}{9} & \frac{1}{3} & \frac{2}{9} & \frac{1}{9} \end{bmatrix}$. Applying this filter across rows and columns of $g_1(n,m)$ will result in

$$g_2(n,m) = \begin{bmatrix}
0.69 & 0.95 & 1.23 & 1.54 & 1.83 & 1.89 & 1.73 & 1.48 & 1.31 & 1.19 & 0.99 & 0.65 \\
1.02 & 1.27 & 1.49 & 1.77 & 2.17 & 2.41 & 2.43 & 2.29 & 2.13 & 1.9 & 1.51 & 0.95 \\
1.28 & 1.52 & 1.63 & 1.76 & 2.07 & 2.41 & 2.69 & 2.88 & 2.93 & 2.67 & 2.07 & 1.27 \\
1.49 & 1.80 & 1.83 & 1.71 & 1.72 & 1.92 & 2.35 & 2.92 & 3.34 & 3.22 & 2.54 & 1.53 \\
1.8 & 2.24 & 2.26 & 1.9 & 1.57 & 1.49 & 1.83 & 2.54 & 3.21 & 3.32 & 2.74 & 1.70 \\
2.10 & 2.58 & 2.56 & 2.07 & 1.6 & 1.39 & 1.55 & 2.09 & 2.71 & 2.94 & 2.56 & 1.66 \\
2.15 & 2.54 & 2.41 & 1.93 & 1.63 & 1.61 & 1.81 & 2.17 & 2.52 & 2.61 & 2.26 & 1.48 \\
1.90 & 2.14 & 1.93 & 1.57 & 1.64 & 2.10 & 2.64 & 2.97 & 3.01 & 2.71 & 2.12 & 1.31 \\
1.68 & 1.85 & 1.63 & 1.41 & 1.80 & 2.72 & 3.62 & 3.98 & 3.73 & 3.02 & 2.12 & 1.21 \\
1.64 & 1.87 & 1.71 & 1.57 & 2.09 & 3.15 & 4.10 & 4.34 & 3.85 & 2.94 & 1.97 & 1.09 \\
1.56 & 1.84 & 1.77 & 1.66 & 2.10 & 2.98 & 3.69 & 3.70 & 3.12 & 2.30 & 1.53 & 0.86 \\
1.14 & 1.38 & 1.36 & 1.28 & 1.54 & 2.07 & 2.45 & 2.35 & 1.90 & 1.37 & 0.93 & 0.54
\end{bmatrix}$$

Now we downsample the signal $g_2(n,m)$ by a factor of 3.

$$g(n,m) = \begin{bmatrix}
0.69 & 1.54 & 1.73 & 1.19 \\
1.49 & 1.71 & 2.35 & 3.22 \\
2.15 & 1.93 & 1.81 & 2.61 \\
1.64 & 1.57 & 4.10 & 2.94
\end{bmatrix}$$

FIGURE 2.39   256 × 256 LENA image.

FIGURE 2.40   Upsampled LENA image.

Example 2.17

In this example, the LENA image is upsampled by a factor of 2 and downsampled by a factor 3. The linear interpolation filters are used for both the upsampling and downsampling operations. The original image is shown in Figure 2.39. The upsampled and the downsampled images are shown in Figures 2.40 and 2.41, respectively.

FIGURE 2.41   LENA image at a resolution of $\frac{2}{3}$ of the original image.

2.8   IMAGE ENHANCEMENT

The goal of image enhancement is to provide a more instinctively pleasing image. Enhancements are also used to simplify visual interpretation and understanding of images. There are different techniques used in image enhancement such as unsharp masking and histogram equalization [5,19–21].

2.8.1   UNSHARP MASKING

The operation of unsharp masking is used to sharpen an image by enhancing its high-frequency components. The process consists of the following three steps.

1. Image $f(n, m)$ is slightly blurred to create a smooth (low-pass) image $\bar{f}(n, m)$.
2. Difference between the original image and the smooth image is $d(n, m) = f(n, m) - \bar{f}(n, m)$. This image consists mainly of high frequencies of the original image.
3. Difference image is boosted by a factor of β and added back to the original image to create a sharper image. The degree of sharpness is controlled by the boost factor and the size of the LPF

$$f_{UM}(n, m) = f(n, m) + \beta(f(n, m) - \bar{f}(n, m)) \qquad (2.121)$$

The block diagram representing the unsharp masking process is shown in Figure 2.42. In digital processing, the entire process of linear unsharp masking can be compactly represented as a single high-pass FIR filter operation.

Consider performing unsharp masking using an FIR filter $h_{LP}(n, m)$ as the blurring operator and a boost factor of β. The filter used in this process is denoted by $\delta(n, m)$ and has all components as zeros except for the center tap (which is one),

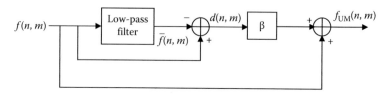

FIGURE 2.42 Unsharp masking.

the equivalent FIR filter starting from the original image $f(n,m)$ to the output unsharp masked image $f_{UM}(n,m)$ is given by

$$h_{UM}(n,m) = (1 + \beta)\delta(n,m) - \beta h_{LP}(n,m) \tag{2.122}$$

For example, if the LPF $h_{LP}(n,m)$ is a simple 5×5 moving average filter, then

$$h_{UM}(n,m) = \frac{1}{25}\begin{bmatrix} -\beta & -\beta & -\beta & -\beta & -\beta \\ -\beta & -\beta & -\beta & -\beta & -\beta \\ -\beta & -\beta & 24+25\beta & -\beta & -\beta \\ -\beta & -\beta & -\beta & -\beta & -\beta \\ -\beta & -\beta & -\beta & -\beta & -\beta \end{bmatrix} \tag{2.123}$$

The following two unsharp masked filters are designed with $\beta = \frac{25}{11} = 2.2727$ and $\beta = \frac{25}{4} = 6.25$.

$$h_{UM}(n,m)|_{\beta=2.2727} = \frac{1}{11}\begin{bmatrix} -1 & -1 & -1 & -1 & -1 \\ -1 & -1 & -1 & -1 & -1 \\ -1 & -1 & 35 & -1 & -1 \\ -1 & -1 & -1 & -1 & -1 \\ -1 & -1 & -1 & -1 & -1 \end{bmatrix} \tag{2.124}$$

and

$$h_{UM}(n,m)|_{\beta=6.25} = \frac{1}{4}\begin{bmatrix} -1 & -1 & -1 & -1 & -1 \\ -1 & -1 & -1 & -1 & -1 \\ -1 & -1 & 28 & -1 & -1 \\ -1 & -1 & -1 & -1 & -1 \\ -1 & -1 & -1 & -1 & -1 \end{bmatrix} \tag{2.125}$$

In general, higher values of β result in more image sharpening. Similarly, in the above filters, a smaller center tap implies more sharpening since it increases the relative contribution of the negative taps on the filtered output.

2.8.2 IMAGE HISTOGRAM

Consider an $M \times M$, 8 bit image $f(n,m)$. The image histogram $N(f)$, for each code value f is the number of times the code value f appears in the image. The PDF of image $f(n,m)$ is the normalized histogram defined by

$$p(f) = \frac{N(f)}{M^2} \quad f = 0, 1, \ldots, 255 \tag{2.126}$$

The cumulative distribution function CDF(f) is defined as

$$CDF(f) = \sum_{i=0}^{f} p(i) \qquad (2.127)$$

As an example, the histogram, PDF and CDF for the LENA image are shown in Figures 2.43 through 2.45.

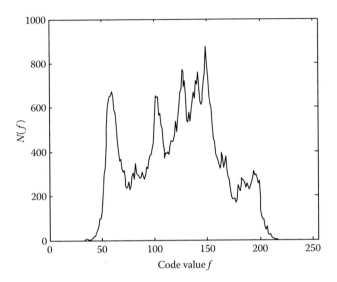

FIGURE 2.43   Histogram of LENA image.

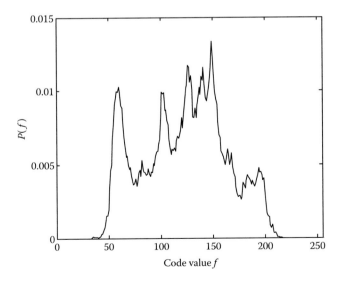

FIGURE 2.44   PDF of LENA Image.

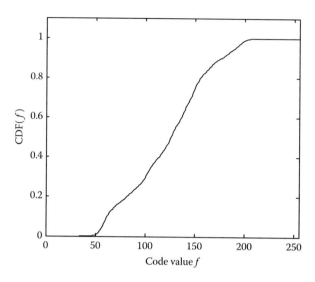

FIGURE 2.45 CDF of LENA image.

2.8.3 HISTOGRAM EQUALIZATION

Assume that r is a continuous random variable taking values between 0 and 1 with a PDF $p_r(r)$. Let $T(r)$ be a single-valued and monotonically increasing function of its argument such that

$$0 \leq T(r) \leq 1 \quad \text{for } 0 \leq r \leq 1 \tag{2.128}$$

Then, according to the theory of functions of one random variable, the random variable $s = T(r)$ has a PDF given by

$$p_s(s) = \frac{p_r(r)}{\frac{dT}{dr}}\bigg|_{r=T^{-1}(s)} \tag{2.129}$$

Now if we want the random variable s to be uniformly distributed over the interval $[0\ \ 1]$, then we set $p_s(s) = 1$. Substituting this into Equation 2.129 yields

$$1 = \frac{p_r(r)}{\frac{dT}{dr}} \tag{2.130}$$

or

$$\frac{dT}{dr} = p_r(r) \tag{2.131}$$

Integrating both sides of Equation 2.131, we have

$$s = T(r) = \int_0^r p_r(x)dx \tag{2.132}$$

Example 2.18

Image f with gray levels between 0 and 1 has a PDF as shown in Figure 2.46. Find the transformation that changes the image f to a new image having a uniform PDF.

The transformation that changes the PDF of f to a uniform distribution is given by Equation 2.133.

$$s = T(r) = \int_0^r p_r(x)dx = \int_0^r 2x\,dx = r^2 \tag{2.133}$$

This means that each gray level in the output image is the square of the corresponding gray level in the input image.

In histogram equalization, the objective is to obtain a uniform histogram for the output image. The transformation that maps the input image to the output image is obtained by discrete approximation to the integral Equation 2.133. Assuming that the input image has L levels, then

$$s = \left[\frac{CDF(r) - a}{1 - a}(L - 1) + 0.5 \right] \quad r = 0, 1, \ldots, L - 1 \tag{2.134}$$

where $CDF(r)$ is the cumulative distribution function of the input image, a is the minimum value of $CDF(r)$ and $[x]$ stands for the largest integer less than or equal to x.

Example 2.19

The low contrast LENA image and its histogram are shown in Figure 2.47. The equalized image and its histogram are shown in Figure 2.48.

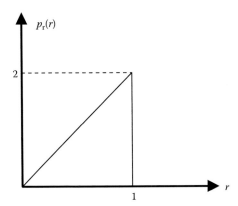

FIGURE 2.46   PDF of image f.

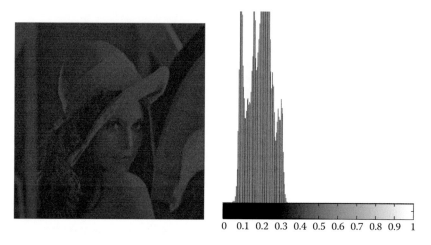

FIGURE 2.47   Low contrast LENA image and its histogram.

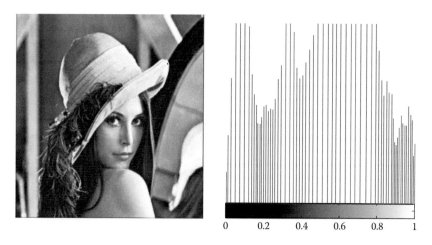

FIGURE 2.48   Equalized image and its histogram.

2.9   IMAGE RESTORATION

The goal in image restoration is to reconstruct the original image which has been subjected to degradation and noise [21,22]. It is usually done by compensating for the image noise and blur, assuming certain degradation models. The closeness of the restored image to the original image depends on the accuracy of the model and the availability of the degradation operator and additive noise statistics. Wiener filtering, weighted least square, and maximum likelihood are common techniques for image restoration. Figure 2.49 shows the block diagram of an image degradation and restoration system.

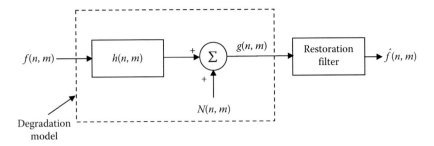

FIGURE 2.49 Image degradation and restoration filter.

The observed image $g(n, m)$ is given by

$$g(n, m) = f(n, m) * h(n, m) + N(n, m) \qquad (2.135)$$

where $f(n, m)$ is the original image, $g(n, m)$ is the degraded noisy image, $N(n, m)$ is zero mean stationary additive noise process, $h(n, m)$ is a linear filter representing the blur and $\hat{f}(n, m)$ is the restored image. The restoration filter is designed by minimizing a metric that is a measure of closeness of $\hat{f}$ to f according to some criterion. Here, we discuss an approach based on minimizing the mean-square error between the original image and the restored image. This is known as MMSE filter or Wiener Filter.

2.9.1 WIENER FILTER RESTORATION

The Wiener filter is designed by minimizing the following objective function

$$J = E(|e(n, m)|^2) \qquad (2.136)$$

where $e(n, m) = f(n, m) - \hat{f}(n, m)$ is the estimation error and E is the expected value operator. The solution to the above minimization problem can be obtained using calculus of variation and is given by

$$W(\omega_x, \omega_y) = \frac{H^*(\omega_x, \omega_y)}{|H(\omega_x, \omega_y)|^2 + \frac{S_N(\omega_x, \omega_y)}{S_f(\omega_x, \omega_y)}} \qquad (2.137)$$

In Equation 2.137, $W(\omega_x, \omega_y)$ is the frequency response of the Wiener filter (restoration filter), $H(\omega_x, \omega_y)$ is the blur (degradation kernel) frequency response, $S_N(\omega_x, \omega_y)$ is the noise power spectrum and $S_f(\omega_x, \omega_y)$ is the power spectrum of the original image. If there is no additive noise $(S_N(\omega_x, \omega_y) = 0)$, then the Wiener filter becomes

$$W(\omega_x, \omega_y) = \frac{H^*(\omega_x, \omega_y)}{|H(\omega_x, \omega_y)|^2} = \frac{1}{H(\omega_x, \omega_y)} \qquad (2.138)$$

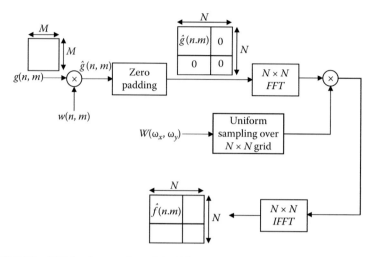

FIGURE 2.50 FFT implementation of the Wiener filter.

This is called inverse filter. If there is no degradation except for additive noise, that is, $H(\omega_x, \omega_y) = 1$, then

$$W(\omega_x, \omega_y) = \frac{1}{1 + \frac{S_N(\omega_x, \omega_y)}{S_f(\omega_x, \omega_y)}} \qquad (2.139)$$

This is the Wiener filter for noise removal. In practice, the noise and the image power spectra are not known and the ratio $\frac{S_N(\omega_x, \omega_y)}{S_f(\omega_x, \omega_y)}$ is replaced by the constant α.

$$W(\omega_x, \omega_y) = \frac{H^*(\omega_x, \omega_y)}{|H(\omega_x, \omega_y)|^2 + \alpha} \qquad (2.140)$$

The constant α is a measure of noise power to image signal power (inverse of SNR). It is generally chosen by trial and error. The Wiener filter is normally implemented in frequency domain using the FFT algorithm. The block diagram of FFT-based implementation of the Wiener filter is shown in Figure 2.50.

As indicated in Figure 2.50, the blur frequency response is sampled and multiplied by the FFT of the zero-padded and windowed input image. The window $w(n, m)$ is a separable 2-D window given as product of two 1-D windows.

$$w(n, m) = w_1(n)w_2(m) \qquad (2.141)$$

Example 2.20

The LENA image is blurred by a 5×5 moving average filter and white Gaussian noise is added to the resulting image. The original image is shown in Figure 2.51.

FIGURE 2.51 Original LENA image.

FIGURE 2.52 Blurred noisy LENA image.

The blurred and noisy image is shown in Figure 2.52. The restored image using the inverse filter is shown in Figure 2.53. As shown, the inverse filter amplifies the noise to an extent that the image is not visible. The restored image using the Wiener filter is shown in Figure 2.54. The Wiener filter reduces the noise effect and the resulting image is closer to the original image than the inverse filtered image. The best result is obtained when $\alpha = 0.01$ in the Wiener filter.

2.10 IMAGE HALFTONING

Digital image halftoning converts a gray scale image to a binary image for display and printing on binary devices. Most of the printed images in this book are halftone images. Halftone images appear as if they are gray scale images. This is due to local

FIGURE 2.53   Restored image using inverse filter.

FIGURE 2.54   Restored image using the Wiener filter.

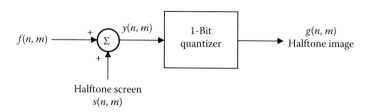

FIGURE 2.55   Halftone image generation.

spatial averaging (low-pass filtering) performed by the eye. A simple technique for generating a halftone image is shown in Figure 2.55. A halftone screen value $s(n, m)$ is added to each continuous tone image sample $f(n, m)$. The resulting signal is denoted by $y(n, m)$ and is given by

$$y(n, m) = f(n, m) + s(n, m) \tag{2.142}$$

The halftone image is obtained by quantizing $y(n, m)$ to 1 bit, that is,

$$g(n, m) = \begin{cases} 1 & y(n, m) \geq T \\ 0 & \text{otherwise} \end{cases} \tag{2.143}$$

The threshold T for an 8 bit image is 126.

Equation 2.144 is a typical halftone pattern. This halftone pattern is repeated periodically to generate a halftone screen $s(n, m)$ of the same size as the original continuous tone image $f(n, m)$.

$$\text{Halftone pattern} = \begin{bmatrix} 52 & 44 & 36 & 124 & 132 & 140 & 148 & 156 \\ 60 & 4 & 28 & 116 & 200 & 228 & 236 & 164 \\ 68 & 12 & 20 & 108 & 212 & 252 & 244 & 172 \\ 76 & 84 & 92 & 100 & 204 & 196 & 188 & 180 \\ 132 & 140 & 148 & 156 & 52 & 44 & 36 & 124 \\ 200 & 228 & 236 & 164 & 60 & 4 & 28 & 116 \\ 212 & 252 & 244 & 172 & 68 & 12 & 20 & 108 \\ 204 & 196 & 188 & 180 & 76 & 84 & 92 & 100 \end{bmatrix} \tag{2.144}$$

is an example, the 512×512 LENA image shown in Figure 2.56 is the input to the halftone algorithm shown in Figure 2.55. The halftone screen $s(n, m)$ is shown in Figure 2.57 and is generated using the halftone pattern shown in Equation 2.144. The resulting halftone image is shown in Figure 2.58.

Another common halftoning technique is error diffusion technique discussed in Section 2.10.1.

FIGURE 2.56   Original image.

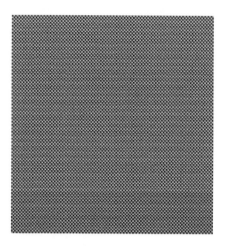

FIGURE 2.57   Halftone screen.

FIGURE 2.58   Halftone image.

2.10.1   ERROR DIFFUSION ALGORITHM

The block diagram of error diffusion algorithm is shown in Figure 2.59. The halftone image is obtained by quantizing $y(n, m)$ to 1 bit. The quantization error signal $e(n, m)$ is diffused to the neighboring pixels by the LPF $h(n, m)$.

The signal $y(n, m)$ is given as

$$y(n, m) = f(n, m) - e(n, m) * h(n, m)$$

$$= f(n, m) - \sum_j \sum_i h(i, j) e(n - i, m - j) \qquad (2.145)$$

where $e(n, m) = g(n, m) - y(n, m)$.

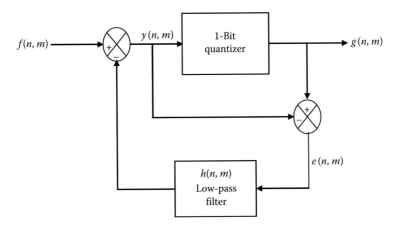

FIGURE 2.59 Error diffusion halftoning.

The halftone image $g(n, m)$ is generated by 1 bit quantization of $y(n, m)$ and is given by

$$g(n, m) = \begin{cases} 1 & y(n, m) \geq 126 \\ 0 & \text{otherwise} \end{cases} \tag{2.146}$$

An example of a low-pass diffusion filter commonly used is

$$h = \frac{1}{16} \begin{bmatrix} 0 & 0 & 7 \\ 3 & 5 & 1 \end{bmatrix} \tag{2.147}$$

As indicated by the diffusion filter, the quantization error is diffused to the four neighboring pixels.

Example 2.21

The gray scale LENA image is halftoned using the error diffusion algorithm with the filter given in Equation 2.147. The original image and the halftone image are shown in Figures 2.60 and 2.61, respectively.

Example 2.22

The RGB color LENA image is halftoned using the error diffusion algorithm with the filter given in Equation 2.147. The original image and the halftone image are shown in Figures 2.62 and 2.63, respectively.

FIGURE 2.60   LENA Image.

FIGURE 2.61   Halftone LENA image.

FIGURE 2.62 (See color insert following page 428.)   RGB LENA image.

FIGURE 2.63 (See color insert following page 428.)   Halftone LENA image.

PROBLEMS

2.1 Consider a digital camera with a focal length of 50 mm, an f number of 5.6, and a CCD sensor pixel dimension of 16 μm × 16 μm. Consider two objects at distances of 1 and 3 m from the camera, respectively. We wish to take a picture of these two objects and keep both of them in focus as much as possible.

    a. Find the distance at which the camera should be focused so that the PSF resulting from the defocus of both of these objects have the same radius. Is this distance halfway in between the two objects?

    b. Find the radius of blur in terms of the number of pixels of the CCD sensor for the above set of parameters.

    c. Repeat part (b) for an f number of 2 and an f number of 16.

2.2 Consider an out-of-focus camera with a blur radius of R. Assume that the object moves horizontally with uniform motion during the exposure period such that the resulting PSF on the film has a width of a. Find the horizontal profile of the resulting cascaded PSF for a given vertical distance of y on the film.

2.3 Find the DFT of the checkerboard image defined by

$$f(n, m) = 0.5 + 0.5(-1)^{n+m} \quad n, m = 0, 1, \ldots, N - 1$$

2.4 A circularly symmetric imaging system has a PSF given by

$$h(r) = e^{-|r|}$$

    a. Derive an expression for the OTF and MTF of this imaging system.

    b. At what spatial radial frequency does the system MTF fall to $\frac{1}{\sqrt{2}}$?

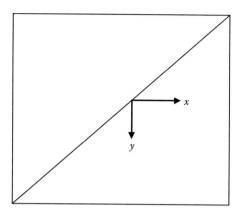

FIGURE 2.64

2.5 Consider a linear, shift-invariant image degradation system with a PSF $h(x, y) = e^{-2(|x|+|y|)}$. Suppose the input to the system is an image consisting of one line as shown in Figure 2.64. What is the output image $g(x, y)$?

2.6 Suppose that an image $f(x, y)$ undergoes planer motion, and let $x_0(t)$ and $y_0(t)$ be the time-varying components of motion in the x- and y-directions, respectively. The total exposure at any point of the recording medium (e.g., film) is obtained by integrating the instantaneous exposure over the time interval during which the shutter is open. Then, if T is the duration of exposure, we have

$$g(x, y) = \frac{1}{T} \int_0^T f(x - x_0(t),\ y - y_0(t)) dt$$

where $g(x, y)$ is the output image.
a. Show that the OTF of the system is

$$H(\omega_x, \omega_y) = \frac{1}{T} \int_0^T e^{-j(\omega_x x_0(t) + \omega_y y_0(t))} dt$$

b. Suppose that the image undergoes motion in the x-direction only, at the rate of $x_0(t) = 0.5at^2$. Sketch the resulting MTF.

2.7 A PSF has no spatial frequencies greater than 400 cycles/mm. What values would you assign to the sampling interval Δx and the DFT length so as to obtain samples of the MTF in which aliasing is negligible and samples are spaced no further than 5 cycles/mm apart? The DFT length must be a power of 2.

2.8 Find the Fourier transform of a line oriented at an angle θ as shown in Figure 2.65.

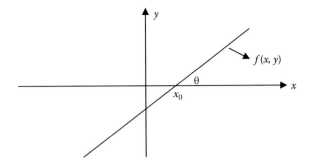

FIGURE 2.65

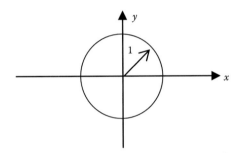

FIGURE 2.66

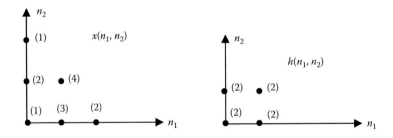

FIGURE 2.67

2.9 Find the Fourier transform of the 2-D image $f(x,y) = \delta(x^2 + y^2 - 1)$. The image is shown in Figure 2.66. Display the transform as an image.

2.10 Consider the two sequences $x(n_1, n_2)$ and $h(n_1, n_2)$ shown in Figure 2.67.
  a. Determine $y(n_1, n_2) = x(n_1, n_2) * h(n_1, n_2)$, the linear convolution of $x(n_1, n_2)$ and $h(n_1, n_2)$.
  b. Develop a procedure to compute $y(n_1, n_2) = x(n_1, n_2) * h(n_1, n_2)$ using DFT.

2.11 Determine the approximate number of multiplications required for the 2-D convolution of $f(n, m)$ and $h(n, m)$ for the three cases given below. Assume $f(n, m)$ to be $N \times N$ and $h(n, m)$ to be $L \times L$.
a. Both $f(n, m)$ and $h(n, m)$ are nonseparable.
b. Both $f(n, m)$ and $h(n, m)$ are separable.
c. $f(n, m)$ is not separable but $h(n, m)$ is separable.

2.12 Determine the convolution of $f_x(n, m)$ with $h(n, m)$, where

$$x(n, m) = \begin{bmatrix} 1 & 3 & 2 & 3 \\ 4 & 7 & 1 & 0 \\ 3 & 0 & 5 & 8 \\ 2 & 0 & 2 & 6 \end{bmatrix} \quad \text{and} \quad h(n, m) = \begin{bmatrix} 4 & 12 \\ 2 & 6 \end{bmatrix}$$

2.13 An image of size 64×64 is to be filtered using a 5×5 FIR LPF. Find
a. The number of real multiplications if the convolution is performed in the pixel domain.
b. The number of real multiplications if the overlap–add technique is used with an 8-point FFT algorithm.

2.14 The signal $f(x)$ is band limited so that its spectrum, $F(\omega)$ (i.e., Fourier transform), is zero outside the interval $-2\pi B < \omega < 2\pi B$.
a. Expand the spectrum $F(\omega)$ in a Fourier series on the interval $-2\pi B < \omega < 2\pi B$, having the form

$$F(\omega) = \sum_{k=-\infty}^{\infty} f_k \exp\left(-j\frac{k\omega}{2B}\right)$$

b. Show that the coefficients, f_k, in this expansion are proportional to samples of $f(x)$ at the interval of $\frac{1}{2B}$ unit of length. Using these values for f_k in the above equation, take the inverse Fourier transform of $F(\omega)$ to give $f(x)$ and show that the result expresses the signal $f(x)$ in terms of the sample values $f(\frac{k}{2B})$ and an "interpolation function" of the form

$$\frac{\sin 2\pi B\left(x - \frac{k}{2B}\right)}{2\pi B\left(x - \frac{k}{2B}\right)}$$

This last result is, then, actually a statement of the most popular form of the sampling theorem, in that it expresses the band-limited signal $f(x)$ in terms of discrete samples taken at intervals of $\frac{1}{2B}$ units of length.

2.15 Let $P(x) = xe^{-0.5x^2}u(x)$ and let the number of quantizer levels be 8 (3 bit).
a. What are the decision and reconstruction levels?
b. Find the resulting MSE.

2.16 Design a 2-D 11×11 zero-phase FIR filter to approximate the desired frequency response shown in Figure 2.68.

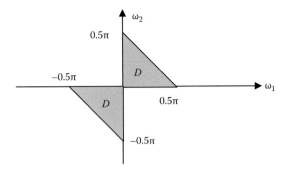

FIGURE 2.68

$$H_d(\omega_1, \omega_2) = \begin{cases} 1 & (\omega_1, \omega_2) \in D \\ 0 & \text{otherwise} \end{cases}$$

2.17 Consider the 8×8 image $f(n, m)$ shown below.

$$f(n, m) = \begin{bmatrix}
123 & 125 & 130 & 123 & 140 & 145 & 150 & 15 \\
128 & 129 & 135 & 124 & 145 & 140 & 129 & 19 \\
24 & 150 & 120 & 140 & 123 & 134 & 29 & 35 \\
129 & 180 & 109 & 150 & 155 & 160 & 29 & 65 \\
190 & 185 & 98 & 100 & 162 & 89 & 68 & 34 \\
100 & 23 & 88 & 136 & 123 & 35 & 187 & 43 \\
86 & 12 & 0 & 5 & 135 & 34 & 45 & 56 \\
23 & 14 & 4 & 8 & 128 & 56 & 25 & 23
\end{bmatrix}$$

a. Assume this image is filtered by a separable filter having the following horizontal and vertical frequency responses

$$H_h(\omega_x) = \tfrac{1}{3}(1 + 2\cos(\omega_x)),$$
$$H_v(\omega_y) = 3 - 2\cos(\omega_y)$$

What are the gray level values of the output image at the following locations?

$$g(2, 3) \quad \text{and} \quad g(5, 5)$$

b. Repeat part (1), for a nonseparable filter with the following frequency response.

$$H(\omega_x, \omega_y) = 4 - \cos(\omega_x) - \cos(\omega_y)$$

2.18 Consider the 3×3 FIR filter

$$h(n, m) = \frac{1}{8} \begin{bmatrix} -2 & -4 & -2 \\ 6 & 12 & 6 \\ -2 & -4 & -2 \end{bmatrix}$$

a. What is the DC gain of this filter $(H(0,0))$?
b. What is the high-frequency gain $H(\pi, \pi)$?
c. Is this a zero-phase filter?
d. Is it separable?
e. Find the output image $g(n, m)$.

$$f(n, m) = 2 + \cos(\pi n + \pi m) \rightarrow \boxed{h(n, m)} \mapsto g(n, m)$$

2.19 The histogram of a 32×32, 3 bit gray level image is given in Table 2.4. We wish to modify the gray scale of this image such that the histogram of the processed image is as close as possible to being constant. Determine a transformation that will achieve this objective.

2.20 Consider the following 3 bit gray level image.

$$f(n, m) = \begin{bmatrix} 2 & 4 & 3 & 3 & 1 & 1 & 0 & 0 \\ 0 & 1 & 2 & 3 & 1 & 3 & 2 & 6 \\ 3 & 2 & 5 & 5 & 7 & 3 & 2 & 3 \\ 3 & 6 & 6 & 2 & 4 & 3 & 3 & 6 \\ 2 & 1 & 5 & 2 & 3 & 1 & 4 & 3 \\ 5 & 1 & 3 & 3 & 4 & 1 & 2 & 7 \\ 2 & 0 & 7 & 0 & 6 & 7 & 5 & 7 \\ 2 & 3 & 1 & 3 & 4 & 2 & 4 & 3 \end{bmatrix}$$

a. Find the histogram of this image.
b. We wish to modify the gray scale of this image such that the histogram of the processed image is as close as possible to being constant. Determine a transformation that will achieve this objective.

2.21 A motion-blurred image (uniform motion in the x-direction with $x_0(t) = a\frac{t}{T}$) is observed in the presence of additive white noise with a power spectrum of $S_n(\omega_x, \omega_y) = N_0$.
a. What is the transfer function of the Wiener filter in terms of the power spectrum of the image?
b. Give an algorithm for FFT implementation of the Wiener filter.

TABLE 2.4

Histogram of Gray Scale Level

Gray level	0	1	2	3	4	5	6	7
Histogram	98	145	107	180	152	95	116	131

REFERENCES

1. Jahne, B. *Digital Image Processing: Concepts, Algorithms, and Scientific Applications*, Springer-Verlag, New York, 1997.
2. Petrou, M. and Bosdogianni, P. *Image Processing: The Fundamentals*, John Wiley & Sons, United Kingdom, 1999.
3. Jain, A.K. *Digital Image Processing: Concepts, Algorithms, and Scientific Applications*, Springer-Verlag, New York, 1988.
4. Lee, H.-C. Review of imaging-blur models in a photographic system using the principles of optics, *J. Opt. Eng.*, 29, 405–421, 1990.
5. Lim, J.S. *Two-Dimensional Signal and Image Processing*, Prentice Hall, Upper Saddle River, NJ, 1990.
6. Max, J. Quantizing for minimum distortion, *IRE Trans. Info. Theory*, IT-6, 7–12, 1960.
7. Gray, R.M. Vector quantization, *IEEE Trans. Acous. Speech Signal Processing*, ASSP-1 (2), 4–29, 1984.
8. Equitz, W.H. A new vector quantization clustering algorithm, *IEEE Trans. Acous. Speech Signal Processing*, ASSP-37(10), 1568–1575, 1989.
9. Bracewell, R.N. *The Fourier Transform and its Applications*, 3rd ed., McGraw-Hill, New York, 2000.
10. Cooley, J.W., Lewis, P.A.W., and Welch, P.D. Historical notes on the fast Fourier transform, *IEEE Trans. Audio Electroacoustics*, AU-15(2), 76–79, 1967.
11. Brigham, E.O. *The Fast Fourier Transform and its Applications*, Prentice Hall, Upper Saddle River, NJ, 1988.
12. Ahmed, N., Natarajan, T., and Rao, K.R. Discrete cosine transforms, *IEEE Trans. Comp.*, C-23, 90–93, 1974.
13. Clarke, R.J. *Transform Coding of Images*, Academic Press, New York, 1985.
14. Lu, W.-S. and Antoniou, A. *Two-Dimensional Digital Filters*, Marcel Dekker, New York, 1992.
15. Marr, D. and Hildreth, E. Theory of edge detection, *Proc. R. Soc. Lond.*, B207, 187–217, 1980.
16. Martelli, A. Edge detection using Heuristic search methods, *Comput. Graphics Image Proc.*, 1, 169–182, 1972.
17. Martelli, A. An application of Heuristic search methods to edge and contour detection, *Comm. ACM*, 19(2), 73–83, 1976.
18. Unser, M., Aldroubi, A., and Eden, M. Enlargement or reduction of digital images with minimum loss of information, *IEEE Trans. Image Processing*, 4(5), 247–257, 1995.
19. Pratt, W.K. *Digital Image Processing*, 3rd ed., John Wiley & Sons, New York, 2001.
20. Hummel, R.A. Histogram modification techniques. Technical Report TR-329. F-44620–72C-0062, Computer Science Center, University of Maryland, College Park, MD, 1974.
21. Gonzalez, R.C. and Woods, R.E. *Digital Image Processing*, Prentice Hall, Upper Saddle River, NJ, 2007.
22. MacAdam, D.P. Digital image restoration by constrained deconvolution, *J. Opt. Soc. Am.*, 60, 1617–1627, 1970.

3 Mathematical Foundations

3.1 INTRODUCTION

The motivation for this chapter is to present mathematical tools for analysis and design of open- and closed-loop continuous- and discrete-time control systems. A broad class of linear time-invariant (LTI) systems can be represented by linear differential equations (DEs) with constant coefficients in case of continuous-time and linear difference equations with constant coefficients in case of discrete-time systems. The Fourier and Laplace transforms play important roles in analyzing and designing such systems. The z-transform is a tool for analyzing and designing discrete-time systems and is the counterpart of Laplace transform that is used for continuous-time systems. In this chapter, we introduce LTI continuous-time systems, Laplace transform, discrete-time systems, and z-transform. Matrices and linear algebra are also important tools in analyzing control systems in the state-space form. Eigenvalues and eigenvectors, singular value decomposition (SVD), and functions of matrices are other mathematical tools used to design control systems with state-space approach. The rest of this chapter is devoted to this important subject.

3.2 GENERAL CONTINUOUS-TIME SYSTEM DESCRIPTION

Consider a single-input single-output (SISO) continuous-time system with input $u(t)$ and output $y(t)$ that can be represented by constant-coefficients linear DE of the form

$$\frac{d^N y(t)}{dt^N} + a_{N-1}\frac{d^{N-1} y(t)}{dt^{N-1}} + \cdots + a_0 y(t) = b_M \frac{d^M u(t)}{dt^M} + b_{M-1}\frac{d^{M-1} u(t)}{dt^{M-1}} + \cdots + b_0 u(t)$$

(3.1)

where
   N is the order of the DE
   M is typically less than or equal to N

Given the input signal $u(t)$ and N initial conditions $y(0)$, $y'(0)$, $\ldots$, $y^{(N-1)}(0)$, the output signal $y(t)$ can be determined uniquely by solving the DE given by Equation 3.1 either in time domain or in the transform domain using Fourier or Laplace transform. We will first solve the equation in time domain, and then we will use Laplace transform to solve the equation in s-domain.

3.2.1 Solution of Constant-Coefficients Linear Differential Equations

We first consider zero-input response that is the response of the system defined by Equation 3.1 when the input signal $u(t)$ is zero. Consider an Nth-order constant-coefficients linear DE with zero input driven by initial conditions $y(0)$, $y'(0), \ldots$, and $y^{(N-1)}(0)$:

$$\frac{\mathrm{d}^N y(t)}{\mathrm{d}t^N} + a_{N-1} \frac{\mathrm{d}^{N-1} y(t)}{\mathrm{d}t^{N-1}} + \cdots + a_0 y(t) = 0 \tag{3.2}$$

Assuming a solution of the form $y(t) = Ce^{Dt}$, we would have

$$C\left[D^N + a_{N-1}D^{N-1} + a_{N-2}D^{N-2} + \cdots + a_0\right]e^{Dt} = 0 \tag{3.3}$$

Since Ce^{Dt} is nonzero, then

$$D^N + a_{N-1}D^{N-1} + a_{N-2}D^{N-2} + \cdots + a_0 = 0 \tag{3.4}$$

This polynomial is called characteristic equation of the DE. It is a polynomial of degree N that has N roots $D_1, D_2, \ldots$, and D_N. These roots are generally complex and are called characteristic roots of the DE. The zero-input solution is given by

$$y(t) = C_1 e^{D_1 t} + C_2 e^{D_2 t} + \cdots + C_N e^{D_N t} \tag{3.5}$$

The constants $C_1, C_2, \ldots$, and C_N are found by applying the N initial conditions.

The stability of the system is defined in terms of the zero-input response. The discrete LTI system described by DE given in Equation 3.1 is stable if the zero-input response decays to zero as $t \to \infty$. This implies that the characteristic roots must have real parts less than zero, that is,

$$\mathrm{Re}(D_i) < 0 \quad \text{for } i = 1, 2, \ldots, N \tag{3.6}$$

Example 3.1

Solve the following second-order DE with initial conditions $y(0) = -1$ and $\frac{\mathrm{d}y(t)}{\mathrm{d}t}\big|_{t=0} = 4$:

$$\frac{\mathrm{d}^2 y(t)}{\mathrm{d}t^2} + 3\frac{\mathrm{d}y(t)}{\mathrm{d}t} + 2y(t) = 0$$

Solution

The characteristic equation is $D^2 + 3D + 2 = 0$, which can be written as

$$D^2 + 3D + 2 = (D+1)(D+2) = 0$$

Therefore, the roots are $D_1 = -1$ and $D_2 = -2$. The zero-input response is

$$y(t) = C_1 e^{D_1 t} + C_2 e^{D_2 t} = C_1 e^{-t} + C_2 e^{-2t}$$

The constants C_1 and C_2 are found by applying the initial conditions

$$y(0) = C_1 + C_2 = -1$$

$$y'(0) = -C_1 - 2C_2 = 4$$

The solutions are $C_1 = 2$ and $C_2 = -3$. Therefore,

$$y(t) = 2e^{-t} - 3e^{-2t} \quad \text{for } t \geq 0$$

Next, we consider the homogenous solution. The homogenous solution is the result of driving the system with an input with zero initial conditions. The homogenous solution is typically similar to the input signal; for example, if the input signal is an exponential function, the response will be exponential and if the input signal is a sinusoid, the output is also sinusoid with different amplitude and phase. This is due to the properties of linear systems. A simple example that shows the process of obtaining the homogenous solution is given below.

Example 3.2

Solve the following second-order DE with initial conditions $y(0) = -1$ and $\frac{dy(t)}{dt}\big|_{t=0} = 4$:

$$\frac{d^2 y(t)}{dt^2} + 3\frac{dy(t)}{dt} + 2y(t) = e^{-4t}u(t) \qquad (3.7)$$

where $u(t)$ is the unit step function.

SOLUTION

Since the input is an exponential function, the homogenous solution will be another exponential, that is,

$$y_h(t) = Ae^{-4t} \qquad (3.8)$$

Substituting Equation 3.8 into Equation 3.7 yields

$$16Ae^{-4t} - 12Ae^{-4t} + 2Ae^{-4t} = e^{-4t}$$

Solving for A yields

$$A = \tfrac{1}{6}$$

Therefore, the total solution is the sum of the zero-input response and the homogeneous solution

$$y(t) = C_1 e^{-t} + C_2 e^{-2t} + \tfrac{1}{6}e^{-4t}$$

The constants C_1 and C_2 are found by applying the initial conditions:

$$y(0) = C_1 + C_2 + \tfrac{1}{6} = -1$$

$$y'(0) = -C_1 - 2C_2 - \tfrac{4}{6} = 4$$

The solutions are $C_1 = \tfrac{7}{3}$ and $C_2 = -\tfrac{7}{2}$. Therefore,

$$y(t) = \tfrac{7}{3}e^{-t} - \tfrac{7}{2}e^{-2t} + \tfrac{1}{6}e^{-4t} \quad \text{for } t \geq 0$$

Another approach for solving constant-coefficients linear DEs is use of Fourier or Laplace transform. In the next section, we briefly cover Laplace transform with its properties. Interested readers can refer to Ref. [1] for more details.

3.3 LAPLACE TRANSFORM

The Laplace transform of a continuous signal $x(t)$ is a mapping from time domain to complex frequency domain defined as

$$X(s) = \int_{-\infty}^{\infty} x(t)e^{-st}dt \tag{3.9}$$

where $s = \sigma + j\omega$ is the complex frequency. The integral in Equation 3.9 may not converge for all values of s. The region in complex plain s, where the complex function $X(s)$ converges, is known as the region of convergence (ROC) of $X(s)$. The transform defined by Equation 3.9 is referred to as double-sided Laplace transform. In control applications, the signals are defined over the time interval of 0 to ∞. This is due to the fact that in control systems, we are mainly concerned with transient response of the system. Therefore, we are interested in one-sided Laplace transform defined as

$$X(s) = \int_{0}^{\infty} x(t)e^{-st}dt \tag{3.10}$$

Example 3.3

Find the Laplace transform and ROC of the signal $x(t) = e^{-at}u(t)$.

SOLUTION

$$X(s) = \int_{0}^{\infty} x(t)e^{-st}dt = \int_{0}^{\infty} e^{-at}e^{-st}dt = \int_{0}^{\infty} e^{-(s+a)t}dt = \frac{-1}{s+a}e^{-(s+a)t}\Big|_{0}^{\infty} = \frac{1}{s+a}$$

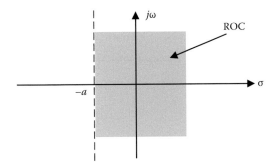

FIGURE 3.1   ROC of $X(s)$.

The convergence of $X(s)$ requires that $\lim_{t \to \infty} e^{-(s+a)t} \to 0$. Thus, the ROC is the set of points in the complex s-plane for which $\mathrm{Re}(s) > -a$, as shown in Figure 3.1. Notice that the function $X(s)$ has a single pole located at $s = -a$, which is outside ROC of $X(s)$.

Example 3.4

Find the Laplace transform and ROC of the signal $x(t) = u(t)$, where $u(t)$ is the unit step function.

SOLUTION

This is a special case of Example 3.3 where $a = 0$. Therefore,

$$X(s) = \frac{1}{s} \quad \text{ROC: Re}(s) > 0$$

Example 3.5

Find the Laplace transform and ROC of the unit impulse signal $x(t) = \delta(t)$.

SOLUTION

$$X(s) = \int_0^\infty x(t)e^{-st}dt = \int_0^\infty \delta(t)e^{-st}dt = \delta(t)e^{-st}\big|_{t=0} = 1$$

Since the above integral converges for all values of s, the ROC is the entire complex s-plane.

The Laplace transform of some elementary signals and their ROCs are listed in Table 3.1. Properties of Laplace transform are given in Table 3.2. Interested readers are referred to Ref. [1] for proofs of these properties.

TABLE 3.1

Laplace Transform of Elementary Functions

Signal	Transform	ROC
$\delta(t)$	1	s-Plane
$u(t)$	$\dfrac{1}{s}$	$\mathrm{Re}(s) > 0$
$e^{-at}u(t)$	$\dfrac{1}{s+a}$	$\mathrm{Re}(s) > -\mathrm{Re}(a)$
$te^{-at}u(t)$	$\dfrac{1}{(s+a)^2}$	$\mathrm{Re}(s) > -\mathrm{Re}(a)$
$t^n e^{-at}u(t)$	$\dfrac{n!}{(s+a)^{n+1}}$	$\mathrm{Re}(s) > -\mathrm{Re}(a)$
$\cos \omega_0 t\, u(t)$	$\dfrac{1}{s^2+\omega_0^2}$	$\mathrm{Re}(s) > 0$
$\sin \omega_0 t\, u(t)$	$\dfrac{s}{s^2+\omega_0^2}$	$\mathrm{Re}(s) > 0$
$e^{-at}\cos \omega_0 t\, u(t)$	$\dfrac{1}{(s+a)^2+\omega_0^2}$	$\mathrm{Re}(s) > -\mathrm{Re}(a)$
$e^{-at}\sin \omega_0 t\, u(t)$	$\dfrac{s+a}{(s+a)^2+\omega_0^2}$	$\mathrm{Re}(s) > -\mathrm{Re}(a)$
$\delta(t-T)$	e^{-sT}	s-Plane

3.3.1 INVERSE LAPLACE TRANSFORM

An important application of Laplace transform is in analysis of LTI continuous systems. This analysis involves computing the response of the system to a given input using Laplace transform. Once the Laplace transform of the output signal is determined, inverse Laplace transform is used to find the corresponding time-domain function. There are different techniques such as inversion integral and partial-fraction expansion to compute the inverse Laplace transform from a given algebraic expression. In this section, we only consider partial-fraction approach. Interested readers are referred to Ref. [2] for a complete coverage of this subject.

Partial Fraction Expansion
If $X(s)$ is a rational function, that is,

$$X(s) = \frac{\sum_{k=0}^{M} b_k s^k}{\sum_{k=0}^{N} a_k s^k} = \frac{\sum_{k=0}^{M} b_k s^k}{\prod_{i=1}^{Q} (s - p_i)^{m_i}} \tag{3.11}$$

where

p_i, $i = 1, 2, \ldots, Q$ are the poles of $X(s)$ or roots of polynomial $\sum_{k=1}^{N} a_k s^k$
m_i is the multiplicity of the ith pole

TABLE 3.2
Properties of the Laplace Transform

Property	$x(t) \xrightarrow{L} X(s)$
	$y(t) \xrightarrow{L} Y(s)$
Linearity	$ax(t) + by(t) \xrightarrow{L} aX(s) + bY(s)$
Differentiation (first derivative)	$\dfrac{dx(t)}{dt} \xrightarrow{L} sX(s) - x(0)$
Differentiation (second derivative)	$\dfrac{d^2x(t)}{dt^2} \xrightarrow{L} s^2X(s) - sx(0) - x'(0)$
Multiplication by t	$tx(t) \xrightarrow{L} -\dfrac{dX(s)}{ds}$
Shift	$x(t - t_0) \xrightarrow{L} e^{-st_0}X(s)$
Multiplication by t^n	$t^n e^{-at} u(t) \xrightarrow{L} \dfrac{n!}{(s + a)^{n+1}}$
Scaling	$x(at) \xrightarrow{L} \dfrac{1}{|a|}X\left(\dfrac{s}{a}\right)$
Integration	$\displaystyle\int_0^t x(\tau)d\tau \xrightarrow{L} \dfrac{X(s)}{s}$
Convolution	$x(t) * y(t) \xrightarrow{L} X(s)Y(s)$

Note that $m_1 + m_2 + \cdots + m_Q = N$. Assuming that $N > M$, partial fraction of $X(s)$ yields

$$X(s) = \left[\frac{A_1}{s - p_1} + \frac{A_2}{(s - p_1)^2} + \cdots + \frac{A_{m_1}}{(s - p_1)^{m_1}}\right]$$

$$+ \left[\frac{B_1}{s - p_2} + \frac{B_2}{(s - p_2)^2} + \cdots + \frac{B_{m_2}}{(s - p_2)^{m_2}}\right] + \cdots \qquad (3.12)$$

The residues $A_1, A_2, \ldots, A_{m_1}$ corresponding to the pole p_1 are computed using

$$A_k = \frac{d^{k-m_i}}{dz^{k-m_i}}(s - p_1)^{m_1} X(s)\big|_{s=p_1} \quad k = m_i, m_{i-1}, \ldots, 2, 1 \qquad (3.13)$$

Similarly, the other set of residues corresponding to other poles are computed. Once the partial fraction is completed, the time-domain function $x(t)$ is

$$x(t) = \left[A_1 e^{p_1 t} + A_2 \frac{t e^{p_1 t}}{1!} + A_3 \frac{t^2 e^{p_1 t}}{2!} + \cdots + A_{m_1} \frac{t^{m_1-1} e^{p_1 t}}{(m_1 - 1)!}\right]$$

$$+ \left[B_1 e^{p_2 t} + B_2 \frac{t e^{p_2 t}}{1!} + B_3 \frac{t^2 e^{p_2 t}}{2!} + \cdots + B_{m_2} \frac{t^{m_2-1} e^{p_2 t}}{(m_2 - 1)!}\right] + \cdots \qquad (3.14)$$

Example 3.6

Find the inverse Laplace transform of the following function of the right-sided signal $x(t)$ if

$$X(s) = \frac{s}{(s+1)^2(s+2)}$$

SOLUTION

Using partial-fraction expansion, we would have

$$X(s) = \frac{s}{(s+1)^2(s+2)} = \frac{A_1}{s+1} + \frac{A_2}{(s+1)^2} + \frac{B_1}{s+2}$$

where

$$B_1 = \lim_{s \to -2} (s+2)X(s) = \lim_{s \to -2} \frac{s}{(s+1)^2} = -2$$

$$A_2 = \lim_{s \to -1} (s+1)^2 X(s) = \lim_{s \to -1} \frac{s}{s+2} = -1$$

$$A_1 = \lim_{s \to -1} \frac{d(s+1)^2 X(s)}{ds} = \lim_{s \to -1} \frac{d}{ds}\frac{s}{s+2} = \lim_{s \to -1} \frac{2}{(s+2)^2} = 2$$

Therefore,

$$X(s) = \frac{2}{s+1} - \frac{1}{(s+1)^2} - \frac{2}{s+2}$$

Hence,

$$x(t) = [2e^{-t} - te^{-t} - 2e^{-2t}]u(t)$$

Example 3.7

Solve the following second-order DE using Laplace transform. The initial conditions are $y(0) = -1$ and $\frac{dy(t)}{dt}\big|_{t=0} = 4$.

$$\frac{d^2 y(t)}{dt^2} + 3\frac{dy(t)}{dt} + 2y(t) = e^{-4t}u(t)$$

where $u(t)$ is the unit step function.

SOLUTION

Taking Laplace transform from both sides of the above DE, we have

$$s^2 Y(s) - sy(0) - y'(0) + 3[sY(s) - y(0)] + 2Y(s) = \frac{1}{s+4}$$

Using the initial conditions, we have

$$(s^2 + 3s + 2)Y(s) + s - 1 = \frac{1}{s+4}$$

Therefore,

$$Y(s) = \frac{-s^2 - 3s + 5}{(s+4)(s^2+3s+2)}$$

Using partial-fraction expansion, we have

$$Y(s) = \frac{-s^2 - 3s + 5}{(s+4)(s^2+3s+2)} = \frac{-s^2 - 3s + 5}{(s+4)(s+1)(s+2)}$$

$$Y(s) = \frac{A_1}{s+1} + \frac{A_2}{s+2} + \frac{A_3}{s+4}$$

The residues A_1, A_2, and A_3 are computed as

$$A_1 = \lim_{s \to -1} (s+1)Y(s) = \frac{-s^2 - 3s + 5}{(s+2)(s+4)} = \frac{7}{3}$$

$$A_2 = \lim_{s \to -2} (s+2)Y(s) = \frac{-s^2 - 3s + 5}{(s+1)(s+4)} = -\frac{7}{2}$$

$$A_3 = \lim_{s \to -4} (s+4)Y(s) = \frac{-s^2 - 3s + 5}{(s+1)(s+2)} = \frac{1}{6}$$

Therefore,

$$y(t) = \left(\tfrac{7}{3}e^{-t} - \tfrac{7}{2}e^{-2t} + \tfrac{1}{6}e^{-4t} \right) u(t)$$

3.4 GENERAL LINEAR DISCRETE-TIME SYSTEMS

Consider a SISO discrete-time system with input $u(n)$ and output $y(n)$ that can be represented by DE of the form

$$y(n) = -\sum_{i=1}^{N} a_i y(n-i) + \sum_{i=0}^{M} b_i u(n-i) \qquad (3.15)$$

This defines an Nth-order constant-coefficients DE. This equation describes a general LTI discrete-time system. This system can be analyzed in time domain or in transform domain using z-transform. In the next section, we solve the DE in time domain. The theory of z-transform with its properties and inverse z–transform are covered in the subsequent sections.

3.4.1 SOLUTION OF CONSTANT-COEFFICIENTS DIFFERENCE EQUATIONS

We first consider zero-input response that is the response of a system defined by Equation 3.15 when the input signal $u(n)$ is zero. Consider an Nth-order constant-coefficients DE with zero input driven by initial conditions $y(-1)$, $y(-2)$, ..., and $y(-N)$:

$$y(n) = -\sum_{i=1}^{N} a_i y(n-i) \tag{3.16}$$

Assuming a solution of the form $y(n) = CD^n$, we would have

$$CD^n = -\sum_{i=1}^{N} a_i CD^{n-i} \tag{3.17}$$

or

$$CD^n \left(1 + \sum_{i=1}^{N} a_i D^{-i} \right) = 0 \tag{3.18}$$

Since CD^n is nonzero, then

$$1 + \sum_{i=1}^{N} a_i D^{-i} = 0 \tag{3.19}$$

This is called characteristic equation of the DE. It is a polynomial of degree N that has N roots D_1, D_2, ..., and D_N. These roots are generally complex and are called characteristic roots of the DE. The zero-input solution is given by

$$y(n) = C_1 D_1^n + C_2 D_2^n + \cdots + C_N D_N^n \tag{3.20}$$

The constants C_1, C_2, ..., and C_N are found by applying the N initial conditions.

The stability of the system is defined in terms of the zero-input response. The discrete LTI system described by DE given in Equation 3.15 is stable if the zero-input response decays to zero as $n \to \infty$. This implies that the characteristic roots must have magnitude less than one, that is,

$$|D_i| < 1 \quad \text{for } i = 1, 2, \ldots, N \tag{3.21}$$

Example 3.8

Solve the following second-order DE with initial conditions $y(-2) = 1$ and $y(-1) = -1$:

$$y(n) = 0.75 y(n-1) - 0.125 y(n-2) \tag{3.22}$$

SOLUTION

The characteristic equation $1 - 0.75D^{-1} + 0.125D^{-2} = 0$ can be written as

$$D^2 - 0.75D + 0.125 = (D - 0.5)(D - 0.25) = 0 \qquad (3.23)$$

Therefore, the roots are $D_1 = 0.5$ and $D_2 = 0.25$. The zero-input response is

$$y(n) = C_1(0.5)^n + C_2(0.25)^n \qquad (3.24)$$

The constants C_1 and C_2 are found by applying the initial conditions

$$y(-1) = C_1(0.5)^{-1} + C_2(0.25)^{-1} = 2C_1 + 4C_2 = -1 \qquad (3.25)$$

$$y(-2) = C_1(0.5)^{-2} + C_2(0.25)^{-2} = 4C_1 + 16C_2 = 1 \qquad (3.26)$$

The solutions are $C_1 = -1.25$ and $C_2 = 0.375$. Therefore,

$$y(n) = -1.25(0.5)^n + 0.375(0.25)^n \quad n \geq -2 \qquad (3.27)$$

Next, we consider the homogenous solution. The homogenous solution is found by driving the system with an input but zero initial conditions. The homogenous solution is similar to input signal; for example, if the input is an exponential function, the response will be exponential, if the input is a sinusoid, the output is also a sinusoid.

Example 3.9

Consider the following second-order DE:

$$y(n) = 0.75y(n - 1) - 0.125y(n - 2) + u(n) \qquad (3.28)$$

Find the output signal $y(n)$ if the input signal is $u(n) = 2(0.4)^n$ for $n \geq 0$. Assume zero initial conditions, that is, let $y(-2) = y(-1) = 0$.

SOLUTION

Since the input is an exponential function, the homogenous solution will be another exponential, that is,

$$y_h(n) = A(0.4)^n \qquad (3.29)$$

Substituting Equation 3.29 into Equation 3.28 yields

$$A(0.4)^n = 0.75A(0.4)^{n-1} - 0.125A(0.4)^{n-2} + 2(0.4)^n \qquad (3.30)$$

Simplifying the right side of Equation 3.30, we have

$$A(0.4)^n = [0.75(0.4)^{-1}A - 0.125(0.4)^{-2}A + 2](0.4)^n \qquad (3.31)$$

Therefore,

$$A = 0.75(0.4)^{-1}A - 0.125(0.4)^{-2}A + 2 \qquad (3.32)$$

Solving for A yields

$$A = \frac{2}{1 - 0.75(0.4)^{-1} + 0.125(0.4)^{-2}} = -\frac{64}{3} \qquad (3.33)$$

Therefore,

$$y_h(n) = -\frac{64}{3}(0.4)^n \quad n \geq 0 \qquad (3.34)$$

The total solution is the sum of the homogenous and the zero-input response. Therefore,

$$y(n) = C_1(0.5)^n + C_2(0.25)^n + y_h(n) = C_1(0.5)^n + C_2(0.25)^n - \frac{64}{3}(0.4)^n \quad (3.35)$$

The constants C_1 and C_2 are found by applying the initial conditions

$$y(-1) = C_1(0.5)^{-1} + C_2(0.25)^{-1} - \frac{160}{3} = 2C_1 + 4C_2 - \frac{160}{3} = 0 \qquad (3.36)$$

$$y(-2) = C_1(0.5)^{-2} + C_2(0.25)^{-2} - \frac{400}{3} = 4C_1 + 16C_2 - \frac{400}{3} = 0 \qquad (3.37)$$

Solving Equations 3.36 and 3.37, we have $C_1 = 20$ and $C_2 = \frac{10}{3}$. Therefore,

$$y(n) = 20(0.5)^n + \frac{10}{3}(0.25)^n - \frac{64}{3}(0.4)^n \quad n \geq -2 \qquad (3.38)$$

DE can also be solved using z-transform. In the next section, we define z-transform, inverse z-transform, and properties associated with the transform. We will then use z-transform to solve discrete-time linear systems.

3.5   z-TRANSFORM

The z-transform of discrete-time signal $x(n)$ is defined as

$$X(z) = \sum_{n=-\infty}^{\infty} x(n)z^{-n} \qquad (3.39)$$

where z is a complex variable. The infinite sum in Equation 3.39 may not converge for all values of z. The region where the complex function $X(z)$ converges is known as the region of convergence. This is referred to as double-sided z-transform. If the signal $x(n) = 0$ for $n < 0$, then we have one-sided z-transform, which is defined as

$$X(z) = \sum_{n=0}^{\infty} x(n)z^{-n} \tag{3.40}$$

Convergence of the z-transform requires that

$$\sum_{n=-\infty}^{\infty} |x(n)r^{-n}| < \infty \tag{3.41}$$

for some positive values of r. This means absolute summability of the exponentially weighted sequence $x(n)$. For example, the following sequences do not have z-transform, since neither of these sequences multiplied by r^{-n} would be absolutely summable for any value of r. They are not satisfying the condition given above by Equation 3.41.

$$x_1(n) = A \frac{\sin(\omega_0 n)}{\omega_0 n} \tag{3.42}$$

$$x_2(n) = A \cos(\omega_0 n + \theta) \tag{3.43}$$

$$x_3(n) = a^n \qquad -\infty < n < \infty \tag{3.44}$$

Example 3.10

Find the z-transform of the one-sided signal $x(n) = a^n u(n)$.

SOLUTION

$$X(z) = \sum_{n=-\infty}^{\infty} x(n)z^{-n} = \sum_{n=0}^{\infty} a^n z^{-n} = \sum_{n=0}^{\infty} (az^{-1})^n = \frac{1}{1 - az^{-1}} = \frac{z}{z - a}$$

The convergence of $X(z)$ requires that $|az^{-1}| < 1$. Thus, the ROC is the set of points in complex z-plane for which $|z| > |a|$, as shown in Figure 3.2. Notice that the function $X(z)$ has a single pole located at $z = a$, which is outside ROC of $X(z)$.

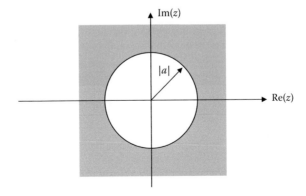

FIGURE 3.2 ROC for Example 3.10.

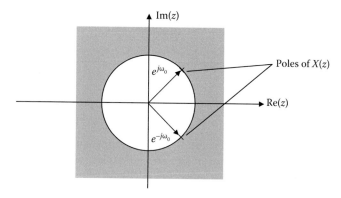

FIGURE 3.3   ROC for Example 3.11.

Example 3.11

Consider the sequence $x(n) = \cos(\omega_0 n)u(n)$. The z-transform of this sequence is

$$X(z) = \sum_{n=-\infty}^{\infty} x(n)z^{-n} = \sum_{n=0}^{\infty} \cos(\omega_0 n)z^{-n} = 0.5\sum_{n=0}^{\infty}(e^{jn\omega_0} + e^{-jn\omega_0})z^{-n}$$

$$= 0.5\sum_{n=0}^{\infty}(e^{j\omega_0}z^{-1})^n + 0.5\sum_{n=0}^{\infty}(e^{-j\omega_0}z^{-1})^n \qquad (3.45)$$

If $|e^{j\omega_0}z^{-1}| < 1$ and $|e^{-j\omega_0}z^{-1}| < 1$ or $|z| > 1$, the sums in Equation 3.45 converge and

$$X(z) = 0.5\frac{1}{1 - e^{j\omega_0}z^{-1}} + 0.5\frac{1}{1 - e^{-j\omega_0}z^{-1}} = \frac{0.5z}{z - e^{j\omega_0}} + \frac{0.5z}{z - e^{-j\omega_0}}$$

$$= \frac{z(z - \cos\omega_0)}{z^2 - 2\cos\omega_0 z + 1} \qquad (3.46)$$

The ROC of $X(z)$ is the set of points outside the unit circle, as shown in Figure 3.3. The function $X(z)$ has two complex conjugate poles at $p_1 = e^{j\omega_0}$ and $p_2 = e^{-j\omega_0}$. These poles are outside ROC of $X(z)$. In general, any discrete-time sequence, which is the linear combination of exponential functions, has a z-transform, which is the ratio of two polynomials in z (rational function). These polynomials can be represented by a constant gain, a set of poles, and zeros.

Example 3.12

Consider the sequence $x(n) = a^{|n|}$, where $0 < a < 1$. Find the z-transform of the sequence.

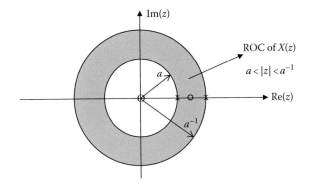

FIGURE 3.4 Pole–zero plot and ROC of Example 3.12.

SOLUTION

$$X(z) = \sum_{n=-\infty}^{\infty} x(n)z^{-n} = \sum_{n=-\infty}^{-1} (a)^{-n}z^{-n} + \sum_{n=0}^{\infty} a^n z^{-n} = \sum_{n=-\infty}^{-1} (az)^{-n} + \sum_{n=0}^{\infty} (az^{-1})^n$$

This can be simplified as

$$X(z) = \sum_{n=1}^{\infty} (az)^n + \sum_{n=0}^{\infty} (az^{-1})^n = -1 + \sum_{n=0}^{\infty} (az)^n + \sum_{n=0}^{\infty} (az^{-1})^n$$

$$= -1 + \frac{1}{1-az} + \frac{1}{1-az^{-1}} \tag{3.47}$$

or

$$X(z) = \frac{z(z+a-a^{-1})}{(z-a)(z-a^{-1})}$$

For convergence of $X(z)$, both sums in Equation 3.47 must converge. This requires that $|az| < 1$ and $|az^{-1}| < 1$, or equivalently, $|z| < a^{-1}$ and $|z| > a$. Therefore, the ROC of $X(z)$ is $a < |z| < a^{-1}$. The pole–zero plot as well as ROC of $X(z)$ is shown in Figure 3.4.

The z-transforms of some elementary functions and their ROCs are listed in Table 3.3.

3.5.1 PROPERTIES OF z-TRANSFORM

Most common properties of z-transform are listed below:

(a) **Linearity**: z-Transform is a linear transform, that is, if

$$x_1(n) \xrightarrow{z} X_1(z), \quad \text{ROC} = R_1 \tag{3.48}$$

TABLE 3.3

z-Transform of Elementary Sequences

x(n)	X(z)	ROC
$\delta(n)$	1	Entire z-plane
$\delta(n-k)$	z^{-k}	Entire z-plane
$u(n)$	$\dfrac{z}{z-1}$	$|z| > 1$
$a^n u(n)$	$\dfrac{z}{z-a}$	$|z| > |a|$
$a^n u(-n-1)$	$-\dfrac{z}{z-a}$	$|z| < |a|$
$a^{|n|}$	$\dfrac{z(z+a-a^{-1})}{(z-a)(z-a^{-1})}$	$a < |z| < \dfrac{1}{a}$ $0 < a < 1$
$nu(n)$	$\dfrac{z}{(z-1)^2}$	$|z| > 1$
$na^n u(n)$	$\dfrac{az}{(z-a)^2}$	$|z| > |a|$
$n^2 a^n u(n)$	$\dfrac{az(z+a)}{(z-a)^3}$	$|z| > |a|$
$\dfrac{n(n-1)\cdots(n-m+1)}{m!} a^{n-m} u(n-m+1)$	$\dfrac{z}{(z-a)^{m+1}}$	$|z| > |a|$
$\cos(\omega_0 n) u(n)$	$\dfrac{z(z-\cos\omega_0)}{z^2 - 2z\cos\omega_0 + 1}$	$|z| > 1$
$\sin(\omega_0 n) u(n)$	$\dfrac{z\sin\omega_0}{z^2 - 2z\cos\omega_0 + 1}$	$|z| > 1$
$a^n \cos(\omega_0 n) u(n)$	$\dfrac{z(z-a\cos\omega_0)}{z^2 - 2az\cos\omega_0 + a^2}$	$|z| > |a|$
$a^n \sin(\omega_0 n) u(n)$	$\dfrac{za\sin\omega_0}{z^2 - 2za\cos\omega_0 + a^2}$	$|z| > |a|$
$a^{|n|}$	$\dfrac{z(z+a-a^{-1})}{(z-a)(z-a^{-1})}$	$a < |z| < \dfrac{1}{a}$ $0 < a < 1$

and

$$x_2(n) \xrightarrow{Z} X_2(z), \quad \text{ROC} = R_2 \qquad (3.49)$$

then

$$\alpha x_1(n) + \beta x_2(n) \xrightarrow{Z} \alpha X_1(z) + \beta X_2(z), \quad \text{ROC} = R = R_1 \cap R_2 \qquad (3.50)$$

This means that the z-transform of linear combination of two sequences is the linear combination of their respective z-transforms and the ROC is the intersection of the individual regions of convergence. The proof is straightforward and is based on the fact that sum of two sequences is equal to sum of individual sequences.

Example 3.13

Find the z-transform of the following sequence:

$$x(n) = 2\left(\tfrac{1}{3}\right)^n u(n) + 3\left(\tfrac{1}{2}\right)^n u(n)$$

SOLUTION

To find $X(z)$, we use the linearity property and the z-transform pair $a^n u(n) \rightarrow \frac{z}{z-a}$, $|z| > |a|$. Then

$$X(z) = 2\,\frac{z}{z - \tfrac{1}{3}} + 3\,\frac{z}{z - \tfrac{1}{2}} = \frac{z(5z - 2)}{\left(z - \tfrac{1}{3}\right)\left(z - \tfrac{1}{2}\right)}$$

and the ROC is

$$R = \text{ROC} = \left(|z| > \tfrac{1}{3}\right) \cap \left(|z| > \tfrac{1}{2}\right) = |z| > \tfrac{1}{2}$$

(b) **Delay Property of the z-Transform**: If

$$x(n)u(n) \rightarrow X(z), \quad \text{ROC} = R \tag{3.51}$$

then

$$x(n - k)u(n - k) \rightarrow z^{-k}X(z), \quad \text{ROC} = R \tag{3.52}$$

The delay k is an integer.

Proof:

$$\sum_{n=-\infty}^{\infty} x(n - k)u(n - k)z^{-n} = \sum_{n=k}^{\infty} x(n - k)z^{-n} = \sum_{l=0}^{\infty} x(l)z^{-l-k}$$

$$= z^{-k} \sum_{l=0}^{\infty} x(l)z^{-l} = z^{-k}X(z) \tag{3.53}$$

Example 3.14

Find the z-transform of the following sequence:

$$x(n) = \begin{cases} 0 & n \le 3 \\ (0.6)^n & n \ge 4 \end{cases}$$

SOLUTION

The solution is easily found using the delay property of the z-transform, note that

$$x(n) = (0.6)^n u(n-4) = (0.6)^4 (0.6)^{n-4} u(n-4)$$

Therefore,

$$X(z) = (0.6)^4 z^{-4} \frac{z}{z - 0.6} = \frac{(0.6)^4}{z^3(z - 0.6)}$$

(c) **Convolution Property**: Convolution property of z-transform states that convolution in time domain is multiplication in z-domain. This means that the z-transform of convolution of two functions is the product of their corresponding z-transforms. If

$$y(n) = x(n) * h(n) \tag{3.54}$$

then

$$Y(z) = X(z)H(z) \tag{3.55}$$

Proof:

$$Y(z) = \sum_{n=-\infty}^{\infty} y(n)z^{-n} = \sum_{n=-\infty}^{\infty} \sum_{k=-\infty}^{\infty} x(k)h(n-k)z^{-n}$$

$$= \sum_{k=-\infty}^{\infty} x(k) \sum_{n=-\infty}^{\infty} h(n-k)z^{-n} \tag{3.56}$$

Using shifty property, we have

$$Y(z) = \sum_{k=-\infty}^{\infty} x(k)H(z)z^{-k} = H(z) \sum_{k=-\infty}^{\infty} x(k)z^{-k} = H(z)X(z) \tag{3.57}$$

As a result of this important theorem, the z-transform of the output of a linear shift-invariant system is the product of the z-transform of the input signal and the z-transform of the impulse response of the system. The z-transform of the system impulse response $h(n)$ is the system transfer function $H(z)$.

Example 3.15

Convolve the following two sequences:

$$x(n) = \begin{bmatrix} 2 & 3 & 1 \end{bmatrix} \quad \text{and} \quad h(n) = \begin{bmatrix} 1 & -2 & 3 & 2 \end{bmatrix}$$

SOLUTION

The z-transforms of $x(n)$ and $h(n)$ are

$$X(z) = 2 + 3z^{-1} + z^{-2}$$
$$H(z) = 1 - 2z^{-1} + 3z^{-2} + 2z^{-3}$$

Using the convolution property of z-transform, we have

$$Y(z) = X(z)H(z) = (2 + 3z^{-1} + z^{-2})(1 - 2z^{-1} + 3z^{-2} + 2z^{-3})$$
$$= 2 - z^{-1} + z^{-2} + 11z^{-3} + 9z^{-4} + 2z^{-5}$$

Therefore,

$$y(n) = x(n) * h(n) = \begin{bmatrix} 2 & -1 & 1 & 11 & 9 & 2 \end{bmatrix}$$

Example 3.16

Convolve the following two sequences using z-transform:

$$x(n) = (0.5)^n u(n)$$
$$h(n) = (0.7)^n u(n)$$

SOLUTION

The z-transform of $y(n)$, the convolution of $x(n)$ and $h(n)$ is

$$Y(z) = X(z)H(z) = \frac{z}{z - 0.5} \frac{z}{z - 0.7} = \frac{z^2}{(z - 0.5)(z - 0.7)}$$

Performing partial fraction, we have

$$Y(z) = \frac{3.5z}{z - 0.7} - \frac{2.5z}{z - 0.5}$$

Taking the inverse z-transform yields

$$y(n) = 3.5(0.7)^n u(n) - 2.5(0.5)^n u(n)$$

(d) **Multiplication by n or Differentiation of $X(z)$**: The z-transform of $nx(n)$ is $-z\frac{dX(z)}{dz}$, where $X(z)$ is the z-transform of $x(n)$.

Proof: The z-transform of $x(n)$ is

$$X(z) = \sum_{n=0}^{\infty} x(n)z^{-n} \tag{3.58}$$

Differentiating both sides of Equation 3.58 with respect to the complex variable z results in

$$\frac{dX(z)}{dz} = \sum_{n=-\infty}^{\infty} x(n)(-n)z^{-n-1} = -z^{-1}\sum_{n=-\infty}^{\infty} nx(n)z^{-n} \tag{3.59}$$

Multiplying both sides of Equation 3.59 by $-z$ yields

$$-z\frac{dX(z)}{dz} = \sum_{n=-\infty}^{\infty} nx(n)z^{-n} \tag{3.60}$$

The following example uses the differentiation property.

Example 3.17

Find the z-transform of the sequence

$$x(n) = na^n u(n) \tag{3.61}$$

SOLUTION

Since the z-transform of $a^n u(n)$ is $\frac{z}{z-a}$, then the z-transform of $x(n) = na^n u(n)$ is

$$X(z) = -z\frac{d}{dz}\left(\frac{z}{z-a}\right) = -z\frac{z-a-z}{(z-a)^2} = \frac{az}{(z-a)^2} \tag{3.62}$$

(e) **Initial Value Theorem**: If $x(n) = 0$ for $n < 0$, then

$$x(0) = \lim_{z\to\infty} X(z) \tag{3.63}$$

Proof: To derive this property, take the limit on both sides of Equation 3.64 as z approaches infinity:

$$X(z) = \sum_{n=0}^{\infty} x(n)z^{-n} = x(0) + x(1)z^{-1} + x(2)z^{-2} + \cdots \tag{3.64}$$

The result is

$$\lim_{z\to\infty} X(z) = x(0) + 0 + 0 + \cdots = x(0) \qquad (3.65)$$

(f) **Final Value Theorem**: If $x(n) = 0$ for $n < 0$ and $X(z)$ does not have any poles on the boundary of unit circle, then

$$x(\infty) = \lim_{z\to 1} (z - 1)X(z) \qquad (3.66)$$

To derive this property, we use the delay and linear properties of the z-transform, mainly

$$(z - 1)X(z) = zX(z) - X(z) = z\text{-transform of } [x(n+1) - x(n)] \qquad (3.67)$$

Therefore,

$$(z - 1)X(z) = \sum_{n=-1}^{\infty} [x(n+1) - x(n)]z^{-n} \qquad (3.68)$$

Taking the limit as $z \to 1$ results in

$$\lim_{z\to 1} (z - 1)X(z) = \sum_{n=-1}^{\infty} [x(n+1) - x(n)]$$
$$= x(0) - x(-1) + x(1) - x(0) + x(2) - x(1) + \cdots \qquad (3.69)$$

Since $x(-1) = 0$, then:

$$\lim_{z\to 1} (z - 1)X(z) = x(\infty) \qquad (3.70)$$

Example 3.18

Find the initial and the final values of the one-sided signal x(n), if

$$X(z) = \frac{z(z - 0.7)}{z^2 - 1.25z + 0.25}$$

SOLUTION
Using the initial and final value theorems, we have

$$x(0) = \lim_{z\to\infty} X(z) = \lim_{z\to\infty} \frac{z(z - 0.7)}{z^2 - 1.25z + 0.25} = 1$$

and

$$x(\infty) = \lim_{z \to 1} (z - 1)X(z) = \lim_{z \to 1} (z - 1)\frac{z(z - 0.7)}{(z - 1)(z - 0.25)} = \lim_{z \to 1} \frac{z(z - 0.7)}{z - 0.25} = 0.4$$

3.5.2 INVERSE z-TRANSFORM

An important application of the z-transform is in the analysis of linear discrete-time systems. This analysis involves computing the response of the systems to a given input using z-transform. Once the z-transform of the output signal is determined, inverse z-transform is used to find the corresponding time-domain sequence. There are different techniques to find the inverse z-transform from a given algebraic expression. In this section, we consider techniques such as inversion integral, power series expansion, and partial fraction.

(a) **Inversion Integral**: The inverse z-transform of $X(z)$ is given by the integral

$$x(n) = \frac{1}{j2\pi} \oint_C X(z)z^{n-1}dz \tag{3.71}$$

where C is any closed contour in the ROC of $X(z)$ excluding the origin, as shown in Figure 3.5. If $X(z)$ is a rational function of its argument z, then using the residue theorem, we have

$$x(n) = \sum \text{Residues of } X(z)z^{n-1} \text{at poles of } X(z) \text{ inside } C \tag{3.72}$$

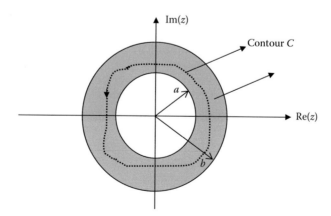

FIGURE 3.5　Closed contour C.

It can be shown that

$$x(n) = \begin{cases} \sum \text{Residues of } X(z)z^{n-1} \text{ at poles of } X(z) \text{ inside } C & n \geq 0 \\ -\sum \text{Residues of } X(z)z^{n-1} \text{ at poles of } X(z) \text{ outside } C & n \leq -1 \end{cases}$$

(3.73)

Example 3.19

Find the inverse z-transform of the following function:

$$X(z) = \frac{z^2}{(z-0.2)(z-0.3)}, \quad \text{ROC: } |z| > 0.3$$

SOLUTION

The ROC and the contour C for this example are shown in Figure 3.6. Since there are no poles outside C, we have

$$x(n) = 0 \quad \text{for } n < 0$$

For $n \geq 0$, we have

$$x(n) = \sum \text{Residues of } \frac{z^2}{(z-0.2)(z-0.3)} z^{n-1} \text{ at poles of } X(z) \text{ inside } C = A_1 + A_2$$

The residue A_1 corresponding to pole $p_1 = 0.2$ is

$$A_1 = (z-0.2)\frac{z^2}{(z-0.2)(z-0.3)} z^{n-1}\Big|_{z=0.2} = \frac{0.2^2(0.2)^{n-1}}{(0.2-0.3)} = -2(0.2)^n$$

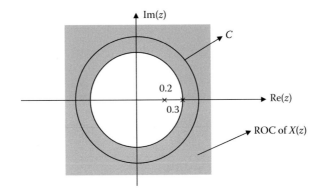

FIGURE 3.6 ROC and contour C for Example 3.19.

and the residue A_2 corresponding to the second pole is

$$A_2 = (z - 0.3)\frac{z^2}{(z - 0.2)(z - 0.3)}z^{n-1}\Big|_{z=0.3} = \frac{0.3^2(0.3)^{n-1}}{(0.3 - 0.2)} = 3(0.3)^n$$

Therefore,

$$x(n) = A_1 + A_2 = -2(0.2)^n u(n) + 3(0.3)^n u(n)$$

(b) **Inversion by Power Series Expansion**: In this case, we expand $X(z)$ as a power series of z^{-1}, that is, $X(z) = \sum_{n=-\infty}^{\infty} a_n z^{-n}$. The expansion coefficients are $x(n)$, that is, $x(n) = a_n$. If the function $X(z)$ is a rational function, the power series can be obtained using long division as illustrated by the following example.

Example 3.20

Find the inverse z-transform of the function $X(z) = \frac{z}{z-0.2}$ using power series expansion.

SOLUTION

The power series expansion of $X(z)$ is

$$X(z) = \frac{z}{z - 0.2} = \frac{1}{1 - 0.2z^{-1}} = 1 + 0.2z^{-1} + (0.2)^2 z^{-2} + (0.2)^3 z^{-3} + \cdots$$

Therefore,

$$x(n) = \begin{bmatrix} 1 & 0.2 & (0.2)^2 & (0.2)^3 & (0.2)^4 & \cdots \end{bmatrix} = (0.2)^n u(n)$$

Example 3.21

Find the inverse z-transform of

$$X(z) = \frac{z^3 - 2.3z^2 + 0.84z}{z^3 - 1.4z^2 + 0.63z - 0.09}$$

SOLUTION

Expanding $X(z)$, we have

$$X(z) = \frac{z^3 - 2.3z^2 + 0.84z}{z^3 - 1.4z^2 + 0.63z - 0.09}$$
$$= 1 - 0.9z^{-1} + 1.05z^{-2} - 0.813z^{-3} - 0.5577z^{-4} + \cdots$$

Therefore,

$$x(0) = 1,\ x(1) = -0.9,\ x(2) = 1.05,\ x(3) = -0.813,\ x(4) = -0.5577,\ldots$$

In general, it is not possible to obtain closed-form solution using long-division technique.

Example 3.22

Find the inverse z-transform of the irrational function

$$X(z) = \log(1 + az^{-1}) \quad \text{ROC: } |z| > a$$

SOLUTION

Using Taylor series expansion, we have

$$\log(1 + az^{-1}) = \sum_{n=1}^{\infty} (-1)^{n+1} \frac{a^n z^{-n}}{n}$$

Therefore,

$$x(n) = (-1)^{n+1} \frac{a^n}{n} u(n-1)$$

(c) **Inversion by Partial Fraction Expansion**: If $X(z)$ is a rational function, that is,

$$X(z) = \frac{\sum_{k=0}^{M} b_k z^k}{\sum_{k=0}^{M} a_k z^k} = \frac{\sum_{k=0}^{M} b_k z^k}{\prod_{i=1}^{Q}(z - p_i)^{m_i}} \tag{3.74}$$

where

$p_i,\ i = 1, 2, \ldots, Q$ are the poles of $X(z)$ or roots of polynomial $\sum_{k=1}^{M} a_k z^k$
m_i is the multiplicity of the ith pole. Note that $m_1 + m_2 + \cdots + m_Q = M$

Partial fraction of $\frac{X(z)}{z}$ yields

$$\frac{X(z)}{z} = \frac{A_0}{z} + \left[\frac{A_1}{z - p_1} + \frac{A_2}{(z - p_1)^2} + \cdots + \frac{A_{m_1}}{(z - p_1)^{m_1}} \right]$$
$$+ \left[\frac{B_1}{z - p_2} + \frac{B_2}{(z - p_2)^2} + \cdots + \frac{B_{m_2}}{(z - p_2)^{m_2}} \right] + \cdots \tag{3.75}$$

or

$$
X(z) = A_0 + \left[\frac{A_1 z}{z - p_1} + \frac{A_2 z}{(z - p_1)^2} + \cdots + \frac{A_{m_1} z}{(z - p_1)^{m_1}} \right]
$$
$$
+ \left[\frac{B_1 z}{z - p_2} + \frac{B_2 z}{(z - p_2)^2} + \cdots + \frac{B_{m_2} z}{(z - p_2)^{m_2}} \right] + \cdots \qquad (3.76)
$$

where

$$
A_0 = z \left. \frac{X(z)}{z} \right|_{z=0} = X(0) \qquad (3.77)
$$

The residues $A_1, A_2, \ldots, A_{m_1}$ corresponding to the pole p_1 are computed using

$$
A_k = \frac{d^{k-m_i}}{dz^{k-m_i}} (z - p_1)^{m_1} \left. \frac{X(z)}{z} \right|_{z=p_1} \qquad k = m_i, m_{i-1}, \ldots, 2, 1 \qquad (3.78)
$$

Similarly, the other set of residues corresponding to other poles are computed. Once the residues are computed, the time-domain function is

$$
x(n) = A_0 \delta(n)
$$
$$
+ \left[A_1 (p_1)^n u(n) + A_1 n(p_1)^{n-1} u(n) + \cdots \right.
$$
$$
+ A_{m_1} \frac{n(n-1)(n-2)\cdots(n-m_1+1)}{m_1!} (p_1)^{n-m_1} u(n - m_1 + 1) \Big]
$$
$$
+ \left[B_1 (p_2)^n u(n) + B_1 n(p_2)^{n-1} u(n) + \cdots \right.
$$
$$
+ B_{m_2} \frac{n(n-1)(n-2)\cdots(n-m_2+1)}{m_2!} (p_2)^{n-m_2} u(n - m_2 + 1) \Big] + \cdots
$$
$$
(3.79)
$$

Example 3.23

Find the inverse z-transform of the following function:

$$
X(z) = \frac{z^2}{(z - 0.2)^2 (z - 0.3)}, \qquad \text{ROC: } |z| > 0.3
$$

SOLUTION

Using partial-fraction expansion, we would have

$$
X(z) = \frac{-30z}{z - 0.2} - \frac{2z}{(z - 0.2)^2} + \frac{30z}{z - 0.3}
$$

Taking inverse z-transform yields

$$x(n) = -30(0.2)^n u(n) - \frac{2n(0.2)^n}{0.2} u(n) + 30(0.3)^n u(n)$$

or

$$x(n) = [-30(0.2)^n - 10n(0.2)^n + 30(0.3)^n] u(n)$$

Example 3.24

Consider a SISO system described by the second-order DE:

$$y(n + 2) + 0.4y(n + 1) + 0.03y(n) = x(n)$$

Find the output of this system if the input signal is a unit step function $x(n) = u(n)$ with the initial conditions $y(1) = y(0) = 0$.

SOLUTION

Taking z-transform from both sides of the above DE yields

$$z^2 Y(z) + 0.4zY(z) + 0.03Y(z) = X(z) = \frac{z}{z - 1}$$

Therefore,

$$Y(z) = \frac{z}{(z^2 + 0.4z + 0.03)(z - 1)} = \frac{z}{(z + +0.1)(z + 0.3)(z - 1)}$$

Partial fraction of $\frac{Y(z)}{z}$ results in

$$\frac{Y(z)}{z} = \frac{1}{(z + +0.1)(z + 0.3)(z - 1)} = \frac{0.6993}{z - 1} - \frac{4.5454}{z + 0.1} + \frac{3.8461}{z + 0.3}$$

$$Y(z) = \frac{0.6993z}{z - 1} - \frac{4.5454z}{z + 0.1} + \frac{3.8461z}{z + 0.3}$$

Hence,

$$y(n) = [0.6993 - 4.5454(-0.1)^n + 3.8461(-0.3)^n] u(n)$$

3.5.3 RELATION BETWEEN THE z-TRANSFORM AND THE LAPLACE TRANSFORM

Consider analog signal $x_a(t)$ being sampled at the rate of $f_s = \frac{1}{T}$ as shown in Figure 3.7.

The output of the sampler could be considered to be either the discrete signal $x(n)$ or the continuous signal $x^*(t)$ defined as

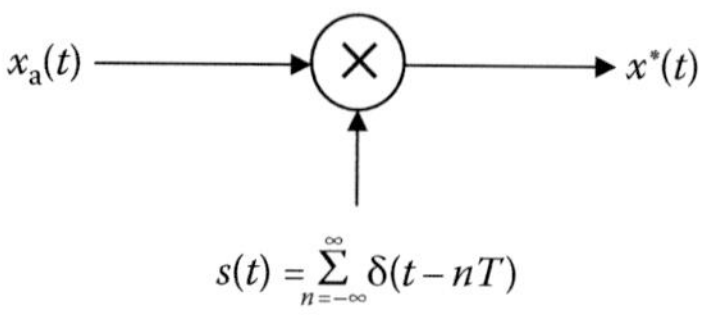

$$x_a(t) \longrightarrow \boxed{\times} \longrightarrow x^*(t)$$

$$s(t) = \sum_{n=-\infty}^{\infty} \delta(t - nT)$$

FIGURE 3.7 Sampling process.

$$x(n) = x_a(nT) \tag{3.80}$$

$$x^*(t) = \sum_{n=-\infty}^{\infty} x_a(nT)\delta(t - nT) \tag{3.81}$$

The Laplace transform of $x^*(t)$ is

$$X^*(s) = \sum_{n=-\infty}^{\infty} x_a(nT)e^{-nTs} \tag{3.82}$$

The z-transform of the discrete signal $x(n)$ is

$$X(z) = \sum_{n=-\infty}^{\infty} x(n)z^{-n} \tag{3.83}$$

Comparing Equations 3.81 and 3.82, we have

$$X^*(s) = X(z)\big|_{z=e^{Ts}} \tag{3.84}$$

Thus, the z-transform of a discrete signal $x(n)$ is the Laplace transform of the sampled signal $x^*(t)$ with the change of variable

$$z = e^{Ts} \tag{3.85}$$

The above equation defines a mapping from complex s-plane to complex z-plane, as shown in Figure 3.8.

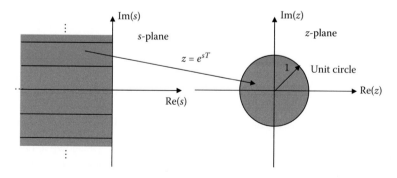

FIGURE 3.8 Mapping from the complex s-plane to the complex z-plane.

3.6 DISCRETE-TIME FOURIER TRANSFORM

The discrete-time Fourier transform (DTFT) of discrete signal $x(n)$ is defined as

$$X(j\omega) = \sum_{n=-\infty}^{\infty} x(n)e^{-jn\omega} \tag{3.86}$$

This is the z-transform $X(z)$ evaluated on the boundary of unit circle in z-plane, that is,

$$X(j\omega) = X(z)\big|_{z=e^{j\omega}} \tag{3.87}$$

Therefore, discrete signal $x(n)$ has a DTFT if the ROC of $X(z)$ contains the unit circle. Since the DTFT is closely related to the z-transform, its properties are very similar to the properties of the z-transform. For this reason, we briefly cover some important properties of DTFT without proof.

3.6.1 PROPERTIES OF DISCRETE-TIME FOURIER TRANSFORM

(a) **Linearity**: DTFT is a linear transform, that is, if

$$x(n) \xrightarrow{\text{DTFT}} X(j\omega) \tag{3.88}$$

$$y(n) \xrightarrow{\text{DTFT}} Y(j\omega) \tag{3.89}$$

then

$$ax(n) + by(n) \xrightarrow{\text{DTFT}} aX(j\omega) + bY(j\omega) \tag{3.90}$$

(b) **Periodicity**: $X(j\omega)$ is periodic with a period of 2π, that is,

$$X(j\omega) = X[j(\omega + 2\pi)] \tag{3.91}$$

(c) **Delay Property**: If

$$x(n) \xrightarrow{\text{DTFT}} X(j\omega) \tag{3.92}$$

then

$$x(n - n_0) \xrightarrow{\text{DTFT}} e^{-jn_0\omega}X(j\omega) \tag{3.93}$$

Example 3.25

Find the DTFT of the five-point discrete signal:

$$x(n) = \begin{bmatrix} 2 & -3 & 7 & -3 & 2 \end{bmatrix}$$

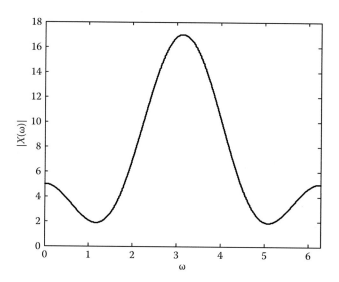

FIGURE 3.9   Magnitude of DTFT of $x(n)$.

SOLUTION

The DTFT of $x(n)$ is

$$X(j\omega) = \sum_{n=-\infty}^{\infty} x(n)e^{-jn\omega} = 2 - 3e^{-j\omega} + 7e^{-2j\omega} - 3e^{-3j\omega} + 2e^{-4j\omega}$$
$$= 2e^{-2j\omega}(e^{2j\omega} + e^{-2j\omega}) - 3e^{-2j\omega}(e^{j\omega} + e^{-j\omega}) + 7e^{-2j\omega}$$
$$= e^{-2j\omega}[4\cos 2\omega - 6\cos\omega + 7]$$

The magnitude of $X(j\omega)$ is

$$|X(j\omega)| = 7 - 6\cos\omega + 4\cos 2\omega$$

The magnitude plot is shown in Figure 3.9.

3.6.2   INVERSE DTFT

The inverse DTFT can be computed using the inversion integral used in computing the inverse z-transform, that is,

$$x(n) = \frac{1}{j2\pi} \oint_C X(z)z^{n-1}dz \qquad (3.94)$$

Let the integration contour C be the unit circle $z = e^{j\omega}$ in the complex z-plane. We can use this closed contour since $x(n)$ has Fourier transform and its ROC contains the unit circle. Therefore,

$$x(n) = \frac{1}{j2\pi} \oint_C X(z) z^{n-1} dz = \frac{1}{j2\pi} \int_{-\pi}^{\pi} X(j\omega) e^{j(n-1)\omega} je^{j\omega} d\omega = \frac{1}{2\pi} \int_{-\pi}^{\pi} X(j\omega) e^{jn\omega} d\omega$$

$$(3.95)$$

Example 3.26

Find the inverse DTFT of $X(j\omega)$ given by

$$X(j\omega) = e^{-|\omega|} \quad \text{for } -\pi < \omega < \pi$$

SOLUTION

Using the inversion integral, we have

$$x(n) = \frac{1}{2\pi} \int_{-\pi}^{\pi} X(j\omega) e^{jn\omega} d\omega = \frac{1}{2\pi} \int_{-\pi}^{\pi} e^{-|\omega|} e^{jn\omega} d\omega$$

The above integral is decomposed into two integrals:

$$x(n) = \frac{1}{2\pi} \int_{-\pi}^{0} e^{\omega} e^{jn\omega} d\omega + \frac{1}{2\pi} \int_{0}^{\pi} e^{-\omega} e^{jn\omega} d\omega$$

Hence,

$$x(n) = \frac{1}{2\pi} \int_{-\pi}^{0} e^{(1+jn)\omega} d\omega + \frac{1}{2\pi} \int_{0}^{\pi} e^{-(1-jn)\omega} d\omega = \frac{1}{2\pi} \frac{e^{(1+jn)\omega}}{1+jn} \Big|_{-\pi}^{0} - \frac{1}{2\pi} \frac{e^{-(1-jn)\omega}}{1-jn} \Big|_{0}^{\pi}$$

$$= \frac{1}{2\pi} \frac{1 - e^{-(1+jn)\pi}}{1+jn} - \frac{1}{2\pi} \frac{e^{-(1-jn)\pi} - 1}{1-jn} = \frac{1}{2\pi} \frac{1 - e^{-\pi} e^{-jn\pi}}{1+jn} - \frac{1}{2\pi} \frac{e^{-\pi} e^{jn\pi} - 1}{1-jn}$$

Therefore,

$$x(n) = \frac{1}{2\pi} \frac{2 - 2e^{-\pi} \cos n\pi}{1+n^2} = \frac{1 - e^{-\pi} \cos n\pi}{(1+n^2)\pi}$$

3.7 TWO-DIMENSIONAL z-TRANSFORM

The two-dimensional (2-D) z-transform of the 2-D sequence $x(n_1, n_2)$ is defined as

$$X(z_1, z_2) = \sum_{n_1=-\infty}^{\infty} \sum_{n_2=-\infty}^{\infty} x(n_1, n_2) z_1^{-n_1} z_2^{-n_2}$$

$$(3.96)$$

where z_1 and z_2 are complex variables. The region in four-dimensional (4-D) $\{z_1, z_2\}$ space where $X(z_1, z_2)$ converges is defined as ROC of $X(z_1, z_2)$. The ROC depends on $|z_1|$ and $|z_2|$. It is typically shown as a set in the 2-D $|z_1| - |z_2|$ plane. Two-dimensional z-transform has applications in the design of 2-D infinite impulse-response filters and stability analysis of numerical solution of partial DEs.

Example 3.27

Find the z-transform of the sequence

$$x(n_1, n_2) = \begin{cases} a^{n_1} b^{n_2} & n_1, n_2 \geq 0 \\ 0 & \text{otherwise} \end{cases}$$

SOLUTION

$$X(z_1, z_2) = \sum_{n_1=-\infty}^{\infty} \sum_{n_2=-\infty}^{\infty} x(n_1, n_2) z_1^{-n_1} z_2^{-n_2} = \sum_{n_1=0}^{\infty} (az_1^{-1})^{n_1} \sum_{n_2=0}^{\infty} (bz_2^{-1})^{n_2} \quad (3.97)$$

For convergence of $X(z_1, z_2)$, both sums in Equation 3.97 must converge. This requires that $|az_1^{-1}| < 1$ and $|bz_2^{-1}| < 1$ or equivalently, $|z_1| > |a|$ and $|z_2| > |b|$. Therefore,

$$X(z_1, z_2) = \frac{1}{1 - az_1^{-1}} \frac{1}{1 - bz_2^{-1}} = \frac{z_1 z_2}{(z_1 - a)(z_2 - b)}$$

The ROC of $X(z_1, z_2)$ is shown in Figure 3.10.

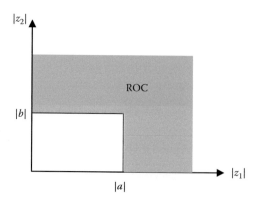

FIGURE 3.10   ROC of Example 3.23.

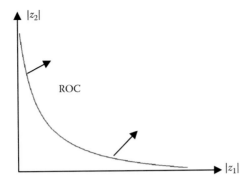

FIGURE 3.11   ROC of Example 3.24.

Example 3.28

Find the z-transform of the 2-D sequence

$$x(n_1, n_2) = \begin{cases} 1 & n_1 = n_2 \geq 0 \\ 0 & \text{otherwise} \end{cases} \tag{3.98}$$

SOLUTION

The 2-D z-transform of $x(n_1, n_2)$ is

$$X(z_1, z_2) = \sum_{n_1=-\infty}^{\infty} \sum_{n_2=-\infty}^{\infty} x(n_1, n_2) z_1^{-n_1} z_2^{-n_2} = \sum_{n_1=0}^{\infty} (z_1 z_2)^{-n_1} \tag{3.99}$$

If $|z_1^{-1} z_2^{-1}| < 1$ or $|z_1||z_2| > 1$, the sum in Equation 3.99 convergences to

$$X(z_1, z_2) = \frac{1}{1 - z_1^{-1} z_2^{-1}} = \frac{z_1 z_2}{z_1 z_2 - 1} \tag{3.100}$$

The ROC is shown in Figure 3.11

3.8   TWO-DIMENSIONAL DISCRETE-SPACE FOURIER TRANSFORM

The 2-D discrete-space Fourier transform (DSFT) of 2-D discrete signal $x(n_1, n_2)$ is defined as

$$X(j\omega_1, j\omega_2) = \sum_{n_2=-\infty}^{\infty} \sum_{n_1=-\infty}^{\infty} x(n_1, n_2) e^{-jn_1\omega_1} e^{-jn_2\omega_2} \tag{3.101}$$

This is the z-transform $X(z_1, z_2)$ evaluated on the boundary of unit circles $z_1 = e^{j\omega_1}$ and $z_2 = e^{j\omega_2}$ in the 4-D complex space (z_1, z_2), that is,

$$X(j\omega_1, j\omega_2) = X(z_1, z_2)\big|_{z_1=e^{j\omega_1}, z_2=e^{j\omega_2}} \tag{3.102}$$

3.8.1 PROPERTIES OF 2-D DSFT

Properties of 2-D DSFT are similar to the properties of 1-D DTFT. Here are some important properties of 2-D DSFT:

(a) **Linearity**: 2-D DSFT is a linear transform, that is, if

$$x(n_1, n_2) \xrightarrow{\text{DSFT}} X(j\omega_1, j\omega_2) \tag{3.103}$$

$$y(n_1, n_2) \xrightarrow{\text{DSFT}} Y(j\omega_1, j\omega_2) \tag{3.104}$$

then

$$ax(n_1, n_2) + by(n_1, n_2) \xrightarrow{\text{DSFT}} aX(j\omega_1, j\omega_2) + bY(j\omega_1, j\omega_2) \tag{3.105}$$

(b) **Periodicity**: $X(j\omega_1, j\omega_2)$ is periodic with a period of 2π, that is,

$$X(j\omega_1, j\omega_2) = X[j(\omega_1 + 2\pi), j(\omega_2 + 2\pi)] \tag{3.106}$$

(c) **Delay Property**: If

$$x(n_1, n_2) \xrightarrow{\text{DSFT}} X(j\omega_1, j\omega_2) \tag{3.107}$$

then

$$x(n_1 - \alpha, n_2 - \beta) \xrightarrow{\text{DSFT}} e^{-j\alpha\omega_1} e^{-j\beta\omega_2} X(j\omega_1, j\omega_2) \tag{3.108}$$

3.8.2 INVERSE 2-D DSFT

The inverse 2-D DSFT can be computed using the inversion integral given by

$$x(n_1, n_2) = \frac{1}{4\pi^2} \int_{-\pi}^{\pi} \int_{-\pi}^{\pi} X(j\omega_1, j\omega_2) e^{jn_1\omega_1} e^{jn_2\omega_2} d\omega_1 d\omega_2 \tag{3.109}$$

Example 3.29

Find the inverse DSFT of $X(j\omega_1, j\omega_2)$ given by

$$X(j\omega_1, j\omega_2) = \begin{cases} 1 & \text{if } (\omega_1, \omega_2) \in D \\ 0 & \text{otherwise} \end{cases}$$

where D is the dashed area shown in Figure 3.12.

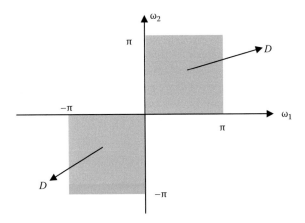

FIGURE 3.12   Support of $X(j\omega_1, j\omega_2)$.

SOLUTION

Using inverse 2-D DSFT, we have

$$x(n_1, n_2) = \frac{1}{4\pi^2} \int\limits_{-\pi}^{\pi} \int\limits_{-\pi}^{\pi} X(j\omega_1, j\omega_2) e^{jn_1\omega_1} e^{jn_2\omega_2} \, d\omega_1 \, d\omega_2$$

The above integral is

$$x(n_1, n_2) = \frac{1}{4\pi^2} \int\limits_{-\pi}^{0} \int\limits_{-\pi}^{0} e^{jn_1\omega_1} e^{jn_2\omega_2} \, d\omega_1 \, d\omega_2 + \frac{1}{4\pi^2} \int\limits_{0}^{\pi} \int\limits_{0}^{\pi} e^{jn_1\omega_1} e^{jn_2\omega_2} \, d\omega_1 \, d\omega_2$$

Upon integration, we have

$$x(n_1, n_2) = \frac{1}{4\pi^2} \frac{1 - e^{-j\pi n_1}}{jn_1} \frac{1 - e^{-j\pi n_2}}{jn_2} + \frac{1}{4\pi^2} \frac{e^{j\pi n_1} - 1}{jn_1} \frac{e^{j\pi n_2} - 1}{jn_2}$$

which can be simplified to

$$x(n_1, n_2) = \frac{1}{4\pi^2} \frac{1 - e^{-j\pi n_1}}{jn_1} \frac{1 - e^{-j\pi n_2}}{jn_2} + \frac{1}{4\pi^2} \frac{e^{j\pi n_1} - 1}{jn_1} \frac{e^{j\pi n_2} - 1}{jn_2}$$

Therefore,

$$x(n_1, n_2) = \begin{cases} 0.5 & n_1 = n_2 = 0 \\ 0 & n_1 = 0, \, n_2 \neq 0 \\ 0 & n_1 \neq 0, \, n_2 = 0 \\ \frac{(1 - \cos \, n_1 \pi)(1 - \cos \, n_2 \pi)}{2\pi^2 n_1 n_2} & \text{otherwise} \end{cases}$$

3.9 EIGENVALUES AND EIGENVECTORS

Spectral analysis of matrices through use of eigenvalues/eigenvectors and SVD plays an important role in analysis and design of control systems using state-space approach. Techniques such as state feedback by pole placement and design of state estimators are all based on eigenvalues/eigenvectors decomposition. Solution of LTI continuous and discrete systems is also directly related to functions of matrices that is computed using eigenvalue and eigenvector decomposition.

3.9.1 DEFINITION OF EIGENVALUE AND EIGENVECTOR

The nonzero vector x is an eigenvector of square $n \times n$ matrix A if there is a scale factor λ such that

$$Ax = \lambda x \tag{3.110}$$

The scale factor λ is called the eigenvalue corresponding to the eigenvector x. The above equation can be considered as an operator operating on x. The eigenvectors of A are vectors that are not changed by the operator, they are only scaled by λ. This means that eigenvectors are invariant with respect to operator A. This is similar to the concept of eigenfunctions of LTI systems. For example the steady-state response of an LTI system to an input of a sinusoidal signal is a sinusoidal signal with same frequency as that of the input, but different magnitude and phase. Therefore, sinusoidal signals are eigenfunctions of LTI systems. The equation $Ax = \lambda x$ can be written as

$$(\lambda I - A)x = 0 \tag{3.111}$$

This equation has a nontrivial solution if and only if matrix $\lambda I - A$ is singular, that is,

$$\det (\lambda I - A) = 0 \tag{3.112}$$

The above determinant is a polynomial of degree n and is denoted by $P(\lambda)$. This polynomial is called the characteristic polynomial of matrix A. The characteristic polynomial has n roots that are eigenvalues of matrix A. Corresponding to each eigenvalue there is an eigenvector. The eigenvalues can be repeated eigenvalues and also they may be complex.

Example 3.30

Find the eigenvalues and eigenvectors of the 2×2 matrix A:

$$A = \begin{bmatrix} -2 & 2 \\ -24 & 12 \end{bmatrix}$$

SOLUTION

First we compute the two eigenvalues λ_1 and λ_2. The characteristic polynomial of matrix A is

$$P(\lambda) = |\lambda I - A| = \begin{vmatrix} \lambda + 2 & -2 \\ 24 & \lambda - 12 \end{vmatrix} = \lambda^2 - 10\lambda + 24 = (\lambda - 4)(\lambda - 6) = 0$$

Therefore, $\lambda_1 = 4$ and $\lambda_2 = 6$. Now we compute the eigenvectors corresponding to the two eigenvalues. Eigenvector corresponding to $\lambda_1 = 4$ is computed as

$$Ax_1 = \lambda_1 x_1$$

$$\begin{bmatrix} -2 & 2 \\ -24 & 12 \end{bmatrix} \begin{bmatrix} a \\ b \end{bmatrix} = 4 \begin{bmatrix} a \\ b \end{bmatrix} \rightarrow \begin{cases} -a + b = 2a \\ -12a + 6b = 2b \end{cases} \rightarrow \quad b = 3a$$

Let $a = 1$, then $b = 3$ and

$$x_1 = \begin{bmatrix} a \\ b \end{bmatrix} = \begin{bmatrix} 1 \\ 3 \end{bmatrix}$$

Eigenvector corresponding to $\lambda_2 = 6$ is given by

$$Ax_2 = \lambda_2 x_2$$

or

$$\begin{bmatrix} -1 & 1 \\ -12 & 6 \end{bmatrix} \begin{bmatrix} a \\ b \end{bmatrix} = 3 \begin{bmatrix} a \\ b \end{bmatrix} \rightarrow \begin{cases} -a + b = 3a \\ -12a + 6b = 3b \end{cases} \rightarrow \quad b = 4a$$

Let $a = 1$, then $b = 4$ and

$$x_2 = \begin{bmatrix} a \\ b \end{bmatrix} = \begin{bmatrix} 1 \\ 4 \end{bmatrix}$$

Example 3.31

Find the eigenvalues and eigenvectors of the 3×3 matrix A:

$$A = \begin{bmatrix} 2.6 & 1.3 & -2.5 \\ 0.8 & 5.4 & -5 \\ 0.8 & 1.4 & -1 \end{bmatrix}$$

SOLUTION

The characteristic polynomial of matrix A is

$$P(\lambda) = |\lambda I - A| = \lambda^3 - 7\lambda^2 + 14\lambda - 8 = (\lambda - 1)(\lambda - 2)(\lambda - 4) = 0$$

The eigenvalues are $\lambda_1 = 1$, $\lambda_2 = 2$, and $\lambda_3 = 4$. The corresponding eigenvectors are computed as

$$Ax_1 = \lambda_1 x_1 \rightarrow \begin{bmatrix} 2.6 & 1.3 & -2.5 \\ 0.8 & 5.4 & -5 \\ 0.8 & 1.4 & -1 \end{bmatrix} \begin{bmatrix} a \\ b \\ c \end{bmatrix} = \begin{bmatrix} a \\ b \\ c \end{bmatrix} \rightarrow \begin{cases} 2.6a + 1.3b - 2.5c = a \\ 0.8a + 5.4b - 5c = b \\ 0.8a + 1.4b - c = c \end{cases}$$

These three equations are not linearly independent. Only two of them are linearly independent. Using the first two equations with $a = 1$, we have

$$\begin{cases} 1.3b - 2.5c = -1.6 \\ 4.4b - 5c = -0.8 \end{cases} \rightarrow \begin{cases} b = \frac{4}{3} \\ c = \frac{4}{3} \end{cases} \rightarrow x_1 = \frac{1}{3} \begin{bmatrix} 3 \\ 4 \\ 4 \end{bmatrix}$$

Similarly,

$$Ax_2 = \lambda_2 x_2 \rightarrow \begin{bmatrix} 2.6 & 1.3 & -2.5 \\ 0.8 & 5.4 & -5 \\ 0.8 & 1.4 & -1 \end{bmatrix} \begin{bmatrix} a \\ b \\ c \end{bmatrix} = 2 \begin{bmatrix} a \\ b \\ c \end{bmatrix}$$

Therefore,

$$\begin{cases} 2.6a + 1.3b - 2.5c = 2a \\ 0.8a + 5.4b - 5c = 2b \\ 0.8a + 1.4b - c = 2c \end{cases}$$

$$\begin{cases} 1.3b - 2.5c = -0.6 \\ 3.4b - 5c = -0.8 \end{cases} \rightarrow \begin{cases} b = 0.5 \\ c = 0.5 \end{cases} \rightarrow x_2 = \begin{bmatrix} 1 \\ 0.5 \\ 0.5 \end{bmatrix}$$

The third eigenvector is

$$Ax_3 = \lambda_3 x_3 \rightarrow \begin{bmatrix} 2.6 & 1.3 & -2.5 \\ 0.8 & 5.4 & -5 \\ 0.8 & 1.4 & -1 \end{bmatrix} \begin{bmatrix} a \\ b \\ c \end{bmatrix} = 4 \begin{bmatrix} a \\ b \\ c \end{bmatrix}$$

Hence,

$$\begin{cases} 2.6a + 1.3b - 2.5c = 4a \\ 0.8a + 5.4b - 5c = 4b \\ 0.8a + 1.4b - c = 4c \end{cases}$$

$$\begin{cases} 1.3b - 2.5c = -1.4 \\ 1.4b - 5c = -0.8 \end{cases} \rightarrow \begin{cases} b = 3 \\ c = 1 \end{cases} \rightarrow x_3 = \begin{bmatrix} 1 \\ 3 \\ 1 \end{bmatrix}$$

3.9.2 PRODUCT AND SUM OF EIGENVALUES

The product and sum of eigenvalues of any matrix are equal to the determinant and the trace of that matrix, respectively. We first show that the product of eigenvalues is equal to the determinant of the matrix. Let A be an $n \times n$ matrix with characteristic polynomial $P(\lambda)$, then

$$P(\lambda) = |\lambda I - A| = (\lambda - \lambda_1)(\lambda - \lambda_2) \cdots (\lambda - \lambda_n) \qquad (3.113)$$

Set $\lambda = 0$, then

$$|-A| = (-\lambda_1)(-\lambda_2) \cdots (-\lambda_n) = (-1)^n \lambda_1 \lambda_2 \cdots \lambda_n \qquad (3.114)$$

or

$$(-1)^n |A| = (-1)^n \lambda_1 \lambda_2 \cdots \lambda_n \qquad (3.115)$$

Hence, we have

$$\lambda_1 \lambda_2 \cdots \lambda_n = |A| \qquad (3.116)$$

Now we show that the sum of eigenvalues is equal to the trace of the matrix:

$$P(\lambda) = |\lambda I - A| = \begin{bmatrix} \lambda - a_{11} & -a_{12} & \cdots & -a_{1n} \\ -a_{21} & \lambda - a_{22} & \cdots & -a_{2n} \\ \vdots & \vdots & \ddots & \vdots \\ -a_{n1} & -a_{n2} & \cdots & \lambda - a_{nn} \end{bmatrix} \qquad (3.117)$$

By expanding the above determinant along the first column, we have

$$P(\lambda) = (\lambda - a_{11})|M_{11}| - \sum_{i=2}^{n} (-1)^{i+1} a_{i1} |M_{i1}| \qquad (3.118)$$

where M_{ij} is the determinant of matrix obtained by deleting the ith row and the jth column of A. Expanding determinant of M_{11}, we have

$$|M_{11}| = (\lambda - a_{22})|M'_{11}| - \sum_{i=3}^{n} (-1)^{i+1} a_{i1} |M'_{i1}| \qquad (3.119)$$

Continuing these expansions, we will have

$$P(\lambda) = (\lambda - a_{11})(\lambda - a_{22}) \cdots (\lambda - a_{nn}) + P'(\lambda) \qquad (3.120)$$

where $P'(\lambda)$ is a polynomial of degree $n - 2$. Therefore, the leading coefficient of $P(\lambda)$ is one and the second leading coefficient is $a_{11} + a_{22} + \cdots + a_{nn}$. Also from Equation 3.113, the second leading coefficient of $P(\lambda)$ is $\lambda_1 + \lambda_2 + \cdots + \lambda_n$. Therefore,

$$\lambda_1 + \lambda_2 + \cdots + \lambda_n = a_{11} + a_{22} + \cdots + a_{nn} = \text{Trace}(A) \qquad (3.121)$$

3.9.3 Finding Characteristic Polynomial of a Matrix

Let A be an $n \times n$ matrix. The characteristic polynomial of matrix A can be found by using the following recursive algorithm. Let $W_k = \text{Trace}(A^k)$, $k = 1, 2, \ldots, n$, then the coefficients of the characteristic equation are [3]

$$
\begin{aligned}
\alpha_1 &= -W_1 \\
\alpha_2 &= -\tfrac{1}{2}(a_1 W_1 + W_2) \\
\alpha_3 &= -\tfrac{1}{3}(a_2 W_1 + a_1 W_2 + W_3) \\
&\vdots \\
\alpha_n &= -\frac{1}{n}(a_{n-1} W_1 + a_{n-2} W_2 + \cdots + a_1 W_{n-1} + W_n)
\end{aligned}
\tag{3.122}
$$

and

$$
P(\lambda) = \lambda^n + \alpha_1 \lambda^{n-1} + \alpha_2 \lambda^{n-2} + \cdots + \alpha_{n-1}\lambda + \alpha_n
\tag{3.123}
$$

The above algorithm is known as Bocher's formula [4].

Example 3.32

Find the characteristic polynomial of the following 3×3 matrix:

$$
A = \begin{bmatrix} -2 & 4 & 2 \\ 1 & 6 & 3 \\ 1 & -1 & -5 \end{bmatrix}
$$

Solution

The trace of A, A^2, and A^3 are

$$
A^2 = A \times A = \begin{bmatrix} -2 & 4 & 2 \\ 1 & 6 & 3 \\ 1 & -1 & -5 \end{bmatrix} \begin{bmatrix} -2 & 4 & 2 \\ 1 & 6 & 3 \\ 1 & -1 & -5 \end{bmatrix} = \begin{bmatrix} 10 & 14 & -2 \\ 7 & 37 & 5 \\ -8 & 3 & 24 \end{bmatrix}
$$

$$
A^3 = A^2 \times A = \begin{bmatrix} 10 & 14 & -2 \\ 7 & 37 & 5 \\ -8 & 3 & 24 \end{bmatrix} \begin{bmatrix} -2 & 4 & 2 \\ 1 & 6 & 3 \\ 1 & -1 & -5 \end{bmatrix} = \begin{bmatrix} -8 & 126 & 72 \\ 28 & 245 & 100 \\ 43 & -38 & -127 \end{bmatrix}
$$

$$
W_1 = \text{Trace}(A) = -1
$$

$$
W_2 = \text{Trace}(A^2) = 71
$$

$$
W_3 = \text{Trace}(A^3) = 110
$$

Hence,

$$\alpha_1 = -W_1 = 1$$

$$\alpha_2 = -\tfrac{1}{2}(a_1 W_1 + W_2) = -\tfrac{1}{2}(-1 \times 1 + 71) = -35$$

$$\alpha_3 = -\tfrac{1}{3}(\alpha_2 W_1 + \alpha_1 W_2 + W_3) = -\tfrac{1}{3}(35 \times 1 + 1 \times 71 + 110) = -72$$

Therefore,

$$P(\lambda) = \lambda^3 + 8\lambda^2 + 9\lambda - 2$$

3.9.4   MODAL MATRIX

Let $x_1, x_2, \ldots, x_n$ be the n independent eigenvectors of $n \times n$ matrix A. The $n \times n$ matrix M formed by side-by-side stacking of eigenvectors is called the modal matrix of A:

$$M = \begin{bmatrix} x_1 & x_2 & \cdots & x_n \end{bmatrix} \tag{3.124}$$

Since the n columns of M are linearly independent, it is full rank and hence invertible.

Example 3.33

Find the modal matrix of

$$A = \begin{bmatrix} 2.6 & 1.3 & -2.5 \\ 0.8 & 5.4 & -5 \\ 0.8 & 1.4 & -1 \end{bmatrix}$$

SOLUTION

Matrix A has three independent eigenvectors (see Example 3.31):

$$x_1 = \begin{bmatrix} 1 \\ 1.3333 \\ 1.3333 \end{bmatrix}, \quad x_2 = \begin{bmatrix} 1 \\ 0.5 \\ 0.5 \end{bmatrix}, \quad \text{and} \quad x_3 = \begin{bmatrix} 1 \\ 3 \\ 1 \end{bmatrix}$$

Therefore, the modal matrix is

$$M = \begin{bmatrix} x_1 & x_2 & \cdots & x_n \end{bmatrix} = \begin{bmatrix} 1 & 1 & 1 \\ 1.333 & 0.5 & 3 \\ 1.333 & 0.5 & 1 \end{bmatrix}$$

3.9.5   MATRIX DIAGONALIZATION

Let A be an $n \times n$ matrix with n distinct eigenvalues $\lambda_1, \lambda_2, \ldots, \lambda_n$. Assume that the corresponding independent eigenvectors are $x_1, x_2, \ldots, x_n$. Therefore, we have

$$Ax_1 = \lambda_1 x_1 \tag{3.125}$$

$$Ax_2 = \lambda_2 x_2 \tag{3.126}$$

$$\vdots$$

$$Ax_n = \lambda_n x_n \tag{3.127}$$

These equations can be put together in matrix form to obtain

$$\begin{bmatrix} Ax_1 & Ax_2 & \cdots & Ax_n \end{bmatrix} = \begin{bmatrix} \lambda_1 x_1 & \lambda_2 x_2 & \cdots & \lambda_n x_n \end{bmatrix} \tag{3.128}$$

Equation 3.128 can be written as

$$A\begin{bmatrix} x_1 & x_2 & \cdots & x_n \end{bmatrix} = \begin{bmatrix} x_1 & x_2 & \cdots & x_n \end{bmatrix} \begin{bmatrix} \lambda_1 & 0 & \cdots & 0 \\ 0 & \lambda_2 & \cdots & 0 \\ \vdots & \vdots & \ddots & \vdots \\ 0 & 0 & \cdots & \lambda_n \end{bmatrix} \tag{3.129}$$

Define Λ to be the diagonal matrix of eigenvalues, then the above matrix equation can be written in terms of Λ and modal matrix M as

$$AM = M\Lambda \tag{3.130}$$

This equation is valid regardless of whether the eigenvectors are linearly independent or not. However, if the eigenvectors are linearly independent, then M is full rank and has an inverse and in this case, we can post-multiply the above equation by M^{-1} to obtain

$$A = M\Lambda M^{-1} \tag{3.131}$$

or

$$\Lambda = M^{-1}AM \tag{3.132}$$

If matrix A has repeated eigenvalues, it is diagonalizable if and only if the eigenvectors are linearly independent.

Example 3.34

Diagonalize 3×3 matrix A given by

$$A = \begin{bmatrix} 2.6 & 1.3 & -2.5 \\ 0.8 & 5.4 & -5 \\ 0.8 & 1.4 & -1 \end{bmatrix}$$

SOLUTION

The eigenvalues and modal matrix of A are

$$\lambda_1 = 1, \quad \lambda_2 = 2, \quad \lambda_3 = 4, \quad \text{and} \quad M = \begin{bmatrix} 1 & 1 & 1 \\ 1.333 & 0.5 & 3 \\ 1.333 & 0.5 & 1 \end{bmatrix}$$

Therefore,

$$
\begin{aligned}
M^{-1}AM &= \begin{bmatrix} 1 & 1 & 1 \\ 1.333 & 0.5 & 3 \\ 1.333 & 0.5 & 1 \end{bmatrix}^{-1} \begin{bmatrix} 2.6 & 1.3 & -2.5 \\ 0.8 & 5.4 & -5 \\ 0.8 & 1.4 & -1 \end{bmatrix} \begin{bmatrix} 1 & 1 & 1 \\ 1.333 & 0.5 & 3 \\ 1.333 & 0.5 & 1 \end{bmatrix} \\
&= \begin{bmatrix} -0.6 & -0.3 & 1.5 \\ 1.6 & -0.2 & -1 \\ 0 & 0.5 & -0.5 \end{bmatrix} \begin{bmatrix} 2.6 & 1.3 & -2.5 \\ 0.8 & 5.4 & -5 \\ 0.8 & 1.4 & -1 \end{bmatrix} \begin{bmatrix} 1 & 1 & 1 \\ 1.333 & 0.5 & 3 \\ 1.333 & 0.5 & 1 \end{bmatrix} \\
&= \begin{bmatrix} 1 & 0 & 0 \\ 0 & 2 & 0 \\ 0 & 0 & 4 \end{bmatrix}
\end{aligned}
$$

Example 3.35

Diagonalize matrix A:

$$A = \begin{bmatrix} 5 & 1 & 0 \\ 2 & 4 & 0 \\ -1 & 1 & 6 \end{bmatrix}$$

SOLUTION

The eigenvalues and eigenvectors of A are

$$\lambda_1 = 3 \quad \text{and} \quad \lambda_2 = \lambda_3 = 6$$

$$x_1 = \begin{bmatrix} 1 \\ -2 \\ 1 \end{bmatrix}, \quad x_2 = \begin{bmatrix} 0 \\ 0 \\ 1 \end{bmatrix}, \quad \text{and} \quad x_3 = \begin{bmatrix} 23 \\ 23 \\ 18 \end{bmatrix}$$

The repeated eigenvalues have independent eigenvectors, therefore, matrix A is diagonalizable. Hence,

$$M^{-1}AM = \begin{bmatrix} 1 & 0 & 23 \\ -2 & 0 & 23 \\ 1 & 1 & 18 \end{bmatrix}^{-1} \begin{bmatrix} 5 & 1 & 0 \\ 2 & 4 & 0 \\ -1 & 1 & 6 \end{bmatrix} \begin{bmatrix} 1 & 0 & 23 \\ -2 & 0 & 23 \\ 1 & 1 & 18 \end{bmatrix} = \begin{bmatrix} 3 & 0 & 0 \\ 0 & 6 & 0 \\ 0 & 0 & 6 \end{bmatrix}$$

Note that if matrix A has repeated eigenvalues, it is not always possible to diagonalize it as is shown in the following example.

Example 3.36

Diagonalize the 2×2 matrix A:

$$A = \begin{bmatrix} 0 & 1 \\ -4 & -4 \end{bmatrix}$$

SOLUTION

The eigenvalues of A are $\lambda_1 = \lambda_2 = -2$. The eigenvectors of A are computed as

$$Ax = \lambda x \rightarrow \begin{bmatrix} 0 & 1 \\ -4 & -4 \end{bmatrix} \begin{bmatrix} a \\ b \end{bmatrix} = -2 \begin{bmatrix} a \\ b \end{bmatrix} \rightarrow \begin{cases} b = -2a \\ -4a - 4b = -2b \end{cases} \rightarrow b = -2a$$

Therefore, there is one eigenvector $x_1 = \begin{bmatrix} 1 \\ -2 \end{bmatrix}$ and the matrix A is not diagonalizable.

3.9.6 DEFINITE MATRICES

Positive and negative definite (semi-definite) matrices are an important class of matrices with applications in signal processing, image processing, and control systems. They are particularly useful matrices in optimization problems. We now define four types of definite matrices:

(a) **Positive Definite Matrices**: The $n \times n$ Hermitian matrix A is said to be positive definite if for any nonzero vector $x \in R^n$, the quantity $x^H A x > 0$. Here H stands for conjugate transpose. Matrix A is said to be Hermitian if $A = A^H$.

(b) **Positive Semi-Definite Matrices**: The $n \times n$ Hermitian matrix A is said to be positive semi-definite if for any nonzero vector $x \in R^n$, the quantity $x^H A x \geq 0$.

(c) **Negative Definite Matrices**: The $n \times n$ Hermitian matrix A is said to be negative definite if for any nonzero vector $x \in R^n$, the quantity $x^H A x < 0$.

(d) **Negative Semi-Definite Matrices**: The $n \times n$ Hermitian matrix A is said to be negative semi-definite if for any nonzero vector $x \in R^n$, the quantity $x^H A x \leq 0$.

The following theorem states the condition for positiveness of a Hermitian matrix.

THEOREM 3.1

An $n \times n$ Hermitian matrix is positive definite if and only if all its eigenvalues are positive.

Proof: First assume that A is Hermitian and all of its eigenvalues are positive, then

$$x^H A x = x^H M \Lambda M^H x \tag{3.133}$$

Let $y = M^H x$, then Equation 3.133 can be written as

$$x^H A x = x^H M \Lambda M^H x = y^H \Lambda y = \sum_{i=1}^{n} \lambda_i |y_i|^2 > 0 \tag{3.134}$$

Next we need to show that if A is Hermitian and positive definite, then all of its eigenvalues are positive.

Let x_i be the eigenvector corresponding to the ith eigenvalue λ_i of A, then

$$0 < x_i^H A x_i = x_i^H \lambda_i x_i = \lambda_i \|x_i\|^2 \tag{3.135}$$

Therefore, $\lambda_i > 0$.

Similar theorems can be stated for positive semi-definite, negative definite, and negative semi-definite matrices. Here we state a summary of the results of these theorems:

(a) Positive definite: All eigenvalues are positive.
(b) Negative definite: All eigenvalues are negative.
(c) Positive semi-definite: All eigenvalues are nonnegative (zero or positive).
(d) Negative semi-definite: All eigenvalues are nonpositive (zero or negative).

Example 3.37

Check the following symmetric matrices for their definiteness:

(a) $A = \begin{bmatrix} 5 & -2 \\ -2 & 5 \end{bmatrix}$

(b) $B = \begin{bmatrix} 2 & 1 \\ 1 & -2 \end{bmatrix}$

(c) $C = \begin{bmatrix} -6 & -4 \\ -4 & -6 \end{bmatrix}$

SOLUTION

The eigenvalues of A are 3 and 7. Both are positive, hence matrix A is positive definite matrix. Eigenvalues of matrix B are -2.2361 and 2.2361, therefore B is indefinite; and finally, the eigenvalues of matrix C are -2 and -10 that makes it negative definite.

3.10 SINGULAR VALUE DECOMPOSITION

One of the most important tools in signal processing and numerical linear algebra is the SVD. The SVD was discovered for square matrices by Beltrami and Jordan in the eighteenth century. The theory for general matrices was established by Eckart and Young. We first state the SVD theorem and then we look at its applications.

THEOREM 3.2

Let A be an $m \times n$ real or complex matrix with rank r, then there exist unitary matrices $U(m \times m)$ and $V(n \times n)$ such that

$$A = U\Sigma V^{\mathrm{H}} \tag{3.136}$$

where Σ is an $m \times n$ matrix with entries

$$\Sigma_{ij} = \begin{cases} \sigma_i & \text{if } i = j \\ 0 & \text{if } i \neq j \end{cases} \tag{3.137}$$

The quantities $\sigma_1 \geq \sigma_2 \geq \cdots \geq \sigma_r > \sigma_{r+1} = \sigma_{r+2} = \cdots = \sigma_n = 0$ are called singular values of A.

Proof: Let $S = A^{\mathrm{H}}A$. Matrix S is $n \times n$ Hermitian and positive semi-definite with rank r. Therefore, it has nonnegative eigenvalues. Let the eigenvalues and eigenvectors of S be $\sigma_1^2 \geq \sigma_2^2 \geq \cdots \geq \sigma_r^2 > \sigma_{r+1}^2 = \sigma_{r+2}^2 = \cdots = \sigma_n^2 = 0$ and $v_1, v_2, \ldots, v_n$. These eigenvectors form an orthonormal set. Let $V_1 = \begin{bmatrix} v_1 & v_2 & \cdots & v_r \end{bmatrix}$, $V_2 = \begin{bmatrix} v_{r+1} & v_{r+2} & \cdots & v_n \end{bmatrix}$, and $\Lambda = \mathrm{diag}(\sigma_1, \sigma_2, \ldots, \sigma_r)$, then

$$A^{\mathrm{H}}AV_1 = V_1\Lambda^2 \tag{3.138}$$

Pre-multiply both sides of Equation 3.138 by V_1^{H} followed by post- and pre-multiplication by Λ^{-1} results in

$$\Lambda^{-1}V_1^{\mathrm{H}}A^{\mathrm{H}}AV_1\Lambda^{-1} = I \tag{3.139}$$

Choose $U_1 = AV_1\Lambda^{-1}$, then by Equation 3.139 we have $U_1^{\mathrm{H}}U_1 = I$. Notice that U_1 is a unitary matrix of size $m \times r$. Choose U_2 to be another unitary matrix of size $m \times (m - r)$ orthogonal to U_1. Then

$$U^H AV = [U_1 \quad U_2]^H A[V_1 \quad V_2] = \begin{bmatrix} U_1^H AV_1 & U_1^H AV_2 \\ U_2^H AV_1 & U_2^H AV_2 \end{bmatrix} = \begin{bmatrix} \Lambda & 0 \\ 0 & 0 \end{bmatrix} = \Sigma \quad (3.140)$$

Pre- and post-multiplying both sides of Equation 3.140 by U and V^H, we have

$$A = U\Sigma V^H \qquad (3.141)$$

The nonnegative numbers $\sigma_1 \geq \sigma_2 \geq \cdots \geq \sigma_r > \sigma_{r+1} = \sigma_{r+2} = \cdots = \sigma_n = 0$ are called singular values of A and they are square roots of eigenvalues of $A^H A$ or AA^H.

Example 3.38

Find the SVD of matrix A given by

$$A = \begin{bmatrix} 2 & 4 & 6 \\ -2 & 6 & -4 \end{bmatrix}$$

SOLUTION

We first form $S = A^T A$

$$S = A^T A = \begin{bmatrix} 8 & -4 & 20 \\ -4 & 52 & 0 \\ 20 & 0 & 52 \end{bmatrix}$$

The eigenvalues of S are $\sigma_1^2 = 60$, $\sigma_2^2 = 52$, and $\sigma_3^2 = 0$. The corresponding eigenvectors are

$$v_1 = \begin{bmatrix} -0.3651 \\ 0.1826 \\ -0.9129 \end{bmatrix}, \quad v_2 = \begin{bmatrix} 0 \\ -0.9806 \\ 0.1961 \end{bmatrix}, \quad \text{and} \quad v_3 = \begin{bmatrix} -0.9309 \\ -0.0716 \\ 0.3581 \end{bmatrix}$$

Therefore,

$$V_1 = \begin{bmatrix} -0.3651 & 0 \\ 0.1826 & -0.9806 \\ -0.9129 & 0.1961 \end{bmatrix} \quad \text{and} \quad V_2 = \begin{bmatrix} -0.9309 \\ -0.0716 \\ 0.3581 \end{bmatrix}$$

and

$$U_1 = AV_1 \Lambda^{-1} = \begin{bmatrix} 2 & 4 & 6 \\ -2 & 6 & -4 \end{bmatrix} \begin{bmatrix} -0.3651 & 0 \\ 0.1826 & -0.9806 \\ -0.9129 & 0.1961 \end{bmatrix} \begin{bmatrix} \sqrt{60} & 0 \\ 0 & \sqrt{52} \end{bmatrix}^{-1}$$

$$= \begin{bmatrix} -0.707 & -0.707 \\ 0.707 & -0.707 \end{bmatrix}$$

Since $m = 2$ and $r = 2$, we have $U = U_1$. Therefore, we have

$$A = U\Sigma V^H = \begin{bmatrix} -0.707 & -0.707 \\ 0.707 & -0.707 \end{bmatrix} \begin{bmatrix} \sqrt{60} & 0 & 0 \\ 0 & \sqrt{52} & 0 \end{bmatrix} \begin{bmatrix} -0.3651 & 0 & -0.9309 \\ 0.1826 & -0.9806 & -0.0716 \\ -0.9129 & 0.1961 & 0.3581 \end{bmatrix}^T$$

Example 3.39

Consider the 256×256 cameraman image. If we decompose the image into its SVD components, each component is an image of size 256×256. The ith component of the image is $\sigma_i^2 u_i v_i$ and the whole image is

$$A = U\Sigma V^T = \sum_{i=1}^{256} \sigma_i u_i v_i^T$$

Plot of the 256 singular values of the image normalized with respect to the largest singular value is shown in Figure 3.13. As can be seen, most of the singular values are small. The original image and the images reconstructed using the first 20 and 50 singular values are shown in Figures 3.14 through 3.16, respectively.

3.10.1 MATRIX NORM

Matrix norm like other vector-space norms must satisfy the properties of vector norm. Let A be an $m \times n$ matrix mapping vector space R^n to vector space R^m. By definition, the p matrix norm is defined by

$$\|A\|_p = \sup_{\substack{x \in R^n \\ \|x\|_p = 1}} \|Ax\|_p \qquad (3.142)$$

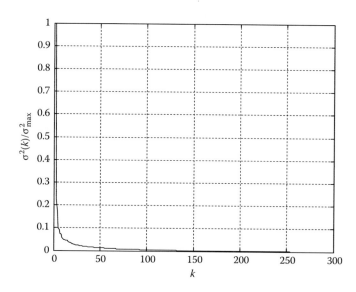

FIGURE 3.13   Singular values normalized with respect to the largest singular value.

FIGURE 3.14   Cameraman image.

FIGURE 3.15   Image reconstructed using the first 10 eigenimages.

FIGURE 3.16   Image reconstructed using the first 50 eigenimages.

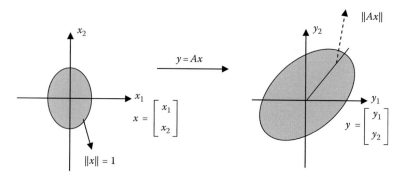

FIGURE 3.17 Matrix norm.

Matrix norm is a measure of boundness of that matrix. The concept of matrix norm for a 2×2 matrix is illustrated in Figure 3.17.

The matrix norm depends on vector norm, for example, if the vector norm is l_1 norm then the matrix norm is based on vector l_1 norm. We consider three cases corresponding to $p = 1$, $p = 2$, and $p = \infty$.

Case I: $p = 1$
In this case, the matrix norm becomes

$$\|A\|_1 = \max_{\|x\|_1 = 1} \|Ax\|_1 = \max_j \sum_{i=1}^{m} |a_{ij}| \tag{3.143}$$

Therefore, $\|A\|_1$ is equal to the longest column sum, which means that to find the $p = 1$ norm, compute the sum of absolute values of each column and pick the maximum.

Case II: $p = \infty$
In this case, the matrix norm becomes

$$\|A\|_\infty = \max_{\|x\|_\infty = 1} \|Ax\|_\infty = \max_i \sum_{j=1}^{m} |a_{ij}| \tag{3.144}$$

Therefore, $\|A\|_\infty$ is equal to the longest row sum, which means that to find the $p = \infty$ norm, compute the sum of absolute values of each row and pick the maximum.

Case III: $p = 2$
In this case, the matrix norm becomes

$$\|A\|_2 = \max_{\|x\|_2 = 1} \|Ax\|_2 \tag{3.145}$$

To find the $p = 2$ matrix norm, we must solve the following optimization problem:

$$\underset{x}{\text{maximize}} \ \|Ax\|_2 = x^H A^H Ax$$
$$\text{subject to:} \quad x^H x = 1 \tag{3.146}$$

Using Lagrange multiplier technique, this is equivalent to the following optimization problem:

$$\underset{x}{\text{maximize}} \ J = x^H A^H Ax - \lambda(x^H x - 1) \tag{3.147}$$

Setting the gradient of J with respect to x equal to zero, we obtain the equation

$$\frac{\partial J}{\partial x} = 2A^H Ax - 2\lambda x = 0 \rightarrow A^H Ax = \lambda x \tag{3.148}$$

Therefore, the solution for vector x must be an eigenvector of square matrix $A^H A$ corresponding to eigenvalue λ and the resulting norm is

$$\|Ax\|_2 = x^H A^H Ax = \lambda x^H x = \lambda \tag{3.149}$$

Since we are maximizing the norm, λ must be chosen to be the maximum eigenvalue of the positive definite matrix $A^H A$, therefore,

$$\|A\|_2 = \max \lambda(A^H A) \tag{3.150}$$

The p norm has the property that for any two matrices A and B, the following inequality holds:

$$\|Ax\|_p \leq \|A\|_p \|x\|_p \tag{3.151}$$

and

$$\|AB\|_p \leq \|A\|_p \|B\|_p \tag{3.152}$$

Frobenius Norm
Frobenius norm is another matrix norm that is not a p norm. It is defined as

$$\|A\|_F = \left(\sum_{i=1}^m \sum_{j=1}^n |a_{ij}|^2 \right)^{\frac{1}{2}} \tag{3.153}$$

Frobenius norm is also called Euclidean norm. As a simple example, the Frobenius norm of an $n \times n$ identity matrix is $\|I\|_F = \sqrt{n}$. The Frobenius norm can also be expressed as

$$\|A\|_F = \sqrt{\text{Trace}(A^H A)} \tag{3.154}$$

Example 3.40

Find the $p = 1$, $p = 2$, $p = \infty$, and Frobenius norm of the following matrix:

$$A = \begin{bmatrix} 4 & 2 \\ -6 & 7 \end{bmatrix}$$

SOLUTION

The different matrix norms are computed as

$$\|A\|_1 = \max_j \sum_{i=1}^{m} |a_{ij}| = \max_j \left[\sum_{i=1}^{m} |a_{i1}| \quad \sum_{i=1}^{m} |a_{i2}| \right] = \max_j \begin{bmatrix} 10 & 9 \end{bmatrix} = 10$$

$$\|A\|_\infty = \max_i \sum_{j=1}^{m} |a_{ij}| = \max_i \left[\sum_{j=1}^{m} |a_{1j}| \quad \sum_{j=1}^{m} |a_{2j}| \right] = \max_i \begin{bmatrix} 6 & 13 \end{bmatrix} = 13$$

$$\|A\|_2 = \sqrt{\max \lambda(A^H A)} = \sqrt{\max \lambda\left(\begin{bmatrix} 4 & -6 \\ 2 & 7 \end{bmatrix} \begin{bmatrix} 4 & 2 \\ -6 & 7 \end{bmatrix} \right)}$$

$$= \sqrt{\max \lambda\left(\begin{bmatrix} 52 & -34 \\ -34 & 53 \end{bmatrix} \right)} = \sqrt{86.5037} = 9.3007$$

$$\|A\|_F = \left(\sum_{i=1}^{2} \sum_{j=1}^{2} |a_{ij}|^2 \right)^{\frac{1}{2}} = \sqrt{16 + 4 + 36 + 49} = \sqrt{105} = 10.247$$

3.10.2   PRINCIPAL COMPONENTS ANALYSIS

Principal component analysis (PCA) is a statistical technique to extract patterns, similarities, and differences hidden in data. The PCA is extremely useful in case of data of higher dimensions since illustrative techniques are not applicable in case of data of dimension greater than three. An important application of PCA is data dimensionality reduction or data compression. This is achieved by reducing the data dimension with a small loss of information. Let X be an m-dimensional random vector, and let $X_1, X_2, \ldots, X_N$ be N observations of random vector X. The $m \times m$ covariance matrix R of these data is estimated using the sample covariance by

$$R = \frac{1}{N} \sum_{n=1}^{N} (X_n - \mu)(X_n - \mu)^T \tag{3.155}$$

where μ is the sample mean vector given by

$$\mu = \frac{1}{N} \sum_{n=1}^{N} X_n \tag{3.156}$$

Let $q_1, q_2, \ldots, q_r$ be r vectors of size $m \times 1$ obtained as linear combinations of $X_1, X_2, \ldots, X_N$, that is,

$$q_j = \sum_{i=1}^{N} a_{ij} X_i = a_j^T X \quad j = 1, 2, \ldots, r \tag{3.157}$$

where
$a_j = \begin{bmatrix} a_{1j} & a_{2j} & \cdots & a_{Nj} \end{bmatrix}^T$ is a unit norm vector
$X = \begin{bmatrix} X_1 & X_2 & \cdots & X_N \end{bmatrix}$ is a data matrix

The sample variance of q_j is given by

$$\sigma_j^2 = \frac{1}{N} \left(q_i - a_j^T \mu \right)^T \left(q_j - a_j^T \mu \right) = \frac{1}{N} \left(a_j^T X - a_j^T \mu \right)^T \left(a_j^T X - a_j^T \mu \right)$$

$$= a_j^T \frac{1}{N} (X - \mu)^T (X - \mu) a_j = a_j^T \frac{1}{N} \sum_{i=1}^{N} (X_i - \mu)^T (X_i - \mu) a_j$$

$$= a_j^T R a_j \tag{3.158}$$

We now maximize $\sigma_j^2 = a_j^T R a_j$ with respect to unit norm vector a_j. The solution is the normalized eigenvector corresponding to the largest eigenvalue of the sample covariance matrix R. Therefore, the r principal components $q_1, q_2, \ldots, q_r$ are the normalized eigenvectors corresponding to the r largest eigenvalues of R. Note that the principal components are pairwise orthogonal. Once the principal components are established, any random vector in this class can be approximated by linear combinations of these principal components, that is,

$$\hat{X} = \sum_{i=1}^{r} \alpha_i q_i \tag{3.159}$$

where

$$\alpha_i = X^T q_i \quad i = 1, 2, \ldots, r \tag{3.160}$$

Example 3.41

Figure 3.18 shows 2-D scatter data obtained by plotting two neighboring pixels $x = f(i, j)$ and $y = f(i + 1, j)$ of the LENA image. The estimated covariance matrix of this data set is

$$R = \begin{bmatrix} 0.2529 & 0.2512 \\ 0.2512 & 0.2525 \end{bmatrix}$$

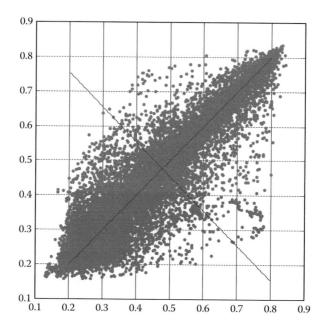

FIGURE 3.18   LENA image scatter data with the two principal components.

The two eigenvectors and the corresponding eigenvalues of covariance matrix R are

$$q_1 = \begin{bmatrix} 0.7074 \\ 0.7078 \end{bmatrix}, \quad \lambda_1 = 0.5039$$

and

$$q_2 = \begin{bmatrix} 0.7068 \\ -0.7074 \end{bmatrix}, \quad \lambda_2 = 0.0014$$

The two principal components are also shown in Figure 3.18.

Example 3.42

In this example, the gray-scale cameraman image is used to obtain the principal components (eigenimages). These eigenimages are used as basis functions for compressing the image. The image is partitioned into 8×8 block and the 64×64 sample covariance matrix is formed. There are 64 basis functions. Each basis function is an 8×8 image. To compress the image, we use 16 of these principal component images corresponding to the 16 largest singular values. The original image, the basis functions, the compressed image, and the error image are shown in Figures 3.19 through 3.22, respectively. The peak signal to noise ratio (PSNR) of the compressed image is 65 dB. Note that the basis functions are not quantized. If we quantize the basis

FIGURE 3.19   Original image.

FIGURE 3.20   Eigenimages.

FIGURE 3.21   Compressed image.

FIGURE 3.22   Error image.

functions and the weights of the basis function, the PSNR will decrease and the compression loss will be more visible.

Example 3.43

To compress an RGB image, we transform the image from RGB color space to YC_bC_r color space [5]. The three channels are independently partitioned into 8×8 blocks and the 64×64 sample covariance matrix is estimated for each channel. There are 64 basis functions per channel. Each basis function is an 8×8 image. To compress the image, we use 8 of these principal component images corresponding to the 8 largest singular values. The original image, the compressed image, and the error image are shown in Figures 3.23 through 3.25, respectively. The PSNR of the

FIGURE 3.23 (See color insert following page 428.)   Original RGB image.

FIGURE 3.24 (See color insert following page 428.)   Compressed RGB image.

FIGURE 3.25 (See color insert following page 428.)   Error RGB image.

compressed image is 55 dB. Again no quantization is performed on the basis of functions and weights.

3.11   MATRIX POLYNOMIALS AND FUNCTIONS OF SQUARE MATRICES

Matrix polynomials and functions of matrices have application in different engineering discipline such as modern control theory, solution of simultaneous linear DEs, queuing theory, and communication systems. In this section, we focus on matrix polynomials and some special function of square matrices with their analytical computations. They are useful for high-quality color imaging systems.

3.11.1 MATRIX POLYNOMIAL

Consider the monic polynomial $f(x)$ of degree m given by

$$f(x) = x^m + a_1 x^{m-1} + a_2 x^{m-2} + \cdots + a_{m-1} x + a_m \qquad (3.161)$$

If the scalar variable x is replaced by the $n \times n$ matrix A, then the corresponding matrix polynomial $f(A)$ is defined by

$$f(A) = A^m + a_1 A^{m-1} + a_2 A^{m-2} + \cdots + a_{m-1} A + a_m I \qquad (3.162)$$

where

$$A^m = A \times \overset{m \text{ times}}{A} \times \cdots \times A$$

I is an $n \times n$ identity matrix

The polynomial $f(x)$ can be written in factor form

$$f(x) = (x - \alpha_1)(x - \alpha_2) \cdots (x - \alpha_n) \qquad (3.163)$$

where $\alpha_1, \alpha_2, \ldots, \alpha_n$ are the roots of the polynomial $f(x)$. The matrix polynomial $f(A)$ can be factored as

$$f(A) = (A - \alpha_1 I)(A - \alpha_2 I) \cdots (A - \alpha_n I) \qquad (3.164)$$

3.11.2 INFINITE SERIES OF MATRICES

An infinite series of matrix A is defined as

$$S(A) = a_0 I + a_1 A + a_2 A^2 + \cdots = \sum_{k=0}^{\infty} a_k A^k \qquad (3.165)$$

It can be shown that the matrix infinite series $S(A)$ converges if and only if the scalar infinite series $S(\lambda_i)$ converges for all values of i, where λ_i is the ith eigenvalue of matrix A. For example, the geometric matrix series

$$S(A) = I + aA + a^2 A^2 + \cdots = \sum_{k=0}^{\infty} a^k A^k \qquad (3.166)$$

is a convergence series and converges to

$$S(A) = (I - aA)^{-1} \qquad (3.167)$$

If and only if the scalar geometric series

$$S(\lambda) = 1 + a\lambda + a^2 \lambda^2 + \cdots = \sum_{k=0}^{\infty} (a\lambda)^k \qquad (3.168)$$

converges for all eigenvalues of A. With the assumption that eigenvalues of A are distinct, we must have

$$|a\lambda_i| < 1 \quad \text{or} \quad -\frac{1}{|a|} < \lambda_i < \frac{1}{|a|} \quad \text{for } i = 1, 2, \dots, n \tag{3.169}$$

Another example is the exponential matrix polynomial defined by

$$e^A = I + A + \frac{A^2}{2!} + \frac{A^3}{3!} + \cdots = \sum_{k=0}^{\infty} \frac{A^k}{k!} \tag{3.170}$$

3.11.3 CAYLEY–HAMILTON THEOREM

THEOREM 3.3

Any $n \times n$ square matrix satisfies its own characteristic polynomial, that is,

$$P(A) = 0 \tag{3.171}$$

where $P(\lambda) = |\lambda I - A|$ is the characteristic polynomial of matrix A.

Proof: We prove the theorem for a special case when matrix A has n distinct eigenvalues. Let $P(\lambda)$ be the characteristic polynomial of A. Then $P(\lambda)$ is a polynomial of degree n in λ and is given by

$$P(\lambda) = \lambda^n + \alpha_1 \lambda^{n-1} + \alpha_2 \lambda^{n-2} + \cdots + \alpha_{n-1}\lambda + \alpha_n \tag{3.172}$$

Then

$$P(A) = A^n + \alpha_1 A^{n-1} + \alpha_2 A^{n-2} + \cdots + \alpha_{n-1}A + \alpha_n I = \sum_{i=0}^{n} \alpha_i A^{n-i} \tag{3.173}$$

Since $A^k = M\Lambda^k M^{-1}$, therefore,

$$P(A) = \sum_{i=0}^{n} \alpha_i A^{n-i} = \sum_{i=0}^{n} \alpha_i M\Lambda^{n-i} M^{-1} = M \sum_{i=0}^{n} \alpha_i \Lambda^{n-i} M^{-1}$$

$$= M \begin{bmatrix} P(\lambda_1) & 0 & \cdots & 0 \\ 0 & P(\lambda_2) & \cdots & 0 \\ \vdots & \vdots & \ddots & \vdots \\ 0 & 0 & \cdots & P(\lambda_n) \end{bmatrix} M^{-1} \tag{3.174}$$

Since $P(\lambda_1) = P(\lambda_2) = \cdots = P(\lambda_n) = 0$, then

$$P(A) = 0 \qquad (3.175)$$

Example 3.44

Show that matrix A satisfies its characteristic polynomial:

$$A = \begin{bmatrix} 2 & -7 \\ 5 & 1 \end{bmatrix}$$

SOLUTION

The characteristic polynomial of A is

$$P(\lambda) = |\lambda I - A| = \begin{vmatrix} \lambda - 2 & 7 \\ -5 & \lambda - 1 \end{vmatrix} = \lambda^2 - 3\lambda + 37$$

Then

$$
\begin{aligned}
P(A) &= A^2 - 3A + 37I \\
&= \begin{bmatrix} 2 & -7 \\ 5 & 1 \end{bmatrix}\begin{bmatrix} 2 & -7 \\ 5 & 1 \end{bmatrix} - 3\begin{bmatrix} 2 & -7 \\ 5 & 1 \end{bmatrix} + 37\begin{bmatrix} 1 & 0 \\ 0 & 0 \end{bmatrix} \\
&= \begin{bmatrix} -31 & -21 \\ 15 & -34 \end{bmatrix} + \begin{bmatrix} -6 & 21 \\ -15 & -3 \end{bmatrix} + \begin{bmatrix} 37 & 0 \\ 0 & 37 \end{bmatrix} = \begin{bmatrix} 0 & 0 \\ 0 & 0 \end{bmatrix} = 0
\end{aligned}
$$

Cayley–Hamilton theorem is used to reduce any matrix polynomial of $n \times n$ matrix A to a polynomial of degree $n - 1$. This is proved in the following theorem.

THEOREM 3.4

Consider an $n \times n$ matrix A and a matrix polynomial in A of degree $m \geq n$. That is,

$$f(A) = a_m A^m + a_{m-1}A^{m-1} + a_{m-2}A^{m-2} + \cdots + a_1 A + a_0 I \qquad (3.176)$$

Then $f(A)$ is reducible to a polynomial of degree $n - 1$ given by

$$R(A) = r_{n-1}A^{n-1} + r_{n-2}A^{n-2} + r_{n-3}A^{n-3} + \cdots + r_1 A + r_0 I \qquad (3.177)$$

Proof: Consider the scalar polynomial $f(\lambda)$. Divide this polynomial by $P(\lambda)$, the characteristic polynomial of matrix A, then

$$f(\lambda) = Q(\lambda)P(\lambda) + R(\lambda) \qquad (3.178)$$

where $Q(\lambda)$ and $R(\lambda)$ are the quotient and remainder polynomial after dividing $f(\lambda)$ by $P(\lambda)$, respectively. The degree of the remainder polynomial is at most $n-1$ since the degree of characteristic polynomial is n. Since $P(A) = 0$ by the Cayley–Hamilton theorem, then we have

$$f(A) = Q(A)P(A) + R(A) = R(A) \qquad (3.179)$$

Example 3.45

Find the matrix polynomial $f(A) = A^8 - 8A^6 + 2A^5 + 2A^3 - 3A^2 + A + 7I$ if

$$A = \begin{bmatrix} 4 & 2 \\ -7 & 0 \end{bmatrix}$$

SOLUTION

The characteristic polynomial of A is

$$P(\lambda) = |\lambda I - A| = \begin{vmatrix} \lambda - 4 & 2 \\ -3 & \lambda \end{vmatrix} = \lambda^2 - 4\lambda + 6$$

Dividing $f(\lambda)$ by $P(\lambda)$, we have

$$\frac{\lambda^8 - 8\lambda^6 + 2\lambda^5 + 2\lambda^3 - 3\lambda^2 + \lambda + 7}{\lambda^2 - 4\lambda + 6} = \lambda^6 + 4\lambda^5 + 2\lambda^4 - 14\lambda^3 - 68\lambda^2 - 186\lambda$$
$$- 339 + \frac{-239\lambda + 2041}{\lambda^2 - 4\lambda + 6}$$

Therefore,

$$f(A) = R(A) = -239A + 2041I = -239 \begin{bmatrix} 4 & -2 \\ 3 & 0 \end{bmatrix} + 2041 \begin{bmatrix} 1 & 0 \\ 0 & 1 \end{bmatrix}$$
$$= \begin{bmatrix} -956 & 478 \\ -717 & 0 \end{bmatrix} + \begin{bmatrix} 2041 & 0 \\ 0 & 2041 \end{bmatrix} = \begin{bmatrix} 1085 & 478 \\ -71 & 2041 \end{bmatrix}$$

3.11.4 FUNCTION OF MATRICES

There are several techniques to find function of matrices. They include Cayley–Hamilton and matrix diagonalization techniques.

3.11.4.1 Cayley–Hamilton Technique

First we assume that the eigenvalues of matrix A are distinct. Using the results of the Cayley–Hamilton theorem, we have

$$f(A) = R(A) = r_{n-1}A^{n-1} + r_{n-2}A^{n-2} + r_{n-3}A^{m-3} + \cdots + r_1A + r_0I \qquad (3.180)$$

where

$$f(\lambda) = Q(\lambda)P(\lambda) + R(\lambda) \qquad (3.181)$$

To get the coefficients of the polynomial $R(\lambda)$, we set $\lambda = \lambda_i$, $i = 1, 2, \ldots, n$ in the above equation:

$$f(\lambda_i) = Q(\lambda_i)P(\lambda_i) + R(\lambda_i) = R(\lambda_i) \qquad (3.182)$$

This yield a set of n simultaneous linear equations that can be solved for $r_0, r_1, \ldots, r_{n-1}$:

$$
\begin{bmatrix}
1 & \lambda_1 & \lambda_1^2 & \cdots & \lambda_1^{n-1} \\
1 & \lambda_2 & \lambda_2^2 & \cdots & \lambda_2^{n-1} \\
1 & \lambda_3 & \lambda_3^2 & \cdots & \lambda_3^{n-1} \\
\vdots & \vdots & \vdots & \vdots & \vdots \\
1 & \lambda_n & \lambda_n^2 & \cdots & \lambda_n^{n-1}
\end{bmatrix}
\begin{bmatrix}
r_0 \\ r_1 \\ r_2 \\ \vdots \\ r_{n-1}
\end{bmatrix}
=
\begin{bmatrix}
f(\lambda_1) \\ f(\lambda_2) \\ f(\lambda_3) \\ \vdots \\ f(\lambda_n)
\end{bmatrix}
\qquad (3.183)
$$

Now let us assume that we have repeated eigenvalues. Without loss of generality, we assume that the $n \times n$ matrix A has one eigenvalue of multiplicity m and $n - m$ distinct eigenvalues. That is, the eigenvalues of A are

$$\underbrace{\lambda_1, \lambda_1, \ldots, \lambda_1}_{m}, \lambda_{m+1}, \lambda_{m+2}, \ldots, \lambda_n \qquad (3.184)$$

Since λ_1 is an eigenvalue with multiplicity of m, then

$$\left. \frac{d^k Q(\lambda)P(\lambda)}{d\lambda^k} \right|_{\lambda_1} = 0 \quad \text{for } k = 1, 2, \ldots, m \qquad (3.185)$$

Therefore, we have the following n equations for n unknown $r_0, r_1, \ldots, r_{n-1}$:

$$
\begin{aligned}
f(\lambda_1) &= R(\lambda_1) \\
f'(\lambda_1) &= R'(\lambda_1) \\
f''(\lambda_1) &= R''(\lambda_1) \\
\vdots \quad &= \quad \vdots \\
f^{m-1}(\lambda_1) &= R^{m-1}(\lambda_1) \\
f(\lambda_{m+1}) &= R(\lambda_{m+1}) \\
f(\lambda_{m+2}) &= R(\lambda_{m+2}) \\
\vdots \quad &= \quad \vdots \\
f(\lambda_n) &= R(\lambda_n)
\end{aligned}
\qquad (3.186)
$$

Example 3.46

Find $f(A) = e^{At}$ if

$$A = \begin{bmatrix} 0 & 1 \\ -2 & -3 \end{bmatrix}$$

SOLUTION

The characteristic polynomial of A is

$$|\lambda I - A| = \begin{bmatrix} \lambda & -1 \\ 2 & \lambda + 3 \end{bmatrix} = (\lambda + 1)(\lambda + 2) = 0$$

Hence the eigenvalues are

$$\lambda_1 = -1 \quad \text{and} \quad \lambda_2 = -2$$

Since they are simple eigenvalues, we have

$$f(\lambda_1) = R(\lambda_1) = r_0 + b_1\lambda_1$$
$$f(\lambda_2) = R(\lambda_2) = r_1 + b_1\lambda_2$$

or in matrix form

$$\begin{bmatrix} 1 & -1 \\ 1 & -2 \end{bmatrix}\begin{bmatrix} r_0 \\ r_1 \end{bmatrix} = \begin{bmatrix} e^{-t} \\ e^{-2t} \end{bmatrix}$$

Solving for r_0 and r_1, we have

$$\begin{bmatrix} r_0 \\ r_1 \end{bmatrix} = \begin{bmatrix} 1 & -1 \\ 1 & -2 \end{bmatrix}^{-1}\begin{bmatrix} e^{-t} \\ e^{-2t} \end{bmatrix} = \begin{bmatrix} 2 & -1 \\ 1 & -1 \end{bmatrix}\begin{bmatrix} e^{-t} \\ e^{-2t} \end{bmatrix} = \begin{bmatrix} 2e^{-t} - e^{-2t} \\ e^{-t} - e^{-2t} \end{bmatrix}$$

Therefore,

$$e^{At} = r_0 I + r_1 A = \begin{bmatrix} r_0 & 0 \\ 0 & r_0 \end{bmatrix} + \begin{bmatrix} 0 & r_1 \\ -2r_1 & -3r_1 \end{bmatrix} = \begin{bmatrix} r_0 & r_1 \\ -2r_1 & r_0 - 3r_1 \end{bmatrix}$$

Substituting for r_0 and r_1

$$e^{At} = \begin{bmatrix} r_0 & r_1 \\ -2r_1 & r_0 - 3r_1 \end{bmatrix} = \begin{bmatrix} 2e^{-t} - e^{-2t} & 2e^{-t} - e^{-2t} \\ 2e^{-t} - e^{-2t} & 2e^{-t} - e^{-2t} \end{bmatrix}$$

3.11.4.2 Modal-Matrix Technique

Assume that matrix A has distinct eigenvalues and let M be the modal matrix of A, then

$$A = M\Lambda M^{-1} \tag{3.187}$$

Let the Taylor series expansion of the function $f(A)$ be

$$f(A) = \sum_{k=0}^{\infty} \alpha_k A^k \tag{3.188}$$

Now substituting Equation 3.187 into Equation 3.188 results in

$$f(A) = \sum_{k=0}^{\infty} \alpha_k (M \Lambda M^{-1})^k = \sum_{k=0}^{\infty} \alpha_k M \Lambda^k M^{-1} = M \sum_{k=0}^{\infty} \alpha_k \Lambda^k M^{-1} \tag{3.189}$$

Since

$$
\sum_{k=0}^{\infty} \alpha_k \Lambda^k =
\begin{bmatrix}
\sum_{k=0}^{\infty} \alpha_k \lambda_1^k & 0 & \cdots & 0 & 0 \\
0 & \sum_{k=0}^{\infty} \alpha_k \lambda_2^k & \cdots & 0 & 0 \\
\vdots & \vdots & \ddots & \vdots & \vdots \\
0 & 0 & \cdots & \sum_{k=0}^{\infty} \alpha_k \lambda_{n-1}^k & 0 \\
0 & 0 & \cdots & 0 & \sum_{k=0}^{\infty} \alpha_k \lambda_n^k
\end{bmatrix}
$$
$$
=
\begin{bmatrix}
f(\lambda_1) & 0 & \cdots & 0 & 0 \\
0 & f(\lambda_2) & \cdots & 0 & 0 \\
\vdots & \vdots & \ddots & \vdots & \vdots \\
0 & 0 & \cdots & f(\lambda_{n-1}) & 0 \\
0 & 0 & \cdots & 0 & f(\lambda_n)
\end{bmatrix}
\tag{3.190}
$$

Therefore,

$$
f(A) = M
\begin{bmatrix}
f(\lambda_1) & 0 & \cdots & 0 & 0 \\
0 & f(\lambda_2) & \cdots & 0 & 0 \\
\vdots & \vdots & \ddots & \vdots & \vdots \\
0 & 0 & \cdots & f(\lambda_{n-1}) & 0 \\
0 & 0 & \cdots & 0 & f(\lambda_n)
\end{bmatrix}
M^{-1}
\tag{3.191}
$$

Example 3.47

Find $f(A) = A^k$ if

$$
A =
\begin{bmatrix}
0.75 & 0.5 & -0.75 \\
0.5 & 3 & -3 \\
0.5 & 2 & -2
\end{bmatrix}
$$

Mathematical Foundations

Solution

The characteristic polynomial of A is

$$P(\lambda) = |\lambda I - A| = \begin{vmatrix} \lambda - 0.75 & -0.5 & 0.75 \\ -0.5 & \lambda - 3 & 3 \\ -0.5 & -2 & \lambda + 2 \end{vmatrix} = \lambda^3 - 1.75\lambda^2 + 0.875\lambda - 0.125$$

$$= (\lambda - 1)(\lambda - 0.5)(\lambda - 0.25)$$

The eigenvalues are $\lambda_1 = 1$, $\lambda_2 = 0.5$, and $\lambda_3 = 0.25$. The corresponding eigenvectors are

$$x_1 = \begin{bmatrix} 0 \\ 3 \\ 2 \end{bmatrix}, \quad x_2 = \begin{bmatrix} 1 \\ 1 \\ 1 \end{bmatrix}, \quad \text{and} \quad x_3 = \begin{bmatrix} 1 \\ 2 \\ 2 \end{bmatrix}$$

Therefore,

$$f(A) = Mf(\Lambda)M^{-1} = \begin{bmatrix} 0 & 1 & 1 \\ 3 & 1 & 2 \\ 2 & 1 & 2 \end{bmatrix} \begin{bmatrix} (1)^k & 0 & 0 \\ 0 & (0.5)^k & 0 \\ 0 & 0 & (0.25)^k \end{bmatrix} \begin{bmatrix} 0 & 1 & 1 \\ 3 & 1 & 2 \\ 2 & 1 & 2 \end{bmatrix}^{-1}$$

$$= \begin{bmatrix} 0 & (0.5)^k & (0.25)^k \\ 3 & (0.5)^k & 2(0.55)^k \\ 2 & (0.5)^k & 2(0.25)^k \end{bmatrix} \begin{bmatrix} 0 & 1 & -1 \\ 2 & 2 & -3 \\ -1 & -2 & 3 \end{bmatrix}$$

$$= \begin{bmatrix} 2(0.5)^k - (0.25)^k & 2(0.5)^k - 2(0.25)^k & -3(0.5)^k + 3(0.25)^k \\ 2(0.5)^k - 2(0.25)^k & 3 + 2(0.5)^k - 4(0.25)^k & -3 - 3(0.5)^k + 6(0.25)^k \\ 2(0.5)^k - 2(0.25)^k & 2 + 2(0.5)^k - 4(0.25)^k & -2 - 3(0.5)^k + 6(0.25)^k \end{bmatrix}$$

Hence,

$$A^k = \begin{bmatrix} 2(0.5)^k - (0.25)^k & 2(0.5)^k - 2(0.25)^k & -3(0.5)^k + 3(0.25)^k \\ 2(0.5)^k - 2(0.25)^k & 3 + 2(0.5)^k - 4(0.25)^k & -3 - 3(0.5)^k + 6(0.25)^k \\ 2(0.5)^k - 2(0.25)^k & 2 + 2(0.5)^k - 4(0.25)^k & -2 - 3(0.5)^k + 6(0.25)^k \end{bmatrix}$$

3.11.5 Matrix Exponential Function e^{At}

The matrix exponential function e^{At} is a very important function with applications in analysis and design of continuous time-control systems. It is defined as an infinite series:

$$e^{At} = I + At + \frac{A^2 t^2}{2!} + \frac{A^3 t^3}{3!} + \cdots + \frac{A^n t^n}{n!} + \cdots \tag{3.192}$$

This series is a convergent series for all values of t and can be computed using different techniques such as Cayley–Hamilton and Laplace transform. The derivative of e^{At} with respect to t is

$$\frac{de^{At}}{dt} = A + \frac{2A^2t^2}{1!} + \frac{A^3t^2}{2!} + \cdots + \frac{A^n t^{n-1}}{(n-1)!} + \cdots$$

$$= A\left(I + At + \frac{A^2t^2}{2!} + \frac{A^3t^3}{3!} + \cdots\right)$$

$$= Ae^{At} = e^{At}A \tag{3.193}$$

The integral of the exponential function e^{At} is

$$\int_0^t e^{A\tau}d\tau = A^{-1}(e^{At} - I) \tag{3.194}$$

The above integral is valid if and only if matrix A is nonsingular. If A is singular, there is no simple closed-form solution.

Example 3.48

Find $\int_0^t e^{A\tau}d\tau$ if

(a) $A = \begin{bmatrix} 0 & 1 \\ -2 & -3 \end{bmatrix}$

(b) $A = \begin{bmatrix} 2 & -2 \\ 1 & -1 \end{bmatrix}$

SOLUTION

(a) First we find e^{At} using Cayley–Hamilton technique:

$$|\lambda I - A| = \begin{vmatrix} \lambda & -1 \\ 2 & \lambda + 3 \end{vmatrix} = (\lambda + 1)(\lambda + 2) = 0 \;\rightarrow\; \lambda_1 = -1, \;\; \lambda_2 = -2$$

The corresponding eigenvectors are

$$Ax_1 = \lambda_1 x_1, \quad \text{therefore } x_1 = [\,1 \quad -1\,]^T$$

$$Ax_2 = \lambda_2 x_2, \quad \text{therefore } x_2 = [\,1 \quad -2\,]^T$$

Therefore,

$$
e^{At} = M\Lambda(t)M^{-1} = \begin{bmatrix} 1 & 1 \\ -1 & -2 \end{bmatrix} \begin{bmatrix} e^{-t} & 0 \\ 0 & e^{-2t} \end{bmatrix} \begin{bmatrix} 1 & 1 \\ -1 & -2 \end{bmatrix}
$$

$$
= \begin{bmatrix} 2e^{-t} - e^{-2t} & e^{-t} - e^{-2t} \\ -2e^{-t} + 2e^{-2t} & -e^{-t} + 2e^{-2t} \end{bmatrix}
$$

Since matrix A is nonsingular, we have

$$
\int_0^t e^{A\tau}d\tau = A^{-1}(e^{At} - I) = \begin{bmatrix} 0 & 1 \\ -2 & -3 \end{bmatrix}^{-1} \left(\begin{bmatrix} 2e^{-t} - e^{-2t} & e^{-t} - e^{-2t} \\ -2e^{-t} + 2e^{-2t} & -e^{-t} + 2e^{-2t} \end{bmatrix} - \begin{bmatrix} 1 & 0 \\ 0 & 1 \end{bmatrix} \right)
$$

$$
= \begin{bmatrix} -1.5 & -0.5 \\ 1 & 0 \end{bmatrix} \begin{bmatrix} 2e^{-t} - e^{-2t} - 1 & e^{-t} - e^{-2t} \\ -2e^{-t} + 2e^{-2t} & -e^{-t} + 2e^{-2t} - 1 \end{bmatrix}
$$

$$
= \begin{bmatrix} -2e^{-t} + 0.5e^{-2t} + 1.5 & -e^{-t} - 0.5e^{-2t} \\ 2e^{-t} - e^{-2t} - 1 & e^{-t} - e^{-2t} \end{bmatrix}
$$

(b) In this case, matrix A is singular, and

$$
e^{At} = \begin{bmatrix} -1 + 2e^t & 2 - 2e^t \\ -1 + e^t & 2 - e^t \end{bmatrix}
$$

Therefore, we use direct integration:

$$
\int_0^t e^{A\tau}d\tau = \begin{bmatrix} \int_0^t (-1 + 2e^\tau)d\tau & \int_0^t (2 - 2e^\tau)d\tau \\ \int_0^t (-1 + e^\tau)d\tau & \int_0^t (2 - e^\tau)d\tau \end{bmatrix} = \begin{bmatrix} -t - 2 + 2e^t & 2t + 2 - 2e^t \\ -t - 1 + e^t & 2t + 1 - e^t \end{bmatrix}
$$

3.11.6 COMPUTING e^{At} USING LAPLACE TRANSFORM

Laplace transform can be used to compute e^{At}. In this method, e^{At} is computed by solving the DE describing e^{At} using the Laplace transform. Let $\varphi(t) = e^{At}$, then

$$
\frac{d\varphi(t)}{dt} = Ae^{At} = A\varphi(t) \tag{3.195}
$$

with $\varphi(0) = e^{A0} = I$

Taking the Laplace transform from both sides of Equation 3.195, we have

$$
s\Phi(s) - \varphi(0) = A\Phi(s) \tag{3.196}
$$

Since $\varphi(0) = I$, then

$$
(sI - A)\Phi(s) = I \tag{3.197}
$$

and

$$\Phi(s) = (sI - A)^{-1} \qquad (3.198)$$

Therefore,

$$\varphi(t) = L^{-1}[(sI - A)^{-1}] \qquad (3.199)$$

Example 3.49

Find e^{At} if

$$A = \begin{bmatrix} -3 & -1 \\ 0 & -4 \end{bmatrix}$$

SOLUTION

Using Laplace transform, we have

$$\Phi(s) = (sI - A)^{-1} = \begin{bmatrix} s+3 & 1 \\ 0 & s+4 \end{bmatrix}^{-1} = \frac{1}{s^2 + 7s + 12} \begin{bmatrix} s+4 & -1 \\ 0 & s+3 \end{bmatrix}$$

or

$$\Phi(s) = \begin{bmatrix} \dfrac{s+4}{(s+3)(s+4)} & \dfrac{-1}{(s+3)(s+4)} \\ 0 & \dfrac{s+3}{(s+3)(s+4)} \end{bmatrix}$$

Partial fraction of the entries of matrix $\Phi(s)$ yields

$$\Phi(s) = \begin{bmatrix} \dfrac{1}{s+3} & -\dfrac{1}{s+3} + \dfrac{1}{s+4} \\ 0 & \dfrac{1}{s+4} \end{bmatrix}$$

Hence,

$$e^{At} = L^{-1}(sI - A)^{-1} = \begin{bmatrix} e^{-3t} & -e^{-3t} + e^{-4t} \\ 0 & e^{-4t} \end{bmatrix}$$

3.11.7 MATRIX EXPONENTIAL FUNCTION A^k

Matrix exponential A^k has applications in design of discrete time-control systems. It is the state-transition matrix of LTI discrete systems in state-space form. It can be computed using either Cayley–Hamilton technique or z-transform. To use z-transform, let

$$\varphi(k) = A^k \tag{3.200}$$

Then it is obvious that

$$\varphi(k + 1) = A^{k+1} = A\varphi(k)$$
$$\varphi(0) = I \tag{3.201}$$

Taking the z-transform from both sides of Equation 3.201, we have

$$z\Phi(z) - z\varphi(0) = A\Phi(z) \tag{3.202}$$

Since $\varphi(0) = I$, then we have

$$(zI - A)\Phi(z) = zI \tag{3.203}$$

Solving Equation 3.203 for $\Phi(z)$,

$$\Phi(z) = z(zI - A)^{-1} \tag{3.204}$$

Therefore,

$$\varphi(k) = Z^{-1}[z(zI - A)^{-1}] \tag{3.205}$$

Example 3.50

Find A^k, if

$$A = \begin{bmatrix} 0.1 & 0.4 & -0.1 \\ -0.15 & 0.6 & -0.05 \\ -0.2 & 0.4 & 0.2 \end{bmatrix}$$

SOLUTION

Using z-transform, we have

$$\Phi(z) = z(zI - A)^{-1} = z \begin{bmatrix} z - 0.1 & -0.4 & 0.1 \\ 0.15 & z - 0.6 & 0.05 \\ 0.2 & -0.4 & z - 0.2 \end{bmatrix}^{-1}$$

After taking the inverse, we have

$$
\Phi(z) = \frac{z}{(z - 0.2)(z - 0.3)(z - 0.4)}
$$

$$
\times \begin{bmatrix} (z - 0.6)(z - 0.2) + 0.02 & 0.4(z - 0.2) - 0.04 & -0.1(z - 0.4) - 0.02 \\ -0.15(z - 0.2) + 0.01 & (z - 0.1)(z - 0.2) - 0.02 & -0.05(z - 0.1) - 0.015 \\ -0.2(z - 0.6) - 0.02 & 0.4(z - 0.1) - 0.08 & (z - 0.1)(z - 0.6) + 0.06 \end{bmatrix}
$$

Performing partial-fraction expansion, we have

$$
\Phi(z) = \begin{bmatrix} \frac{z}{z - 0.2} + \frac{z}{z - 0.3} - \frac{z}{z - 0.4} & -\frac{2z}{z - 0.2} + \frac{2z}{z - 0.4} & -\frac{2z}{z - 0.2} + \frac{3z}{z - 0.3} - \frac{z}{z - 0.4} \\ \frac{0.5z}{z - 0.2} + \frac{0.5z}{z - 0.3} - \frac{z}{z - 0.4} & -\frac{z}{z - 0.2} + \frac{2z}{z - 0.4} & \frac{0.5z}{z - 0.2} - \frac{0.5z}{z - 0.3} \\ \frac{3z}{z - 0.2} - \frac{4z}{z - 0.3} + \frac{z}{z - 0.4} & -\frac{2z}{z - 0.2} + \frac{2z}{z - 0.4} & -\frac{5z}{z - 0.2} + \frac{12z}{z - 0.3} - \frac{6z}{z - 0.4} \end{bmatrix}
$$

Therefore,

$$
\varphi(k) = A^k
$$

$$
= \begin{bmatrix} (0.2)^k + (0.3)^k - (0.4)^k & -2(0.2)^k + 2(0.4)^k & -2(0.2)^k + 3(0.3)^k - (0.4)^k \\ 0.5(0.2)^k + 0.5(0.3)^k - (0.4)^k & -(0.2)^k + 2(0.4)^k & 0.5(0.2)^k - 0.5(0.3)^k \\ 3(0.2)^k - 4(0.3)^k + (0.4)^k & -2(0.2)^k + 2(0.4)^k & -5(0.2)^k + 12(0.3)^k - 6(0.4)^k \end{bmatrix}
$$

3.12 FUNDAMENTALS OF MATRIX CALCULUS

In a variety of multivariate optimization problems, it is necessary to differentiate a scalar function with respect to a vector. In this section, we look at derivatives of functions with respect to vectors.

3.12.1 DERIVATIVES OF A SCALAR FUNCTION WITH RESPECT TO A VECTOR

Let $f(x)$ be a scalar function of vector $x \in R^n$, then by definition, the derivative or gradient of $f(x)$ with respect to vector x is an $n \times 1$ vector, which is defined as

$$
\frac{\partial f(x)}{\partial x} = \begin{bmatrix} \dfrac{\partial f(x)}{\partial x_1} \\ \dfrac{\partial f(x)}{\partial x_2} \\ \vdots \\ \dfrac{\partial f(x)}{\partial x_n} \end{bmatrix} \tag{3.206}
$$

The second derivative of $f(x)$ with respect to $x \in R^n$ is an $n \times n$ matrix, which is defined by

$$\frac{\partial^2 f(x)}{\partial x^2} = \begin{bmatrix} \dfrac{\partial^2 f(x)}{\partial x_1^2} & \dfrac{\partial f(x)}{\partial x_1 \partial x_2} & \cdots & \dfrac{\partial^2 f(x)}{\partial x_1 \partial x_n} \\[3mm] \dfrac{\partial^2 f(x)}{\partial x_2 \partial x_1} & \dfrac{\partial^2 f(x)}{\partial x_2^2} & \cdots & \dfrac{\partial^2 f(x)}{\partial x_2 \partial x_n} \\[3mm] \vdots & \vdots & \vdots & \vdots \\[3mm] \dfrac{\partial^2 f(x)}{\partial x_n \partial x_1} & \dfrac{\partial^2 f(x)}{\partial x_n \partial x_2} & \cdots & \dfrac{\partial^2 f(x)}{\partial x_n^2} \end{bmatrix} \tag{3.207}$$

Example 3.51

Let $x \in R^3$ and $f(x) = x_1^3 + 3x_1^2 + 2x_2 - 7x_2 x_3 + 4$. Find $\frac{\partial f(x)}{\partial x}$.

SOLUTION

$$\frac{\partial f(x)}{\partial x} = \begin{bmatrix} \dfrac{\partial f(x)}{\partial x_1} \\[3mm] \dfrac{\partial f(x)}{\partial x_2} \\[3mm] \dfrac{\partial f(x)}{\partial x_3} \end{bmatrix} = \begin{bmatrix} 3x_1^2 + 6x_1 \\[2mm] 2 - 7x_3 \\[2mm] -7x_2 \end{bmatrix}$$

The second derivative is

$$\frac{\partial^2 f(x)}{\partial x^2} = \begin{bmatrix} 6x_1 + 6 & 0 & 0 \\ 0 & 0 & -7 \\ 0 & -7 & 0 \end{bmatrix}$$

Example 3.52

Let $x \in R^n$ and $f(x) = c^T x$, where c is an $n \times 1$ column vector, then

$$f(x) = c^T x = c_1 x_1 + c_2 x_2 + \cdots + c_n x_n$$

and

$$\frac{\partial f(x)}{\partial x} = \begin{bmatrix} \dfrac{\partial f(x)}{\partial x_1} \\[2mm] \dfrac{\partial f(x)}{\partial x_2} \\[2mm] \vdots \\[2mm] \dfrac{\partial f(x)}{\partial x_n} \end{bmatrix} = \begin{bmatrix} c_1 \\ c_2 \\ \vdots \\ c_n \end{bmatrix}$$

Hence,

$$\frac{\partial c^{\mathrm{T}} x}{\partial x} = c \qquad (3.208)$$

Similarly,

$$\frac{\partial x^{\mathrm{T}} c}{\partial x} = c \qquad (3.209)$$

3.12.2 DERIVATIVES OF QUADRATIC FUNCTIONS

Let $x \in R^n$ and $f(x) = x^{\mathrm{T}} A x$, where A is an $n \times n$ symmetric matrix. The multivariate quadratic function $f(x)$ can be written as

$$f(x) = x^{\mathrm{T}} A x = \sum_{i=1}^{n} \sum_{j=1}^{n} a_{ij} x_i x_j \qquad (3.210)$$

Consider the first component of vector $\frac{\partial f(x)}{\partial x}$, that is,

$$\begin{aligned}
\frac{\partial f(x)}{\partial x_1} &= \frac{\partial}{\partial x_1} \sum_{i=1}^{n} \sum_{j=1}^{n} a_{ij} x_i x_j = \frac{\partial}{\partial x_1} \sum_{i=2}^{n} \sum_{j=2}^{n} a_{ij} x_i x_j + \frac{\partial}{\partial x_1} a_{11} x_1 x_1 \\
&\quad + \frac{\partial}{\partial x_1} \sum_{j=2}^{n} a_{1j} x_1 x_j + \frac{\partial}{\partial x_1} \sum_{i=2}^{n} a_{i1} x_i x_1 \\
&= \frac{\partial}{\partial x_1} \sum_{i=2}^{n} \sum_{j=2}^{n} a_{ij} x_i x_j + \frac{\partial}{\partial x_1} a_{11} x_1^2 + \frac{\partial}{\partial x_1} x_1 \sum_{j=2}^{n} a_{1j} x_j + \frac{\partial}{\partial x_1} x_1 \sum_{i=2}^{n} a_{i1} x_i \qquad (3.211)
\end{aligned}$$

The first term in the above expansion is independent of x_1, therefore it does not contribute to the derivative. Hence,

$$\frac{\partial f(x)}{\partial x_1} = 0 + 2 a_{11} x_1 + \sum_{j=2}^{n} a_{1j} x_j + \sum_{i=2}^{n} a_{i1} x_i = \sum_{j=1}^{n} a_{1j} x_j + \sum_{i=1}^{n} a_{i1} x_i = 2 \sum_{j=1}^{n} a_{1j} x_j$$

$$(3.212)$$

This is equal to the first element of the product $2Ax$. Similarly, it can be shown that $\frac{\partial f(x)}{\partial x_i}$ is equal to the ith element of the product. Therefore,

$$\frac{\partial x^T A x}{\partial x} = 2Ax \tag{3.213}$$

Similarly, it can be shown that the second derivative of $f(x)$ is given by

$$\frac{\partial^2 x^T A x}{\partial x^2} = 2A \tag{3.214}$$

Example 3.53

Let $x \in R^3$ and $f(x) = 13x_1^2 + 8x_2^2 - 4x_3^2 - 3x_1x_2 + 5x_1x_3 + 6x_2x_3$. Find $\frac{\partial f(x)}{\partial x}$ and $\frac{\partial^2 f(x)}{\partial x^2}$.

SOLUTION

$$f(x) = 13x_1^2 + 8x_2^2 - 4x_3^2 - 3x_1x_2 + 5x_1x_3 + 6x_2x_3$$

$$= \begin{bmatrix} x_1 & x_2 & x_3 \end{bmatrix} \begin{bmatrix} 13 & -1.5 & 2.5 \\ -1.5 & 8 & 3 \\ 2.5 & 3 & -4 \end{bmatrix} \begin{bmatrix} x_1 \\ x_2 \\ x_3 \end{bmatrix} = x^T A x$$

The first derivative of $f(x)$ is

$$\frac{\partial f(x)}{\partial x} = 2Ax = 2 \begin{bmatrix} 13 & -1.5 & 2.5 \\ -1.5 & 8 & 3 \\ 2.5 & 3 & -4 \end{bmatrix} \begin{bmatrix} x_1 \\ x_2 \\ x_3 \end{bmatrix} = \begin{bmatrix} 26x_1 - 3x_2 + 5x_3 \\ -3x_1 + 16x_2 + 6x_3 \\ 5x_1 + 6x_2 - 8x_3 \end{bmatrix}$$

The second derivative of $f(x)$ is

$$\frac{\partial^2 f(x)}{\partial x^2} = \frac{\partial 2Ax}{\partial x} = 2A = \begin{bmatrix} 26 & -3 & 5 \\ -3 & 16 & 6 \\ 5 & 6 & -8 \end{bmatrix}$$

Example 3.54

Let $x \in R^n$, A be an $n \times n$ symmetric matrix, b be an $n \times 1$ column vector, and c a scalar. Define $f(x) = \frac{1}{2}x^T A x + b^T x + c$. Find $\frac{\partial f(x)}{\partial x}$ and $\frac{\partial^2 f(x)}{\partial x^2}$.

SOLUTION
The first derivative of $f(x)$ is

$$\frac{\partial f(x)}{\partial x} = Ax + b$$

And the second derivative is

$$\frac{\partial^2 f(x)}{\partial x^2} = \frac{\partial(Ax + b)}{\partial x} = A$$

3.12.3 DERIVATIVE OF A VECTOR FUNCTION WITH RESPECT TO A VECTOR

Let $f(x) \in R^m$ be a vector function of vector $x \in R^n$, then by definition

$$\frac{\partial f(x)}{\partial x} = \begin{bmatrix} \dfrac{\partial f_1(x)}{\partial x_1} & \dfrac{\partial f_2(x)}{\partial x_1} & \cdots & \dfrac{\partial f_m(x)}{\partial x_1} \\[2ex] \dfrac{\partial f_1(x)}{\partial x_2} & \dfrac{\partial f_2(x)}{\partial x_2} & \cdots & \dfrac{\partial f_m(x)}{\partial x_2} \\[2ex] \vdots & \vdots & \vdots & \vdots \\[2ex] \dfrac{\partial f_1(x)}{\partial x_n} & \dfrac{\partial f_2(x)}{\partial x_n} & \cdots & \dfrac{\partial f_m(x)}{\partial x_n} \end{bmatrix} \tag{3.215}$$

Example 3.55

Let $x \in R^3$ and $f(x) = \begin{bmatrix} x_1^2 x_2 + 3x_2 + 4x_3 - 5 \\ x_1 x_2 + x_2 - 2x_1 x_2 x_3 \end{bmatrix}$, then

$$\frac{\partial f(x)}{\partial x} = \begin{bmatrix} \dfrac{\partial f_1(x)}{\partial x_1} & \dfrac{\partial f_2(x)}{\partial x_1} \\[2ex] \dfrac{\partial f_1(x)}{\partial x_2} & \dfrac{\partial f_2(x)}{\partial x_2} \\[2ex] \dfrac{\partial f_1(x)}{\partial x_3} & \dfrac{\partial f_2(x)}{\partial x_3} \end{bmatrix} = \begin{bmatrix} 2x_1 x_2 & x_2 - 2x_2 x_3 \\ x_1^2 + 3 & x_1 - 2x_1 x_3 + 1 \\ 4 & -2x_1 x_2 \end{bmatrix}$$

PROBLEMS

3.1 (a) Let

$$x(n) = \delta(n) + 2\delta(n - 1) + 3\delta(n - 2) \quad \text{and}$$
$$h(n) = \delta(n) + \delta(n - 1) + 2\delta(n - 2)$$

Find $x(n) * h(n)$.

(b) Consider the LTI system with impulse response

$$h(n) = \left(\tfrac{1}{2}\right)^n u(n)$$

Using z-transform, find the response $y(n)$ of this system to the input

$$x(n) = \left(\tfrac{1}{4}\right)^n u(n)$$

3.2 Determine the ROCs and z-transforms of each of the following sequences:
(a) $x(n) = (0.3)^n u(n) + 2^n u(-n-1)$
(b) $y(n) = (0.2)^n u(n-2)$
(c) $z(n) = u(n) + \delta(n) + 3^n u(-n)$

3.3 Evaluate the inverse z-transform of the following function:

(a) $X(z) = \dfrac{z^2 - z + \frac{1}{6}}{z^2 - \frac{5}{6}z + \frac{1}{6}}$,   ROC: $\frac{1}{3} < z < \frac{1}{2}$

(b) $X(z) = \log(1 - 2z^{-1})$,   ROC: $|z| > 2$

3.4 Determine the z-transform and its ROC for each of the following 2-D sequences:
(a) $x(n_1, n_2) = \left(\tfrac{1}{2}\right)^{n_1} u(n_1)u(n_2)$
(b) $y(n_1, n_2) = \left(\tfrac{1}{3}\right)^{n_1+n_2} u(n_1)u(n_2)$

3.5 Find the eigenvalues and eigenvectors of matrix A given below:

$$A = \begin{bmatrix} 1 & 2 \\ 3 & -1 \end{bmatrix}$$

3.6 Find the eigenvalues and eigenvectors of the following matrices:

$$A = \begin{bmatrix} 2 & -1 \\ 0 & 3 \end{bmatrix}, \quad B = \begin{bmatrix} a & -b \\ b & a \end{bmatrix}, \quad \text{and} \quad C = \begin{bmatrix} 0 & 1 & 3 \\ 0 & 3 & 0 \\ -3 & 0 & 0 \end{bmatrix}$$

3.7 Let A be a 4×4 matrix with eigenvalues $-1, -2, -3$, and -4. Find (if possible) the following quantities:
(a) $\det(A^T)$
(b) $\text{Trace}(A^{-1})$
(c) $\det(A - 8I)$

3.8 Compute the eigenvalues and eigenvectors of the matrix

$$A = \begin{bmatrix} 0 & 2 & 2 \\ 2 & 0 & 2 \\ 2 & 2 & 0 \end{bmatrix}$$

Can matrix A be diagonalized? If yes, find the transformation to diagonalize matrix A.

3.9 Compute the eigenvalues and eigenvectors of the matrix

$$A = \begin{bmatrix} 1 & 1 & 1 \\ 0 & 2 & 1 \\ 0 & 0 & 3 \end{bmatrix}$$

Find, if possible, the matrix M such that $M^{-1}AM$ is a diagonal matrix.

3.10 Show that the eigenvalues of $A + \alpha I$ are related to the eigenvalues of A by

$$\lambda(A + \alpha I) = \lambda(A) + \alpha$$

3.11 Find the SVD of the following matrices:

$$A = \begin{bmatrix} 1 & -2 \\ -2 & 4 \end{bmatrix}, \quad B = \begin{bmatrix} 1 & -1 & 0 \\ 3 & -2 & 0 \\ 1 & -3 & 5 \end{bmatrix}, \quad \text{and} \quad C = \begin{bmatrix} 1 & -1 \\ 2 & -3 \\ 3 & 4 \end{bmatrix}$$

3.12 Consider the symmetric square matrix

$$A = \begin{bmatrix} b & 1 \\ 1 & a \end{bmatrix} \begin{bmatrix} 4 & 0 \\ 0 & -2 \end{bmatrix} \begin{bmatrix} b & 1 \\ 1 & a \end{bmatrix}^{-1}$$

(a) Find eigenvalues of $A + 3I$.
(b) For what value of α, the matrix $A + \alpha I$ is singular.

3.13 Determine whether the following matrices are (a) positive definite, (b) positive semi-definite, (c) negative definite, and (d) negative semi-definite:

$$A = \begin{bmatrix} 2 & 1 & 1 \\ 1 & 4 & 0 \\ 1 & 0 & 1 \end{bmatrix}, \quad B = \begin{bmatrix} -1 & -1 & -1 \\ -1 & -4 & -1 \\ -1 & -1 & -2 \end{bmatrix}, \quad \text{and} \quad C = \begin{bmatrix} 7 & -6 \\ 6 & 2 \end{bmatrix}$$

3.14 Consider the following square matrix:

$$A = \begin{bmatrix} 2 & -1 \\ 1 & 2 \end{bmatrix}$$

(a) Compute A^k and e^{At} using the Cayley–Hamilton theorem.
(b) Find the square roots of the matrix, that is, find all matrices B such that $B^2 = A$ using the Cayley–Hamilton theorem method.

3.15 Consider the following square matrix:

$$A = \begin{bmatrix} 2 & 2 & 0 \\ 0 & 0 & 2 \\ 0 & 0 & 2 \end{bmatrix}$$

(a) Compute A^7 and A^{73} using the Cayley–Hamilton theorem.
(b) Compute e^{At} using the Cayley–Hamilton theorem.

3.16 Assume that the following matrix is nonsingular:

$$C = \begin{bmatrix} a & 1 & 0 \\ 0 & b & 0 \\ 0 & 0 & c \end{bmatrix}$$

Compute the natural logarithm of the matrix C; that is, identify the matrix $B = \ln(C)$ that satisfies the equation $C = e^B$. Is the assumption of non-singularity of C really needed for the solution of this problem?

3.17 Given the 3×3 matrix

$$A = \begin{bmatrix} -3 & 2 & 4 \\ 1 & 0 & -1 \\ -2 & 1 & 5 \end{bmatrix}$$

Express the matrix polynomial $f(A) = A^4 + 2A^3 + A^2 - A + 3I$ as a linear combination of the matrices A^2, A, and I only.

3.18 Let the 2×2 matrix A be defined as

$$A = \begin{bmatrix} 0 & 7 \\ -1 & 0 \end{bmatrix}$$

Compute A^k and e^{At}.

3.19 Let the 3×3 matrix A be defined as

$$A = \begin{bmatrix} 0 & -3 & 0 \\ -3 & 0 & -3 \\ 0 & -3 & 0 \end{bmatrix}$$

Compute A^k and e^{At}.

3.20 Let the 3×3 matrix A be defined as

$$A = \begin{bmatrix} -1 & -1 & -1 \\ 0 & -1 & -1 \\ 0 & 0 & -1 \end{bmatrix}$$

Compute A^k and e^{At}.

REFERENCES

1. A. V. Oppenheim, A. S. Willsky, and S. H. Nawab, *Signals and Systems*, 2nd edn., Prentice Hall, Upper Saddle River, NJ, 1996.
2. J. L. Schiff, *The Laplace Transform: Theory and Applications*, Springer-Verlag New York Inc., New York, 1999.
3. P. M. DeRusso, R. J. Roy, C. M. Close, and A. Desrochers, *State Variables for Engineers*, 2nd edn., Wiley-Interscience, New York, 1997.
4. M. Bocher, *Introduction to Higher Algebra*, Dover Publications, 2004.
5. G. A. Baxes, *Digital Image Processing: Principles and Applications*, John Wiley & Sons, New York, 1994.

4 State-Variable Representation

4.1 INTRODUCTION

Modem control system theory is based on the state-space formulation of an underlying system. It provides a unified approach for system modeling and design that is applicable to both linear and nonlinear systems, time-varying and time-invariant systems as well as single-input single-output (SISO) and multiple-input multiple-output (MIMO) systems [1]. In this chapter, we present system modeling in state space, solution of state equations, and controllability and observability of linear time-invariant (LTI) systems.

4.2 CONCEPT OF STATES

The state-space approach is a unified approach for representation of both continuous- and discrete-time dynamical systems. It covers a broad range of systems from nonlinear to linear, time-invariant to time-varying systems. It can also be used in modeling stochastic dynamic systems. In general terms, the states of a dynamic system is the minimum number of variables called state variables such that knowledge of these states at any given time together with the input at that time and future time uniquely determines the behavior of the system past that time [2].

4.3 STATE-SPACE REPRESENTATION OF CONTINUOUS-TIME SYSTEMS

4.3.1 DEFINITION OF STATE

The state of a continuous-time dynamic system is the minimum number of variables called state variables such that the knowledge of these variables at time $t = t_0$ together with the input for $t \geq t_0$ uniquely determines the behavior of the system for $t \geq t_0$. If N variables are needed, then these N variables are considered as components of an N-dimensional vector x called the state vector. The N-dimensional space whose coordinates are the states of the system is called state space. The state of a system at time t is a point in the N-dimensional state space [2,3].

177

4.3.2 STATE EQUATIONS OF CONTINUOUS-TIME SYSTEMS

The state-space equations describing a MIMO continuous-time dynamic system have a general form given by

$$\frac{dx(t)}{dt} = f(x(t), u(t), t) \tag{4.1}$$

$$y(t) = h(x(t), u(t), t) \tag{4.2}$$

where

$x(t) = \begin{bmatrix} x_1(t) & x_2(t) & \cdots & x_N(t) \end{bmatrix}^T$ is the $N \times 1$ state vector
$u(t)$ is the $M \times 1$ input vector
$y(t)$ is the $P \times 1$ output vector

The vector functions $f(.)$ and $h(.)$ are in general nonlinear functions of the state and input vectors and are given by the following equations:

$$f(x(t), u(t), t) = \begin{bmatrix} f_1(x_1(t), x_2(t), \ldots, x_N(t), u_1(t), u_2(t), \ldots, u_M(t), t) \\ f_2(x_1(t), x_2(t), \ldots, x_N(t), u_1(t), u_2(t), \ldots, u_M(t), t) \\ \vdots \\ f_N(x_1(t), x_2(t), \ldots, x_N(t), u_1(t), u_2(t), \ldots, u_M(t), t) \end{bmatrix} \tag{4.3}$$

and

$$h(x(t), u(t), t) = \begin{bmatrix} h_1(x_1(t), x_2(t), \ldots, x_N(t), u_1(t), u_2(t), \ldots, u_M(t), t) \\ h_2(x_1(t), x_2(t), \ldots, x_N(t), u_1(t), u_2(t), \ldots, u_M(t), t) \\ \vdots \\ h_P(x_1(t), x_2(t), \ldots, x_N(t), u_1(t), u_2(t), \ldots, u_M(t), t) \end{bmatrix} \tag{4.4}$$

The state equations given by Equations 4.3 and 4.4 are applicable to both linear and nonlinear, time-invariant, and time-varying systems. If the system is linear, then the state equations can be written as

$$\frac{dx(t)}{dt} = A(t)x(t) + B(t)u(t) \tag{4.5}$$

$$y(t) = C(t)x(t) + D(t)u(t) \tag{4.6}$$

where $A(t)$, $B(t)$, $C(t)$, and $D(t)$ are time-varying matrices of appropriate dimensions. If the system is LTI, then A, B, C, and D are constant matrices and the state equation becomes

$$\frac{dx(t)}{dt} = Ax(t) + Bu(t) \tag{4.7}$$

The output equation is

$$y(t) = Cx(t) + Du(t) \tag{4.8}$$

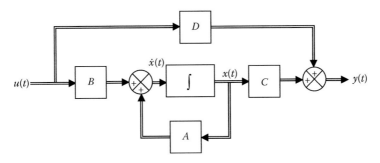

FIGURE 4.1   Block diagram of a continuous LTI system in state space.

where A, B, C, and D are $N \times N$, $N \times M$, $P \times N$, and $P \times M$ constant matrices, respectively. The block diagram of a linear continuous time-invariant control system in state-space form is shown in Figure 4.1.

In the next section, we show how the state equations can be derived for electrical and mechanical systems.

4.3.3 STATE-SPACE EQUATIONS OF ELECTRICAL SYSTEMS

Example 4.1

Consider the series RLC circuit shown in Figure 4.2 with the input $u(t)$ and output $y(t)$.

To determine the output of this system, one needs to know the input signal, the initial charge on the capacitor, and the initial current through the inductor; therefore, the system has two states. The first state is defined as the voltage across the capacitor and the second state the current through the inductor. Therefore, the first state is defined as

$$x_1(t) = v_C(t) \tag{4.9}$$

and the second state is defined as

$$x_2(t) = i_L(t) \tag{4.10}$$

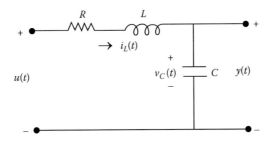

FIGURE 4.2   Series RLC circuit.

Differentiating both sides of Equation 4.9 yields

$$\dot{x}_1(t) = \frac{dv_C(t)}{dt} = \frac{i_L(t)}{C} = \frac{1}{C}x_2(t) \qquad (4.11)$$

Similarly, we differentiae both sides of Equation 4.10:

$$\dot{x}_2(t) = \frac{di_L(t)}{dt} = \frac{v_L(t)}{L} = \frac{1}{L}[u(t) - v_R(t) - v_C(t)] = \frac{1}{L}[u(t) - Rx_2(t) - x_1(t)]$$
$$= \frac{u(t)}{L} - \frac{R}{L}x_2(t) - \frac{1}{L}x_1(t) \qquad (4.12)$$

The output equation is given by

$$y(t) = x_1(t) \qquad (4.13)$$

The above equations can be written in matrix form as

$$\begin{bmatrix} \dot{x}_1(t) \\ \dot{x}_2(t) \end{bmatrix} = \begin{bmatrix} \frac{1}{C} & 0 \\ -\frac{1}{L} & -\frac{R}{L} \end{bmatrix} \begin{bmatrix} x_1(t) \\ x_2(t) \end{bmatrix} + \begin{bmatrix} 0 \\ \frac{1}{L} \end{bmatrix} u(t)$$
$$y(t) = \begin{bmatrix} 1 & 0 \end{bmatrix} \begin{bmatrix} x_1(t) \\ x_2(t) \end{bmatrix} \qquad (4.14)$$

Therefore, the A, B, C, and D matrices of the system are

$$A = \begin{bmatrix} \frac{1}{C} & 0 \\ -\frac{1}{L} & -\frac{R}{L} \end{bmatrix}, \quad B = \begin{bmatrix} 0 \\ \frac{1}{L} \end{bmatrix}, \quad C = \begin{bmatrix} 1 & 0 \end{bmatrix}, \quad \text{and} \quad D = 0 \qquad (4.15)$$

As it can be seen from this simple example, the number of states needed to model an electric circuit in state-space form is equal to the total number of independent energy-storage elements in the circuit. In the circuit shown in Figure 4.2, there are two elements that can store energy, the capacitor and the inductor; hence, two states are needed. In Example 4.2, we consider an electric circuit with three independent energy-storage elements.

Example 4.2

Consider the circuit shown in Figure 4.3.
The three states of the circuit are defined as follows:

$$x_1(t) = v_{C_2}(t)$$
$$x_2(t) = v_{C_1}(t) \qquad (4.16)$$
$$x_3(t) = i_L(t)$$

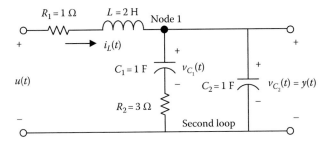

FIGURE 4.3 Electric circuit with three states.

Kirchhoff's voltage law (KVL) around the second loop is

$$x_1(t) = v_{C_2}(t) = v_{C_1}(t) + v_{R_2}(t) = x_2(t) + R_2 C_1 \dot{x}_2(t) = x_2(t) + 3\dot{x}_2(t) \qquad (4.17)$$

Therefore, we have

$$\dot{x}_2(t) = \tfrac{1}{3} x_1(t) - \tfrac{1}{3} x_2(t) \qquad (4.18)$$

Kirchhoff's current law (KCL) at Node 1 yields

$$x_3(t) = C_1 \dot{x}_2(t) + C_2 \dot{x}_1(t) = \dot{x}_2(t) + \dot{x}_1(t) \qquad (4.19)$$

Therefore,

$$\dot{x}_1(t) = x_3(t) - \dot{x}_2(t) = -\tfrac{1}{3} x_1(t) + \tfrac{1}{3} x_2(t) + x_3(t) \qquad (4.20)$$

Finally, writing the KVL around the first loop results in

$$R_1 x_3(t) + L\dot{x}_3(t) + x_2(t) + R_2 \dot{x}_2(t) = u(t) \qquad (4.21)$$

Hence, we have

$$\dot{x}_3(t) = -\tfrac{1}{2} x_3(t) - \tfrac{1}{2} x_2(t) - \tfrac{3}{2} \dot{x}_2(t) + \tfrac{1}{2} u(t) = -\tfrac{1}{2} x_1(t) - \tfrac{1}{2} x_3(t) + \tfrac{1}{2} u(t) \qquad (4.22)$$

Equations 4.18, 4.20, and 4.22 can be written in matrix form as

$$\begin{bmatrix} \dot{x}_1(t) \\ \dot{x}_2(t) \\ \dot{x}_3(t) \end{bmatrix} = \begin{bmatrix} -\tfrac{1}{3} & \tfrac{1}{3} & 1 \\ \tfrac{1}{3} & -\tfrac{1}{3} & 0 \\ -\tfrac{1}{2} & 0 & -\tfrac{1}{2} \end{bmatrix} \begin{bmatrix} x_1(t) \\ x_2(t) \\ x_3(t) \end{bmatrix} + \begin{bmatrix} 0 \\ 0 \\ \tfrac{1}{2} \end{bmatrix} u(t) \qquad (4.23)$$

The output equation is given by

$$y(t) = x_1(t) = [1 \quad 0 \quad 0] \begin{bmatrix} x_1(t) \\ x_2(t) \\ x_3(t) \end{bmatrix} \tag{4.24}$$

Note that the states of a system are not unique. Any linear combinations of a set of states can be used as a new set of states for a given system. This means that if the N-dimensional vector x is a state of a dynamic system with state and output equations given by

$$\frac{dx(t)}{dt} = Ax(t) + Bu(t)$$
$$y(t) = Cx(t) + Du(t) \tag{4.25}$$

then the N-dimensional vector z is defined as

$$z = Px \tag{4.26}$$

where P is a nonsingular $N \times N$ transformation, which is another state vector for the same system. The new state and output equations are given by

$$\frac{dz(t)}{dt} = PAP^{-1}z(t) + PBu(t)$$
$$y(t) = CPz(t) + Du(t) \tag{4.27}$$

4.3.4   STATE-SPACE EQUATIONS OF MECHANICAL SYSTEMS

The state-space modeling of mechanical systems can be derived from the first principles similar to the electrical circuits. The main equation in this case is Newton's law of motion.

Example 4.3

As the first example, consider the ideal spring–mass system shown in Figure 4.4.
Applying Newton's law, $F = ma$, where m is the mass, a is the acceleration, and F is the overall forcing function with the assumption of linear spring, we have

$$m\frac{d^2x(t)}{dt^2} = u(t) - kx(t) \tag{4.28}$$

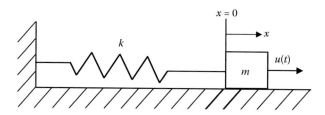

FIGURE 4.4   Spring–mass system.

Define the two states of the system to be the mass displacement $x(t)$ and its velocity $\dot{x}(t)$. Then, we have

$$x_1(t) = x(t)$$
$$x_2(t) = \dot{x}(t)$$

(4.29)

Differentiating both sides of Equations 4.29 results in

$$\dot{x}_1(t) = \frac{dx(t)}{dt} = x_2(t)$$
$$\dot{x}_2(t) = \frac{d^2x(t)}{dt^2} = -\frac{k}{m}x(t) + \frac{1}{m}u(t) = -\frac{k}{m}x_1(t) + \frac{1}{m}u(t)$$

(4.30)

The above equations can be written in matrix form as

$$\begin{bmatrix} \dot{x}_1(t) \\ \dot{x}_2(t) \end{bmatrix} = \begin{bmatrix} 0 & 1 \\ -\frac{k}{m} & 0 \end{bmatrix} \begin{bmatrix} x_1(t) \\ x_2(t) \end{bmatrix} + \begin{bmatrix} 0 \\ \frac{1}{m} \end{bmatrix} u(t)$$

(4.31)

Example 4.4

As another example, consider the same ideal spring–mass system with friction that is assumed to be proportional to velocity. Applying Newton's law of motion to the system results in

$$m\frac{d^2x(t)}{dt^2} = u(t) - kx(t) - b\frac{dx(t)}{dt}$$

(4.32)

Using the same states, the state equations become

$$\begin{bmatrix} \dot{x}_1(t) \\ \dot{x}_2(t) \end{bmatrix} = \begin{bmatrix} 0 & 1 \\ -\frac{k}{m} & -\frac{b}{m} \end{bmatrix} \begin{bmatrix} x_1(t) \\ x_2(t) \end{bmatrix} + \begin{bmatrix} 0 \\ \frac{1}{m} \end{bmatrix} u(t)$$

(4.33)

Example 4.5 is the classical nonlinear control problem known as inverted pendulum.

Example 4.5

Consider a cart with an inverted pendulum as shown in Figure 4.5. It is assumed that the cart and the pendulum move in only one plane and the friction is negligible. The goal is to maintain the pendulum at the vertical position.

The horizontal (F_x) and vertical (F_y) forces applied by the cart on the pendulum are given by

$$F_x = m\frac{d^2}{dt^2}(x + l\sin\theta) = m\frac{d}{dt}(\dot{x} + l\dot{\theta}\cos\theta) = m(\ddot{x} + l\ddot{\theta}\cos\theta - l\dot{\theta}^2\sin\theta)$$

(4.34)

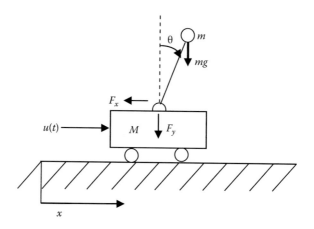

FIGURE 4.5   Inverted pendulum.

$$F_y = mg - m\frac{d^2}{dt^2}(l\cos\theta) = mg + m\frac{d}{dt}(l\dot\theta\sin\theta) = mg + ml(\ddot\theta\sin\theta + \dot\theta^2\cos\theta)$$

(4.35)

Appling Newton's law of motion to the cart results in

$$M\frac{d^2x}{dt^2} = u(t) - F_x = u(t) - m(\ddot x + l\ddot\theta\cos\theta - l\dot\theta^2\sin\theta)$$ (4.36)

Equation 4.36 can be written as

$$(M + m)\ddot x + ml\ddot\theta\cos\theta - ml\dot\theta^2\sin\theta = u(t)$$ (4.37)

Newton's law applied to the motion of the pendulum yields

$$ml^2\ddot\theta = mgl\sin\theta + F_yl\sin\theta - F_xl\cos\theta$$ (4.38)

Substituting F_x and F_y into Equation 4.38 results in

$$ml^2\ddot\theta = mgl\sin\theta + m(g + l\ddot\theta\sin\theta + l\dot\theta^2\cos\theta)l\sin\theta$$
$$- l\cos\theta\, m(\ddot x + l\ddot\theta\cos\theta - l\dot\theta^2\sin\theta)$$ (4.39)

This equation can be simplified as

$$l(1 + \cos 2\theta)\ddot\theta - l(\sin 2\theta)\dot\theta^2 + \cos\theta\ddot x = 2g\sin\theta$$ (4.40)

Equations 4.38 and 4.40 are used to solve for $\ddot x$ and $\ddot\theta$. The results are

$$\ddot x = \frac{-mg\sin 2\theta}{(2M + m)\cos\theta} + \frac{1 + \cos 2\theta}{(2M + m)\cos\theta}u(t)$$ (4.41)

and

$$\ddot{\theta} = (\sin\theta)\dot{\theta}^2 + \frac{2g(M+m)\sin\theta}{(2M+m)l\cos\theta} - \frac{1}{2M+m}u(t) \tag{4.42}$$

Define the state variables as $x_1 = x$, $x_2 = \dot{x}$, $x_3 = \theta$, and $x_4 = \dot{\theta}$, then we have

$$\begin{aligned}
\dot{x}_1 &= x_2 \\
\dot{x}_2 &= \frac{-mg\sin 2x_3}{(2M+m)\cos x_3} + \frac{1+\cos 2x_3}{(2M+m)\cos x_3}u(t) \\
\dot{x}_3 &= x_4 \\
\dot{x}_4 &= x_4^2\sin x_3 + \frac{2g(M+m)\sin x_3}{(2M+m)l\cos x_3} - \frac{1}{2M+m}u(t)
\end{aligned} \tag{4.43}$$

The above state equations are nonlinear. To obtain linear equations, we linearize them about $\theta = 0$. The results of linearization are

$$\begin{aligned}
\dot{x}_1 &= x_2 \\
\dot{x}_2 &= \frac{-2mg}{2M+m}x_3 + \frac{2}{2M+m}u(t) \\
\dot{x}_3 &= x_4 \\
\dot{x}_4 &= \frac{2g(M+m)}{(2M+m)l}x_3 - \frac{1}{2M+m}u(t)
\end{aligned} \tag{4.44}$$

The linear equation given by Equation 4.44 can be written in matrix form as

$$\begin{bmatrix} \dot{x}_1(t) \\ \dot{x}_2(t) \\ \dot{x}_3(t) \\ \dot{x}_4(t) \end{bmatrix} = \begin{bmatrix} 0 & 1 & 0 & 0 \\ 0 & 0 & -\dfrac{2mg}{2M+m} & 0 \\ 0 & 0 & 0 & 1 \\ 0 & 0 & \dfrac{2g(M+m)}{2M+m} & 0 \end{bmatrix} \begin{bmatrix} x_1(t) \\ x_2(t) \\ x_3(t) \\ x_4(t) \end{bmatrix} + \begin{bmatrix} 0 \\ \dfrac{2}{2M+m} \\ 0 \\ -\dfrac{1}{2M+m} \end{bmatrix} u(t) \tag{4.45}$$

4.4 STATE-SPACE REPRESENTATION OF GENERAL CONTINUOUS LTI SYSTEMS

Consider a SISO system described by the Nth order constant coefficients differential equation with input $u(t)$ and output $y(t)$:

$$\frac{d^N y(t)}{dt^N} + a_1\frac{d^{N-1}y(t)}{dt^{N-1}} + \cdots + a_N y(t) = b_0\frac{d^N u(t)}{dt^N} + b_1\frac{d^{N-1}u(t)}{dt^{N-1}} + \cdots + b_N u(t) \tag{4.46}$$

The state-space representation of the system defined by differential equation given in Equation 4.46 is not unique. We define two useful canonical forms that are suitable for state feedback and state estimator design. The first one is controllable canonical form and the second one is observable canonical form.

4.4.1 Controllable Canonical Form

The controllable canonical form is given by

$$
\begin{bmatrix} \dot{x}_1(t) \\ \dot{x}_2(t) \\ \vdots \\ \dot{x}_{N-1}(t) \\ \dot{x}_N(t) \end{bmatrix} = \begin{bmatrix} 0 & 1 & 0 & \cdots & 0 \\ 0 & 0 & 1 & \cdots & 0 \\ \vdots & \vdots & \vdots & \vdots & \vdots \\ 0 & 0 & 0 & \cdots & 1 \\ -a_N & -a_{N-1} & -a_{N-2} & \cdots & -a_1 \end{bmatrix} \begin{bmatrix} x_1(t) \\ x_2(t) \\ \vdots \\ x_{N-1}(t) \\ x_N(t) \end{bmatrix} + \begin{bmatrix} 0 \\ 0 \\ \vdots \\ 0 \\ 1 \end{bmatrix} u(t) \quad (4.47)
$$

with the output equation

$$
y(t) = [b_N - a_N b_0 \quad b_{N-1} - a_{N-1} b_0 \quad b_{N-2} - a_{N-2} b_0 \quad \cdots \quad b_1 - a_1 b_0] \begin{bmatrix} x_1(t) \\ x_2(t) \\ \vdots \\ x_{N-1}(t) \\ x_N(t) \end{bmatrix} + b_0 u(t)
$$

$$(4.48)$$

4.4.2 Observable Canonical Form

The observable canonical form is given by

$$
\begin{bmatrix} \dot{x}_1(t) \\ \dot{x}_2(t) \\ \vdots \\ \dot{x}_{N-1}(t) \\ \dot{x}_N(t) \end{bmatrix} = \begin{bmatrix} 0 & 0 & \cdots & 0 & -a_N \\ 1 & 0 & \cdots & 0 & -a_{N-1} \\ 0 & 1 & \cdots & 0 & a_{N-3} \\ \vdots & \vdots & & \vdots & \vdots \\ 0 & 0 & \cdots & 1 & -a_1 \end{bmatrix} \begin{bmatrix} x_1(t) \\ x_2(t) \\ \vdots \\ x_{N-1}(t) \\ x_N(t) \end{bmatrix} + \begin{bmatrix} b_N - a_N b_0 \\ b_{N-1} - a_{N-1} b_0 \\ b_{N-2} - a_{N-2} b_0 \\ \vdots \\ b_1 - a_1 b_0 \end{bmatrix} u(t) \quad (4.49)
$$

with the output equation

$$
y(t) = [0 \quad 0 \quad 0 \quad \cdots \quad 1] \begin{bmatrix} x_1(t) \\ x_2(t) \\ \vdots \\ x_{N-1}(t) \\ x_N(t) \end{bmatrix} + b_0 u(t) \quad (4.50)
$$

Example 4.6

Consider the following SISO system:

$$\frac{d^3y(t)}{dt^3} + 5\frac{d^2y(t)}{dt^2} + 2\frac{dy(t)}{dt} + 8y = 9\frac{d^2u(t)}{dt^2} + 7\frac{du(t)}{dt} + 13u(t) \qquad (4.51)$$

Obtain a state-space representation of this system in (a) controllable and (b) observable canonical forms.

SOLUTION

(a) Controllable canonical form:

$$\begin{bmatrix} \dot{x}_1(t) \\ \dot{x}_2(t) \\ \dot{x}_3(t) \end{bmatrix} = \begin{bmatrix} 0 & 1 & 0 \\ 0 & 0 & 1 \\ -8 & -2 & -5 \end{bmatrix} \begin{bmatrix} x_1(t) \\ x_2(t) \\ x_3(t) \end{bmatrix} + \begin{bmatrix} 0 \\ 0 \\ 1 \end{bmatrix} u(t) \text{ and } y(t) = \begin{bmatrix} 13 & 7 & 9 \end{bmatrix} \begin{bmatrix} x_1(t) \\ x_2(t) \\ x_3(t) \end{bmatrix} \qquad (4.52)$$

(b) Observable canonical form:

$$\begin{bmatrix} \dot{x}_1(t) \\ \dot{x}_2(t) \\ \dot{x}_3(t) \end{bmatrix} = \begin{bmatrix} 0 & 0 & -8 \\ 1 & 0 & -2 \\ 0 & 1 & -5 \end{bmatrix} \begin{bmatrix} x_1(t) \\ x_2(t) \\ x_3(t) \end{bmatrix} + \begin{bmatrix} 13 \\ 7 \\ 9 \end{bmatrix} u(t) \text{ and } y(t) = \begin{bmatrix} 0 & 0 & 1 \end{bmatrix} \begin{bmatrix} x_1(t) \\ x_2(t) \\ x_3(t) \end{bmatrix}$$

$$(4.53)$$

4.4.3 TRANSFER FUNCTION (MATRIX) FROM STATE-SPACE EQUATIONS

The transfer function (matrix) can be derived from state and output equations. For a SISO system, the transfer function is 1×1 and for a MIMO system with M inputs and P outputs, the transfer matrix is $P \times M$. Let the state equations of a MIMO system be

$$\frac{dx(t)}{dt} = Ax(t) + Bu(t) \qquad (4.54)$$

Taking Laplace transform from both sides of Equation 4.54 yields

$$sX(s) = AX(s) + BU(s) \qquad (4.55)$$

Hence,

$$X(s) = (sI - A)^{-1}BU(s) \qquad (4.56)$$

The output equation in Laplace transform domain is

$$Y(s) = CX(s) + DU(s) = [C(sI - A)^{-1}B + D]U(s) \qquad (4.57)$$

Therefore, the transfer matrix is

$$H(s) = C(sI - A)^{-1}B + D \tag{4.58}$$

Example 4.7

Find the transfer function corresponding to the state equations given by

$$\begin{bmatrix} \dot{x}_1(t) \\ \dot{x}_2(t) \end{bmatrix} = \begin{bmatrix} 0 & 1 \\ -5 & -6 \end{bmatrix} \begin{bmatrix} x_1(t) \\ x_2(t) \end{bmatrix} + \begin{bmatrix} 2 \\ 3 \end{bmatrix} u(t) \quad \text{and} \quad y(t) = \begin{bmatrix} 4 & 5 \end{bmatrix} \begin{bmatrix} x_1(t) \\ x_2(t) \end{bmatrix} \tag{4.59}$$

SOLUTION
The transfer function is

$$H(s) = C(sI - A)^{-1}B + D = \begin{bmatrix} 41 & 5 \end{bmatrix} \begin{bmatrix} s & -1 \\ 5 & s+6 \end{bmatrix}^{-1} \begin{bmatrix} 2 \\ 3 \end{bmatrix} \tag{4.60}$$

Hence,

$$
\begin{aligned}
H(s) &= \frac{1}{s^2 + 6s + 5} \begin{bmatrix} 4 & 5 \end{bmatrix} \begin{bmatrix} s+6 & 1 \\ -5 & s \end{bmatrix} \begin{bmatrix} 2 \\ 3 \end{bmatrix} \\
&= \frac{1}{s^2 + 6s + 5} \begin{bmatrix} 4 & 5 \end{bmatrix} \begin{bmatrix} 2s + 15 \\ 3s - 10 \end{bmatrix} = \frac{23s + 10}{s^2 + 6s + 5}
\end{aligned}
\tag{4.61}
$$

4.5   SOLUTION OF LTI CONTINUOUS-TIME STATE EQUATIONS

Consider an LTI system described by the state-space equations:

$$\dot{x}(t) = Ax(t) + Bu(t) \tag{4.62}$$

with the output equation,

$$y(t) = Cx(t) + Du(t) \tag{4.63}$$

To obtain the output of this system for a given input and an initial state, we first solve the state equation (Equation 4.62) and then the solution is substituted into algebraic equation (Equation 4.63) in order to find the output $y(t)$. The solution to the state equation has two parts, homogeneous solution and particular solution [4]. We first consider the homogeneous solution.

4.5.1   SOLUTION OF HOMOGENEOUS STATE EQUATION

The homogeneous solution is the solution of state equations to an arbitrary initial condition with zero input. The homogeneous state equation is given by

$$\dot{x}(t) = Ax(t) \tag{4.64}$$

Assuming that the initial state is $x(0)$, then

$$x(t) = e^{At}x(0) \tag{4.65}$$

where e^{At} is the matrix exponential function. To verify that this is the solution to the homogeneous equation, we need to show that it satisfies the initial condition as well as the differential equation. The initial condition is satisfied since

$$x(0) = e^{A0}x(0) = Ix(0) = x(0) \tag{4.66}$$

Differentiating both sides of Equation 4.64, we have

$$\dot{x}(t) = \frac{de^{At}x(0)}{dt} = Ae^{At}x(0) = Ax(t) \tag{4.67}$$

Therefore, the homogeneous part of the solution is $x(t) = e^{At}x(0)$. The matrix exponential e^{At} is called state-transition matrix and is denoted by $\varphi(t)$. The state-transition matrix $\varphi(t)$ is an $N \times N$ matrix and is the solution to the homogeneous equation

$$\dot{\varphi}(t) = A\varphi(t) \tag{4.68}$$

with initial condition $\varphi(0) = I$. The state-transition matrix is given by

$$\varphi(t) = e^{At} \tag{4.69}$$

The state-transition matrix satisfies the following properties. These properties are stated without proof:

(a) $\varphi(0) = I$ $\qquad\qquad$ (4.70)

(b) $\varphi(-t) = e^{-At} = (e^{At})^{-1} = \varphi^{-1}(t)$ $\qquad\qquad$ (4.71)

(c) $\varphi(t_1 + t_2) = \varphi(t_1)\varphi(t_2) = \varphi(t_2)\varphi(t_1)$ $\qquad\qquad$ (4.72)

(d) $\varphi(t_3 - t_2)\varphi(t_2 - t_1) = \varphi(t_3 - t_1)$ $\qquad\qquad$ (4.73)

The proof is left as an exercise (see Problem 4.5).

4.5.2 COMPUTING STATE-TRANSITION MATRIX

There are several methods for computing the state-transition matrix. Here, we review two methods, which are very useful.

(a) **Modal Matrix Approach**: Let $\{\lambda_i, x_i\}_{i=1}^{N}$ be the set of eigenvalues and their corresponding eigenvectors of matrix A. Under the assumption of independent eigenvectors, the state-transition matrix is given by

$$\varphi(t) = M\Lambda(t)M^{-1} \qquad (4.74)$$

where

$M = \begin{bmatrix} x_1 & x_2 & \cdots & x_N \end{bmatrix}$ is the $N \times N$ matrix of eigenvectors known as modal matrix

$\Lambda(t)$ is an $N \times N$ diagonal matrix given by

$$\Lambda(t) = \begin{bmatrix} e^{\lambda_1 t} & 0 & \cdots & 0 \\ 0 & e^{\lambda_2 t} & \cdots & 0 \\ \vdots & \vdots & \ddots & \vdots \\ 0 & 0 & \cdots & e^{\lambda_n t} \end{bmatrix} \qquad (4.75)$$

The second approach which is more general is based on Laplace Transform.

(b) **Laplace Transform Technique**: In this method, $\varphi(t)$ is computed using the Laplace transform. Taking the Laplace transform from both sides of Equation 4.62, we have

$$s\Phi(s) - \varphi(0) = A\Phi(s) \qquad (4.76)$$

Since $\varphi(0) = I$, then

$$\Phi(s) = (sI - A)^{-1} \qquad (4.77)$$

Therefore,

$$\varphi(t) = L^{-1}[(sI - A)^{-1}] \qquad (4.78)$$

where L^{-1} stands for inverse the Laplace transform.

Example 4.8

Find the state-transition matrix of the following dynamic system using the two methods described above:

$$\dot{x}(t) = \begin{bmatrix} -1 & 1 \\ -1 & -1 \end{bmatrix} x(t)$$

SOLUTION

(a) **Modal Matrix Technique**: The eigenvalues of matrix A are

$$|\lambda I - A| = \begin{vmatrix} \lambda + 1 & 1 \\ 1 & \lambda + 1 \end{vmatrix} = (\lambda + 1)^2 + 1 = 0 \;\rightarrow\; \lambda_1 = -1 + j, \lambda_2 = -1 - j$$

The corresponding eigenvectors are

$$Ax_1 = \lambda_1 x_1, \quad \text{therefore } x_1 = \begin{bmatrix} 1 & j \end{bmatrix}^T$$

$$Ax_2 = \lambda_2 x_2, \quad \text{therefore } x_2 = \begin{bmatrix} 1 & -j \end{bmatrix}^T$$

Hence,

$$\varphi(t) = M\Lambda(t)M^{-1} = \begin{bmatrix} 1 & 1 \\ j & -j \end{bmatrix} \begin{bmatrix} e^{-(1-j)t} & 0 \\ 0 & e^{-(1+j)t} \end{bmatrix} \begin{bmatrix} 1 & 1 \\ j & -j \end{bmatrix}^{-1}$$

$$= \begin{bmatrix} e^{-(1-j)t} & e^{-(1+j)t} \\ je^{-(1-j)t} & -je^{-(1+j)t} \end{bmatrix} \begin{bmatrix} 0.5 & -0.5j \\ 0.5 & 0.5j \end{bmatrix}$$

$$= \begin{bmatrix} 0.5e^{-(1-j)t} + 0.5e^{-(1+j)t} & -0.5je^{-(1-j)t} + 0.5je^{-(1+j)t} \\ 0.5je^{-(1-j)t} - 0.5je^{-(1+j)t} & 0.5e^{-(1-j)t} + 0.5e^{-(1+j)t} \end{bmatrix}$$

$$= \begin{bmatrix} e^{-t}\cos t & e^{-t}\sin t \\ -e^{-t}\sin t & e^{-t}\cos t \end{bmatrix}$$

(b) **Use of Laplace Transform**: Using Laplace transform, we have

$$\Phi(s) = (sI - A)^{-1} = \begin{bmatrix} s+1 & -1 \\ 1 & s+1 \end{bmatrix}^{-1} = \frac{1}{(s+1)^2 + 1} \begin{bmatrix} s+1 & 1 \\ -1 & s+1 \end{bmatrix}$$

or

$$\Phi(s) = \begin{bmatrix} \dfrac{s+1}{(s+1)^2 + 1} & \dfrac{1}{(s+1)^2 + 1} \\ \dfrac{-1}{(s+1)^2 + 1} & \dfrac{s+1}{(s+1)^2 + 1} \end{bmatrix}$$

Taking inverse Laplace transform, we have

$$\varphi(t) = \begin{bmatrix} e^{-t}\cos t & e^{-t}\sin t \\ -e^{-t}\sin t & e^{-t}\cos t \end{bmatrix}$$

4.5.3 COMPLETE SOLUTION OF STATE EQUATION

The complete solution of the state equation x (Equation 4.62) is the sum of the homogeneous and particular solutions and is given by

$$x(t) = \varphi(t)x(0) + \int_0^t \varphi(t - \tau)Bu(\tau)d\tau \tag{4.79}$$

To verify that this is the total solution, we need to show that it satisfies the state equation and the initial condition. The initial condition is satisfied since

$$x(0) = \varphi(0)x(0) + \int_0^0 \varphi(0-\tau)Bu(\tau)d\tau = I \times x(0) + 0 = x(0) \tag{4.80}$$

To show that it satisfies the state equation, differentiate both sides of Equation 4.79:

$$\dot{x}(t) = \dot{\varphi}(t)x(0) + \frac{d}{dt}\int_0^t \varphi(t-\tau)Bu(\tau)d\tau$$

$$= A\varphi(t)x(0) + \varphi(t-t)Bu(t) + \int_0^t \dot{\varphi}(t-\tau)Bu(\tau)d\tau$$

$$= A\varphi(t)x(0) + \varphi(0)Bu(t) + \int_0^t A\varphi(t-\tau)Bu(\tau)d\tau$$

$$= A\varphi(t)x(0) + Bu(t) + A\int_0^t \varphi(t-\tau)Bu(\tau)d\tau$$

$$= A\left[\varphi(t)x(0) + \int_0^t \varphi(t-\tau)Bu(\tau)d\tau\right] + Bu(t)$$

$$= Ax(t) + Bu(t) \tag{4.81}$$

The first term in the total solution, that is $\varphi(t)x(0)$, is the homogeneous solution or the zero-input response. The second term is the particular solution, which is also referred to as zero-state response.

Example 4.9

Find the output of the dynamic system described by the state equation:

$$\dot{x}(t) = \begin{bmatrix} -1 & 1 \\ -1 & -1 \end{bmatrix} x(t) + \begin{bmatrix} 1 \\ 0 \end{bmatrix} u(t)$$

$$y(t) = [\,1 \quad -1\,]x(t)$$

With the input and initial conditions given by,

$$u(t) = \begin{cases} 1 & t > 0 \\ 0 & t < 0 \end{cases} \quad \text{and} \quad x(0) = \begin{bmatrix} 1 \\ 0 \end{bmatrix}$$

SOLUTION

The homogeneous solution or the zero-state response is given by

$$x_h(t) = \varphi(t)x(0) = \begin{bmatrix} e^{-t}\cos t & e^{-t}\sin t \\ -e^{-t}\sin t & e^{-t}\cos t \end{bmatrix}\begin{bmatrix} 1 \\ 0 \end{bmatrix} = \begin{bmatrix} e^{-t}\cos t \\ -e^{-t}\sin t \end{bmatrix}$$

The zero-state response or the particular solution is

$$x_p(t) = \int_0^t \varphi(t-\tau)Bu(\tau)d\tau = \int_0^t \begin{bmatrix} e^{-t+\tau}\cos(t-\tau) & e^{-t+\tau}\sin(t-\tau) \\ -e^{-t+\tau}\sin(t-\tau) & e^{-t+\tau}\cos(t-\tau) \end{bmatrix}\begin{bmatrix} 1 \\ 0 \end{bmatrix}d\tau$$

$$= \int_0^t \begin{bmatrix} e^{-t+\tau}\cos(t-\tau) \\ -e^{-t+\tau}\sin(t-\tau) \end{bmatrix}d\tau$$

$$= \begin{bmatrix} e^{-t}\cos t\int_0^t e^{\tau}\cos\tau\,d\tau + e^{-t}\sin t\int_0^t e^{\tau}\sin\tau\,d\tau \\ -e^{-t}\sin t\int_0^t e^{\tau}\cos\tau\,d\tau + e^{-t}\cos t\int_0^t e^{\tau}\sin\tau\,d\tau \end{bmatrix}$$

$$= \begin{bmatrix} e^{-t}\cos t[0.5e^{\tau}\cos\tau + 0.5e^{\tau}\sin\tau]|_0^t + e^{-t}\sin t[-0.5e^{\tau}\cos\tau + 0.5e^{\tau}\sin\tau]|_0^t \\ -e^{-t}\sin t[0.5e^{\tau}\cos\tau + 0.5e^{\tau}\sin\tau]|_0^t + e^{-t}\cos t[-0.5e^{\tau}\cos\tau + 0.5e^{\tau}\sin\tau]|_0^t \end{bmatrix}$$

$$= \begin{bmatrix} e^{-t}\cos t[0.5e^{t}\cos t + 0.5e^{t}\sin t - 0.5] + e^{-t}\sin t[-0.5e^{t}\cos t + 0.5e^{t}\sin t + 0.5] \\ -e^{-t}\sin t[0.5e^{t}\cos t + 0.5e^{t}\sin t - 0.5] + e^{-t}\cos t[-0.5e^{t}\cos t + 0.5e^{t}\sin t + 0.5] \end{bmatrix}$$

$$= \begin{bmatrix} -0.5e^{-t}\cos t + 0.5e^{-t}\sin t + 0.5 \\ 0.5e^{-t}\cos t - 0.5e^{-t}\sin t - 0.5 \end{bmatrix}$$

Therefore,

$$x(t) = x_h(t) + x_p(t) = \begin{bmatrix} e^{-t}\cos t \\ -e^{-t}\sin t \end{bmatrix} + \begin{bmatrix} -0.5e^{-t}\cos t + 0.5e^{-t}\sin t + 0.5 \\ 0.5e^{-t}\cos t - 0.5e^{-t}\sin t - 0.5 \end{bmatrix}$$

$$= \begin{bmatrix} 0.5e^{-t}\cos t + 0.5e^{-t}\sin t + 0.5 \\ 0.5e^{-t}\cos t - 1.5e^{-t}\sin t - 0.5 \end{bmatrix}$$

The output $y(t)$ is

$$y(t) = \begin{bmatrix} 1 & -1 \end{bmatrix}x(t) = \begin{bmatrix} 1 & -1 \end{bmatrix}\begin{bmatrix} 0.5e^{-t}\cos t + 0.5e^{-t}\sin t + 0.5 \\ 0.5e^{-t}\cos t - 1.5e^{-t}\sin t - 0.5 \end{bmatrix}$$

$$= 2e^{-t}\sin t + 1$$

4.6   STATE-SPACE REPRESENTATION OF DISCRETE-TIME SYSTEMS

4.6.1   DEFINITION OF STATE

The state of a discrete-time dynamic system is the minimum number of variables called state variables such that the knowledge of these variables at time $k = k_0$ together with input for $k \geq k_0$ uniquely determines the behavior of the system for $k \geq k_0$. If N variables are needed, then these N variables are considered as components of an N-dimensional vector x called state vector. The N-dimensional space whose coordinates are the states of the system is called state space. The state of a system at time n is a point in the state space [5].

4.6.2 State Equations

The general expressions for state-space equations of a MIMO dynamic system are given by

$$x(k+1) = f(x(k), u(k), k) \tag{4.82}$$
$$y(k) = h(x(k), u(k), k) \tag{4.83}$$

where
$x(k) = \begin{bmatrix} x_1(k) & x_2(k) & \cdots & x_N(k) \end{bmatrix}^T$ is the $N \times 1$ state vector
$u(k)$ is the $M \times 1$ input vector
$y(n)$ is the $P \times 1$ output vector

The functions f and h are nonlinear and are defined as

$$f(x(k), u(k), k) = \begin{bmatrix} f_1(x_1(k), x_2(k), \ldots, x_N(k), u_1(k), u_2(k), \ldots, u_M(k), k) \\ f_2(x_1(k), x_2(k), \ldots, x_N(k), u_1(k), u_2(k), \ldots, u_M(k), k) \\ \vdots \\ f_N(x_1(k), x_2(k), \ldots, x_N(k), u_1(k), u_2(k), \ldots, u_M(k), k) \end{bmatrix}$$

$$\tag{4.84}$$

$$h(x(n), u(n), n) = \begin{bmatrix} h_1(x_1(k), x_2(k), \ldots, x_N(k), u_1(k), u_2(k), \ldots, u_M(k), k) \\ h_2(x_1(k), x_2(k), \ldots, x_N(k), u_1(k), u_2(k), \ldots, u_M(k), k) \\ \vdots \\ h_P(x_1(k), x_2(k), \ldots, x_N(k), u_1(k), u_2(k), \ldots, u_M(k), k) \end{bmatrix}$$

$$\tag{4.85}$$

If the system is linear, then the state equations can be written as

$$x(k+1) = A(k)x(k) + B(k)u(k) \tag{4.86}$$
$$y(k) = C(k)x(k) + D(k)u(k) \tag{4.87}$$

If the system is LTI, then matrices A, B, C, and D are constant matrices and the state equations become

$$x(k+1) = Ax(k) + Bu(k) \tag{4.88}$$
$$y(k) = Cx(k) + Du(k) \tag{4.89}$$

where A, B, C, and D are $N \times N$, $N \times M$, $P \times N$, and $P \times M$ constant matrices, respectively. The block diagram of an LTI control system in state-space form is shown in Figure 4.6.

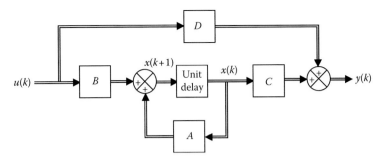

FIGURE 4.6   Block diagram of a discrete LTI system in state space.

4.7   STATE-SPACE REPRESENTATION OF DISCRETE-TIME LTI SYSTEMS

Consider the LTI discrete-time dynamic system described by the difference equation:

$$y(k) = -\sum_{i=1}^{N} a_i y(k-i) + \sum_{i=0}^{M} b_i u(k-i) \tag{4.90}$$

The state-space realizations of the above system in two most important standard forms, i.e., the controllable and observable canonical forms, are discussed below.

4.7.1   CONTROLLABLE CANONICAL FORM

The controllable canonical form is given by

$$
\begin{bmatrix} x_1(k+1) \\ x_2(k+1) \\ \vdots \\ x_{N-1}(k+1) \\ x_N(k+1) \end{bmatrix}
=
\begin{bmatrix} 0 & 1 & 0 & \cdots & 0 \\ 0 & 0 & 1 & \cdots & 0 \\ \vdots & \vdots & \vdots & \vdots & \vdots \\ 0 & 0 & 0 & \cdots & 1 \\ -a_N & -a_{N-1} & -a_{N-2} & \cdots & -a_1 \end{bmatrix}
\begin{bmatrix} x_1(k) \\ x_2(k) \\ \vdots \\ x_{N-1}(k) \\ x_N(k) \end{bmatrix}
+
\begin{bmatrix} 0 \\ 0 \\ \vdots \\ 0 \\ 1 \end{bmatrix} u(k) \tag{4.91}
$$

with the output equation

$$
y(k) = [b_N - a_N b_0 \quad b_{N-1} - a_{N-1} b_0 \quad b_{N-2} - a_{N-2} b_0 \quad \cdots \quad b_1 - a_1 b_0]
\begin{bmatrix} x_1(k) \\ x_2(k) \\ \vdots \\ x_{N-1}(k) \\ x_N(k) \end{bmatrix}
+ b_0 u(k)
$$

$$\tag{4.92}$$

4.7.2 Observable Canonical Form

The observable canonical form is given by

$$
\begin{bmatrix} x_1(k+1) \\ x_2(k+1) \\ \vdots \\ x_{N-1}(k+1) \\ x_N(k+1) \end{bmatrix} = \begin{bmatrix} 0 & 0 & \cdots & 0 & -a_N \\ 1 & 0 & \cdots & 0 & -a_{N-1} \\ 0 & 1 & \cdots & 0 & a_{N-3} \\ \vdots & \vdots & \vdots & \vdots & \vdots \\ 0 & 0 & \cdots & 1 & -a_1 \end{bmatrix} \begin{bmatrix} x_1(k) \\ x_2(k) \\ \vdots \\ x_{N-1}(k) \\ x_N(k) \end{bmatrix} + \begin{bmatrix} b_N - a_N b_0 \\ b_{N-1} - a_{N-1} b_0 \\ b_{N-2} - a_{N-2} b_0 \\ \vdots \\ b_1 - a_1 b_0 \end{bmatrix} u(k) \quad (4.93)
$$

with the output equation given as

$$
y(k) = \begin{bmatrix} 0 & 0 & 0 & \cdots & 1 \end{bmatrix} \begin{bmatrix} x_1(k) \\ x_2(k) \\ \vdots \\ x_{N-1}(k) \\ x_N(k) \end{bmatrix} + b_0 u(k) \quad (4.94)
$$

Example 4.10

Consider the following SISO discrete-time LTI system

$$
y(k+2) + 0.75y(k+1) + 0.25y(k) = u(k+2) + 3u(k+1) - 4u(k)
$$

Obtain a state-space representation of this system in (a) controllable and (b) observable canonical forms.

Solution

(a) Controllable canonical form: The state equations are

$$
\begin{bmatrix} x_1(k+1) \\ x_2(k+1) \end{bmatrix} = \begin{bmatrix} 0 & 1 \\ -0.25 & -0.75 \end{bmatrix} \begin{bmatrix} x_1(k) \\ x_2(k) \end{bmatrix} + \begin{bmatrix} 0 \\ 1 \end{bmatrix} u(k)
$$

and the output equation is

$$
y(k) = \begin{bmatrix} -4.25 & 2.25 \end{bmatrix} \begin{bmatrix} x_1(k) \\ x_2(k) \end{bmatrix} + u(k)
$$

(b) Observable canonical form: The state equations and the output equation are

$$
\begin{bmatrix} x_1(k+1) \\ x_2(k+1) \end{bmatrix} = \begin{bmatrix} 0 & -0.25 \\ 1 & -0.75 \end{bmatrix} \begin{bmatrix} x_1(k) \\ x_2(k) \end{bmatrix} + \begin{bmatrix} -4.25 \\ 2.25 \end{bmatrix} u(k) \quad \text{and}
$$

$$
y(k) = \begin{bmatrix} 0 & 1 \end{bmatrix} \begin{bmatrix} x_1(k) \\ x_2(k) \end{bmatrix} + u(k)
$$

4.7.3 Transfer Function (Matrix) from State-Space Equations

The transfer function (matrix) can be derived from state and output equations. For a SISO system, the transfer function is 1×1 and for a MIMO system with M inputs and P outputs, the transfer matrix is $P \times M$. Let the state equations of a discrete MIMO system be

$$x(k + 1) = Ax(k) + Bu(k) \tag{4.95}$$

Taking z-transform from both sides of Equation 4.95 yields

$$zX(z) = AX(z) + BU(z) \tag{4.96}$$

Hence,

$$X(z) = (zI - A)^{-1}BU(z) \tag{4.97}$$

The output equation in z-domain is

$$Y(z) = CX(z) + DU(z) = [C(zI - A)^{-1}B + D]U(z) \tag{4.98}$$

Therefore, the transfer function matrix is

$$H(z) = C(zI - A)^{-1}B + D \tag{4.99}$$

Example 4.11

Find the transfer function corresponding to the state equation given by

$$\begin{bmatrix} x_1(k + 1) \\ x_2(k + 1) \end{bmatrix} = \begin{bmatrix} 0 & 1 \\ -0.6 & -0.5 \end{bmatrix} \begin{bmatrix} x_1(k) \\ x_2(k) \end{bmatrix} + \begin{bmatrix} 2 \\ 3 \end{bmatrix} u(k)$$

$$y(k) = \begin{bmatrix} 1 & -2 \end{bmatrix} \begin{bmatrix} x_1(k) \\ x_2(k) \end{bmatrix}$$

Solution
The transfer function is

$$H(z) = C(zI - A)^{-1}B + D = \begin{bmatrix} 1 & -2 \end{bmatrix} \begin{bmatrix} z & -1 \\ 0.6 & z + 0.5 \end{bmatrix}^{-1} \begin{bmatrix} 2 \\ 3 \end{bmatrix}$$

or

$$H(z) = \frac{1}{z^2 + 0.5z + 0.6} \begin{bmatrix} 1 & -2 \end{bmatrix} \begin{bmatrix} z + 0.5 & 1 \\ -0.6 & z \end{bmatrix} \begin{bmatrix} 2 \\ 3 \end{bmatrix}$$

Therefore,

$$H(z) = \frac{-4z + 6.4}{z^2 + 0.5z + 0.6}$$

4.8 SOLUTION OF LTI DISCRETE-TIME STATE EQUATIONS

Consider a discrete LTI system described by the state-space equation

$$x(k + 1) = Ax(k) + Bu(k) \tag{4.100}$$

and the output equation

$$y(k) = Cx(k) + Du(k) \tag{4.101}$$

To obtain the output of this system for a given input and an initial state, we first solve the state equation (Equation 4.100) and then the solution is substituted into the algebraic equation (Equation 4.101) in order to find the output $y(k)$. The solution to state equations has two parts, a homogeneous solution and a particular solution. We first consider the homogeneous solution.

4.8.1 SOLUTION OF HOMOGENEOUS STATE EQUATION

The homogeneous solution is the solution of state equations to an arbitrary initial condition with zero input. The homogeneous state equation is given by

$$x(k + 1) = Ax(k) \tag{4.102}$$

Assuming that the initial state is $x(0)$, we have

$$
\begin{aligned}
x(1) &= Ax(0) \\
x(2) &= Ax(1) = A \times Ax(0) = A^2 x(0) \\
x(3) &= Ax(2) = A \times A^2 x(0) = A^3 x(0) \\
&\vdots \\
x(k) &= Ax(k - 1) = A \times A^{k-1} x(0) = A^k x(0)
\end{aligned}
\tag{4.103}
$$

Therefore, in matrix form, the solution to the homogeneous equation is

$$x(k) = A^k x(0) \tag{4.104}$$

Therefore, the homogeneous part of the solution is $x(k) = A^k x(0)$. The matrix exponential A^k is called state-transition matrix and is denoted by $\varphi(k)$. The state-transition matrix $\varphi(k)$ is an $N \times N$ matrix and is the solution to the homogeneous equation

$$\varphi(k+1) = A\varphi(k)$$
$$\varphi(0) = I$$
(4.105)

The state-transition matrix is given by

$$\varphi(k) = A^k$$
(4.106)

The state transition matrix satisfies the following properties. These properties are stated without proof.

(a) $\varphi(0) = I$ (4.107)

(b) $\varphi(-k) = A^{-k} = (A^k)^{-1} = \varphi^{-1}(k)$ (4.108)

(c) $\varphi(k_1 + k_2) = \varphi(k_1)\varphi(k_2) = \varphi(k_2)\varphi(k_1)$ (4.109)

(d) $\varphi(k_3 - k_2)\varphi(k_2 - k_1) = \varphi(k_3 - k_1)$ (4.110)

The proof is left as an exercise (see Problem 4.6).

4.8.2 COMPUTING STATE-TRANSITION MATRIX

Similar to the continuous case, the discrete state-transition matrix can be computed using modal matrix or z-transform. Here, we discuss these two methods.

(a) **Modal Matrix Approach**: Let $\{\lambda_i, V_i\}_{i=1}^{N}$ be the set of eigenvalues and their corresponding eigenvectors of matrix A. Then the state-transition matrix is computed using

$$\varphi(k) = M\Lambda(k)M^{-1}$$
(4.111)

where
$M = [V_1 \quad V_2 \quad \cdots \quad V_N]$ is the $N \times N$ matrix of eigenvectors known as modal matrix
$\Lambda(k)$ is an $N \times N$ diagonal matrix given by

$$\Lambda(k) = \begin{bmatrix} \lambda_1^k & 0 & \cdots & 0 \\ 0 & \lambda_2^k & \cdots & 0 \\ \vdots & \vdots & \ddots & \vdots \\ 0 & 0 & \cdots & \lambda_N^k \end{bmatrix}$$
(4.112)

(b) z-**Transform Approach**: In this method, $\varphi(k)$ is computed using the z-transform. Taking the z-transform on both sides of Equation 4.105, we have

$$z\Phi(z) - z\varphi(0) = A\Phi(z)$$
(4.113)

Since $\varphi(0) = I$, then

$$(zI - A)\Phi(z) = zI \tag{4.114}$$

and

$$\Phi(z) = z(zI - A)^{-1} \tag{4.115}$$

Therefore,

$$\Phi(k) = Z^{-1}[z(zI - A)^{-1}] \tag{4.116}$$

Example 4.12

Find the state-transition matrix of the following dynamic system using the two methods presented previously:

$$x(k + 1) = \begin{bmatrix} 1 & -0.25 \\ 1.5 & -0.25 \end{bmatrix} x(k)$$

SOLUTION

(a) **Modal Matrix Approach**: The characteristic polynomial of matrix A is

$$|\lambda I - A| = \begin{vmatrix} \lambda - 1 & 0.25 \\ -1.5 & \lambda + 0.25 \end{vmatrix} = \lambda^2 - 0.75\lambda + 0.125 = (\lambda - 0.5)(\lambda - 0.25) = 0$$

Therefore, the eigenvalues of A are

$$\lambda_1 = 0.5 \quad \lambda_2 = 0.25$$

The corresponding eigenvectors are

$$AV_1 = \lambda_1 V_1, \quad \text{therefore } V_1 = \begin{bmatrix} 1 & 2 \end{bmatrix}^T$$

$$AV_2 = \lambda_2 V_2, \quad \text{therefore } V_2 = \begin{bmatrix} 1 & 3 \end{bmatrix}^T$$

Using the modal matrix approach,

$$\varphi(k) = M\Lambda(k)M^{-1} = \begin{bmatrix} 1 & 1 \\ 2 & 3 \end{bmatrix} \begin{bmatrix} (0.5)^k & 0 \\ 0 & (0.25)^k \end{bmatrix} \begin{bmatrix} 1 & 1 \\ 2 & 3 \end{bmatrix}^{-1}$$

$$= \begin{bmatrix} (0.5)^k & (0.25)^k \\ 2(0.5)^k & 3(0.25)^k \end{bmatrix} \begin{bmatrix} 3 & -1 \\ -2 & 1 \end{bmatrix}$$

Therefore,

$$\varphi(k) = \begin{bmatrix} 3(0.5)^k - 2(0.25)^k & -(0.5)^k + (0.25)^k \\ 6(0.5)^k - 6(0.25)^k & -2(0.5)^k + 3(0.25)^k \end{bmatrix}$$

(b) **z-Transform Approach**: Using z-transform, we have

$$\Phi(z) = z(zI - A)^{-1} = z \begin{bmatrix} z - 1 & 0.25 \\ -1.5 & z + 0.25 \end{bmatrix}^{-1} = \frac{z}{z^2 - 0.75z + 0.125} \begin{bmatrix} z + 0.25 & -0.25 \\ 1.5 & z - 1 \end{bmatrix}$$

$$\Phi(z) = z \begin{bmatrix} \dfrac{z + 0.25}{(z - 0.5)(z - 0.25)} & -\dfrac{0.25}{(z - 0.5)(z - 0.25)} \\ \dfrac{1.5}{(z - 0.5)(z - 0.25)} & \dfrac{z - 1}{(z - 0.5)(z - 0.25)} \end{bmatrix}$$

Partial fraction expansion yields

$$\Phi(z) = z \begin{bmatrix} \dfrac{3}{z - 0.5} + \dfrac{-2}{z - 0.25} & \dfrac{-1}{z - 0.5} + \dfrac{1}{z - 0.25} \\ \dfrac{6}{z - 0.5} + \dfrac{-6}{z - 0.25} & \dfrac{-2}{z - 0.5} + \dfrac{3}{z - 0.25} \end{bmatrix} = \begin{bmatrix} \dfrac{3z}{z - 0.5} + \dfrac{-2z}{z - 0.25} & \dfrac{-z}{z - 0.5} + \dfrac{z}{z - 0.25} \\ \dfrac{6z}{z - 0.5} + \dfrac{-6z}{z - 0.25} & \dfrac{-2z}{z - 0.5} + \dfrac{3z}{z - 0.25} \end{bmatrix}$$

Hence,

$$\varphi(k) = Z^{-1}[\Phi(z)] = \begin{bmatrix} 3(0.5)^k - 2(0.25)^k & -(0.5)^k + (0.25)^k \\ 6(0.5)^k - 6(0.25)^k & -2(0.5)^k + 3(0.25)^k \end{bmatrix}$$

4.8.3 COMPLETE SOLUTION OF STATE EQUATIONS

Assuming that the initial state is $x(0)$, we have

$$x(1) = Ax(0) + Bu(0)$$
$$x(2) = Ax(1) + Bu(1) = A^2 x(0) + ABu(0) + Bu(1)$$
$$x(3) = Ax(2) + Bu(2) = A^3 x(0) + A^2 Bu(0) + ABu(1) + Bu(2)$$

$$\vdots$$

$$x(k) = Ax(k - 1) + Bu(k - 1) = A^k x(0) + A^{k-1} Bu(0) + A^{k-2} Bu(1) + \cdots + Bu(k - 1)$$

$$(4.117)$$

Therefore,

$$x(k) = A^k x(0) + \sum_{n=0}^{k-1} A^{k-1-n} Bu(n) \qquad (4.118)$$

In terms of state-transition matrix, the total solution is given by

$$x(k) = \varphi(k)x(0) + \sum_{n=0}^{k-1} \varphi(k-1-n)Bu(n) \qquad (4.119)$$

The total solution consists of two terms, the first term $\varphi(k)x(0)$ is referred to as the zero-input response and the second term $\sum_{n=0}^{k-1} \varphi(k-1-n)Bu(n)$ is called zero-state response. The zero-input response is the response of the system due to the initial conditions only and the zero-state response is the response due to the input with zero initial conditions. In the following example, we compute the zero-state and zero-input responses of a second-order system.

Example 4.13

Find the output of the following dynamic system if $u(k) = \begin{cases} 1 & k \geq 0 \\ 0 & k < 0 \end{cases}$ and
$x(0) = \begin{bmatrix} 1 \\ -1 \end{bmatrix}$:

$$x(k+1) = \begin{bmatrix} 1 & -0.25 \\ 1.5 & -0.25 \end{bmatrix} x(k) + \begin{bmatrix} -1 \\ 1 \end{bmatrix} u(k)$$

$$y(k) = \begin{bmatrix} 1 & 2 \end{bmatrix} x(k)$$

SOLUTION

First we find the zero-input response:

$$\varphi(k)x(0) = \begin{bmatrix} 3(0.5)^k - 2(0.25)^k & -(0.5)^k + (0.25)^k \\ 6(0.5)^k - 6(0.25)^k & -2(0.5)^k + 3(0.25)^k \end{bmatrix} \begin{bmatrix} 1 \\ -1 \end{bmatrix} = \begin{bmatrix} 4(0.5)^k - 3(0.25)^k \\ 8(0.5)^k - 9(0.25)^k \end{bmatrix}$$

Next, we find the zero-state response:

$$\sum_{n=0}^{k-1} \varphi(k-1-n)Bu(n) = \sum_{n=0}^{k-1} \begin{bmatrix} \varphi_{11}(k-1-n) & \varphi_{12}(k-1-n) \\ \varphi_{21}(k-1-n) & \varphi_{22}(k-1-n) \end{bmatrix} \begin{bmatrix} 1 \\ -1 \end{bmatrix}$$

$$= \sum_{n=0}^{k-1} \begin{bmatrix} \varphi_{11}(k-1-n) - \varphi_{12}(k-1-n) \\ \varphi_{21}(k-1-n) - \varphi_{22}(k-1-n) \end{bmatrix}$$

$$= \sum_{n=0}^{k-1} \begin{bmatrix} 4(0.5)^{k-1-n} - 3(0.25)^{k-1-n} \\ 8(0.5)^{k-1-n} - 9(0.25)^{k-1-n} \end{bmatrix}$$

or

$$
\sum_{n=0}^{k-1} \varphi(k-1-n)Bu(n) = \begin{bmatrix} 4(0.5)^{k-1}\sum_{n=0}^{k-1}2^n - 3(0.25)^{k-1}\sum_{n=0}^{k-1}4^n \\ 8(0.5)^{k-1}\sum_{n=0}^{k-1}2^n - 9(0.25)^{k-1}\sum_{n=0}^{k-1}4^n \end{bmatrix}
$$

$$
= \begin{bmatrix} 4(0.5)^{k-1}\dfrac{1-2^k}{1-2} - 3(0.25)^{k-1}\dfrac{1-4^k}{1-4} \\ 8(0.5)^{k-1}\dfrac{1-2^k}{1-2} - 3(0.25)^{k-1}\dfrac{1-4^k}{1-4} \end{bmatrix}
$$

$$
= \begin{bmatrix} -4(0.5)^{k-1}+4+(0.25)^{k-1} \\ -8(0.5)^{k-1}+12+(0.25)^{k-1} \end{bmatrix} = \begin{bmatrix} 4 - 8(0.5)^k + 4(0.25)^k \\ 12 - 16(0.5)^k + 4(0.25)^k \end{bmatrix}
$$

The total system response is the sum of zero-state and zero-input responses. Therefore, we have

$$
x(k) = \varphi(k)x(0) + \sum_{n=0}^{k-1} \varphi(k-1-n)Bu(n) = \begin{bmatrix} 4(0.5)^k - 3(0.25)^k \\ 8(0.5)^k - 9(0.25)^k \end{bmatrix} + \begin{bmatrix} 4 - 8(0.5)^k + 4(0.25)^k \\ 12 - 16(0.5)^n + 4(0.25)^k \end{bmatrix}
$$

$$
= \begin{bmatrix} 4 - 4(0.5)^k + (0.25)^k \\ 12 - 8(0.5)^k - 5(0.25)^k \end{bmatrix}
$$

The output signal is

$$
y(k) = [1 \quad 2\,]x(k) = [1 \quad 2\,]\begin{bmatrix} 4 - 5(0.5)^k + 2(0.25)^k \\ 12 - 10(0.5)^k - 2(0.25)^k \end{bmatrix} = 28 - 25(0.5)^k - 2(0.25)^k
$$

4.9 CONTROLLABILITY OF LTI SYSTEMS

In general, a system is controllable if there exists an input that can transfer the states of the system from an arbitrary initial state to a final state in a finite interval of time. In this section, we examine the controllability of LTI systems. Since the results obtained are identical for continuous and discrete-time systems, we only consider controllability of discrete-time LTI systems [6,7].

4.9.1 DEFINITION OF CONTROLLABILITY

A discrete-time LTI system is said to be controllable at time k_0 if there exists an input $u(k)$ for $k \geq k_0$ that can transfer the system from any initial state $x(k_0)$ to the origin in finite number of steps.

4.9.2 CONTROLLABILITY CONDITION

The following theorem gives the necessary and sufficient conditions for controllability of a discrete-time LTI system.

THEOREM 4.1

A discrete-time LTI system described by

$$x(k+1) = Ax(k) + Bu(k) \tag{4.120}$$

is completely state controllable if the controllability matrix Q is full rank, where

$$Q = \begin{bmatrix} B & AB & A^2B & \cdots & A^{N-1}B \end{bmatrix} \tag{4.121}$$

Proof: We assume that the system has one input. Let the initial state be $x(k_0)$ at time k_0 and the input for time $k \geq k_0$ be $u(k_0), u(k_0 + 1), \ldots, u(k_0 + N - 1)$, then the state of the system at time $k = N + k_0$ is given by

$$x(N + k_0) = A^N x(k_0) + \sum_{n=k_0}^{N+k_0-1} A^{N+k_0-1-n} Bu(n) \tag{4.122}$$

To drive the system to origin, we must have

$$A^N x(k_0) + A^{N-1} Bu(k_0) + A^{N-2} Bu(k_0 + 1) + \cdots + Bu(N + k_0 - 1) = 0 \tag{4.123}$$

or

$$\begin{bmatrix} B & AB & A^2B & \cdots & A^{N-1} & B \end{bmatrix} \begin{bmatrix} u(N + k_0 - 1) \\ u(N + k_0 - 2) \\ \vdots \\ u(k_0 + 1) \\ u(k_0) \end{bmatrix} = -A^N x(k_0) \tag{4.124}$$

Equation 4.124 has a solution if matrix $Q = \begin{bmatrix} B & AB & A^2B & \cdots & A^{N-1} & B \end{bmatrix}$ is full rank.

Example 4.14

Consider the following system:

$$x(k+1) = \begin{bmatrix} 1 & -0.25 \\ 1.5 & -0.25 \end{bmatrix} x(k) + \begin{bmatrix} -1 \\ 1 \end{bmatrix} u(k)$$

The controllability matrix Q is

$$Q = [B \quad AB] = \begin{bmatrix} -1 & -1.25 \\ 1 & -1.75 \end{bmatrix}$$

Since $\det(Q) = 3 \neq 0$, Q is full rank and the system is completely state controllable.

Example 4.15

Consider the SISO system

$$x(k+1) = \begin{bmatrix} 0.5 & 1 & 0 \\ 0 & 0.75 & 1 \\ 0 & 0 & 0.8 \end{bmatrix} x(k) + \begin{bmatrix} 1 \\ -1 \\ 2 \end{bmatrix} u(k)$$

The controllability matrix Q is

$$Q = \begin{bmatrix} B & AB & A^2B \end{bmatrix} = \begin{bmatrix} 1 & -0.5 & 1 \\ -1 & 1.25 & 2.54 \\ 2 & 1.6 & 1.28 \end{bmatrix}$$

Since $\det(Q) = -9.74 \neq 0$, Q is full rank and the system is completely state controllable.

Example 4.16

Consider the MIMO system

$$x(k+1) = \begin{bmatrix} 1 & 0 & 0 \\ 0 & 1 & 0 \\ 0 & 0 & 1 \end{bmatrix} x(k) + \begin{bmatrix} 1 & 0 \\ 0 & 1 \\ -1 & 2 \end{bmatrix} \begin{bmatrix} u_1(k) \\ u_2(k) \end{bmatrix}$$

The controllability matrix Q is

$$Q = \begin{bmatrix} B & AB & A^2B \end{bmatrix} = \begin{bmatrix} 1 & 0 & 1 & 0 & 1 & 0 \\ 0 & 1 & 0 & 1 & 0 & 1 \\ -1 & 2 & -1 & 2 & -1 & 2 \end{bmatrix}$$

Since $\text{rank}(Q) = 2$, Q is not full rank and the system is not completely state controllable.

4.10   OBSERVABILITY OF LTI SYSTEMS

In general, a system is completely state observable if the states of the system can be found from the knowledge of inputs and outputs of the system. Since we know the system and the inputs to the system, the only unknown in determining the states of the system is the initial conditions. Therefore, if we can find all the initial conditions from inputs and outputs of the system, then the system is completely state observable. In this section, we examine the observability of LTI system. Since the results obtained are identical to continuous and discrete-time systems, we only consider observability of discrete-time LTI system [6,7].

4.10.1 DEFINITION OF OBSERVABILITY

A discrete-time LTI system is said to be completely state observable if the states of the system can be estimated from the knowledge of inputs and outputs of the system.

4.10.2 OBSERVABILITY CONDITION

The following theorem gives the necessary and sufficient condition for observability of a discrete-time linear system.

THEOREM 4.2

A discrete-time LTI system given by the input and output equations

$$x(k+1) = Ax(k) + Bu(k)$$
$$y(k) = Cx(k) + Du(k)$$

(4.125)

is completely state observable if the observability matrix Q is full rank, where Q is defined as

$$P = \begin{bmatrix} C \\ CA \\ CA^2 \\ \vdots \\ CA^{N-1} \end{bmatrix}$$

(4.126)

Proof: Let the initial state be $x(0)$ at time $k = 0$ and the input and output for time $k \geq 0$ be $u(k)$ and $y(k)$, respectively, then

$$y(k) = Cx(k) + Du(k) = CA^k x(0) + C \sum_{n=0}^{k-1} A^{k-1-n} Bu(n) + Du(k)$$

(4.127)

Equation 4.127 can be written in matrix form as

$$
\begin{bmatrix} y(0) \\ y(1) \\ y(2) \\ \vdots \\ y(N-1) \end{bmatrix} = \begin{bmatrix} C \\ CA \\ CA^2 \\ \vdots \\ CA^{N-1} \end{bmatrix} x(0) + C \begin{bmatrix} 0 \\ AB \\ AB + B \\ \vdots \\ \sum_{n=0}^{N-2} A^{N-2-n}B \end{bmatrix} + D \begin{bmatrix} u(0) \\ u(1) \\ u(2) \\ \vdots \\ u(N-1) \end{bmatrix} \qquad (4.128)
$$

or

$$
\begin{bmatrix} C \\ CA \\ CA^2 \\ \vdots \\ CA^{N-1} \end{bmatrix} x(0) = \begin{bmatrix} y(0) \\ y(1) \\ y(2) \\ \vdots \\ y(N-1) \end{bmatrix} - C \begin{bmatrix} 0 \\ AB \\ AB + B \\ \vdots \\ \sum_{n=0}^{N-2} A^{N-2-n}B \end{bmatrix} - D \begin{bmatrix} u(0) \\ u(1) \\ u(2) \\ \vdots \\ u(N-1) \end{bmatrix} \qquad (4.129)
$$

Equation 4.129 can be used to solve for $x(0)$ if and only if matrix P is full rank. Once the initial state is obtained, the states of the system at any other time can be obtained by solving the state equations.

Example 4.17

Consider the following system

$$
x(k+1) = \begin{bmatrix} 1 & -0.25 \\ 1.5 & -0.25 \end{bmatrix} x(k) + \begin{bmatrix} -1 \\ 1 \end{bmatrix} u(k)
$$

$$
y(k) = [1 \quad 0]x(k)
$$

The observability matrix P is

$$
P = \begin{bmatrix} C \\ CA \end{bmatrix} = \begin{bmatrix} 1 & 0 \\ 1 & -0.25 \end{bmatrix}
$$

Since $\det(P) = -0.25 \neq 0$, P is full rank and the system is completely state observable.

Example 4.18

Consider the SISO system

$$
x(k+1) = \begin{bmatrix} 0.5 & 1 & 0 \\ 1 & 0.75 & 1 \\ 2 & -1 & 0.8 \end{bmatrix} x(k) + \begin{bmatrix} 1 \\ -1 \\ 2 \end{bmatrix} u(k)
$$

$$
y(k) = [1 \quad 0 \quad -2]x(k) + 3u(k)
$$

The observability matrix P is

$$P = \begin{bmatrix} C \\ CA \\ CA^2 \end{bmatrix} = \begin{bmatrix} 1 & 0 & -2 \\ -3.5 & 3 & -1.6 \\ -1.95 & 0.35 & 1.72 \end{bmatrix}$$

Since $\det(P) = -3.53 \neq 0$, P is full rank and the system is completely state observable.

Example 4.19

Consider the MIMO system

$$x(k+1) = \begin{bmatrix} 1 & 0 & 0 \\ 0 & 1 & 0 \\ 0 & 0 & 1 \end{bmatrix} x(k) + \begin{bmatrix} 1 & 0 \\ 0 & 1 \\ -1 & 2 \end{bmatrix} \begin{bmatrix} u_1(k) \\ u_2(k) \end{bmatrix}$$

$$y(k) = \begin{bmatrix} -1 & 0 & 1 \\ 1 & 1 & 0 \end{bmatrix} x(k)$$

The observability matrix P is

$$P = \begin{bmatrix} C \\ CA \\ CA^2 \end{bmatrix} = \begin{bmatrix} -1 & 0 & 1 \\ 1 & 1 & 0 \\ -1 & 0 & 1 \\ 1 & 1 & 0 \\ -1 & 0 & 1 \\ 1 & 1 & 0 \end{bmatrix}$$

Since $\mathrm{rank}(P) = 2$, P is not full rank and the system is not completely state observable.

Example 4.20

Consider the MIMO system

$$x(k+1) = \begin{bmatrix} 1 & -1 & 0 \\ 0 & 0.5 & 1 \\ -1 & 0 & 0.6 \end{bmatrix} x(k) + \begin{bmatrix} 1 & 0 \\ -3 & 1 \\ -1 & 0 \end{bmatrix} \begin{bmatrix} u_1(k) \\ u_2(k) \end{bmatrix}$$

$$y(k) = \begin{bmatrix} -1 & 1 & 1 \\ 1 & 1 & 0 \end{bmatrix} x(k)$$

The observability matrix P is

$$
P = \begin{bmatrix} C \\ CA \\ CA^2 \end{bmatrix} = \begin{bmatrix} -1 & 1 & 1 \\ 1 & 1 & 0 \\ -2 & 1.5 & 1.6 \\ 1 & -0.5 & 1 \\ -3.6 & 2.75 & 2.46 \\ 0 & -1.25 & 0.1 \end{bmatrix}
$$

Since rank$(P) = 3$, P is full rank and the system is completely state observable.

PROBLEMS

4.1 Find the state-space representation of the following dynamical systems in controllable and observable canonical forms:

(a) $\dfrac{d^3y(t)}{dt^3} + 2\dfrac{d^2y(t)}{dt^2} + 3\dfrac{dy(t)}{dt} + y(t) = 4u(t)$

(b) $y(k+2) + y(k+1) + 0.8y(k) = 8u(k+1) + 6u(k)$

4.2 Find the state-space representation of the following dynamical systems:

(a) $\dfrac{d^3y(t)}{dt^3} + 3\dfrac{d^2y(t)}{dt^2} + 6\dfrac{dy(t)}{dt} + 2y(t) = \dfrac{du(t)}{dt} + 8u(t)$

(b) $y(k+3) + 0.5y(k+2) + y(k+1) + 0.89y(k) = 2.4u(k)$

4.3 Obtain the response $y(t)$ of the following system, where $u(t)$ is the unit step function:

$$
\begin{bmatrix} \dot{x}_1(t) \\ \dot{x}_2(t) \end{bmatrix} = \begin{bmatrix} -1 & -0.75 \\ 1 & 0 \end{bmatrix} \begin{bmatrix} x_1(t) \\ x_2(t) \end{bmatrix} + \begin{bmatrix} 0 \\ 1 \end{bmatrix} u(t)
$$

$$
x(0) = \begin{bmatrix} 1 \\ 0 \end{bmatrix}
$$

$$
y(t) = \begin{bmatrix} 1 & 0 \end{bmatrix} \begin{bmatrix} x_1(t) \\ x_2(t) \end{bmatrix}
$$

4.4 Obtain the response $y(k)$ of the following system, when $u(k) = (-1)^k$ for $k \geq 0$:

$$
\begin{bmatrix} x_1(k+1) \\ x_2(k+1) \end{bmatrix} = \begin{bmatrix} -1 & -0.25 \\ 1 & 0 \end{bmatrix} \begin{bmatrix} x_1(k) \\ x_2(k) \end{bmatrix} + \begin{bmatrix} 1 \\ 0 \end{bmatrix} u(k)
$$

$$
x(0) = \begin{bmatrix} 1 \\ 0 \end{bmatrix}
$$

$$
y(k) = \begin{bmatrix} 1 & -1 \end{bmatrix} \begin{bmatrix} x_1(k) \\ x_2(k) \end{bmatrix}
$$

4.5 Show that the continuous-time state-transition matrix

$$\varphi(t) = e^{At}$$

satisfies the following properties:
(a) $\varphi(0) = I$
(b) $\varphi(-t) = e^{-At} = (e^{At})^{-1} = \varphi^{-1}(t)$
(c) $\varphi(t_1 + t_2) = \varphi(t_1)\varphi(t_2) = \varphi(t_2)\varphi(t_1)$
(d) $\varphi(t_3 - t_2)\varphi(t_2 - t_1) = \varphi(t_3 - t_1)$

4.6 Show that the discrete-time state-transition matrix

$$\varphi(k) = A^k$$

satisfies the following properties:
(a) $\varphi(0) = I$
(b) $\varphi(-k) = A^{-k} = (A^k)^{-1} = \varphi^{-1}(k)$
(c) $\varphi(k_1 + k_2) = \varphi(k_1)\varphi(k_2) = \varphi(k_2)\varphi(k_1)$
(d) $\varphi(k_3 - k_2)\varphi(k_2 - k_1) = \varphi(k_3 - k_1)$

4.7 The state-transition matrix of a time-varying system is the solution to the following partial differential equation:

$$\frac{\partial\varphi(t, \tau)}{\partial t} = A(t)\varphi(t, \tau), \text{ with boundary condition } \varphi(\tau, \tau) = I$$

Find the state-transition matrix of the following systems:

(a) $$\begin{bmatrix} \dot{x}_1(t) \\ \dot{x}_2(t) \end{bmatrix} = \begin{bmatrix} e^{-t} & 0 \\ 0 & -2t \end{bmatrix} \begin{bmatrix} x_1(t) \\ x_2(t) \end{bmatrix}$$

(b) $$\begin{bmatrix} \dot{x}_1(t) \\ \dot{x}_2(t) \end{bmatrix} = \begin{bmatrix} -1 & t \\ 0 & -1 \end{bmatrix} \begin{bmatrix} x_1(t) \\ x_2(t) \end{bmatrix}$$

4.8 An LTI system is described by the following differential equations:

$$\begin{bmatrix} \dot{x}_1(t) \\ \dot{x}_2(t) \end{bmatrix} = \begin{bmatrix} -6 & 4 \\ -2 & 0 \end{bmatrix} \begin{bmatrix} x_1(t) \\ x_2(t) \end{bmatrix} + \begin{bmatrix} 0 \\ 1 \end{bmatrix} u(t)$$

Let $x_1(0) = 1$, $x_2(0) = -1$, and $u(t) = e^{-t}$; derive expressions for $x_1(t)$ and $x_2(t)$.

(a) Find a nonsingular transformation matrix T such that the change in variables $z = Tx$ will lead to a decoupled system of differential equations.

(b) Solve the decoupled system for the new variables $z_1(t)$ and $z_2(t)$, and demonstrate that your results are consistent with the results obtained for $x_1(t)$ and $x_2(t)$.

4.9 A discrete LTI system is described by the following state equations:

$$\begin{bmatrix} x_1(k+1) \\ x_2(k+1) \end{bmatrix} = \begin{bmatrix} 0 & 1 \\ -0.5 & 1.5 \end{bmatrix} \begin{bmatrix} x_1(k) \\ x_2(k) \end{bmatrix} + \begin{bmatrix} -1 \\ 1 \end{bmatrix} u(k)$$

Let $x_1(0) = 1$, $x_2(0) = -1$, and $u(k) = (0.5)^k$, derive expressions for $x_1(k)$ and $x_2(k)$.

(a) Find a nonsingular transformation matrix T such that the change in variables $z(k) = Tx(k)$ will lead to a decoupled system of equations.

(b) Solve the decoupled system for the new variables $z_1(k)$ and $z_2(k)$, and demonstrate that your results are consistent with the results obtained for $x_1(k)$ and $x_2(k)$.

4.10 Consider the continuous transfer function

$$H(s) = \frac{3s + 4}{s^2 + 6s + 8}$$

Obtain the state-space representation of this system in controllable and observable canonical forms.

4.11 Consider the following transfer function

$$H(z) = \frac{3z^2 + 2z + 1}{z^2 + 0.7z + 0.12}$$

Obtain the state-space representation of this system in controllable and observable canonical forms.

4.12 Consider the system defined by

$$\begin{bmatrix} \dot{x}_1(t) \\ \dot{x}_2(t) \end{bmatrix} = \begin{bmatrix} 1 & 2 \\ -3 & -5 \end{bmatrix} \begin{bmatrix} x_1(t) \\ x_2(t) \end{bmatrix} + \begin{bmatrix} -6 \\ 1 \end{bmatrix} u(t)$$

$$y(t) = \begin{bmatrix} 2 & 3 \end{bmatrix} \begin{bmatrix} x_1(t) \\ x_2(t) \end{bmatrix}$$

(a) Transform the system equations into the controllable canonical form.

(b) Transform the system equations into the observable canonical form.

4.13 Consider the system defined by

$$\begin{bmatrix} x_1(k+1) \\ x_2(k+1) \\ x_3(k+1) \end{bmatrix} = \begin{bmatrix} 0 & 1 & 0 \\ 0 & 0 & 1 \\ 0.3 & 0.4 & 1 \end{bmatrix} \begin{bmatrix} x_1(k) \\ x_2(k) \\ x_3(k) \end{bmatrix} + \begin{bmatrix} 0 & 1 \\ 1 & 0 \\ 0 & 1 \end{bmatrix} \begin{bmatrix} u_1(k) \\ u_2(k) \end{bmatrix}$$

$$y(k) = \begin{bmatrix} 1 & 0 & 0 \\ 0 & 1 & 0 \end{bmatrix} \begin{bmatrix} x_1(k) \\ x_2(k) \\ x_3(k) \end{bmatrix}$$

(a) Is the system completely state controllable?
(b) Is the system completely state observable?
(c) Is the system completely output controllable?

4.14 Is the following system completely state controllable and observable?

$$\begin{bmatrix} x_1(k+1) \\ x_2(k+1) \\ x_3(k+1) \end{bmatrix} = \begin{bmatrix} 0 & 1 & 0 \\ 0 & 0 & 1 \\ -3 & -1 & -2 \end{bmatrix} \begin{bmatrix} x_1(k) \\ x_2(k) \\ x_3(k) \end{bmatrix} + \begin{bmatrix} 0 \\ 0 \\ 1 \end{bmatrix} u(k)$$

$$y(k) = \begin{bmatrix} 2 & -3 & -2 \end{bmatrix} \begin{bmatrix} x_1(k) \\ x_2(k) \\ x_3(k) \end{bmatrix}$$

4.15 For the system $x(k+1) = \begin{bmatrix} 0 & 1 \\ -0.125 & 0.75 \end{bmatrix} x(k) + \begin{bmatrix} 0 \\ 2 \end{bmatrix} u(k)$ and $y(k) = \begin{bmatrix} 1 & 1 \end{bmatrix} x(k)$

(a) Find the transfer function $H(z) = \frac{Y(z)}{U(z)}$.
(b) Is the system controllable? Is it observable? If not, find the uncontrollable/unobservable modes.

4.16 Consider the system given by

$$x(k+1) = \begin{bmatrix} 1 & 0 & 0 \\ 0 & 1 & 0 \\ 0 & 0 & 1 \end{bmatrix} x(k) + \begin{bmatrix} b_{11} & b_{12} & b_{13} \\ b_{21} & b_{22} & b_{23} \\ b_{31} & b_{32} & b_{33} \end{bmatrix} u(k)$$

$$y(k) = x_1(k)$$

(a) Find the transfer function $\frac{Y(z)}{U(z)}$.
(b) Find the necessary and sufficient condition for controllability of this system.

4.17 Show that the system given by

$$x(k + 1) = Ax(k) + Bu(k)$$
$$y(k) = Cx(k)$$

where $x \in R^N$, $u \in R^M$, and $y \in R^P$ is completely output controllable if and only if the controllability matrix P given as $P = \begin{bmatrix} CB & CAB. \\ CA^2B & \cdots & CA^{N-1}B \end{bmatrix}$ is of rank M.

REFERENCES

1. K. Ogata, *Modern Control Engineering*, Prentice Hall, Englewood, NJ, 2001.
2. W. L. Brogan, *Modern Control Theory*, Prentice Hall, Englewood, NJ, 1990.
3. B. Friedland, *Control System Design: An Introduction to State-Space Methods*, McGraw-Hill, New York, 1985.
4. K. Ogata, *Discrete-Time Control Systems*, Prentice Hall, Upper Saddle River, NJ, 1994.
5. G. Franklin, J. D. Powell, and A. Emami-Naeini, *Feedback Control of Dynamic Systems*, Prentice-Hall, Upper Saddle River, NJ, 2005.
6. A. Tewari, *Modern Control Design with MATLAB and Simulink*, Wiley, Chichester, United Kingdom, 2002.
7. B.C. Kuo, *Digital Control Systems*, Oxford University Press, New York, 1995.

5 Closed-Loop System Analysis and Design

5.1 INTRODUCTION

This chapter covers the design of closed-loop control systems in state space. The most common technique is state feedback, where the control signal is made to be proportional to the states of the system. The feedback gain is designed through pole placement, where the poles of the closed-loop system are assigned for a specific time response. This approach is valid when all the states are available. In practical applications not all states are measurable and they have to be estimated from control input and measured system output. The state estimation is another topic that will be discussed later in this chapter. We conclude this chapter by introducing optimal control for design of closed-loop control systems. The well-known linear quadratic regulator (LQR) will be used as an example for design of optimal control loops.

5.2 STATE FEEDBACK

5.2.1 BASIC CONCEPT

Consider a single-input single-output (SISO) open-loop LTI discrete-time system in state-space form given by

$$x(k + 1) = Ax(k) + Bu(k) \tag{5.1}$$

where
$A \in R^{N \times N}$
$B \in R^{N}$
$u \in R^{1}$
$x \in R^{N}$

The block diagram of the open-loop system is shown in Figure 5.1.

For an N-state SISO system, the state feedback control signal is generated by using weighted sum of the states, assuming states are accessible. In other words,

$$u(k) = -Kx(k) \tag{5.2}$$

Then, the closed-loop control system will be described by the following equation:

$$x(k + 1) = (A - BK)x(k) \tag{5.3}$$

The block diagram of the closed-loop system with state feedback appears in Figure 5.2.

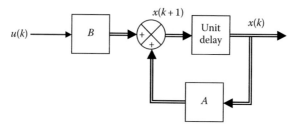

FIGURE 5.1   Block diagram of the open-loop control system.

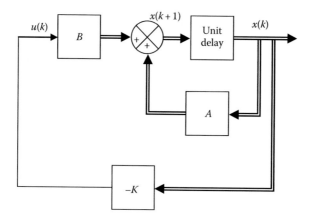

FIGURE 5.2   Block diagram of the closed-loop system with state feedback.

Using techniques described in Chapter 4, the solution to the closed-loop system of Equation 5.3 is

$$x(k) = (A - BK)^k x(0) \qquad (5.4)$$

Therefore, the states of the system will approach zero for any arbitrary initial conditions if the eigenvalues of the closed-loop matrix $A_c = A - BK$ are inside the unit circle in the complex z-plane. The feedback gain K is a vector for SISO system, which can be designed to place the eigenvalues of the closed-loop system inside the unit circle even if the original open-loop system is unstable or does not give an acceptable transient response. The following example illustrates this concept.

Example 5.1

Consider an open-loop control system given by the following state equation:

$$x(k + 1) = \begin{bmatrix} 0 & 1 \\ -0.72 & 1.7 \end{bmatrix} x(k) + \begin{bmatrix} 0 \\ 1 \end{bmatrix} u(k) \qquad (5.5)$$

The poles of the open-loop system or the eigenvalues of the A matrix are $\lambda_1 = 0.9$ and $\lambda_2 = 0.8$. The zero-input response of the system for the initial condition of $x(0) = \begin{bmatrix} 1 & -2 \end{bmatrix}^T$ is given by

$$x(k) = \begin{bmatrix} 0 & 1 \\ -0.72 & 1.7 \end{bmatrix}^k x(0) = \begin{bmatrix} 9(0.8)^k - 8(0.9)^k & -10(0.8)^k + 10(0.9)^k \\ 7(0.8)^k - 7(0.9)^k & -8(0.8)^k + 9(0.9)^k \end{bmatrix} \begin{bmatrix} 1 \\ -2 \end{bmatrix} \tag{5.6}$$

Therefore, the two states of the system are

$$x_1(k) = -11(0.8)^k + 12(0.9)^k \tag{5.7}$$

and

$$x_2(k) = -9(0.8)^k + 11(0.9)^k \tag{5.8}$$

Plot of $x_1(k)$ is shown in Figure 5.3. As is seen in this example, the response of the system is slow and it takes more than 50 time steps to converge to the origin. In order to improve the transient behavior of the system, we use state feedback. The closed-loop system matrix A_c is given by

$$A_c = A - BK = \begin{bmatrix} 0 & 1 \\ -0.72 & 1.7 \end{bmatrix} - \begin{bmatrix} 0 \\ 1 \end{bmatrix} [k_1 \quad k_2] = \begin{bmatrix} 0 & 1 \\ -0.72 - k_1 & 1.7 - k_2 \end{bmatrix} \tag{5.9}$$

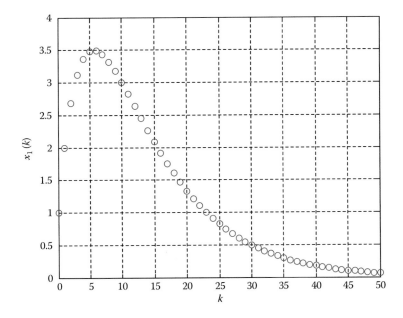

FIGURE 5.3   Transient behavior of the open-loop system.

The characteristic polynomial of the closed-loop system is

$$P(\lambda) = |\lambda I - A_c| = \lambda(\lambda - 1.7 + k_2) + 0.72 + k_1 = \lambda^2 + (k_2 - 1.7)\lambda + 0.72 + k_1$$
(5.10)

To improve the transient response of the closed-loop system, we select the closed-loop poles to be at

$$\lambda_1 = \lambda_2 = 0.2$$
(5.11)

This means that the characteristic polynomial of the closed-loop system should be

$$P(\lambda) = (\lambda - 0.2)(\lambda - 0.2) = \lambda^2 - 0.4\lambda + 0.04$$
(5.12)

Comparing the two equations (Equations 5.10 and 5.12), we have

$$\begin{cases} 0.72 + k_1 = 0.04 \\ k_2 - 1.7 = -0.40 \end{cases}$$
(5.13)

Therefore

$$\begin{cases} k_1 = -0.68 \\ k_2 = 1.3 \end{cases}$$
(5.14)

The closed-loop system response is given by

$$x(k) = (A - BK)^k = \begin{bmatrix} 0 & 1 \\ -0.04 & 0.4 \end{bmatrix}^k x(0)$$

$$= \begin{bmatrix} (1 - k)(0.2)^n & 5k(0.2)^k \\ -0.2k(0.2)^n & (1 + k)(0.2)^k \end{bmatrix} \begin{bmatrix} 1 \\ -2 \end{bmatrix}$$
(5.15)

Therefore the two states of the closed-loop system are

$$x_1(k) = (1 - 11k)(0.2)^k$$
(5.16)

and

$$x_2(k) = -(2 + 2.2k)(0.2)^k$$
(5.17)

The response of the first state of the open-loop and closed-loop systems is shown in Figure 5.4.

As it can be seen, the closed-loop response converges much faster than the open-loop response. Clearly, the state feedback has led to a controllable response to the state, which is one of the reasons why closed-loop systems are preferred over open-loop design. Location of poles directly affects the state response.

In the next section, we discuss the pole-placement design for SISO systems using state feedback [1–3].

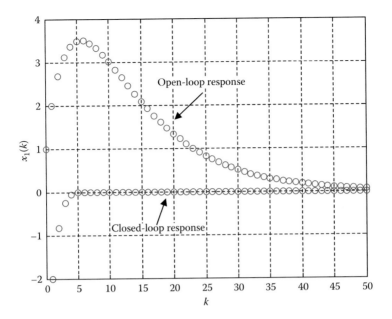

FIGURE 5.4   Open-loop and closed-loop response.

5.2.2   POLE-PLACEMENT DESIGN OF SISO SYSTEMS

Consider the SISO dynamic system described by the following state equation:

$$x(k+1) = Ax(k) + Bu(k) \tag{5.18}$$

With the assumption that the system is completely state controllable using the state feedback $u(k) = -Kx(k)$, we wish to place the poles of the closed-loop system at $\lambda_1, \lambda_2, \ldots, \lambda_N$. This means that the eigenvalues of the closed-loop matrix $A_c = A - BK$ are at the desired locations given by $\lambda_1, \lambda_2, \ldots, \lambda_N$. Therefore the characteristic polynomial of A_c is

$$P(\lambda) = |\lambda I - A_c| = (\lambda - \lambda_1)(\lambda - \lambda_2)\cdots(\lambda - \lambda_N)$$
$$= \lambda^N + \alpha_1\lambda^{N-1} + \alpha_2\lambda^{N-2} + \cdots + \alpha_N \tag{5.19}$$

By Cayley–Hamilton theorem (Section 3.11.3), we have

$$A_c^N + \alpha_1 A_c^{N-1} + \alpha_2 A_c^{N-2} + \cdots + \alpha_N I = P(A_c) = 0 \tag{5.20}$$

But

$$P(A_c) = \alpha_N I + \alpha_{N-1}A_c + \alpha_{N-2}A_c^2 + \cdots + A_c^N$$
$$= \alpha_N I + \alpha_{N-1}(A - BK) + \alpha_{N-2}(A - BK)^2 + \cdots + (A - BK)^N \tag{5.21}$$

Equation 5.21 can be rewritten as

$$P(A_c) = \alpha_N I + \alpha_{N-1}(A - BK) + \alpha_{N-2}(A^2 - ABK - BKA_c)$$
$$+ \alpha_{N-3}(A^3 - A^2BK - ABKA_c - BKA_c^2)$$
$$+ \cdots + (A^N - A^{N-1}BK - A^{N-2}BKA_c - \cdots - BKA_c^{N-1}) \quad (5.22)$$

Simplifying further, this results in

$$P(A_c) = \alpha_N I + \alpha_{N-1}A + \alpha_{N-2}A^2 + \cdots + A^N - \alpha_{N-1}BK - \alpha_{N-2}ABK$$
$$- \alpha_{N-2}BKA_c - \cdots - BKA_c^{N-1}$$
$$= P(A) - B(\alpha_{N-1}K + \alpha_{N-2}KA_c + \cdots + KA_c^{N-1})$$
$$- AB(\alpha_{N-2}K + \alpha_{N-3}KA_c + \cdots + KA_c^{N-2}) - \cdots - A^{N-1}BK \quad (5.23)$$

Equation 5.23 can be written in matrix form as

$$P(A_c) = P(A) - \begin{bmatrix} B & AB & A^2B & \cdots & A^{N-1}B \end{bmatrix} \begin{bmatrix} \alpha_{N-1}K + \alpha_{N-2}KA_c + \cdots KA_c^{N-1} \\ \alpha_{N-2}K + \alpha_{N-3}KA_c + \cdots KA_c^{N-2} \\ \alpha_{N-3}K + \alpha_{N-4}KA_c + \cdots KA_c^{N-3} \\ \vdots \\ K \end{bmatrix}$$

$$(5.24)$$

Since $P(A_c) = 0$, we have

$$P(A) = \begin{bmatrix} B & AB & A^2B & \cdots & A^{N-1}B \end{bmatrix} \begin{bmatrix} \alpha_{N-1}K + \alpha_{N-2}KA_c + \cdots KA_c^{N-1} \\ \alpha_{N-2}K + \alpha_{N-3}KA_c + \cdots KA_c^{N-2} \\ \alpha_{N-3}K + \alpha_{N-4}KA_c + \cdots KA_c^{N-3} \\ \vdots \\ K \end{bmatrix}$$

$$(5.25)$$

Since the system is completely state controllable, the controllability matrix $\begin{bmatrix} B & AB & A^2B & \cdots & A^{N-1}B \end{bmatrix}$ is nonsingular and its inverse exists, therefore we have

$$\begin{bmatrix} \alpha_{N-1}K + \alpha_{N-2}KA_c + \cdots KA_c^{N-1} \\ \alpha_{N-2}K + \alpha_{N-3}KA_c + \cdots KA_c^{N-2} \\ \alpha_{N-3}K + \alpha_{N-4}KA_c + \cdots KA_c^{N-3} \\ \vdots \\ K \end{bmatrix} = \begin{bmatrix} B & AB & A^2B & \cdots & A^{N-1}B \end{bmatrix}^{-1} P(A)$$

$$(5.26)$$

Equation 5.26 can be used to solve for K by premultiplying both sides of this equation by $[0 \quad 0 \quad \cdots \quad 0 \quad 1]$. The resulting gain matrix K is given as

$$K = [0 \quad 0 \quad \cdots \quad 0 \quad 1][B \quad AB \quad A^2B \quad \cdots \quad A^{N-1}B]^{-1}P(A) \qquad (5.27)$$

The above equation is known as Ackermann's formula for pole placement.

Example 5.2

Consider the dynamic system given by

$$x(k+1) = \begin{bmatrix} 0 & 1 \\ -0.72 & 1.7 \end{bmatrix} x(k) + \begin{bmatrix} 0 \\ 1 \end{bmatrix} u(k)$$

Design a state feedback to place the closed-loop poles at

$$\lambda_1 = \lambda_2 = 0.2$$

SOLUTION
The characteristic polynomial of the closed-loop system is

$$P(\lambda) = (\lambda - \lambda_1)(\lambda - \lambda_2) = (\lambda - 0.2)(\lambda - 0.2) = \lambda^2 - 0.4\lambda + 0.04$$

The controllability matrix is

$$[B \quad AB] = \begin{bmatrix} 0 & 1 \\ 1 & 1.7 \end{bmatrix}$$

Since the controllability matrix is full rank, the system is completely state controllable and pole placement is possible. The feedback gain vector is computed using Equation 5.27

$$K = [0 \quad 1][B \quad AB]^{-1}P(A) = [0 \quad 1][B \quad AB]^{-1}(A^2 - 0.4A + 0.04I)$$

$$= [0 \quad 1]\begin{bmatrix} 0 & 1 \\ 1 & 1.7 \end{bmatrix}^{-1}\begin{bmatrix} -0.68 & 1.3 \\ -0.93 & 1.53 \end{bmatrix} = [-0.68 \quad 1.3]$$

There are other algorithms that can be used for pole placement [1, 4]. The second algorithm is summarized below.

a. For the desired poles, find the desired characteristic polynomial of the closed-loop system

$$P_c(\lambda) = |\lambda I - A_c| = (\lambda - \lambda_1)(\lambda - \lambda_2)\cdots(\lambda - \lambda_N)$$
$$= \lambda^N + \alpha_1\lambda^{N-1} + \alpha_2\lambda^{N-2} + \cdots + \alpha_N \qquad (5.28)$$

b. Find the characteristic polynomial of the open-loop system

$$P(\lambda) = |\lambda I - A| = \lambda^N + \beta_1 \lambda^{N-1} + \beta_2 \lambda^{N-2} + \cdots + \beta_N \qquad (5.29)$$

c. Find the transformation T given by

$$T = QW \qquad (5.30)$$

where Q is the controllability matrix and W is given by

$$W = \begin{bmatrix} \beta_{N-1} & \beta_{N-2} & \cdots & \beta_1 & 1 \\ \beta_{N-1} & \beta_{N-1} & \cdots & 1 & 0 \\ \vdots & \vdots & & \vdots & \vdots \\ \beta_1 & 1 & \cdots & 0 & 0 \\ 1 & 0 & \cdots & 0 & 0 \end{bmatrix} \qquad (5.31)$$

d. Find the gain matrix K using the following equation:

$$K = [\alpha_N - \beta_N \quad \alpha_{N-1} - \beta_{N-1} \quad \cdots \quad \alpha_1 - \beta_1]T^{-1} \qquad (5.32)$$

Example 5.3

Consider the dynamic system given by

$$x(k+1) = \begin{bmatrix} 0 & 1 \\ -0.12 & -1 \end{bmatrix} x(k) + \begin{bmatrix} 1 \\ 1 \end{bmatrix} u(k)$$

Design a state feedback controller to place the closed-loop poles at

$$\lambda_1 = 0.3 - j0.4 \quad \text{and} \quad \lambda_2 = 0.3 + j0.4$$

SOLUTION

The characteristic polynomial of the open-loop system is

$$P(\lambda) = |\lambda I - A| = \begin{vmatrix} \lambda & -1 \\ 0.12 & \lambda + 1 \end{vmatrix} = \lambda^2 + \lambda + 0.12 \qquad (5.33)$$

Comparing Equation 5.33 with Equation 5.29, we have $\beta_1 = 1$, and $\beta_2 = 0.12$. The characteristic polynomial of the desired closed-loop system is

$$P_c(\lambda) = (\lambda - 0.3 + j0.4)(\lambda - 0.3 + j0.4) = \lambda^2 - 0.6\lambda + 0.25$$

Hence $\alpha_1 = -0.6$, and $\alpha_2 = 0.25$. The transformation T is given by

$$T = QW = \begin{bmatrix} B & AB \end{bmatrix} \begin{bmatrix} \beta_1 & 1 \\ 1 & 0 \end{bmatrix} = \begin{bmatrix} 1 & 1 \\ 1 & -1.12 \end{bmatrix} \begin{bmatrix} 1 & 1 \\ 1 & 0 \end{bmatrix} = \begin{bmatrix} 2 & 1 \\ -0.12 & 1 \end{bmatrix}$$

The gain matrix K is given as

$$K = [\alpha_2 - \beta_2 \quad \alpha_1 - \beta_1]T^{-1} = [0.13 \quad -1.6]\begin{bmatrix} 2 & 1 \\ -0.12 & 1 \end{bmatrix}^{-1}$$
$$= [-0.0292 \quad -1.5708]$$

Example 5.4

Consider the dynamic system given by

$$x(k+1) = \begin{bmatrix} 0 & 1 \\ -0.12 & -1 \end{bmatrix} x(k) + \begin{bmatrix} 0 \\ 1 \end{bmatrix} u(k)$$

Design a state feedback to place the closed-loop poles at

$$\lambda_1 = \lambda_2 = 0$$

SOLUTION
The characteristic polynomial of the open-loop system is

$$P(\lambda) = \lambda^2 + \lambda + 0.12$$

Hence $\beta_1 = 1$, and $\beta_2 = 0.12$.
    The characteristic polynomial of the desired closed-loop system is

$$P_c(\lambda) = \lambda^2$$

Hence, $\alpha_1 = \alpha_2 = 0$ and the transformation T is given by

$$T = QW = [B \quad AB]\begin{bmatrix} \beta_1 & 1 \\ 1 & 0 \end{bmatrix} = \begin{bmatrix} 0 & 1 \\ 1 & -1 \end{bmatrix}\begin{bmatrix} 1 & 1 \\ 1 & 0 \end{bmatrix} = \begin{bmatrix} 1 & 0 \\ 0 & 1 \end{bmatrix}$$

The gain matrix K is given as

$$K = [\alpha_2 - \beta_2 \quad \alpha_1 - \beta_1]T^{-1} = [-0.12 \quad -1]\begin{bmatrix} 1 & 0 \\ 0 & 1 \end{bmatrix}^{-1} = [-0.12 \quad -1]$$

The closed-loop state equations are

$$x(k+1) = \begin{bmatrix} 0 & 1 \\ -0.12 & -1 \end{bmatrix} x(k) - \begin{bmatrix} 0 \\ 1 \end{bmatrix} Kx(k)$$
$$= \begin{bmatrix} 0 & 1 \\ -0.12 & -1 \end{bmatrix} x(k) - \begin{bmatrix} 0 \\ 1 \end{bmatrix} [-0.12 \quad -1]x(k) = \begin{bmatrix} 0 & 1 \\ 0 & 0 \end{bmatrix} x(k)$$

Now consider an arbitrary initial state $x(0) = [x_1(0) \quad x_2(0)]^T$, then

$$x(1) = \begin{bmatrix} 0 & 1 \\ 0 & 0 \end{bmatrix} x(0) = \begin{bmatrix} 0 & 1 \\ 0 & 0 \end{bmatrix} \begin{bmatrix} x_1(0) \\ x_2(0) \end{bmatrix} = \begin{bmatrix} 0 \\ 0 \end{bmatrix}$$

Therefore any arbitrary initial state is driven to zero in one step. This is called dead beat control.

5.2.3 POLE-PLACEMENT DESIGN OF MULTIPLE-INPUT MULTIPLE-OUTPUT (MIMO) SYSTEMS

Consider an open-loop MIMO system in state-space form given by

$$\begin{aligned} x(k + 1) &= Ax(k) + Bu(k) \\ y(k) &= Cx(k) \end{aligned} \tag{5.34}$$

where

$A \in R^{N \times N}$
$B \in R^{N \times M}$
$u \in R^M$
$C \in R^{P \times N}$
$y \in R^P$

The state feedback control law is given by

$$u(k) = -Kx(k) \tag{5.35}$$

where the gain matrix K is $M \times N$ and is given as

$$K = \begin{bmatrix} K_{11} & K_{12} & \cdots & K_{1N} \\ K_{21} & K_{22} & \cdots & K_{2N} \\ \vdots & \vdots & & \vdots \\ K_{M1} & K_{M2} & \cdots & K_{MN} \end{bmatrix} \tag{5.36}$$

The characteristic polynomial of the closed-loop system is

$$P_c(\lambda) = |\lambda I - A + BK| \tag{5.37}$$

On the other hand, the closed-loop characteristic polynomial in terms of the desired poles is given by

$$\begin{aligned} P_c(\lambda) &= (\lambda - \lambda_1)(\lambda - \lambda_2) \cdots (\lambda - \lambda_N) \\ &= \lambda^N + \alpha_1 \lambda^{N-1} + \alpha_2 \lambda^{N-2} + \cdots + \alpha_N \end{aligned} \tag{5.38}$$

Comparing Equations 5.37 and 5.38, we get N equations with $M \times N$ unknowns. These unknowns are entries of the gain matrix K.

Therefore there are fewer equations than number of unknowns which means that there are infinite number of possible solutions. The MATLAB® function file named "place" implements the Ackermann algorithm for SISO systems. For MIMO systems, the "place" algorithm uses extra degrees of freedom to obtain a robust solution for matrix K. It minimizes the sensitivity of the closed-loop poles to the variations in system's parameters A and B. The command to implement pole placement in MATLAB is $K = \text{place}(A,B,P)$, where A and B are system matrices and P is a vector containing desired poles of the closed-loop system.

Example 5.5

Consider the MIMO dynamic system given by

$$x(k+1) = \begin{bmatrix} 0 & 1 \\ -0.12 & -1 \end{bmatrix} x(k) + \begin{bmatrix} 1 & -1 \\ 0 & 1 \end{bmatrix} \begin{bmatrix} u_1(k) \\ u_2(k) \end{bmatrix}$$

Design a state feedback to place the closed-loop poles at

$$\lambda_1 = 0.3$$
$$\lambda_2 = 0.2$$

SOLUTION

The system matrix of the closed-loop system is

$$
\begin{aligned}
A_c = A - BK &= \begin{bmatrix} 0 & 1 \\ -0.12 & -1 \end{bmatrix} - \begin{bmatrix} 1 & -1 \\ 0 & 1 \end{bmatrix} \begin{bmatrix} K_{11} & K_{12} \\ K_{21} & K_{22} \end{bmatrix} \\
&= \begin{bmatrix} -K_{11}+K_{21} & 1-K_{12}+K_{22} \\ -0.12-K_{21} & -1-K_{22} \end{bmatrix}
\end{aligned}
$$

The characteristic polynomial of the closed-loop system is

$$
\begin{aligned}
P_c(\lambda) &= |\lambda I - A_c| \\
&= \lambda^2 + (K_{11}+K_{22}-K_{21}+1)\lambda + K_{11}(1+K_{22}) + 0.12(1+K_{22}) - K_{12}(0.12+K_{21})
\end{aligned}
$$

Since the desired closed-loop poles are at $\lambda_1 = 0.3$, and $\lambda_2 = 0.2$, then

$$P_c(\lambda) = (\lambda - 0.3)(\lambda - 0.2) = \lambda^2 - 0.5\lambda + 0.06$$

Comparing the two equations of the closed-loop characteristic polynomials, we have

$$
\begin{cases}
K_{11}+K_{22}-K_{21}+1 = -0.5 \\
K_{11}(1+K_{22})+0.12(1+K_{22})-K_{12}(0.12+K_{21}) = 0.06
\end{cases}
$$

As can be seen, we have two equations and four unknowns. Therefore the solution is not unique. For example if we choose $K_{11} = K_{22} = 0$, we have

$$
\begin{cases}
-K_{21}+1 = -0.5 \\
0.12 - K_{12}(0.12+K_{21}) = 0.06
\end{cases}
$$

The solutions to the above equations are

$$
\begin{cases}
K_{21} = 1.5 \\
K_{12} = 0.037
\end{cases}
$$

The resulting gain matrix is

$$K = \begin{bmatrix} K_{11} & K_{12} \\ K_{21} & K_{22} \end{bmatrix} = \begin{bmatrix} 0 & 0.037 \\ 1.5 & 0 \end{bmatrix} \tag{5.39}$$

The MATLAB pole placement yields

$$K = \begin{bmatrix} K_{11} & K_{12} \\ K_{21} & K_{22} \end{bmatrix} = \begin{bmatrix} -0.32 & -0.3 \\ -0.12 & -1.3 \end{bmatrix} \tag{5.40}$$

The MATLAB solution gives a more robust design with respect to variations in the system parameters. For example consider slight perturbation of matrix A by ΔA, that is,

$$A = A + \Delta A = \begin{bmatrix} 0 & 1 \\ -0.12 & -1 \end{bmatrix} + \begin{bmatrix} 0.0039 & -0.0004 \\ 0.0026 & -0.0048 \end{bmatrix} = \begin{bmatrix} 0.0039 & 0.9996 \\ 0.1174 & -1.0048 \end{bmatrix}$$

The closed-loop A matrix using the first design (K matrix given by Equation 5.39) is

$$\begin{aligned} A_C = A - BK &= \begin{bmatrix} 0.0039 & 0.9996 \\ 0.1174 & -1.0048 \end{bmatrix} - \begin{bmatrix} 1 & -1 \\ 0 & 1 \end{bmatrix} \begin{bmatrix} 0 & 0.037 \\ 1.5 & 0 \end{bmatrix} \\ &= \begin{bmatrix} 1.5039 & 0.9626 \\ 1.6174 & -1.0048 \end{bmatrix} \end{aligned}$$

The closed-loop poles are eigenvalues of A_C which are $p_1 = [0.3784 \quad 0.1207]^T$. The percentage change from the desired poles $p = [\lambda_1 \quad \lambda_2]^T = [0.3 \quad 0.2]^T$ is

$$\Delta\lambda = \frac{\|p - p_1\|_2}{\|p\|_2} \times 100 = 30.92\%$$

The closed-loop A matrix using the second design (K matrix given by Equation 5.40) is

$$\begin{aligned} A_C &= \begin{bmatrix} 0.0039 & 0.9996 \\ 0.1174 & -1.0048 \end{bmatrix} - \begin{bmatrix} 1 & -1 \\ 0 & 1 \end{bmatrix} \begin{bmatrix} -0.32 & -0.3 \\ -0.12 & -1.3 \end{bmatrix} \\ &= \begin{bmatrix} 0.2039 & -0.0004 \\ 0.0026 & 0.2952 \end{bmatrix} \end{aligned}$$

The closed-loop poles are $p_2 = [0.2952 \quad 0.2039]^T$ and the percentage change is

$$\Delta\lambda = \frac{\|p - p_2\|_2}{\|p\|_2} \times 100 = 1.72\%$$

Therefore, the second design using the MATLAB pole-placement algorithm is less sensitive to variations in system's parameters.

5.2.4 RELATIONSHIP BETWEEN POLES AND THE CLOSED-LOOP SYSTEM RESPONSE

The closed-loop system response is directly related to the locations of the poles in the complex plane. To illustrate the relationship between the poles and the system response, we consider a second-order dynamic system with zero input. Assume that the closed-loop second-order system is given by

$$x(k+1) = Ax(k) \qquad (5.41)$$

where matrix A is given in the diagonalized form as

$$A = \begin{bmatrix} 1 & -2 \\ 3 & -5 \end{bmatrix} \begin{bmatrix} \lambda_1 & 0 \\ 0 & \lambda_2 \end{bmatrix} \begin{bmatrix} 1 & -2 \\ 3 & -5 \end{bmatrix}^{-1} = \begin{bmatrix} -5\lambda_1 + 6\lambda_2 & 2\lambda_1 - 2\lambda_2 \\ -15\lambda_1 + 15\lambda_2 & 6\lambda_1 - 5\lambda_2 \end{bmatrix} \qquad (5.42)$$

Here λ_1, λ_2 are eigenvalues (poles) of the closed-loop dynamic system which can be assigned arbitrarily. Assume that the system is initially at $x_1(0) = 1$ and $x_2(0) = 2$.
The time response of the system for the given initial condition is given by

$$x(k) = A^k x(0) = \begin{bmatrix} -5\lambda_1 + 6\lambda_2 & 2\lambda_1 - 2\lambda_2 \\ -15\lambda_1 + 15\lambda_2 & 6\lambda_1 - 5\lambda_2 \end{bmatrix}^k \begin{bmatrix} 1 \\ 2 \end{bmatrix} \qquad (5.43)$$

This can be simplified as

$$x_1(k) = [-\lambda_1^k + 2\lambda_2^k]u(k) \qquad (5.44)$$

and

$$x_2(k) = [-3\lambda_1^k + 5\lambda_2^k]u(k) \qquad (5.45)$$

For the system to be stable the poles are chosen to be inside the unit circle in complex plane. This means that $|\lambda_1| < 1$, and $|\lambda_2| < 1$.
The response of the system for different values of λ_1, and λ_2 are shown in Table 5.1.

5.3 LQR DESIGN

5.3.1 INTRODUCTION

The LQR is a well-known optimal control design technique that results in a feedback gain similar to the state feedback design [4,5]. It provides an optimal design based on a specified figure of merit which is quadratic in both states as well as in control. In the derivation of the LQR, we assume that all the states of the system are available to the controller.
Let the plant to be controlled be given in state space as

$$x(k+1) = Ax(k) + Bu(k) \qquad (5.46)$$

TABLE 5.1

System Time Response and Pole Locations

Poles	Time Response	Response Type
$\lambda_1 = 0.25$ $\lambda_2 = 0.25$		Underdamped
$\lambda_1 = 0$ $\lambda_2 = 0$		Deadbeat
$\lambda_1 = -0.42$ $\lambda_2 = -0.42$		Oscillation

The state feedback control law is given by

$$u(k) = -K(k)x(k) \tag{5.47}$$

where $K(k)$ is the controller feedback gain at time k. The closed-loop system dynamic is governed by the state equation given as

$$x(k+1) = (A - BK(k))x(k) \tag{5.48}$$

Let us form a quadratic performance function as follows

$$J = \tfrac{1}{2}x^T(N)Sx(N) + \tfrac{1}{2}\sum_{k=0}^{N-1} x^T(k)Qx(k) + u^T(k)Ru(k) \tag{5.49}$$

where
S and Q are positive or positive semidefinite matrices
R is a positive definite matrix

The first term in the performance index penalizes the deviation of the final state from the origin. The term $x^T(k)Qx(k)$ penalizes deviation of the states from the origin during the control process and the term $u^T(k)Qu(k)$ is a measure of energy used by the control signal. The goal is to minimize performance index J with respect to the control signal $u(k)$ subject to the constraints given by the state equation of the plant (Equation 5.46).

5.3.2 SOLUTION OF THE LQR PROBLEM

The optimal control problem is to minimize the objective function

$$J = \tfrac{1}{2}x^T(N)Sx(N) + \tfrac{1}{2}\sum_{k=0}^{N-1} x^T(k)Qx(k) + u^T(k)Ru(k) \tag{5.50}$$

subject to the constraint given by the state equations

$$x(k+1) = Ax(k) + Bu(k) \tag{5.51}$$

Using Lagrange multipliers, we form a new performance index H as follows

$$H = \tfrac{1}{2}x^T(N)Sx(N) + \tfrac{1}{2}\sum_{k=0}^{N-1} x^T(k)Qx(k) + u^T(k)Ru(k)$$
$$+ \lambda^T(k+1)[x(k+1) - Ax(k) - Bu(k)] \tag{5.52}$$

where $\lambda(k) = \begin{bmatrix} \lambda_1(k) & \lambda_2(k) & \cdots & \lambda_N(k) \end{bmatrix}^T$.

To minimize the functional H, we need to differentiate H with respect to the three variables $x(k)$, $u(k)$, $\lambda(k)$, and $x(N)$ and set the results equal to zero.

$$\frac{\partial H}{\partial x(k)} = Qx(k) + A^{T}\lambda(k + 1) - \lambda(k) = 0 \tag{5.53}$$

$$\frac{\partial H}{\partial u(k)} = Ru(k) + B^{T}\lambda(k + 1) = 0 \tag{5.54}$$

$$\frac{\partial H}{\partial \lambda(k)} = Ax(k - 1) + Bu(k - 1) - x(k) = 0 \tag{5.55}$$

$$\frac{\partial H}{\partial x(N)} = Sx(N) - \lambda(N) = 0 \tag{5.56}$$

The above equations can be solved by assuming that

$$\lambda(k) = P(k)x(k) \tag{5.57}$$

Substituting Equation 5.57 into Equation 5.53 results in

$$Qx(k) + A^{T}P(k + 1)x(k + 1) - P(k)x(k) = 0 \tag{5.58}$$

From Equation 5.54, we have

$$u(k) = -R^{-1}B^{T}\lambda(k + 1) = -R^{-1}B^{T}P(k + 1)x(k + 1) \tag{5.59}$$

By substituting Equation 5.59 into Equation 5.51, we have

$$x(k + 1) = Ax(k) - BR^{-1}B^{T}P(k + 1)x(k + 1) \tag{5.60}$$

Equation 5.60 is now used to solve for $x(k + 1)$. That is

$$x(k + 1) = [I + BR^{-1}B^{T}P(k + 1)]^{-1}Ax(k) \tag{5.61}$$

We now substitute Equation 5.61 into Equation 5.58 to obtain

$$Qx(k) + A^{T}P(k + 1)[I + BR^{-1}B^{T}P(k + 1)]^{-1}Ax(k) - P(k)x(k) = 0 \tag{5.62}$$

Equation 5.62 must be satisfied for all $x(k)$. Therefore, we must have

$$P(k) = A^{T}P(k + 1)[I + BR^{-1}B^{T}P(k + 1)]^{-1}A + Q \tag{5.63}$$

The above equation can be simplified using the matrix inversion lemma which states that for any A, C, and D matrices of appropriate dimensions, we have

$$(A + CD)^{-1} = A^{-1} - A^{-1}C(I + DA^{-1}C)^{-1}DA^{-1} \tag{5.64}$$

Using $A = I$, $C = BR^{-1}$ and $D = B^{T}P(k+1)$ in matrix inversion lemma, we have

$$[I + BR^{-1}B^{T}P(k+1)]^{-1} = I - BR^{-1}(I + B^{T}P(k+1)BR^{-1})B^{T}P(k+1) \quad (5.65)$$

Substituting Equation 5.65 into Equation 5.63, we have

$$P(k) = A^{T}P(k+1)A - A^{T}P(k+1)BR^{-1}(I + B^{T}P(k+1)BR^{-1})B^{T}P(k+1)A + Q \tag{5.66}$$

From Equation 5.56 we have

$$\lambda(N) = Sx(N) = P(N)x(N) \tag{5.67}$$

Therefore

$$P(N) = S \tag{5.68}$$

Equation 5.66 with the boundary condition (Equation 5.68) is known as Riccati equation which can be solved backward in time starting from $k = N$ to $k = 0$. Once $P(k)$ is obtained, the optimal control signal $u(k)$ is given by

$$u(k) = -R^{-1}B^{T}P(k+1)x(k+1)$$
$$= -R^{-1}B^{T}P(k+1)[I + BR^{-1}B^{T}P(k+1)]^{-1}Ax(k) \tag{5.69}$$

Therefore

$$u(k) = -K(k)x(k) \tag{5.70}$$

where the time varying gain matrix $K(k)$ is given by

$$K(k) = R^{-1}B^{T}P(k+1)[I + BR^{-1}B^{T}P(k+1)]^{-1}A \tag{5.71}$$

Example 5.6

Consider the control system given by

$$x(k+1) = 0.5x(k) + u(k)$$

with initial state $x(0) = 1$.
Find the optimal control feedback law to minimize the performance index

$$J = 2x^{2}(8) + \frac{1}{2}\sum_{k=0}^{7}x^{2}(k) + 2u^{2}(k)$$

SOLUTION

In this problem $A = 0.5$, $B = 1$, $S = 4$, $Q = 1$, $R = 2$, and $N = 8$. The Riccati equation is

$$P(k) = 0.25P(k+1) - 0.125P(k+1)(1 + 0.5P(k+1))^{-1}P(k+1) + 1$$

This can be written as

$$P(k) = 0.25P(k+1) - \frac{0.125P^2(k+1)}{1 + 0.5P(k+1)} + 1 = \frac{0.25P(k+1)}{1 + 0.5P(k+1)} + 1 \qquad (5.72)$$

with boundary condition $P(8) = S = 4$

Solving Equation 5.72 backward in time from $k = 8$ to $k = 0$ results in

$$P(8) = 4$$

$$P(7) = \frac{0.25P(8)}{1 + 0.5P(8)} + 1 = \frac{0.25 \times 4}{1 + 0.5 \times 4} + 1 = 1.3333$$

$$P(6) = \frac{0.25P(7)}{1 + 0.5P(7)} + 1 = \frac{0.25 \times 1.3333}{1 + 0.5 \times 1.3333} + 1 = 1.200$$

$$P(5) = \frac{0.25P(6)}{1 + 0.5P(6)} + 1 = \frac{0.25 \times 1.20}{1 + 0.5 \times 1.20} + 1 = 1.1875$$

$$P(4) = \frac{0.25P(5)}{1 + 0.5P(5)} + 1 = \frac{0.25 \times 1.1875}{1 + 0.5 \times 1.1875} + 1 = 1.1863$$

$$P(3) = \frac{0.25P(4)}{1 + 0.5P(4)} + 1 = \frac{0.25 \times 1.1863}{1 + 0.5 \times 1.1863} + 1 = 1.1862$$

$$P(2) = \frac{0.25P(3)}{1 + 0.5P(3)} + 1 = \frac{0.25 \times 1.1862}{1 + 0.5 \times 1.1862} + 1 = 1.1861$$

$$P(1) = \frac{0.25P(2)}{1 + 0.5P(2)} + 1 = \frac{0.25 \times 1.1861}{1 + 0.5 \times 1.1861} + 1 = 1.1861$$

$$P(0) = \frac{0.25P(1)}{1 + 0.5P(1)} + 1 = \frac{0.25 \times 1.1861}{1 + 0.5 \times 1.1861} + 1 = 1.1861$$

The feedback control gain $K(k)$ is computed using Equation 5.71

$$K(k) = 0.25P(k+1)[1 + 0.5P(k+1)]^{-1} = \frac{0.25P(k+1)}{1 + 0.5P(k+1)}$$

Hence we have

$$K(0) = \frac{0.25P(1)}{1 + 0.5P(1)} = \frac{0.25 \times 1.1861}{1 + 0.5 \times 1.1861} = 0.1861$$

$$K(1) = \frac{0.25P(2)}{1 + 0.5P(2)} = \frac{0.25 \times 1.1861}{1 + 0.5 \times 1.1861} = 0.1861$$

$$K(2) = \frac{0.25P(3)}{1 + 0.5P(3)} = \frac{0.25 \times 1.1862}{1 + 0.5 \times 1.1862} = 0.1862$$

$$K(3) = \frac{0.25P(4)}{1 + 0.5P(4)} = \frac{0.25 \times 1.1863}{1 + 0.5 \times 1.1863} = 0.1863$$

$$K(4) = \frac{0.25P(5)}{1 + 0.5P(5)} = \frac{0.25 \times 1.1875}{1 + 0.5 \times 1.1875} = 0.1875$$

$$K(5) = \frac{0.25P(6)}{1 + 0.5P(6)} = \frac{0.25 \times 1.2}{1 + 0.5 \times 1.2} = 0.2$$

$$K(6) = \frac{0.25P(7)}{1 + 0.5P(7)} = \frac{0.25 \times 1.3333}{1 + 0.5 \times 1.3333} = 0.3333$$

$$K(7) = \frac{0.25P(8)}{1 + 0.5P(8)} = \frac{0.25 \times 4}{1 + 0.5 \times 4} = 0.3333$$

5.3.3 STEADY-STATE ALGEBRAIC RICCATI EQUATION

When the time horizon $N \to \infty$, the performance index is given by

$$J = \frac{1}{2} \sum_{k=0}^{\infty} x^{\mathrm{T}}(k)Qx(k) + u^{\mathrm{T}}(k)Ru(k) \tag{5.73}$$

Under this scenario, the gain matrix K approaches a constant and is obtained by solving the steady-state Riccati equation. From Equation 5.66, we have

$$P = A^{\mathrm{T}}PA - A^{\mathrm{T}}PBR^{-1}(I + B^{\mathrm{T}}PBR^{-1})B^{\mathrm{T}}PA + Q \tag{5.74}$$

Once the positive definite matrix P is obtained, the feedback gain matrix K is computed using Equation 5.71

$$K = R^{-1}B^{\mathrm{T}}P(I + BR^{-1}B^{\mathrm{T}}P)^{-1}A \tag{5.75}$$

Example 5.7

Consider the control system given by

$$x(k + 1) = 0.5x(k) + u(k)$$

with initial state $x(0) = 1$.

Find the optimal control feedback law to minimize the performance index

$$J = +\frac{1}{2} \sum_{k=0}^{\infty} x^2(k) + 2u^2(k)$$

SOLUTION

In this problem $A = 0.5$, $B = 1$, $Q = 1$, and $R = 2$. The algebraic Riccati equation is

$$P = 0.25P - 0.125P(1 + 0.5P)^{-1}P + 1 = \frac{0.25P}{1 + 0.5P} + 1$$

or

$$P^2 + 0.5P - 2 = 0$$

The above algebraic quadratic equation has two solutions

$$P = 1.1861 \quad \text{and} \quad P = -1.6861$$

Since matrix P has to be positive definite, the acceptable solution is the positive solution, therefore the constant feedback gain is

$$K = \frac{0.25P}{1 + 0.5P} = \frac{0.25 \times 1.1861}{1 + 0.5 \times 0.1861} = 0.1861$$

In Chapters 7, 8, and 10, we show the benefits of the LQR design technique in designing high-performance color reproduction systems.

5.4 STATE ESTIMATORS (OBSERVERS) DESIGN

5.4.1 INTRODUCTION

In state feedback design using pole placement or optimal control using LQR, the assumption is that all the states are available for feedback. In reality we may not have access to all the states. A state estimator is used to estimate the states of a dynamic system based on the output and the control signals [6,7].

5.4.2 FULL-ORDER OBSERVER DESIGN

In a full-order observer, every state of a dynamic system is estimated. Hence, the order of the observer is the same as the order of the original system. Consider a MIMO system given as

$$x(k + 1) = Ax(k) + Bu(k) \tag{5.76}$$

$$y(k) = Cx(k) \tag{5.77}$$

where
$A \in R^{N \times N}$
$B \in R^{N \times M}$
$u \in R^{M}$
$y \in R^{P}$
$C \in R^{P \times N}$

The block diagram of the system is shown in Figure 5.5.

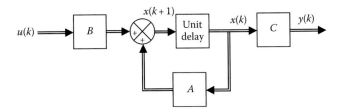

FIGURE 5.5   Block diagram of the system.

The state observer is given by

$$\hat{x}(k+1) = A\hat{x}(k) + Bu(k) + K[y(k) - C\hat{x}(k)] \qquad (5.78)$$

where
   $\hat{x} \in R^{N \times N}$ is an estimate of the state of the system $x(k)$
   K is a gain matrix controlling the dynamic behavior of the observer

Let the estimation error $e(k)$ be

$$e(k) = x(k) - \hat{x}(k) \qquad (5.79)$$

Then

$$e(k+1) = x(k+1) - \hat{x}(k+1) = Ax(k) + Bu(k) - A\hat{x}(k) - Bu(k) - K[y(k) - C\hat{x}(k)]$$
$$= A(x(k) - \hat{x}(k)) - K[Cx(k) - C\hat{x}(k)] = (A - KC)(x(k) - \hat{x}(k))$$
$$= (A - KC)e(k) \qquad (5.80)$$

Equation 5.80 has a solution that is given by

$$e(k) = (A - KC)^k e(0) \qquad (5.81)$$

Therefore if $A - KC$ is a stable matrix (i.e., eigenvalues inside unit circle in complex z-plane), the error signal $e(k)$ will approach zero as $k \to \infty$. Hence we need to choose gain matrix K by assigning eigenvalues to matrix $A - KC$. This is similar to the pole-placement algorithm discussed before, except that in pole placement we assign eigenvalues to $A - BK$. Since $A - KC = (A^T - C^T K^T)^T$, and eigenvalues are invariant under transpose operation, the problem is same as assigning eigenvalues to $A^T - C^T K^T$. Comparing the pole placement with this problem it is obvious that we can use any pole-placement algorithm with the following simple substitutions:

$$A \to A^T$$
$$B \to C^T \qquad (5.82)$$
$$K \to K^T$$

For example the Ackermann's formula becomes

$$K = \left\{ [0 \ 0 \ \cdots \ 0 \ 1] \left[C^T \ A^T C^T \ A^{T^2} C^T \ \cdots \ (A)^{T^{(N-1)}} C^T \right]^{-1} P(A^T) \right\}^T \qquad (5.83)$$

Equation 5.83 can be modified as

$$
K = P(A) \begin{bmatrix} C \\ CA \\ CA^2 \\ \vdots \\ CA^{N-1} \end{bmatrix}^{-1} \begin{bmatrix} 0 \\ 0 \\ \vdots \\ 0 \\ 1 \end{bmatrix} \tag{5.84}
$$

The MATLAB pole-placement command can also be used to compute the observer gain matrix as $G = \mathrm{place}(A^{\mathrm{T}}, B^{\mathrm{T}}, P)$, and then $K = G^{\mathrm{T}}$. The block diagram of the full-order observer is shown in Figure 5.6. Note that the necessary and sufficient condition of the existence of a full-state observer is that observability matrix must have full rank, that is

$$
\mathrm{rank} \begin{bmatrix} C \\ CA \\ CA^2 \\ \vdots \\ CA^{N-1} \end{bmatrix} = N \tag{5.85}
$$

In other words the system must be full-state observable.

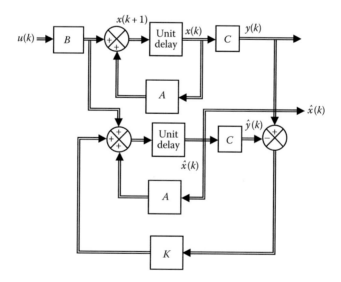

FIGURE 5.6   Block diagram of full-state observer.

Example 5.8

Consider the dynamic system given by

$$x(k + 1) = \begin{bmatrix} 0 & 1 \\ -0.72 & 1.7 \end{bmatrix} x(k) + \begin{bmatrix} 0 \\ 1 \end{bmatrix} u(k)$$

$$y(k) = [1 \quad 0]x(k)$$

Design a full-state observer for the system. Place the poles of the observer at

$$\lambda_1 = \lambda_2 = 0.25$$

SOLUTION

The characteristic polynomial of the observer is

$$P(\lambda) = (\lambda - \lambda_1)(\lambda - \lambda_2) = (\lambda - 0.25)(\lambda - 0.25) = \lambda^2 - 0.5\lambda + 0.0625$$

The observability matrix is

$$\begin{bmatrix} C \\ CA \end{bmatrix} = \begin{bmatrix} 1 & 0 \\ 0 & 1 \end{bmatrix}$$

Since the observability matrix is full rank, the system is completely state observable. The observer gain vector is given by

$$K = P(A) \begin{bmatrix} C \\ CA \\ CA^2 \\ \vdots \\ CA^{N-1} \end{bmatrix}^{-1} \begin{bmatrix} 0 \\ 0 \\ \vdots \\ 0 \\ 1 \end{bmatrix} = [A^2 - 0.5A + 0.0625I] \begin{bmatrix} C \\ CA \end{bmatrix}^{-1} \begin{bmatrix} 0 \\ 1 \end{bmatrix}$$

$$= \begin{bmatrix} -0.6575 & -0.33 \\ 0.2376 & -0.7136 \end{bmatrix} \begin{bmatrix} 1 & 0 \\ 0 & 1 \end{bmatrix} \begin{bmatrix} 0 \\ 1 \end{bmatrix} = \begin{bmatrix} -0.33 \\ -0.7136 \end{bmatrix}$$

The state observer is given by

$$\hat{x}(k + 1) = A\hat{x}(k) + Bu(k) + K[y(k) - C\hat{x}(k)]$$

$$\hat{x}(k + 1) = \begin{bmatrix} 0 & 1 \\ -0.72 & 1.7 \end{bmatrix} \hat{x}(k) + \begin{bmatrix} 0 \\ 1 \end{bmatrix} u(k) + \begin{bmatrix} -0.33 \\ -0.7136 \end{bmatrix} (y(k) - [1 \quad 0]\hat{x}(k))$$

or

$$\hat{x}(k + 1) = \begin{bmatrix} 0.33 & 1 \\ -0.0064 & 0.17 \end{bmatrix} \hat{x}(k) + \begin{bmatrix} 0 \\ 1 \end{bmatrix} u(k) + \begin{bmatrix} -0.33 \\ -0.7136 \end{bmatrix} y(k)$$

5.4.3   REDUCED-ORDER OBSERVER DESIGN

A full-order observer estimates all the states of the system regardless of whether they are measured or not. Consider the case when there are many states where most of them are measured except for a few. In this case, there is no need to estimate the available states. A reduced-order observer is used to estimate the states of a dynamic system, when partial measurement of some states is available.

Assume that the states of a dynamic system can be portioned into two parts, those that are measured directly and those that are not measured directly. That is

$$x(k) = \begin{bmatrix} \text{Measured} \\ \cdots \\ \text{Not measured} \end{bmatrix} = \begin{bmatrix} x_1(k) \\ \cdots \\ x_2(k) \end{bmatrix} \tag{5.86}$$

The state and output equations are portioned accordingly as

$$x_1(k+1) = A_{11}x_1(k) + A_{12}x_2(k) + B_1u(k)$$
$$x_2(k+1) = A_{21}x_1(k) + A_{22}x_2(k) + B_2u(k) \tag{5.87}$$
$$y(k) = Cx_1(k)$$

where

$A_{11}, A_{12}, A_{21}, A_{22}, B_1, B_2,$ and C are matrices of appropriate size
C is a nonsingular square matrix

Since $x_1(k)$ is measured, then its estimate is

$$\hat{x}_1(k) = x_1(k) = C^{-1}y(k) \tag{5.88}$$

The second set of states $x_2(k)$ are estimated as

$$\hat{x}_2(k) = Ly(k) + z(k) \tag{5.89}$$

where

$$z(k+1) = Fz(k) + Gy(k) + Hu(k) \tag{5.90}$$

The estimation error $e(k)$ is defined as

$$e(k) = x_2(k) - \hat{x}_2(k) \tag{5.91}$$

Dynamics of the estimation error $e(k)$ is governed by

$$
\begin{aligned}
e(k+1) &= x_2(k+1) - \hat{x}_2(k+1) = A_{21}x_1(k) + A_{22}x_2(k) + B_2u(k) - Ly(k+1) - z(k+1) \\
&= A_{21}x_1(k) + A_{22}x_2(k) + B_2u(k) - LCx_1(k+1) - Fz(k) - Gy(k) - Hu(k) \\
&= A_{21}x_1(k) + A_{22}x_2(k) + B_2u(k) - LC(A_{11}x_1(k) + A_{12}x_2(k) + B_1u(k)) \\
&\quad - Fz(k) - GCx_1(k) - Hu(k) \\
&= (A_{21} - LCA_{11} - GC)x_1(k) + (A_{22} - LCA_{12})x_2(k) + (B_2 - LCB_1 - H)u(k) - Fz(k) \\
&= (A_{21} - LCA_{11} - GC)x_1(k) + (A_{22} - LCA_{12})x_2(k) + (B_2 - LCB_1 - H)u(k) \\
&\quad - F(\hat{x}_2(k) - Ly(k))
\end{aligned} \tag{5.92}
$$

Equation 5.92 can be simplified further as

$$
\begin{aligned}
e(k+1) &= (A_{21} - LCA_{11} - GC)x_1(k) + (A_{22} - LCA_{12})x_2(k) \\
&\quad + (B_2 - LCB_1 - H)u(k) - F\hat{x}_2(k) + FLCx_1(k) \\
&= (A_{21} - LCA_{11} - GC + FLC)x_1(k) + (A_{22} - LCA_{12})x_2(k) \\
&\quad + (B_2 - LCB_1 - H)u(k) - F(x_2(k) - e(k))
\end{aligned}
\tag{5.93}
$$

or

$$
\begin{aligned}
e(k+1) &= Fe(k) + (A_{21} - LCA_{11} - GC + FLC)x_1(k) \\
&\quad + (A_{22} - LCA_{12} - F)x_2(k) + (B_2 - LCB_1 - H)u(k)
\end{aligned}
\tag{5.94}
$$

For the error to be independent of $x_1(k)$, $x_2(k)$, and $u(k)$, we must have

$$
\begin{aligned}
A_{21} - LCA_{11} - GC + FLC &= 0 \\
A_{22} - LCA_{12} - F &= 0 \\
B_2 - LCB_1 - H &= 0
\end{aligned}
\tag{5.95}
$$

Using the above three equations, we have

$$
F = A_{22} - LCA_{12}
\tag{5.96}
$$

$$
H = B_2 - LCB_1
\tag{5.97}
$$

$$
G = [A_{21} - LCA_{11}]C^{-1} + FL
\tag{5.98}
$$

with these choices for F, G, and H, Equation 5.94 is reduced to

$$
e(k+1) = Fe(k) = (A_{22} - LCA_{12})e(k)
\tag{5.99}
$$

The error signal will approach zero as $k \to \infty$ if the eigenvalues of matrix $A_{22} - LCA_{12}$ are inside unit circle in the complex plain. The gain L matrix is designed by assigning eigenvalues to the matrix $A_{22} - LCA_{12}$. This is similar to the full-observer design with $A_{22} - LCA_{12}$ playing the role of $A - LC$. The following simple example illustrates the design of a reduced-order observer:

Example 5.9

Consider the dynamic system given by

$$
x(k+1) = \begin{bmatrix} 0 & 1 \\ -0.72 & 1.7 \end{bmatrix} x(k) + \begin{bmatrix} 0 \\ 1 \end{bmatrix} u(k)
$$

$$
y(k) = \begin{bmatrix} 1 & 0 \end{bmatrix} x(k)
$$

Design a reduced-order observer to estimate the second state (x_2) of the system. Place the pole of the reduced-order observer at

$$\lambda = 0.1$$

SOLUTION

Since $y(k) = x_1(k)$, we have $C = 1$, $A_{11} = 0$, $A_{12} = 1$, $A_{21} = -0.72$, $A_{22} = 1.7$, $B_1 = 0$, and $B_2 = 1$. Then

$$F = A_{22} - LCA_{12} = 1.7 - L$$
$$H = B_2 - LCB_1 = 1$$
$$G = -0.72 + FL$$

To place the pole of the reduced-order observer at $\lambda = 0.1$, we have

$$1.7 - L = 0.1$$

Solving for L yields

$$L = 1.6$$

and

$$F = 1.7 - L = 0.1$$
$$H = 1$$
$$G = -0.72 + FL = -0.72 + 0.16 = -0.56$$

Therefore, the reduced-order observer is given as

$$\hat{x}_1(k) = x_1(k) = y(k)$$
$$\hat{x}_2(k) = 1.6y(k) + z(k)$$
$$z(k+1) = 0.1z(k) - 0.56y(k) + u(k)$$

5.5 COMBINED STATE ESTIMATION AND CONTROL

5.5.1 INTRODUCTION

Once the states of a dynamic system are estimated, a state feedback can be designed using the estimated states. The two processes of designing the state estimator and state feedback through pole placement are decoupled and are done separately.

5.5.2 COMBINED CONTROLLER AND OBSERVER

The state feedback control law $u(k) = -K_f x(k)$ is replaced by

$$u(k) = -K_f \hat{x}(k) \tag{5.100}$$

where

$\hat{x}(k)$ is the estimate of $x(k)$ at time k

K_f is the state feedback gain matrix obtained through pole placement

The combined controller and observer equations are given by

$x(k+1) = Ax(k) + Bu(k) = Ax(k) - BK_\mathrm{f}\hat{x}(k)$

$\hat{x}(k+1) = A\hat{x}(k) + Bu(k) + K_\mathrm{e}(y(k) - C\hat{x}(k)) = A\hat{x}(k) - BK_\mathrm{f}\hat{x}(k) + K_\mathrm{e}(y(k) - C\hat{x}(k))$

$$(5.101)$$

or

$$\begin{bmatrix} x(k+1) \\ \hat{x}(k+1) \end{bmatrix} = \begin{bmatrix} A & -BK_\mathrm{f} \\ K_\mathrm{e}C & A - BK_\mathrm{f} - K_\mathrm{e}C \end{bmatrix} \begin{bmatrix} x(k) \\ \hat{x}(k) \end{bmatrix} \qquad (5.102)$$

$$y(k) = Cx(k)$$

where K_e is the observer gain matrix. The block diagram of the overall system is shown in Figure 5.7.

The estimation error is controlled by the eigenvalues of $A - K_\mathrm{e}C$. If the (A, C) pair is observable, the eigenvalues can be assigned arbitrarily similar to the pole

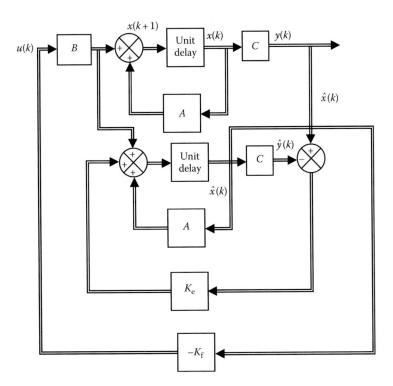

FIGURE 5.7   Combined state estimator and state feedback.

placement. These eigenvalues should be located inside the unit circle closer to the origin than the eigenvalues of $A - BK_f$. With this eigenvalue assignment, the estimation error will approach zero faster than the dynamics of the system.

Example 5.10

Consider the dynamic system given by

$$x(k+1) = \begin{bmatrix} 0 & 1 \\ -0.72 & 1.7 \end{bmatrix} x(k) + \begin{bmatrix} 0 \\ 1 \end{bmatrix} u(k)$$

$$y(k) = [1 \quad 0]x(k)$$

Design a combined state feedback with a state observer for the system. Place the eigenvalues of the closed-loop state feedback system at $\lambda_1 = 0.3$ and $\lambda_2 = 0.2$. Place the observer poles at $\lambda_1 = 0.1$ and $\lambda_2 = 0.1$.

SOLUTION

The characteristic polynomial of the open-loop system is

$$P(\lambda) = |\lambda I - A| = \lambda^2 - 1.7\lambda + 0.72$$

Hence, $\beta_1 = -1.7$ and $\beta_2 = 0.72$.

The characteristic polynomial of the desired closed-loop system is

$$P_c(\lambda) = (\lambda - 0.3)(\lambda - 0.2) = \lambda^2 - 0.5\lambda + 0.06$$

Hence, $\alpha_1 = -0.5$ and $\alpha_2 = 0.06$. The transformation T is given by

$$T = QW = [B \quad AB]\begin{bmatrix} \beta_1 & 1 \\ 1 & 0 \end{bmatrix} = \begin{bmatrix} 0 & 1 \\ 1 & 1.7 \end{bmatrix}\begin{bmatrix} -1.7 & 1 \\ 1 & 0 \end{bmatrix} = \begin{bmatrix} 1 & 0 \\ 0 & 1 \end{bmatrix}$$

The feedback gain matrix K_f is given as

$$K_f = [\alpha_2 - \beta_2 \quad \alpha_1 - \beta_1]T^{-1} = [-0.12 \quad -1]\begin{bmatrix} 1 & 0 \\ 0 & 1 \end{bmatrix}^{-1} = [-0.66 \quad 1.2]$$

We now design a full-state observer for the system. The characteristic polynomial of the observer is

$$P(\lambda) = (\lambda - \lambda_1)(\lambda - \lambda_2) = (\lambda - 0.1)(\lambda - 0.1) = \lambda^2 - 0.2\lambda + 0.01$$

The observability matrix is

$$\begin{bmatrix} C \\ CA \end{bmatrix} = \begin{bmatrix} 1 & 0 \\ 0 & 1 \end{bmatrix}$$

Since the observability matrix is full rank, the system is completely state observable. The observer gain vector K_e is given by

$$K_e = P(A) \begin{bmatrix} C \\ CA \\ CA^2 \\ \vdots \\ CA^{N-1} \end{bmatrix}^{-1} \begin{bmatrix} 0 \\ 0 \\ \vdots \\ 0 \\ 1 \end{bmatrix} = [A^2 - 0.2A + 0.01I] \begin{bmatrix} C \\ CA \end{bmatrix}^{-1} \begin{bmatrix} 0 \\ 1 \end{bmatrix}$$

$$= \begin{bmatrix} -0.71 & 1.5 \\ 1.08 & 1.84 \end{bmatrix} \begin{bmatrix} 1 & 0 \\ 0 & 1 \end{bmatrix} \begin{bmatrix} 0 \\ 1 \end{bmatrix} = \begin{bmatrix} 1.5 \\ 1.84 \end{bmatrix}$$

The dynamics of the state feedback and observer are given by

$$\begin{bmatrix} x(k+1) \\ \hat{x}(k+1) \end{bmatrix} = \begin{bmatrix} A & -BK_f \\ K_eC & A - BK_f - K_eC \end{bmatrix} \begin{bmatrix} x(k) \\ \hat{x}(k) \end{bmatrix}$$

or

$$\begin{bmatrix} x_1(k+1) \\ x_2(k+1) \\ \hat{x}_1(k+1) \\ \hat{x}_2(k+1) \end{bmatrix} = \begin{bmatrix} 0 & 1 & 0 & 0 \\ -0.72 & 1.7 & 0.66 & -1.2 \\ 1.5 & 0 & -1.5 & 1 \\ 1.84 & 0 & -1.9 & 0.5 \end{bmatrix} \begin{bmatrix} x_1(k) \\ x_2(k) \\ \hat{x}_1(k) \\ \hat{x}_2(k) \end{bmatrix}$$

The states of the system with initial conditions of $x(0) = [1 \quad 2]^T$ are shown in Figures 5.8 and 5.9.

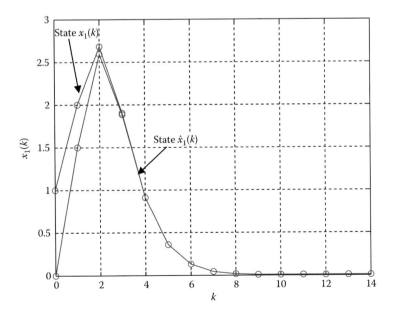

FIGURE 5.8 Combined state estimator and state feedback (first state).

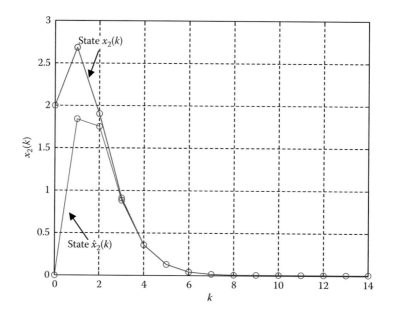

FIGURE 5.9 Combined state estimator and state feedback (second state).

PROBLEMS

5.1 A dynamic system has a plant: $H(z) = \frac{Y(z)}{U(z)} = \frac{10z}{(z+1)(z+0.4)(z+0.7)}$

   a. Define the state variables as $x_1(k) = y(k)$, $x_2(k) = y(k+1)$, and $x_3(k) = y(k+2)$. Derive state equation for this system.

   b. By use of state feedback control $u = -Kx$, obtain the gain matrix K so that the closed-loop poles are at

$$\lambda_1 = 0.1, \quad \lambda_2 = -0.4 + j0.3, \quad \lambda_3 = -0.4 - j0.3$$

5.2 Consider the system defined by

$$\begin{bmatrix} x_1(k+1) \\ x_2(k+1) \end{bmatrix} = \begin{bmatrix} -1 & 1 \\ 0 & 2 \end{bmatrix} \begin{bmatrix} x_1(k) \\ x_2(k) \end{bmatrix} + \begin{bmatrix} 1 \\ 0 \end{bmatrix} u(k)$$

   a. Show that this system cannot be stabilized by the state feedback control $u(k) = -Kx(k)$ for any choice of matrix K.

   b. Is it controllable? If not, determine the uncontrollable mode.

5.3 Prove matrix inversion lemma which states that for any A, C, and D matrices of appropriate dimensions, we have

$$(A + CD)^{-1} = A^{-1} - A^{-1}C(I + DA^{-1}C)^{-1}DA^{-1}$$

5.4 Consider the system defined by

$$x(k + 1) = 0.9x(k) + u(k)$$

Find the state feedback control law $u(k) = -K(k)x(k)$ minimizing the performance index

$$J = 2x^2(10) + \frac{1}{2} \sum_{k=0}^{9} [x^2(k) + 0.5u^2(k)]$$

5.5 Consider the system defined by

$$\begin{bmatrix} x_1(k+1) \\ x_2(k+1) \end{bmatrix} = \begin{bmatrix} -0.6 & 1 \\ 0 & 0.5 \end{bmatrix} \begin{bmatrix} x_1(k) \\ x_2(k) \end{bmatrix} + \begin{bmatrix} 1 \\ 0 \end{bmatrix} u(k)$$

Find the state feedback control law $u(k) = -Kx(k)$ minimizing the performance index

$$J = \frac{1}{2} \sum_{k=0}^{\infty} x_1^2(k) + 2x_2^2(k) + u^2(k)$$

5.6 Consider the system defined by

$$\begin{bmatrix} x_1(k+1) \\ x_2(k+1) \end{bmatrix} = \begin{bmatrix} 0 & 1 \\ 0.4 & 0.5 \end{bmatrix} \begin{bmatrix} x_1(k) \\ x_2(k) \end{bmatrix} + \begin{bmatrix} 1 \\ 1 \end{bmatrix} u(k)$$

$$y(k) = x_1(k)$$

Design a full-order state observer. The desired eigenvalues for the observer matrix are

$$\lambda_1 = 0.1 \quad \text{and} \quad \lambda_2 = 0.2$$

5.7 Consider the system defined by

$$x(k + 1) = Ax(k) + Bu(k)$$

where

$$A = \begin{bmatrix} 0 & 1 & 0 \\ 0 & 0 & 1 \\ 0.1 & 0.25 & -0.4 \end{bmatrix}, \quad B = \begin{bmatrix} 0 \\ 0 \\ 1 \end{bmatrix}$$

Design a linear state feedback for this system. Choose the desired closed-loop poles for deadbeat control.

5.8 Consider the system defined by

$$x(k + 1) = Ax(k) + Bu(k)$$
$$y(k) = Cx(k)$$

where

$$A = \begin{bmatrix} 0 & 1 & 0 \\ 0 & 0 & 1 \\ 0.1 & 0.25 & -0.4 \end{bmatrix}, \quad B = \begin{bmatrix} 0 \\ 0 \\ 1 \end{bmatrix}, \quad C = [1 \ \ 0 \ \ 0]$$

Design a full-order state observer. The desired eigenvalues for the observer matrix are

$$\lambda_1 = \lambda_2 = 2\lambda_3 = 0.2$$

5.9 Consider the system defined by

$$x(k + 1) = Ax(k) + Bu(k)$$
$$y(k) = Cx(k)$$

where

$$A = \begin{bmatrix} 0 & 1 & 0 \\ 0 & 0 & 1 \\ 0.1 & 0.25 & -0.4 \end{bmatrix}, \quad B = \begin{bmatrix} 0 \\ 0 \\ 1 \end{bmatrix}, \quad C = [1 \ \ 0 \ \ 0]$$

Design a reduced-order state observer. The desired eigenvalues for the reduced-order observer matrix are

$$\lambda_1 = \lambda_2 = 0.1$$

5.10 Consider the system

$$x(k + 1) = Ax(k) + Bu(k)$$
$$y(k) = Cx(k)$$

where

$$A = \begin{bmatrix} 0 & 1 & 0 \\ 0 & 0 & 1 \\ 0.1 & 0.25 & -0.4 \end{bmatrix}, \quad B = \begin{bmatrix} 1 \\ 0 \\ 1 \end{bmatrix}, \quad C = \begin{bmatrix} 0 & 0 & 1 \\ 1 & 0 & 0 \end{bmatrix}$$

Design a reduced-order state observer such that the response to the observer error is deadbeat, which is the desired eigenvalue of the reduced-order observer, which is $\lambda = 0$.

REFERENCES

1. K. Ogata, *Modern Control Engineering*, Prentice Hall, Englewood, NJ, 2001.
2. W. L. Brogan, *Modern Control Theory*, Prentice Hall, Englewood, NJ, 1990.
3. B. Friedland, *Control System Design: An Introduction to State-Space Methods*, McGraw-Hill, New York, 1985.
4. K. Ogata, *Discrete-Time Control Systems*, Prentice Hall, Englewood, NJ, 1994.
5. G. Franklin, J. D. Powell, and E. Naeini, *Feedback Control of Dynamic Systems*, Prentice-Hall, Upper Saddle River, NJ, 2005.
6. A. Tewari, *Modern Control Design with MATLAB and Simulink*, Wiley, Chichester, United Kingdom, 2002.
7. B. Kuo, *Digital Control Systems*, Oxford University Press, New York, 1995.

6 Interpolation of Multidimensional Functions

6.1 INTRODUCTION

Multidimensional functions play an important role in digital color imaging systems. For example, a digital color printer can be modeled as a forward map from four-dimensional (4-D) device-dependent *CMYK* color space to three-dimensional (3-D) device-independent *L*a*b** color space. The mapping can be defined by the following three multidimensional functions:

$$L^* = f_1(C, M, Y, K)$$
$$a^* = f_2(C, M, Y, K) \tag{6.1}$$
$$b^* = f_3(C, M, Y, K)$$

Unfortunately, exact closed-form expressions are not easily available for the above equations. Therefore, we have two options. The first option is to have a lookup table (LUT) of all possible combinations of *CMYK*; that is, change each separation from 0 to 255 in steps of one, which requires a LUT size of 255^4 entries that is equivalent to 3×255^4 bytes $= 12.6856$ bytes of *CMYK* – *L*a*b** data. This is not practical. Another approach would be to have a LUT of size smaller than the full size, for example, 17^4 or lesser, that is, *CMYK* – *L*a*b** data and use interpolation to find other colors. Multidimensional interpolation is also extensively used in displays, scanners, and inverse printer maps.

There are two types of interpolation techniques: linear and nonlinear. Generally speaking, linear interpolation techniques are used to interpolate the data, where the color space is uniformly sampled (uniform LUT) and is reasonably linear. If the color space data is nonuniform and is not linear, then nonlinear interpolations are preferred.

In this chapter, we present different methods of multidimensional interpolation. We first consider uniform LUTs and discuss linear interpolation in one-dimensional (1-D), two-dimensional (2-D), and 3-D color spaces. An extension to 4-D color space is straightforward. After this, we consider nonuniform LUTs and describe nonlinear techniques such as the Shepard and moving-matrix approaches. As an example of application, a numerical computation of inverse printer maps and techniques for downsampling the color space LUTs are also presented in some detail.

6.2 INTERPOLATION OF UNIFORMLY SPACED LOOKUP TABLES

Interpolation of uniformly spaced LUTs can be performed using linear, cubic, and other nonlinear techniques [1]. We will first consider linear (1-D) and bilinear (2-D) interpolation techniques. Trilinear interpolation with application to digital printers is discussed in the subsequent section.

6.2.1 LINEAR AND BILINEAR INTERPOLATIONS

We first consider the linear interpolation of a 1-D LUT. Consider the 1-D function $y = f(x)$, shown in Figure 6.1, where x is the independent variable and y is the dependent variable. Assume that a uniform LUT of N data points is available. The LUT is shown in Table 6.1. Suppose that we would like to interpolate point b on the curve. This point is located between the two LUT nodes a and c, that is, $x_i \leq x \leq x_{i+1}$. The interpolated value is located on the straight line connecting the two nodes a and c, as shown in Figure 6.1. The interpolated value y is given by

$$y = \frac{y_{i+1} - y_i}{x_{i+1} - x_i}(x - x_i) + y_i \tag{6.2}$$

for $x_i \leq x \leq x_{i+1}$.

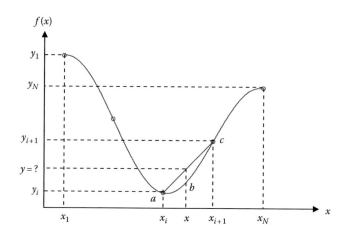

FIGURE 6.1   One-dimensional function $f(x)$.

TABLE 6.1

Uniform LUT of N Data Points

x	$y = f(x)$
x_1	y_1
x_2	y_2
$\vdots$	$\vdots$
x_N	y_N

Example 6.1

Consider the 1-D LUT given in Table 6.2. Use linear interpolation to interpolate the value of the function at $x = 5.4$.

SOLUTION

Figure 6.2 shows the plot of function $y = f(x)$. As seen from this figure, we have

$$y = \frac{y_{i+1} - y_i}{x_{i+1} - x_i}(x - x_i) + y_i = \frac{3.3 - 1.62}{6 - 4}(5.4 - 6) + 3.3 = 2.796$$

In the 2-D case, the underlying function is a function of two variables, $z = f(x, y)$, and the nodes in the LUT are uniformly spaced on grids, as shown in Figure 6.3. Assume that we would like to interpolate the value of the function z at point a with coordinates (x, y). Let point a be within the grids b–e as shown in Figure 6.4. Then the bilinear interpolated value of the function at point a is given by

$$z = f(x, y) = p_{00} + t(p_{01} - p_{00}) + u(p_{10} - p_{00}) + tu(p_{11} - p_{01} - p_{10} - p_{00}) \quad (6.3)$$

where

u and t are the relative distances from the surrounding nodes

p_{00}, p_{01}, p_{11}, and p_{10} are the values of the function $f(x, y)$ at points b, c, d, and e, respectively, as shown in Figure 6.4.

TABLE 6.2
Example of a Uniform LUT

x	0	2	4	6	8	10
y	0.9	0.98	1.62	3.3	6.5	11.8

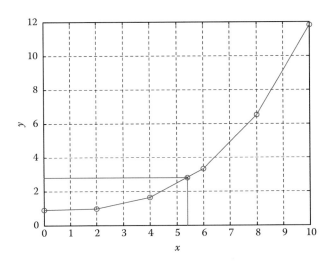

FIGURE 6.2   Plot of function $f(x)$.

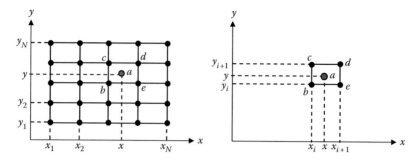

FIGURE 6.3   Two-dimensional uniform grid.

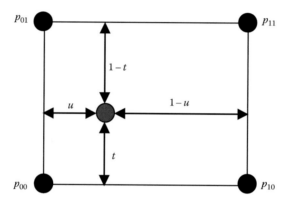

FIGURE 6.4   Relative distances.

The relative distances u and t are given by

$$u = \frac{x - x_i}{x_{i+1} - x_i}, \quad t = \frac{y - y_i}{y_{i+1} - y_i} \tag{6.4}$$

Substituting u and t in Equation 6.3 results in

$$z = f(x, y) = f(x_i, y_i) + \frac{x - x_i}{x_{i+1} - x_i}[f(x_{i+1}, y_i) - f(x_i, y_i)] + \frac{y - y_i}{y_{i+1} - y_i}[f(x_i, y_{i+1}) - f(x_i, y_i)]$$
$$+ \left(\frac{x - x_i}{x_{i+1} - x_i}\right)\left(\frac{y - y_i}{y_{i+1} - y_i}\right)[f(x_{i+1}, y_{i+1}) - f(x_{i+1}, y_i) - f(x_i, y_{i+1}) + f(x_i, y_i)] \tag{6.5}$$

Example 6.2

Consider a static system with two inputs, x_1 and x_2, and two outputs, y_1 and y_2, where $y_1 = f_1(x_1, x_2)$ and $y_2 = f_2(x_1, x_2)$. The LUT representing these two functions is given in Table 6.3 and Figure 6.5. Use bilinear interpolation to find the value of the two functions at point $x = [x_1 \ x_2]$, where $x_1 = 1.5$ and $x_2 = 0.6$.

TABLE 6.3
LUT for a Two-Input
Two-Output System

x_1	x_2	y_1	y_2
0	0	0	0
0	1	1	3
0	2	2	6
1	0	2	1
1	1	3	4
1	2	4	7
2	0	4	2
2	1	5	5
2	2	6	8

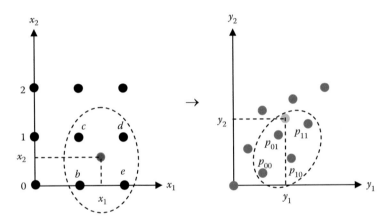

FIGURE 6.5 Range and domain of the two functions $y_1(x_1, x_2)$ and $y_2(x_1, x_2)$.

SOLUTION

The values of u and t are

$$u = \frac{1.5 - 1}{2 - 1} = 0.5 \quad \text{and} \quad t = \frac{0.6 - 0}{1 - 0} = 0.6$$

since x_1 is between 1 and 2, and x_2 is between 0 and 1, the four surrounding points in $x_1 - x_2$ space and their corresponding $\begin{bmatrix} y_1 \\ y_2 \end{bmatrix}$ values are

$$b = \begin{bmatrix} 1 \\ 0 \end{bmatrix}, \quad c = \begin{bmatrix} 1 \\ 1 \end{bmatrix}, \quad d = \begin{bmatrix} 2 \\ 1 \end{bmatrix}, \quad \text{and} \quad e = \begin{bmatrix} 2 \\ 0 \end{bmatrix}$$

$$p_{00} = \begin{bmatrix} 2 \\ 1 \end{bmatrix}, \quad p_{01} = \begin{bmatrix} 3 \\ 4 \end{bmatrix}, \quad p_{11} = \begin{bmatrix} 5 \\ 5 \end{bmatrix}, \quad \text{and} \quad p_{10} = \begin{bmatrix} 4 \\ 2 \end{bmatrix}$$

Therefore,

$$
\begin{bmatrix} y_1 \\ y_2 \end{bmatrix} = p_{00} + t(p_{01} - p_{00}) + u(p_{10} - p_{00}) + tu(p_{11} - p_{01} - p_{10} - p_{00})
$$

$$
= \begin{bmatrix} 2 \\ 1 \end{bmatrix} + 0.6 \begin{bmatrix} 3-2 \\ 4-1 \end{bmatrix} + 0.5 \begin{bmatrix} 4-2 \\ 2-1 \end{bmatrix} + 0.6 \times 0.5 \begin{bmatrix} 5-4-3+2 \\ 5-2-4+1 \end{bmatrix} = \begin{bmatrix} 3.6 \\ 3.3 \end{bmatrix}
$$

We next consider a trilinear interpolation of 3-D LUTs.

6.2.2 TRILINEAR INTERPOLATION

In the 3-D case, the underlying function is a function of three variables, $f(x, y, z)$, and the LUT is uniformly spaced in a 3-D lattice of $N \times M \times L$ points. An example of such a lattice for $N = M = L = 4$ is shown in Figure 6.6. As can be seen, there are a total of 64 grid points uniformly spaced in the 3-D space.

Now assume that we have a 3-D LUT of size $N \times M \times L$ grid points uniformly spaced in the xyz 3-D space and would like to interpolate the grid point (x, y, z). Let the surrounding eight points in the xyz plane be $\begin{bmatrix} n_{000} & n_{001} & n_{010} & n_{011} & n_{100} \\ n_{101} & n_{110} & n_{111} \end{bmatrix}$ and the corresponding points in the f plane be $\begin{bmatrix} p_{000} & p_{001} & p_{010} \\ p_{011} & p_{100} & p_{101} & p_{110} & p_{111} \end{bmatrix}$, as shown in Figure 6.7. Let t, u, and v be the relative distances from the plane and the axes, as shown in Figure 6.7.

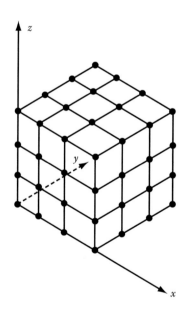

FIGURE 6.6 Uniformly sampled lattice in 3-D space.

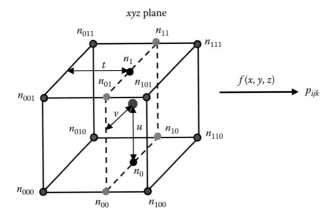

FIGURE 6.7 Interpolation cube and its eight vertices.

The interpolated value of f at location (x, y, z) is computed in three steps as follows:

(1) Use bilinear interpolation to interpolate the corresponding values of the function f at nodes $[\, n_{00} \quad n_{01} \quad n_{10} \quad n_{11}\,]$, as shown by the dotted plane in Figure 6.7:

$$
\begin{aligned}
p_{00} &= p_{000} + t(p_{100} - p_{000}) \\
p_{01} &= p_{001} + t(p_{101} - p_{001}) \\
p_{10} &= p_{010} + t(p_{110} - p_{010}) \\
p_{11} &= p_{011} + t(p_{111} - p_{011})
\end{aligned}
\tag{6.6}
$$

(2) Use linear interpolation to interpolate the corresponding values of the function f at nodes $[n_0 \ n_1]$, as shown in Figure 6.7:

$$
\begin{aligned}
p_0 &= p_{00} + v(p_{10} - p_{00}) \\
p_1 &= p_{01} + v(p_{11} - p_{01})
\end{aligned}
\tag{6.7}
$$

(3) Use linear interpolation to interpolate the corresponding values of the function f at nodes $[x \ y \ z]$, as shown in Figure 6.7:

$$
p = p_0 + u(p_1 - p_0)
\tag{6.8}
$$

The relative distances t, v, and u are given by

$$
t = \frac{x - x_i}{x_{i+1} - x_i}, \quad v = \frac{y - y_i}{y_{i+1} - y_i}, \quad \text{and} \quad u = \frac{z - z_i}{z_{i+1} - z_i}
\tag{6.9}
$$

Example 6.3

Consider a system with three inputs, x, y, and z, and one output, f. The LUT representing the function is given in Table 6.4. Use trilinear interpolation to find the value of the function at points $x = 1.5$, $y = 0.6$, and $z = 1.85$.

The point $p = [x \quad y \quad z]^T = [1.5 \quad 0.6 \quad 1.85]^T$ is inside the cube with vertices $[n_{000} \quad n_{001} \quad n_{010} \quad n_{011} \quad n_{100} \quad n_{101} \quad n_{110} \quad n_{111}]$, as shown in Figure 6.8. The coordinates of these nodes and the corresponding values of the function f are given in Table 6.5.

TABLE 6.4
Example of a Uniformly Sampled 3-D LUT

x	0	1	2	0	1	2	0	1	2	0	1	2	0	1	2	0	1	2	0	1	2	0	1	2	0	1	2
y	0	0	0	1	1	1	2	2	2	0	0	0	1	1	1	2	2	2	0	0	0	1	1	1	2	2	2
z	0	0	0	0	0	0	0	0	0	1	1	1	1	1	1	1	1	1	2	2	2	2	2	2	2	2	2
f	2	3	4	1	5	7	8	3	0	4	3	2	3	2	1	3	7	8	9	2	5	6	4	3	8	6	3

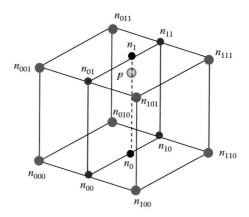

FIGURE 6.8   Eight vertices of the cube surrounding the point p.

TABLE 6.5
LUT for Example 6.3

	n_{000}	n_{001}	n_{010}	n_{011}	n_{100}	n_{101}	n_{110}	n_{111}
x	1	1	1	1	2	2	2	2
y	0	0	1	1	0	0	1	1
z	1	2	1	2	1	2	1	2
f	3	2	2	4	2	5	1	3

SOLUTION

The parameters t, v, and u are

$$t = \frac{x - x_i}{x_{i+1} - x_i} = \frac{1.5 - 1}{2 - 1} = 0.5$$

$$v = \frac{y - y_i}{y_{i+1} - y_i} = \frac{0.6 - 0}{1 - 0} = 0.6$$

$$u = \frac{z - z_i}{z_{i+1} - z_i} = \frac{1.85 - 1}{2 - 1} = 0.85$$

The interpolated values at nodes $\begin{bmatrix} n_{00} & n_{01} & n_{10} & n_{11} \end{bmatrix}$ are

$$p_{00} = p_{000} + t(p_{100} - p_{000}) = 3 + 0.5(2 - 3) = 2.5$$
$$p_{01} = p_{001} + t(p_{101} - p_{001}) = 2 + 0.5(5 - 2) = 3.5$$
$$p_{10} = p_{010} + t(p_{110} - p_{010}) = 2 + 0.5(1 - 2) = 1.5$$
$$p_{11} = p_{011} + t(p_{111} - p_{011}) = 4 + 0.5(3 - 4) = 3.5$$

The interpolated values at nodes $\begin{bmatrix} n_0 & n_1 \end{bmatrix}$ are

$$p_0 = p_{00} + v(p_{10} - p_{00}) = 2.5 + 0.6(1.5 - 2.5) = 1.9$$
$$p_1 = p_{01} + v(p_{11} - p_{01}) = 3.5 + 0.6(3.5 - 3.5) = 3.5$$

Finally, the interpolated value of the function at point p is

$$p = p_0 + u(p_1 - p_0) = 1.9 + 0.85(3.5 - 1.9) = 3.26$$

6.2.3 TETRAHEDRAL INTERPOLATION

Tetrahedral interpolation is another approach for interpolating the regularly sampled LUTs [2]. Now assume that we have a 3-D LUT of size $N \times M \times L$ grid points uniformly spaced in xyz 3-D space and would like to interpolate the grid point (x, y, z). Let the surrounding eight nodes in the xyz plane be $\begin{bmatrix} n_{000} & n_{001} & n_{010} & n_{011} & n_{100} & n_{101} & n_{110} & n_{111} \end{bmatrix}$ and the corresponding points in the f plane be $\begin{bmatrix} p_{000} & p_{001} & p_{010} & p_{011} & p_{100} & p_{101} & p_{110} & p_{111} \end{bmatrix}$, as shown in Figure 6.8. The tetrahedral interpolation divides this cube into six tetrahedrals, as shown in Figure 6.9. The interpolated value is the weighted sum of the values of the function at the four vertices of the tetrahedral enclosing the desired point. That is,

$$p = p_{000} + p_x \frac{x - x_0}{x_1 - x_0} + p_y \frac{y - y_0}{y_1 - y_0} + p_z \frac{z - z_0}{z_1 - z_0} \tag{6.10}$$

The expressions for p_x, p_y, and p_z depend on the location of p with respect to the six tetrahedral and are given as follows:

(1) If $x - x_0 > y - y_0 > z - z_0$, then p is in tetrahedral 1 and

$$p_x = p_{100} - p_{000}, \quad p_y = p_{110} - p_{100}, \quad \text{and} \quad p_z = p_{111} - p_{110} \tag{6.11}$$

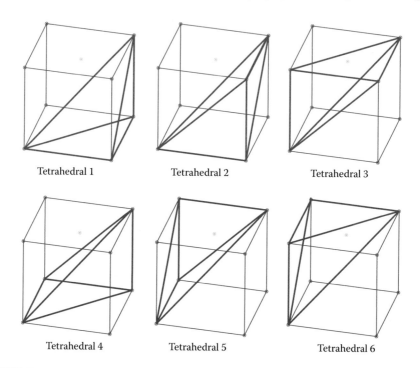

Tetrahedral 1   Tetrahedral 2   Tetrahedral 3

Tetrahedral 4   Tetrahedral 5   Tetrahedral 6

FIGURE 6.9 Dividing the cube into six tetrahedrals.

(2) If $x - x_0 > z - z_0 > y - y_0$, then p is in tetrahedral 2 and

$$p_x = p_{100} - p_{000}, \quad p_y = p_{111} - p_{101}, \quad \text{and} \quad p_z = p_{101} - p_{100} \qquad (6.12)$$

(3) If $z - z_0 > x - x_0 > y - y_0$, then p is in tetrahedral 3 and

$$p_x = p_{101} - p_{001}, \quad p_y = p_{111} - p_{101}, \quad \text{and} \quad p_z = p_{001} - p_{000} \qquad (6.13)$$

(4) If $y - y_0 > x - x_0 > z - z_0$, then p is in tetrahedral 4 and

$$p_x = p_{110} - p_{010}, \quad p_y = p_{010} - p_{000}, \quad \text{and} \quad p_z = p_{111} - p_{110} \qquad (6.14)$$

(5) If $y - y_0 > z - z_0 > x - x_0$, then p is in tetrahedral 5 and

$$p_x = p_{111} - p_{011}, \quad p_y = p_{010} - p_{000}, \quad \text{and} \quad p_z = p_{011} - p_{010} \qquad (6.15)$$

(6) If $z - z_0 > y - y_0 > x - x_0$, then p is in tetrahedral 6 and

$$p_x = p_{111} - p_{011}, \quad p_y = p_{011} - p_{001}, \quad \text{and} \quad p_z = p_{001} - p_{000} \qquad (6.16)$$

Example 6.4

Consider the transformation from device-dependent CMY to device-independent $L^*a^*b^*$ given by the LUT shown in Table 6.6.

Using tetrahedral interpolation, find the $L^*a^*b^*$ values corresponding to the $CMY = [67\ 128\ 54]$ color patch.

SOLUTION

The desired $CMY = [x\ y\ z] = [67\ 128\ 54]$ point is inside the tetrahedral, as shown in Figure 6.10. The CMY values at the eight vertices of the cube are

$$n_{000} = \begin{bmatrix} 0 \\ 85 \\ 0 \end{bmatrix}, \quad n_{001} = \begin{bmatrix} 0 \\ 85 \\ 85 \end{bmatrix}, \quad n_{010} = \begin{bmatrix} 0 \\ 170 \\ 0 \end{bmatrix}, \quad n_{011} = \begin{bmatrix} 0 \\ 170 \\ 85 \end{bmatrix},$$

$$n_{100} = \begin{bmatrix} 85 \\ 85 \\ 0 \end{bmatrix}, \quad n_{101} = \begin{bmatrix} 85 \\ 85 \\ 85 \end{bmatrix}, \quad n_{110} = \begin{bmatrix} 85 \\ 170 \\ 0 \end{bmatrix}, \quad \text{and} \quad n_{111} = \begin{bmatrix} 85 \\ 170 \\ 85 \end{bmatrix}$$

TABLE 6.6
Example of a Printer Forward Map (CMY to $L^*a^*b^*$)

C	M	Y	L*	a*	b*
0	0	0	100.00	−0.04	0.00
0	0	85	97.61	−4.88	35.21
0	0	170	96.02	−7.53	69.72
0	0	255	94.96	−8.59	100.00
0	85	0	82.98	29.86	−7.28
0	85	85	81.49	23.82	24.18
0	85	170	80.94	17.72	87.70
0	85	255	80.51	19.95	53.95
0	170	0	69.10	55.79	−11.76
0	170	85	68.90	48.72	16.33
0	170	170	68.86	44.21	41.85
0	170	255	68.80	41.16	69.85
0	255	0	53.82	86.18	−14.25
0	255	85	54.58	78.25	9.65
0	255	170	55.07	73.16	29.49
0	255	255	55.16	70.02	49.03
85	0	0	83.70	−13.64	−20.44
85	0	85	81.13	−20.69	12.47
85	0	170	79.84	−24.61	43.54
85	0	255	78.66	−26.07	77.41
85	85	0	68.50	15.59	−25.42

(*continued*)

TABLE 6.6 (continued)

Example of a Printer Forward Map (*CMY* to *L*a*b**)

C	M	Y	L*	a*	b*
85	85	85	67.73	7.81	5.02
85	85	170	67.82	2.65	32.37
85	85	255	66.72	1.31	61.35
85	170	0	56.67	40.60	−28.83
85	170	85	57.40	31.75	−0.64
85	170	170	57.65	27.53	23.35
85	170	255	56.89	25.47	47.26
85	255	0	42.76	70.00	−30.95
85	255	85	44.53	62.14	−6.25
85	255	170	45.45	57.65	13.12
85	255	255	45.60	54.93	30.82
170	0	0	66.30	−29.96	−41.98
170	0	85	65.11	−39.47	−9.95
170	0	170	64.37	−44.19	18.92
170	0	255	63.69	−46.42	49.90
170	85	0	54.01	−1.44	−43.99
170	85	85	54.92	−11.56	−14.10
170	85	170	54.96	−16.24	11.53
170	85	255	53.84	−18.73	37.31
170	170	0	44.03	23.42	−45.74
170	170	85	45.67	13.91	−17.88
170	170	170	45.60	9.06	5.44
170	170	255	45.80	6.20	26.89
170	255	0	32.07	53.10	−47.14
170	255	85	34.28	44.77	−22.91
170	255	170	35.64	40.48	−3.57
170	255	255	36.49	37.66	13.50
255	0	0	54.56	−38.18	−55.13
255	0	85	54.79	−51.25	−24.05
255	0	170	54.76	−57.79	4.20
255	0	255	54.54	−61.72	34.77
255	85	0	44.41	−12.50	−55.51
255	85	85	45.39	−24.51	−27.21
255	85	170	45.94	−30.78	−2.37
255	85	255	46.10	−34.92	23.72
255	170	0	35.67	11.46	−56.40
255	170	85	37.20	0.42	−30.48
255	170	170	38.40	−5.71	−7.77
255	170	255	39.12	−10.12	14.22
255	255	0	24.98	42.17	−57.67
255	255	85	26.99	33.01	−35.15
255	255	170	28.80	27.68	−16.30
255	255	255	30.04	23.55	1.02

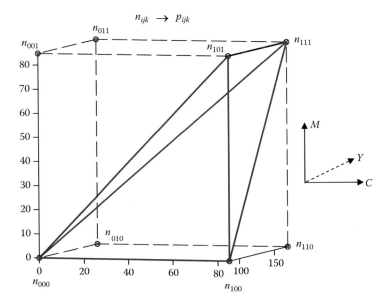

FIGURE 6.10 Tetrahedral surrounding the target color.

The corresponding $L^*a^*b^*$ values are $[p_{000}\ p_{001}\ p_{010}\ p_{011}\ p_{100}\ p_{101}\ p_{110}\ p_{111}]$, which are obtained directly from the LUT:

$$p_{000} = \begin{bmatrix} 82.98 \\ 29.86 \\ -7.28 \end{bmatrix}, \ p_{001} = \begin{bmatrix} 81.49 \\ 23.82 \\ 24.18 \end{bmatrix}, \ p_{010} = \begin{bmatrix} 69.10 \\ 55.79 \\ -11.76 \end{bmatrix}, \ p_{011} = \begin{bmatrix} 68.90 \\ 48.72 \\ 16.33 \end{bmatrix},$$

$$p_{100} = \begin{bmatrix} 68.50 \\ 15.59 \\ -25.42 \end{bmatrix}, \ p_{101} = \begin{bmatrix} 67.73 \\ 7.81 \\ 5.02 \end{bmatrix}, \ p_{110} = \begin{bmatrix} 56.67 \\ 40.60 \\ -28.83 \end{bmatrix}, \ \text{and} \ p_{111} = \begin{bmatrix} 57.40 \\ 31.75 \\ -0.64 \end{bmatrix}$$

The p_x, p_y, and p_z vectors are

$$p_x = p_{100} - p_{000} = \begin{bmatrix} 68.50 \\ 15.59 \\ -25.42 \end{bmatrix} - \begin{bmatrix} 82.98 \\ 29.86 \\ -7.28 \end{bmatrix} = \begin{bmatrix} -14.48 \\ -14.27 \\ -18.14 \end{bmatrix}$$

$$p_y = p_{111} - p_{101} = \begin{bmatrix} 57.40 \\ 31.75 \\ -0.64 \end{bmatrix} - \begin{bmatrix} 67.73 \\ 7.81 \\ 5.02 \end{bmatrix} = \begin{bmatrix} -10.33 \\ 23.94 \\ -5.66 \end{bmatrix}$$

$$p_z = p_{101} - p_{100} = \begin{bmatrix} 67.73 \\ 7.81 \\ 5.02 \end{bmatrix} - \begin{bmatrix} 68.50 \\ 15.59 \\ -25.42 \end{bmatrix} = \begin{bmatrix} -0.77 \\ -7.78 \\ 30.44 \end{bmatrix}$$

Hence,

$$p = p_{000} + p_x \frac{x - x_0}{x_1 - x_0} + p_y \frac{y - y_0}{y_1 - y_0} + p_z \frac{z - z_0}{z_1 - z_0}$$

$$= \begin{bmatrix} 82.98 \\ 29.86 \\ -7.28 \end{bmatrix} + \frac{67 - 0}{85 - 0} \times \begin{bmatrix} -14.48 \\ -14.27 \\ -18.14 \end{bmatrix} + \frac{128 - 85}{170 - 85} \times \begin{bmatrix} -10.33 \\ 23.94 \\ -5.66 \end{bmatrix} + \frac{54 - 0}{85 - 0} \times \begin{bmatrix} -0.77 \\ -7.78 \\ 30.44 \end{bmatrix}$$

Therefore,

$$p = \begin{bmatrix} 65.85 \\ 25.78 \\ -5.10 \end{bmatrix}$$

6.2.4 Sequential Linear Interpolation

Uniformly sampled LUTs do not use the color space efficiently. For example, a 3-D uniformly sampled $L^*a^*b^*$ to CMY LUT used to characterize a digital printer will have many colors that are outside the printer gamut. Therefore, if one can optimally place the grid points to achieve good approximations of the multidimensional functions, then the color space is optimally utilized and the resulting mean-square error (MSE) due to interpolation will be small. The sequential linear interpolation (SLI) is an optimal approach for approximating multidimensional functions by selecting the location of the grid points in a sequential form suitable for sequential interpolation [3]. Here we assume that the grid points have already been selected optimally according to the SLI algorithm and only discuss the SLI. For the 1-D functions, the approach is similar to linear interpolation; therefore, we consider 2-D functions and extend the results to 3-D. Let $y = f(x) = f(x_1, x_2)$ be a nonlinear function of the two variables x_1 and x_2. Figure 6.11 shows an SLI grid structure LUT of size 19.

To interpolate a grid point $p = [x_1 \quad x_2]$ that is not part of the LUT, we first project this point to the two nearest grid lines to the right and left of p, as shown in Figure 6.12.

One-dimensional linear interpolation is used to estimate the values of the function, y_R and y_L, at points p_R and p_L, respectively. They are given by

$$y_R = w_R f(p_{R1}) + (1 - w_R) f(p_{R2}) \tag{6.17}$$

and

$$y_L = w_L f(p_{L1}) + (1 - w_L) f(p_{L2}) \tag{6.18}$$

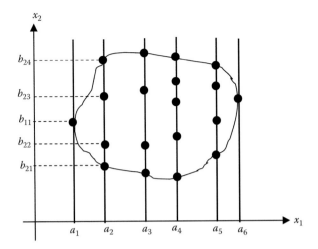

FIGURE 6.11   Example of a 2-D sequential grid structure.

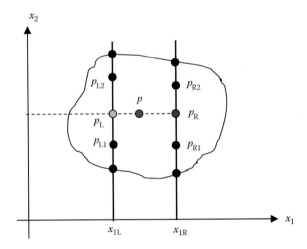

FIGURE 6.12   Interpolation of point p lying between grid line $x_1 = x_{1L}$ and $x_1 = x_{1R}$.

where $w_R = \dfrac{x_2 - x_{2R1}}{x_{2R2} - x_{2R1}}$ and $w_L = \dfrac{x_2 - x_{2L1}}{x_{2L2} - x_{2L1}}$. Once y_R and y_L are determined, the value of the function at point p is obtained by

$$y = u y_L + (1 - u) y_R \qquad (6.19)$$

where

$$u = \frac{x_1 - x_{1R}}{x_{1L} - x_{1R}} \qquad (6.20)$$

The above technique can be easily extended to 3-D and 4-D LUTs with SLI structures.

6.3 NONUNIFORMLY SPACED LOOKUP TABLES

There are several algorithms available to interpolate nonuniformly spaced LUTs. Among them are the Shepard interpolation algorithm, the moving-matrix technique, and the dynamic least-square interpolation algorithm. None of these techniques provide accurate estimations of the values of the underlying functions at grid points that are not part of the LUT. However, they are the only techniques known to us that can be used for the interpolation of nonuniformly spaced LUTs that are not structured. These techniques are discussed in the following sections.

6.3.1 SHEPARD INTERPOLATION

The Shepard interpolation is an approach to interpolate multidimensional irregularly spaced data [4]. The Shepard algorithm is based on a section of a local function to approximate the underlying function at a given point using weighted regression. The function selected is continuously differentiable. To discuss the algorithm, consider a LUT that maps M-dimensional vector y to M-dimensional vector z through a nonlinear transformation, where

$$x \xrightarrow{P} y, \quad y = P(z) \tag{6.21}$$

Assume that a set of N irregularly spaced data points are available. Let the set be given by

$$S = \{(x_1, y_1)\ (x_2, y_2) \cdots (x_N, y_N)\} \tag{6.22}$$

where $x_i \in R^M$ and $y_i \in R^M$. Let x be a point in the input space with the corresponding point y in the output space. Then the interpolated value at x is given by

$$\hat{y} = \begin{cases} \dfrac{\sum_{i=1}^{N} y_i d_i^{-\mu}}{\sum_{i=1}^{N} d_i^{-\mu}} & \text{if } d_i \neq 0 \quad \text{for all } i \\ y_j & \text{if } d_j = 0 \quad \text{for some } j \end{cases} \tag{6.23}$$

where d_i is the distance (based on L_p norm) between vectors x and x_i, that is,

$$d_i = \|x - x_i\|_p \tag{6.24}$$

Note that if x is approaching a data point, then $\hat{y} \to y = P(x)$, indicating that the interpolation formula given in Equation 6.23 is a continuous function, as desired. There are two parameters, p and μ, that can be selected by the user. It can be shown (see Problem 6.6) that if the parameter μ is chosen to be greater than 1, the interpolated function will be differentiable.

Example 6.5

Consider the 1-D function $y = p(x)$ defined by

$$p(x) = \frac{x^2 + x^3}{2} + \frac{\sin(4\pi x)}{24x} \quad 0 \le x \le 1$$

We first nonuniformly sample this function at $N = 10$ points and create the LUT given in Table 6.7.

(a) Use the Shepard interpolation to estimate the value of the function at $x = 0.5$. Use $\mu = 2$ and $p = 2$.
(b) Repeat part a for 100 samples of x that are uniformly spaced between 0 and 1. Plot the results and compare them with the exact values of y.

SOLUTION

(a) $\hat{y} = \dfrac{\sum_{i=1}^{N} y_i d_i^{-\mu}}{\sum_{i=1}^{N} d_i^{-\mu}} = \dfrac{\sum_{i=1}^{10} y_i d_i^{-2}}{\sum_{i=1}^{10} d_i^{-2}} = \dfrac{\sum_{i=1}^{10} y_i |x - x_i|^{-2}}{\sum_{i=1}^{10} |x - x_i|^{-2}} = \dfrac{\sum_{i=1}^{10} y_i |0.5 - x_i|^{-2}}{\sum_{i=1}^{10} |0.5 - x_i|^{-2}} = 0.2292$

(b) The plot of the original function, the irregularly sampled points, and the interpolated function is shown in Figure 6.13.

TABLE 6.7
Uniformly Sampled LUT at $N = 10$ Points

x	0	0.07	0.27	0.34	0.41	0.55	0.66	0.80	0.89	1
y	0.523	0.461	0.0079	−0.0334	0.0266	0.279	0.419	0.545	0.702	1

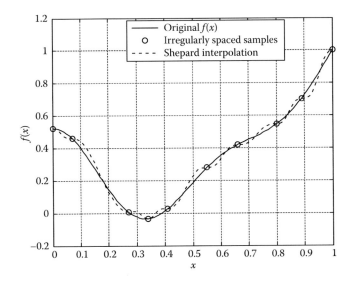

FIGURE 6.13 Shepard interpolation of a 1-D function.

Example 6.6

Consider a *CMY* printer with the 3 × 3 × 3 forward printer map LUT (*CMY* → *L* a* b**) given in Table 6.8. Use the Shepard interpolation algorithm to find

(a) The *CMY* corresponding to $L^* = 50$, $a^* = 30$, and $b^* = 40$. Use $\mu = 2$ and $p = 2$.

(b) The *CMY* corresponding to $L^* = 50$, $a^* = 80$, and $b^* = -14$. Use $\mu = 2$ and $p = 2$.

SOLUTION

(a) $\hat{y} = \dfrac{\sum_{i=1}^{N} y_i d_i^{-\mu}}{\sum_{i=1}^{N} d_i^{-\mu}} = \dfrac{\sum_{i=1}^{27} y_i d_i^{-2}}{\sum_{i=1}^{27} d_i^{-2}} = \dfrac{\sum_{i=1}^{27} CMY_i \| [50\ 30\ 40] - L^*a^*b_i^* \|^{-2}}{\sum_{i=1}^{27} \| [50\ 30\ 40] - L^*a^*b_i^* \|^{-2}}$

$= [98.27 \quad 161.53 \quad 176.64]$

TABLE 6.8
3 × 3 Forward Printer MAP

C	M	Y	L*	a*	b*
0	0	0	100	−0.04	0.00
0	0	127.5	94.49	−5.38	50.30
0	0	255	94.96	−8.59	100.00
0	127.5	0	75.04	43.12	−8.91
0	127.5	127.5	74.41	34.79	34.04
0	127.5	255	74.12	29.53	75.96
0	255	0	53.82	86.18	−14.25
0	255	127.5	55.69	75.09	20.86
0	255	255	55.16	70.02	49.03
127.5	0	0	73.27	−21.68	−29.70
127.5	0	127.5	71.72	−31.65	16.27
127.5	0	255	70.49	−37.09	59.89
127.5	127.5	0	55.41	19.06	−35.43
127.5	127.5	127.5	56.37	8.82	5.92
127.5	127.5	255	55.38	3.27	42.27
127.5	255	0	37.93	59.71	−40.59
127.5	255	127.5	40.09	50.02	−4.48
127.5	255	255	40.70	44.69	24.55
255	0	0	54.56	−38.18	−55.13
255	0	127.5	54.37	−54.87	−9.70
255	0	255	54.54	−61.72	34.77
255	127.5	0	39.35	−2.52	−57.00
255	127.5	127.5	40.61	−15.15	−17.88
255	127.5	255	41.46	−22.62	17.91
255	255	0	24.98	42.17	−57.67
255	255	127.5	26.93	28.03	−26.23
255	255	255	30.04	23.55	1.02

Therefore,

$$[C\ M\ Y] = [98.27\quad 161.53\quad 176.64]$$

(b) $\hat{y} = \dfrac{\sum_{i=1}^{N} y_i d_i^{-\mu}}{\sum_{i=1}^{N} d_i^{-\mu}} = \dfrac{\sum_{i=1}^{27} y_i d_i^{-2}}{\sum_{i=1}^{27} d_i^{-2}} = \dfrac{\sum_{i=1}^{27} CMY_i \|[50\ 80\ -14] - L^*a^*b_i^*\|^{-2}}{\sum_{i=1}^{27} \|[50\ 80\ -14] - L^*a^*b_i^*\|^{-2}}$

$= [27.35\quad 240.57\quad 26.81]$

Therefore,

$$[C\ M\ Y] = [27.35\quad 240.57\quad 26.81]$$

6.3.2 MOVING-MATRIX INTERPOLATION

The moving-matrix approach is another nonlinear technique for interpolation of irregularly spaced or scattered multidimensional data [5]. It is based on weighted least-square regression. The interpolated value $\hat{y}$ at point x is given as

$$\hat{y} = A\tilde{x} \tag{6.25}$$

where
$\tilde{x} = [x\ \ 1]^{\mathrm{T}}$ is the augmented vector x
A is the transformation matrix of size $M \times (M+1)$

If quadratic, cubic, or other terms are included, then the number of terms in the augmented vector and the transformation matrix correspondingly increase.

The transformation matrix A is obtained by minimizing the weighted square error given by

$$E = \sum_{i=1}^{N} W_i \|y_i - A\tilde{x}_i\|^2 \tag{6.26}$$

The above expression for E can be expanded as

$$E = \sum_{i=1}^{N} W_i (y_i - A\tilde{x}_i)(y_i - A\tilde{x}_i)^{\mathrm{T}} = \sum_{i=1}^{N} W_i y_i y_i^{\mathrm{T}} - 2A \sum_{i=1}^{N} W_i \tilde{x}_i y_i^{\mathrm{T}} + A \sum_{i=1}^{N} W_i \tilde{x}_i \tilde{x}_i^{\mathrm{T}} A^{\mathrm{T}} \tag{6.27}$$

Differentiating the above equation with respect to A and setting it equal to zero yields

$$\frac{\partial E}{\partial A} = -2 \sum_{i=1}^{N} W_i y_i \tilde{x}_i^{\mathrm{T}} + 2A \sum_{i=1}^{N} W_i \tilde{x}_i \tilde{x}_i^{\mathrm{T}} = 0 \tag{6.28}$$

Equation 6.28 is used to solve for A. The result is

$$A = SP^{-1} \tag{6.29}$$

where

$$S = \sum_{i=1}^{N} W_i y_i \tilde{x}_i^{\mathrm{T}}$$

$$P = \sum_{i=1}^{N} W_i \tilde{x}_i \tilde{x}_i^{\mathrm{T}}$$

The weight W_i is a function of the distance from the input point x to all the points x_i in the LUT. It is inversely proportional to the distance, which gives more weight to the points closer to the desired point and less weight to points that are farther away from the desired point. The weight W_i is given by

$$W_i = \frac{1}{d_i^{\mu} + \varepsilon} \tag{6.30}$$

where $d_i = \|x - x_i\|$ is the Euclidean norm distance. The variable parameters of this algorithm are μ and ε. These two parameters affect the locality of the regression. Large values of μ and small values of ε give W_i more local behavior, which means that only points closest to the desired point will be considered as significant points. We now consider a sample example.

Example 6.7

Consider the 1-D function $y = P(x)$ given by the LUT shown in Table 6.9.
Use the moving-matrix approach, interpolate, and compute the corresponding value of $P(x)$ at $x = 0.5$. Use $\mu = 2$ and $\varepsilon = 10^{-4}$.

SOLUTION

$$W_i = \frac{1}{d_i^{\mu} + \varepsilon} = \frac{1}{|x - x_i|^2 + 0.0001} = \frac{1}{|0.5 - x_i|^2 + 0.0001} \quad i = 1, 2, \ldots, 10$$

TABLE 6.9

One-Dimensional Function LUT

x	0	0.07	0.27	0.34	0.41	0.55	0.66	0.80	0.89	1
y	0.523	0.461	0.0079	−0.0334	0.0266	0.279	0.419	0.545	0.702	1

TABLE 6.10

Weights as a Function of i

i	1	2	3	4	5	6	7	8	9	10
W_i	15.97	29.16	345.02	1323.87	6038.28	9411.7	1323.87	121.95	43.04	15.97

The weights for different values of i are shown in Table 6.10.

$$S = \sum_{i=1}^{N} W_i y_i x_i^{\mathsf{T}} = \sum_{i=1}^{10} W_i y_i [x_i \ \ 1] = [83.8135 \ \ 144.9269]$$

$$P = \sum_{i=1}^{N} W_i \tilde{x}_i \tilde{x}_i^{\mathsf{T}} = \sum_{i=1}^{10} W_i [x_i \ \ 1]^{\mathsf{T}} [x_i \ \ 1] = \begin{bmatrix} \sum_{i=1}^{10} W_i x_i^2 & \sum_{i=1}^{10} W_i x_i \\ \sum_{i=1}^{10} W_i x_i & \sum_{i=1}^{10} W_i \end{bmatrix} = \begin{bmatrix} 176.0015 & 324.6467 \\ 324.6467 & 634.3268 \end{bmatrix}$$

$$A = SP^{-1} = [0.9789 \ \ -0.2725]$$

$$\hat{y} = A\tilde{x} = [0.9789 \ \ -0.2725] \begin{bmatrix} 0.5 \\ 1 \end{bmatrix} = 0.2169$$

6.3.3 RECURSIVE LEAST-SQUARE IMPLEMENTATION OF MOVING-MATRIX ALGORITHM

The moving-matrix interpolation algorithm given by Equation 6.25 can be implemented using recursive least square (RLS). The transformation matrix A is given by

$$A = SP^{-1} = \left(\sum_{i=1}^{N} W_i y_i \tilde{x}_i^{\mathsf{T}} \right) \left[\sum_{i=1}^{N} W_i \tilde{x}_i \tilde{x}_i^{\mathsf{T}} \right]^{-1} \tag{6.31}$$

As can be seen, matrix inversion is required to compute the matrix A. Since the matrix $P = \sum_{i=1}^{N} W_i \tilde{x}_i \tilde{x}_i^{\mathsf{T}}$ may be ill conditioned, we would like to compute A without using matrix inversion. By using the RLS algorithm (Section 7.4.1.2), it is possible to compute the inverse of P without using any matrix inversion. This can also be achieved by recursive computation of A using the following algorithm:

Step 1: Initialization. Let $B(0) = a^2 I$ and $k = 0$, where $a^2 \gg 0$ and I is the 3×3 identity matrix. Choose ε to be a small positive number.

Step 2: Change $k \rightarrow k + 1$.

Step 3: Compute

$$w(k) = \|\tilde{x} - \tilde{x}_k\| + \varepsilon \tag{6.32}$$

and

$$C(k) = B(k)\widetilde{x}_k$$
$$b(k) = w(k) + \widetilde{x}_k^{\mathrm{T}} C(k)$$
$$B(k+1) = B(k) - C(k)C^{\mathrm{T}}(k)/b(k)$$ (6.33)
$$D(k+1) = D(k) - \widetilde{x}_k y_k^{\mathrm{T}}/w(k)$$

Step 4: Is $k = N - 1$? If the answer is no, go to Step 2; otherwise, stop and compute matrix A using the following equation:

$$A = D(N)B(N)$$ (6.34)

6.4 LOOKUP TABLE INVERSE

6.4.1 INTRODUCTION

A color printer can be seen as a device that is mapping a requested color in the image into a device independent color (or printed color). An approximation to this color mapping can be obtained by constructing a forward map of the printer using experimental data in the form of a LUT. To reproduce colors accurately, we need to build the inverse map of the forward LUT [6,7]. More description about the inverse LUTs can be found in Chapter 7. One form of this LUT associates the input colors in $L^*a^*b^*$ space to printer specific $CMYK$ space. Each entry in the inverse LUT is called a node. It is desirable to have an inverse LUT with input nodes regularly spaced on a sequential plane (i.e., a structured input). While generating the printer forward table, if we select node colors appropriately in the forward LUT, then, theoretically speaking, an inverse LUT is just the reversal of the forward LUT. The reversal can be easily obtained by swapping the data from the forward LUT. The resulting inverse LUT will not conform to the structured specifications required for the input nodes. Also, for colors at the gamut boundary, this type of inverse LUT may not be well defined. It may give multivalued outputs. Multidimensional interpolation approaches, such as tetrahedral, conjugate gradient (CG), and iteratively clustered interpolation (ICI) algorithms, are often used to restructure the inverse LUT when such LUTs are constructed with experimental data such that the inverse LUT finally ends up with a uniformly sampled, structured input grid. However, it is important to note that the various other approaches described in Chapter 7 do not always use these methods.

6.4.2 INVERSE PRINTER MAP

Let us describe the inversion process using the CMY to $L^*a^*b^*$, a three-to-three printer forward map, P, as shown in Figure 6.14. Let $\hat{P}^{-1}$ denote an estimate of the printer inverse map. The CMY values are converted to $CMYK$ using gray-component replacement or under-color removal (GCR/UCR) algorithms before printing on a $CMYK$ to $L^*a^*b^*$ printer (Section 7.5.1). For the purpose of this discussion, the GCR/UCR function is embedded inside the printer map, P, and only in-gamut colors are considered. The inverse printer map $\hat{P}^{-1}$ is a three-to-three map, defined mathematically as $L^*a^*b^* \rightarrow CMY$,

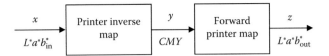

FIGURE 6.14   Forward and inverse printer maps.

where input $L*a*b*$ points are on a 3-D grid of size $h \times h \times h$ having a dynamic range $0 \le L \le 100$, $-127 \le a \le 128$, and $-127 \le b \le 128$, which is created to compensate for the nonlinearity in P, as shown schematically in Figure 6.14.

Another interpretation for the inverse is that, for a given target color $x = L*a*b*_{in}$, we seek an inverse printer map $\hat{P}^{-1}$ such that the printed color z and the target x are as close as possible (or same). When this happens, we say there is a colorimetric match between the target color and the printed color. This inverse printer map $\hat{P}^{-1}$ is a key component of many color control algorithms used in digital printers. The problem of interest is to compute a structured inverse LUT, $\hat{P}^{-1}$, such that the composition $\hat{P}^{-1}[P(\cdot)]$ is as close as possible to the identity matrix in the minimum mean-square error (MMSE) sense. A simple cost function to be minimized can be formulated using the ΔE^*_{ab} color difference formula between the requested input color $x = L*a*b*_{in}$ and the printed output color $z = P(y) = L*a*b*_{out}$:

$$E(y) = \Delta E[P(y), x] = \tfrac{1}{2}\|P(y) - x\|^2 = \tfrac{1}{2}[P(y) - x]^{\mathrm{T}}[P(y) - x] \qquad (6.35)$$

Notice that we do not have an exact model for P; only an approximation given by the forward LUT is available. The interpolation methodology used to solve the three-to-three inverse problem is summarized as follows:

(1) Obtain the forward LUT P for the given color printer. To achieve this, we grid the CMY color space at the input of the GCR algorithm. Process the CMY grids through the GCR algorithm to create print ready $CMYK$ values at each CMY grid node. For this discussion, we can also restrict the printer forward map without the GCR algorithms. That is, simply, a CMY to $L*a*b*$ printer whose structured inverse, $L*a*b*$ to CMY, is required so that the structured inverse conforms to the performance index defined by Equation 6.35. The printed color z_j for each grid node y_j is obtained from experiments on the actual printer.

(2) Select the collection of target colors x_i, $i = 1, 2, \ldots, h$, where $x_i = [L^*_{ix} \;\; a^*_{ix} \;\; b^*_{ix}]^{\mathrm{T}}$. For this, we can uniformly sample the CIELab color space. We assume that every grid point in the sampled CIELab color space is either inside the printer gamut or it has been mapped to a point inside the gamut by an appropriate gamut mapping algorithm (Section 7.6).

(3) Obtain the corresponding value of $y_i = [C_i \; M_i \; Y_i]$ by minimizing the cost function given by Equation 6.35.

An approach to solve the three-to-three optimization problem is described next.

6.4.3 ITERATIVELY CLUSTERED INTERPOLATION

The ICI algorithm is a gradient-based optimization method, with the initial point for the optimization generated through an iterative technique [8]. The ICI algorithm is implemented in the following steps:

Step 1: Select algorithm parameters μ, ε, and k_{max} (their meanings and selections will be discussed shortly). Let $k=0$ and assume the initial condition $y(0) = \lfloor C_y(0) \quad M_y(0) \quad Y_y(0) \rfloor$.

Step 2: Update $y(k)$ using the recursion formula

$$y(k) = y(k+1) - \mu \frac{\partial E}{\partial y(k)} \tag{6.36}$$

where

$$\frac{\partial E}{\partial y(k)} = 2J_k^{\mathrm{T}}\{P[y(k)] - x\} \tag{6.37}$$

Inserting Equation 6.37 into Equation 6.36 results in

$$y(k+1) = y(k) - \mu J_k^{\mathrm{T}}\{P[y(k)] - x\} \tag{6.38}$$

The term J_k is the 3×3 Jacobian matrix computed at the kth iteration. It is given by

$$J_k = \begin{bmatrix} \dfrac{\partial P[y_1(k)]}{\partial C(k)} & \dfrac{\partial P[y_1(k)]}{\partial M(k)} & \dfrac{\partial P[y_1(k)]}{\partial Y(k)} \\[3mm] \dfrac{\partial P[y_2(k)]}{\partial C(k)} & \dfrac{\partial P[y_2(k)]}{\partial M(k)} & \dfrac{\partial P[y_2(k)]}{\partial Y(k)} \\[3mm] \dfrac{\partial P[y_3(k)]}{\partial C(k)} & \dfrac{\partial P[y_3(k)]}{\partial M(k)} & \dfrac{\partial P[y_3(k)]}{\partial Y(k)} \end{bmatrix} \tag{6.39}$$

The Jacobian matrix J_k is computed numerically at each iteration using the forward printer map.

Step 3: Let $k=k+1$. If $E[y(k)] < \varepsilon$ or $k > k_{max}$, then stop; otherwise go to Step 2. The parameter ε is the minimum required ΔE_{ab}^* and can be selected to be any arbitrary small number. The index k_{max} is the maximum number of iterations. The algorithm will stop either when ΔE_{ab}^* is less than μ or when the number of iterations reaches k_{max}. The step size μ controls the rate of convergence and should be selected to achieve fast algorithm convergence and meet accuracy requirements. The upper bound for parameter μ is derived in the next section.

6.4.3.1   Selection of Step Size Parameter μ

The updating equation for $y(k)$ is given by

$$y(k + 1) = y(k) - \mu J_k^T \{P[y(k)] - x\} \tag{6.40}$$

Define the error $e(k)$ as

$$e(k) = P[y(k)] - x \tag{6.41}$$

Use the first-order linear approximation

$$e(k + 1) - e(k) = P[y(k + 1)] - P[y(k)] \approx J_k[y(k + 1) - y(k)] \tag{6.42}$$

Using the updating law given by Equation 6.38, we have

$$e(k + 1) - e(k) \approx -\mu J_k J_k^T e(k) \tag{6.43}$$

or

$$e(k + 1) = \left(I - \mu J_k J_k^T\right) e(k) \tag{6.44}$$

The Jacobian matrix J_k is not changing rapidly from one iteration to another; therefore, we assume that $J_k = J_0$. With this assumption,

$$e(k + 1) = \left(I - \mu J_0 J_0^T\right) e(k) \tag{6.45}$$

Therefore, for the iterations to converge, the error $e(k)$ must approach zero as $k \to \infty$. This requires that all eigenvalues of matrix $I - \mu J_0 J_0^T$ must lie inside the unit circle in the complex plane, that is,

$$\left|\lambda_i\left(I - \mu J_0 J_0^T\right)\right| < 1 \quad i = 1, 2, 3 \tag{6.46}$$

Therefore, the parameter μ should be selected to satisfy the following condition:

$$0 < \mu < \frac{2}{\lambda_{\max}\left(J_0 J_0^T\right)} \tag{6.47}$$

Since $\lambda_{\max}\left(J_0 J_0^T\right) \leq \mathrm{Tr}\left(J_0 J_0^T\right)$, a more conservative, but easier to compute, method for obtaining an upper bound for parameter μ is

$$0 < \mu < \frac{2}{\mathrm{Tr}\left(J_0 J_0^T\right)} \tag{6.48}$$

6.4.3.2   Algorithm Initialization

To find an initial estimate $y(0) = [C(0) \ M(0) \ Y(0)]^{T}$ for the algorithm, we follow the following steps:

(1) Find the auxiliary point $z_{\text{aux}} = \begin{bmatrix} L^*_{\text{aux}} & a^*_{\text{aux}} & b^*_{\text{aux}} \end{bmatrix}^{T}$ in the forward printer LUT that is closest to the target color x in an L_2 norm (Euclidean distance) sense.

(2) Find the point y_{aux}, such that $P(y_{\text{aux}}) = z_{\text{aux}}$. Note that y_{aux} can be found easily, since z_{aux} is a grid point in the forward LUT.

(3) Select N grid points in a neighborhood of y_{aux}; that is, generate a cluster of N points $y_{\text{aux}1}, y_{\text{aux}2}, \dots, y_{\text{aux}N}$ by moving along the C, M, and Y axes around y_{aux}. Use the LUT P and trilinear interpolation to map these N points to obtain the corresponding values in $L^*a^*b^*$ color space. Call them $z_{\text{aux}i} = P(y_{\text{aux}i})$ $i = 1, 2, \dots, N$.

(4) Denote by z_0 the closest point to the set $z_{\text{aux}i}$ $i = 1, 2, \dots, N$. Choose the initial condition to be y_0, where $z_0 = P(y_0)$.

6.4.4   TETRAHEDRAL TECHNIQUE

The tetrahedral interpolation technique covered in Section 6.2.3 can be used to find the inverse printer map. Its low computational complexity makes it an affordable approach for constructing an inverse LUT [2]. For this reason, it is a widely used methodology for inverse computation. Assume a set of *CMY* nodes (y_i) and their corresponding $L^*a^*b^*$ nodes (z_i) are given in the form of a LUT. To find the inverse of $z = L^*a^*b^*$ using the tetrahedral interpolation method, we perform the following steps:

Step 1: Partition the *CMY* color space into tetrahedral segments.

Step 2: Given a target $x = L^*a^*b^*$, find the tetrahedral that contains the z vector. Let $\{z_1 \ z_2 \ z_3 \ z_4\}$ denote the vertices of the tetrahedron that z belongs to in $L^*a^*b^*$ and let $\{y_1 \ y_2 \ y_3 \ y_4\}$ denote their corresponding vertices in *CMY* color space.

Step 3: Compute the corresponding *CMY* value using the following equation:

$$CMY = y = A_{CMY}A_{\text{Lab}}^{-1}(z - z_1) + y_1 \tag{6.49}$$

where

$$A_{CMY} = \begin{bmatrix} y_2 - y_1 & y_3 - y_1 & y_4 - y_1 \end{bmatrix} \tag{6.50}$$

and

$$A_{Lab} = \begin{bmatrix} z_2 - z_1 & z_3 - z_1 & z_4 - z_1 \end{bmatrix} \tag{6.51}$$

6.4.5 Conjugate Gradient Approach

The CG technique is an iterative technique used in solving constrained optimization problems and systems of linear equations of the form $Ax = b$, where matrix A is symmetric and positive definite [9]. The CG algorithm can also be used to solve unconstrained optimization problems such as finding the inverse printer model described in the previous section. Assume a forward printer map $z = P(y)$, where $y \in R^3$ is a *CMY* color patch and $z \in R^3$ is its corresponding $L*a*b*$ value. The goal is to find y for a given $z = x$ by minimizing the ΔE_{ab}^* error between x and $P(y)$, that is,

$$\min_y E(y) = \min_y \Delta E[P(y), x] = \min_y \tfrac{1}{2}\|P(y) - x\|^2$$
$$= \min_y \tfrac{1}{2}[P(y) - x]^T[P(y) - x] \tag{6.52}$$

The iteration to obtain the solution using CG method is given by

$$y(k + 1) = y(k) + \alpha_k d_k \tag{6.53}$$

where d_k and α_k are the search direction vectors and the step size at the kth iteration, respectively. The initialization algorithm to obtain an initial estimate $y(0)$ is similar to the technique used for ICI. The initial search direction is given by

$$d_k = -g_k = -e(k)J_k \tag{6.54}$$

where
$e(k) = P(y(k)) - x(k)$
J_k is the Jacobian matrix defined by Equation 6.39

The updating equations for the search direction and the step size are given as

$$\alpha_k = -\frac{d_k J_k^T e^T(k)}{d_k J_k^T J_k d_k^T} \tag{6.55}$$

$$g_{k+1} = e(k + 1)J_{k+1} \tag{6.56}$$

$$\beta_k = \frac{g_{k+1}^T(g_{k+1} - g_k)}{d_k^T(g_{k+1} - g_k)} \tag{6.57}$$

$$d_{k+1} = -g_{k+1}^T + \beta_k d_k^T \tag{6.58}$$

The criteria for stopping the iterations are given by the threshold parameter ε and the maximum number of iterations K_{max}. The iteration stops when the MSE, ΔE_{ab}^*, is less than ε or the number of iterations exceeds K_{max}. The accuracy of the CG is similar to the ICI algorithm; however, the computational complexity of the CG method far exceeds the ICI approach.

6.4.6 Comparison of Different Inversion Algorithms

We now briefly compare four different mapping algorithms when used for building the inverse map. We modeled a legacy Xerox color digital printer by creating a 13^3 LUT. The LUT is a mapping from a device-dependent CMY to a device-independent $L*a*b*$ color space. The grid points in the CMY color space are uniformly sampled. The inverse map is generated by uniformly sampling the $L*a*b*$ space and finding the inverse of each $L*a*b*$ node on the grid using one of the following algorithms: (a) Shepard, (b) moving matrix, (c) ICI, and (d) CG. Out-of-gamut colors were first mapped to the nearest point in the gamut by a technique that maintains the hue angle.

The quantitative metric used to evaluate the inversion accuracy is the ΔE_{ab}^* accuracy for interpolated colors between the mapped input $L*a*b*$ and the resulting output $L*a*b*$. For 240 in-gamut test colors (Figure 6.15), the ΔE_{ab}^* statistics are simulated. For this simulation, a 13^3 printer forward LUT size was chosen. The 240 in-gamut colors are provided as input to the inverse LUT and the output of the inverse LUT is interpolated through the forward LUT. The interpolation in the forward LUT uses a trilinear algorithm, which is common to all methods. The output of the interpolation is compared through the input to get the error performance. The error results are shown in Table 6.11.

FIGURE 6.15 (See color insert following page 428.)   View of 240 test colors used for comparing inversion algorithms.

TABLE 6.11
Comparison of Four Different Printer Inverse Algorithms

Algorithm	Mean ΔE_{ab}^*	Standard Deviation	Mean $\Delta E_{ab}^* + 2\sigma$	Minimum ΔE_{ab}^*	Maximum ΔE_{ab}^*
Shepard	1.644	0.5448	2.7346	0.151	3.807
Moving matrix	1.149	0.809	2.764	0.023	5.923
CG	0.187	0.5243	1.236	0.001	5.113
ICI	0.044	0.1182	0.2812	0.002	1.681

6.5 COMPRESSION OF LOOKUP TABLES

6.5.1 INTRODUCTION

In general, as indicated above, the inverse map is a LUT that associates points in the printer output color space to points in the printer input color space $(L^*a^*b^* \rightarrow CMYK)$.

An important performance measure of the inverse LUT is the accuracy of inversion. To achieve an improved accuracy for colors inside the printer gamut, often the manufacturers increase the LUT size greatly (e.g., for an $N \times N \times N$ LUT, N may be chosen equal to 33 when compared to 17). To compensate for the printer drift, it is highly desirable to update the inverse printer LUT on a regular basis. If we have to update each node in the inverse LUT through measurements, then this requires printing and measuring a large number of color patches. Hence, a better way to do this is by downsampling the input color space from $N \times N \times N$ to $M \times M \times M$ ($M < N$) by selecting the critical colors for measurement such that, when we upsample the downsampled LUT to the original size, the average ΔE_{ab}^* between the original $L^*a^*b^*$ and the upsampled $L^*a^*b^*$ is minimized. The process is shown schematically in a block diagram in Figure 6.16.

The average ΔE_{ab}^* between the original $L^*a^*b^*$ and $L^*a^*b_1^*$ (the upsampled value from the downsampled LUT corresponding to the patch represented by $L^*a^*b^*$) is defined as

$$\Delta E_{ab}^* = \frac{1}{K} \sum_{i=1}^{K} \| L^*a^*b^* - L^*a^*b_1^*(i) \| \qquad (6.59)$$

These critical colors are selected by minimizing ΔE_{ab}^*. Once these critical colors are identified, the updating of the three-to-three inverse printer map is straightforward. It is obtained simply by first printing and measuring these selected critical colors, and then upsampling the measured $L^*a^*b^*$ values, which are now in the form of a

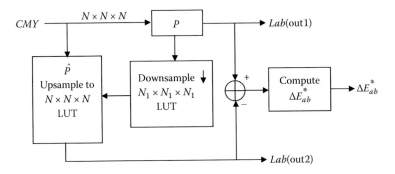

FIGURE 6.16 Optimum color selection method illustrated for obtaining minimal colors for measurement of a three-to-three printer forward map.

reduced-sized forward LUT, to the full size and then applying an inverse algorithm on the upsampled LUT to obtain the full-size inverse map.

There are different techniques available in the literature for selecting these critical colors [3,10,11]. The two most promising techniques for downsampling of multidimensional LUTs are SLI and dynamic optimization (DO) algorithms. In the following sections, we consider both SLI and DO algorithms. Numerical examples are provided for comparison of these two approaches.

6.5.2   DOWNSAMPLING USING SEQUENTIAL LINEAR INTERPOLATION

SLI is a numerical technique to optimally downsample a uniformly spaced LUT to any desired size. Piecewise linear homeomorphism is another technique illustrated in Ref. [10]. These algorithms are very complex and their derivations are not given in this book. Interested readers can refer to Refs. [3,10].

6.5.3   DYNAMIC OPTIMIZATION ALGORITHM

The DO algorithm selects a finite number of colors (or points in $L*a*b$ color space) by minimizing the MSE (ΔE_{ab}^*) between the actual printer output and the upsampled printer output constructed using a finite number of points as defined by

$$\Delta E_{ab}^* = \|L*a*b*(\text{out1}) - L*a*b*(\text{out2})\| \tag{6.60}$$

$$Lab(\text{out1}) = P(CMY) \tag{6.61}$$

$$Lab(\text{out2}) = \hat{P}(CMY) \tag{6.62}$$

The LUT ($\hat{P}$) is obtained by upsampling the smaller LUT containing the finite number of critical colors. This is shown in Figure 6.16. Once the upsampled forward LUT ($\hat{P}$) is constructed for a three-to-three map, its inverse LUT (inverse of $\hat{P}$) can be computed using ICI or the other algorithms mentioned above. The DO algorithm is based on dynamic programming, which uses a multistage decision process and the performance criteria such as minimization of ΔE_{ab}^* error criteria.

To illustrate the algorithm in detail, we first outline the 1-D case and then extend the approach to two and three dimensions. The 3-D approach will be used to find the critical colors for measurement of a three-to-three printer forward map.

6.5.3.1   One-Dimensional DO Algorithm

Consider the 1-D discrete function $f(x)$, where x takes M discrete values $x_1, x_2, \ldots, x_M$. We would like to choose $N < M$ points $\hat{x} = [\hat{x}_1 \ \hat{x}_2 \ \cdots \ \hat{x}_N]$ such that $\hat{x}_1 = x_1$, $\hat{x}_N = x_M$, and $\hat{x} \subset x$ while minimizing the MSE resulting from the piecewise linear approximation defined by $\hat{f}(\hat{x})$ and $f(x)$ over x. The MSE is given by

$$E = \frac{1}{M} \sum_{i=1}^{M} |f(x_j) - L_k(x_i)|^2 \tag{6.63}$$

where $L_k(x)$ is the straight-line segment joining $f(\hat{x}_k)$ to $f(\hat{x}_{k+1})$ and is given by

$$L_k(x) = \frac{f(\hat{x}_{k+1}) - f(\hat{x}_k)}{\hat{x}_{k+1} - \hat{x}_k}(x - \hat{x}_k) + f(\hat{x}_k) \tag{6.64}$$

for $\hat{x}_k \leq x \leq \hat{x}_{k+1}$. Since we always need the first and the last points for interpolation, the number of grid points has to be greater than or equal to three, $N \geq 3$. Now assume that $N = 3$, which means that we need to find one grid point $\hat{x}_2$ between x_1 and x_M such that the total MSE given by Equation 6.63 is minimized. Let the solution be x_j, where $1 < j < N$. The MMSE is given by

$$E^* = \frac{1}{M}\sum_{i=1}^{j}|f(x_i) - L_1(x_i)|^2 + \frac{1}{M}\sum_{i=j+1}^{M}|f(x_i) - L_2(x_i)|^2 \tag{6.65}$$

where $L_1(x)$ and $L_2(x)$ are

$$L_1(x) = \frac{f(x_j) - f(x_1)}{x_j - x_1}(x - x_1) + f(x_1) \tag{6.66}$$

$$L_2(x) = \frac{f(x_M) - f(x_j)}{x_M - x_j}(x - x_j) + f(x_j) \tag{6.67}$$

E^* and the index j are found by the following optimization problem:

$$j = \arg\min_k[E(k)]$$
$$= \arg\min_k \frac{1}{M}\sum_{i=1}^{k}|f(x_i) - L_1(x_i)|^2 + \frac{1}{M}\sum_{i=k+1}^{M}|f(x_i) - L_2(x_i)|^2 \tag{6.68}$$

$$E^* = E(j) \tag{6.69}$$

This means that if we start from x_1 and wish to locate one grid point between x_1 and x_M to approximate $f(x)$, that point will be x_j and the corresponding MMSE will be E^*. We define a new index, $j_1 = j$, and error, $E_1 = E^*$, and assign these two numbers to x_1 and write it as $\{j_1, E_1\}$. We repeat this process for x_2 through x_{N-2} and form the array of numbers as shown in column 1 of Table 6.12. We call this the 1-D single-stage grid allocation algorithm. In this array, E_i is the MSE between the original function and the linearly interpolated function using grid points $\lfloor x_i \quad x_j \quad x_N \rfloor$ over the closed interval of $[x_i \quad x_N]$. These indices and their corresponding MSEs are required for the two-stage optimal search. In the two-stage search, we try to locate two optimal grid points between x_i and x_N such that the total MSE is minimized. Using dynamic programming, we need to minimize

TABLE 6.12

One-Dimensional DO Algorithm

Point	$J = 1$	$J = 2$	$J = \cdots$	$J = N - 2$
x_1	$[j_1, E_1]$	$[k_1, l_1, EE_1]$	$\cdots$	$[m_1, n_1, \ldots, E_{\text{Total}}]$
x_2	$[j_2, E_2]$	$[k_2, l_2, EE_2]$	$\cdots$	—
x_3	$[j_3, E_3]$	$[k_3, l_1, EE_3]$	$\cdots$	—
$\vdots$	$\vdots$	$\vdots$	$\vdots$	$\vdots$
x_{N-4}	$[j_{N-4}, E_{N-4}]$	$[k_{N-4}, l_{N-4}, EE_{N-4}]$	$\cdots$	$\cdots$
x_{N-3}	$[j_{N-3}, E_{N-3}]$	$[N-2, N-1, 0]$	—	—
x_{N-2}	$[N-1, 0]$	—	—	—
x_{N-1}	—	—	—	—
x_N	—	—	—	—

$$k = \arg\ \min_m [E_{im} + E_m], \quad m \geq i + 1$$
$$EE_i = E_{ik_i} + E_{k_i}, \quad l_i = j_{k_i} \tag{6.70}$$

where

E_m is the MMSE corresponding to $j = 1$

E_{im} is the MSE between the original function and the linearly interpolated function using grid points $[x_i, x_m]$ over the interval of $[x_i, \ldots, x_m]$

Then the optimum two-point grid including boundary points will be $[x_1\ x_k\ x_l\ x_N]$, as shown in column 2 of Table 6.12. We call this the 1-D two-stage grid allocation algorithm. Using the same approach, we can now build an n-stage optimization process for the 1-D case with N optimal grid points $[x_1\ x_{m_1}\ x_{n_1} \cdots x_{N_1}]$.

6.5.3.2 Two-Dimensional DO Algorithm

Consider the 2-D discrete function $f(X)$, where $X = [x, y]$ takes $M \times M$ grid points uniformly spaced in the x–y plane, as shown in Figure 6.17. We would like to choose $N \times N$ grid points out of $M \times M$ available data points such that the mean-square interpolation error between the original function and the approximation obtained by upsampling the $N \times N$ LUT is minimized. Since the boundary points should be included for interpolation, degrees of freedom are less than $N \times N$. For example, if A is an internal optimum point, then the four boundary points B, C, D, and E (as shown in Figure 6.17) should also be included in this set. The 2-D DO algorithm is very similar to the 1-D case. In the single-stage 2-D algorithm, assume we start with point $[x_i, y_j]$ and try to optimally select one internal grid point in the set extending from this point to the boundaries of the function $f(X)$. The solution is obtained by direct optimal search and is denoted by the two indices k_i and l_j with its associated four boundary points and MMSE $J_1(i,j) = E_{i,j}$. This process is repeated for each point in the set associated with the domain of the underlying function. With three indices given by

$$[k_i \quad l_j \quad E_{i,j}] \quad i, j = 1, 2, \ldots, M \tag{6.71}$$

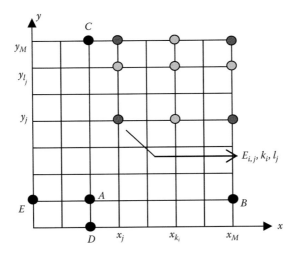

FIGURE 6.17 Grid points in *x–y* plane.

In the second stage, we will try to select two internal points in the same set. If the indices corresponding to the first optimal point are n and m, then the total MSE is

$$E = E_{(i,j),(j,m)} + E_{n,m} \tag{6.72}$$

where $E_{(i,j),(j,m)}$ is the MSE resulting from interpolating all the grid points between indices i to n and j to m using all necessary boundary points. We now minimize E given by Equation 6.72 with respect to i and j. Let the solution be $i = n_i, j = m_j$. Then the indices n_i, m_j, n, m correspond to the two optimum internal points (total of 14 points including boundary points) with the MSE given by

$$E_2(i,j) = E_{(n_i n),(m_j m)} + E_{n,m} = E_{(n_i n),(m_j m)} + E_1(n,m) \tag{6.73}$$

This process is continued until we select $N \times N$ grid points.

6.5.3.3 Three-Dimensional DO Algorithm

Consider a 3-D discrete function $f(X)$ represented by a set of pairs $(X, f(X))$, where X is a 3-D vector. These points are uniformly spaced in the support space of $f(X)$ forming an $M \times M \times M$ LUT. The goal is to downsample this 3-D LUT to a smaller LUT of size $N \times N \times N$ while minimizing the MSE between the original and the upsampled LUT. The 3-D DO algorithm, which is an extension of the 2-D algorithm, can be used for optimally selecting these grid points. Similar to 2-D cases, it is a multistage decision process described by the following steps. In the single-stage decision process, assume that we start from point $\{i\, j\, k\}$ and form a cube extending this point to the boundaries of the underlying function. The goal is to find one point inside the cube with its associated boundary points such that the MSE between the interpolated and original grid point values is minimized. Let the optimal indices be l_i, m_j, and n_k with an MSE of E_{ijk}. In the second stage, we need to find two grid points within the same cube in order to minimize the MSE as in stage one. If the solution for the first point has indices l_{1i}, m_{1j}, and n_{1k}, then the total MSE is given by

$$E_2\left(l_{1i}, m_{1j}, n_{1k}\right) = E(i, j, k) + E_{l_{1i}m_{1j}n_{1k}} \tag{6.74}$$

where $E(i, j, k)$ is the interpolation error between the original and the interpolated functions with the cube extending from grid point $\{i\ j\ k\}$ to grid points l_{1i}, m_{1j}, n_{1k}. Now, we minimize E_2 with respect to the indices l_{1i}, m_{1j}, and n_{1k}. Note that the solution to the first stage is used for finding the optimal solution in the second stage, similar to 1-D and 2-D algorithms. We continue this process until we find the solution to the N-stage problem. Note that while finding the solution to the N-stage grid allocation problem, we obtain the solution to all stages up to and including N.

Example 6.8

Consider the 1-D nonlinear function shown in Figure 6.18. We apply uniform sampling, SLI, and DO algorithms to select 5, 6,..., 20 grid points. We then use linear interpolation and interpolate to the original size of the function and compute the MSE between the original function and the upsampled version of the downsampled function. Figure 6.19 shows the MSE as a function of grid size for three different algorithms (uniformly sampling with reduced samples and SLI). As can be seen, the DO algorithm outperforms the other two algorithms.

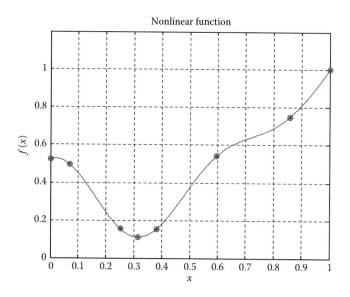

FIGURE 6.18 Nonlinear 1-D function with eight grid points optimally selected by the DO algorithm.

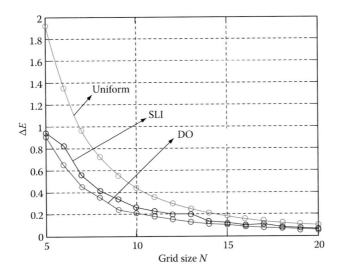

FIGURE 6.19 MSE as a function of grid size.

Example 6.9

Consider a discrete 2-D function defined by the 2-D LUT of size 4×4 as given in Table 6.13. We downsample the LUT to 3×3 using the 2-D DO algorithm.

The optimal grid points (9 points) are shown in Figure 6.20. The resulting MMSE is 1.0869.

TABLE 6.13

Two-Dimensional Function $f(x, y)$

x	y	f
0	0	1
0	0.4	1.48
0	0.8	2.92
0	1.2	5.32
0.4	0	1.134
0.4	0.4	1.678
0.4	0.8	3.182
0.4	1.2	5.646
0.8	0	1.47
0.8	0.4	2.206
0.8	0.8	3.902
0.8	1.2	6.558
1.2	0	1.946
1.2	0.4	3.002
1.2	0.8	5.018
1.2	1.2	7.994

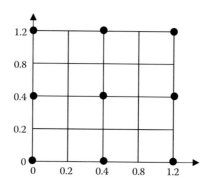

FIGURE 6.20   Grid locations for 2-D example.

Example 6.10

Consider the following 3-D nonlinear function:

$$f(x, y, z) = x + x^2 + xy + 1.5y^{1.6} + xy^4 + 0.1 \quad 0 \le x, y \le 1$$

The function $f(x, y, z)$ is uniformly sampled with $\Delta x = \Delta y = \Delta z = 0.25$ over a cube of size $[0\ 1] \times [0\ 1] \times [0\ 1]$, generating a LUT of size $5 \times 5 \times 5 = 125$ samples. The 3-D DO algorithm is used to downsample the LUT to a smaller LUT of size $3 \times 3 \times 3 = 27$ samples. The resulting MMSE is 1.1107 and the optimal grid points are shown in Table 6.14.

Example 6.11

The overall system for constructing a three-to-three inverse LUT with reduced patch measurements is shown in Figure 6.21 The system performance is measured in terms of the error between the mapped input and output $L^*a^*b^*$ values:

$$E = L^*a^*b^*_{out} - Lab_{in} \text{ (mapped)} \tag{6.75}$$

The forward printer model P is constructed by printing 13^3 uniformly sampled, CMY color patches and measuring their $L^*a^*b^*$ values using a spectrophotometer. This LUT is then downsampled to 3, 4, ..., 13 cubes and then upsampled to the original size (13 cubes). Applying ICI to the newly formed upsampled LUT forms the three-to-three inverse LUT. The 240 test patches that are used to evaluate the overall system performance consist of 216 patches inside the printer gamut and 24 patches inside and on the neutral axis $(a^* = b^* = 0)$ of the printer gamut. The simulation results for a commercial Xerox printer are shown in Table 6.15. As can be seen, a mean $\Delta E_{2000} = 0.76$ is achievable with the reduced measurements of size 125 optimal colors.

TABLE 6.14

Result of Downsampling

x	y	z	f
0	0	0	0.1
0	0	0.75	0.1
0	0	1	0.1
0	0.5	0	0.594
0	0.5	0.75	0.594
0	0.5	1	0.594
0	1	0	1.6
0	1	0.75	1.6
0	1	1	1.6
0.5	0	0	0.85
0.5	0	0.75	1.008
0.5	0	1	1.35
0.5	0.5	0	1.594
0.5	0.5	0.75	1.753
0.5	0.5	1	2.094
0.5	1	0	2.85
0.5	1	0.75	3.008
0.5	1	1	3.35
1	0	0	2.1
1	0	0.75	2.416
1	0	1	3.1
1	0.5	0	3.094
1	0.5	0.75	3.411
1	0.5	1	4.094
1	1	0	4.60
1	1	0.75	4.916
1	1	1	5.6

Example 6.12

Figure 6.21 depicts a chain of events typical for a printing system image path. The input image is in a device-dependent color space such as sRGB. For the present purpose, we will assume it is a standard 24-bit color image that is transformed (T_1) [12] to reference output medium metric *ROMMRGB*. Color space *ROMMRGB* has been developed by Eastman Kodak Company as a standard color space for the storage and manipulation of digital images. Note that the transformations from sRGB to *ROMMRGB* and back to sRGB are nonlinear. Transformation T_2 moves the image from *ROMMRGB* to the device-independent color space *L*a*b**. The results of the simulation for a 512×512 Lena image using a Xerox Phasor 770 digital printer are shown in Table 6.16.

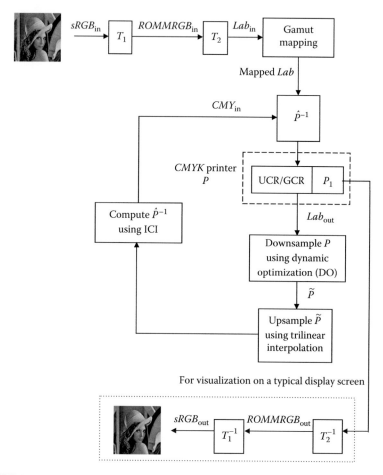

FIGURE 6.21   Printing system image path.

6.6   SMOOTHING ALGORITHM FOR MULTIDIMENSIONAL FUNCTIONS

6.6.1   INTRODUCTION

Data fitting is an important task in printer calibration and characterization. When fitting a function to data, there are two important considerations. The first is the goodness of the fit. Ideally, the fit and the data would have the same value at every point. The goodness of the fit is typically measured by an error metric defined in terms of the difference between the available data and the fit. The L_2 norm (MSE) is a measure normally used for this purpose. Other norms such as L_1 and L_∞ can also be used (Section 3.10.1). The second consideration is the smoothness of the fit. The smoothness of the fit is measured in terms of the first or second derivative of the underlying function. If the function is smooth, then its second derivative must

TABLE 6.15

ΔE Statistics for Downsampled LUTs of Different Sizes

Reduced Measurement Colors	Error Metric	Maximum ΔE	Mean ΔE	95th Percentile ΔE
27	ΔE_{ab}^*	8.56	3.00	4.92
	ΔE_{2000}	6.78	1.27	4.56
48	ΔE_{ab}^*	7.90	1.77	3.51
	ΔE_{2000}	4.42	0.96	3.45
125	ΔE_{ab}^*	7.53	1.37	3.0
	ΔE_{2000}	4.08	0.76	3.0
216	ΔE_{ab}^*	7.41	1.24	3.08
	ΔE_{2000}	4.13	0.74	3.05
343	ΔE_{ab}^*	7.41	1.20	3.19
	ΔE_{2000}	4.14	0.70	2.61
512	ΔE_{ab}^*	7.37	1.1	3.17
	ΔE_{2000}	4.11	0.65	2.48
4913	ΔE_{ab}^*	7.30	0.99	3.12
	ΔE_{2000}	4.18	0.57	1.94

TABLE 6.16

ΔE Statistics for Different LUT Sizes

Downsample LUT Size	Mean ΔE_{ab}^*	Maximum ΔE_{ab}^*	95th Percentile ΔE_{ab}^*
27	3.11	8.84	4.69
64	2.09	8.94	3.93
125	1.46	8.38	3.07
216	1.18	7.05	3.29
2197	1.03	8.80	3.12

be small. We use the second derivative as a measure of the smoothness of fit. If no restriction is made on how smooth the fit must be, then it ends up matching the data exactly. This will create problems if the data is noisy, since we must not fit to the noise in the data. On the other hand, if the fit is too smooth, then it might not approximate the data very well. To find the best fit, both of these factors must be considered. In this section, we consider the problem of smooth curve fitting in 1-D [13] through 4-D cases. As an application, the proposed algorithm can be applied to digital printer calibrations. In the 1-D case, it can be applied to the gray balance calibration [14,15], and for the 2-D case, it can be applied to the 2-D printer calibration [16]. Finally, as a 3-D example, we use the 3-D or 4-D profiling systems. Section 6.6.2 presents the 1-D through 4-D smoothing algorithm. Section 6.6.3 describes applications to printer calibrations with simulation results.

6.6.2 Multidimensional Smoothing Algorithm

The multidimensional smoothing algorithm is covered in this section. We first discuss the 1-D smoothing algorithm and then extend the approach to multidimensional (2-D through 4-D) case.

6.6.2.1 One-Dimensional Smoothing Algorithm

Assume that we have a function of one variable, $f(x)$. The value of this function at n different points is given in a vector $\hat{f}$:

$$\hat{f} = [f(x_0) \quad f(x_1) \quad \cdots \quad f(x_{n-1})]^{\mathrm{T}} \tag{6.76}$$

The objective is to find a vector f that is a smooth approximation of $\hat{f}$. Let the cost function J_1 be a measure of closeness of f to $\hat{f}$, that is,

$$J_1 = \sum_{i=0}^{n-1} (f_i - \hat{f}_i)^2 = \|f - \hat{f}\|^2 = (f - \hat{f})^{\mathrm{T}}(f - \hat{f}) \tag{6.77}$$

The other cost function, J_2, is used to measure the smoothness of the fit f. It is given as the distance between the second derivative of function f and zero:

$$J_2 = \int_{x_0}^{x_{n-1}} \left[\frac{\mathrm{d}^2 f(x)}{\mathrm{d}x^2} \right]^2 \mathrm{d}x \tag{6.78}$$

The cost function J_2 is approximated using

$$J_2 = \sum_{i=0}^{n-3} (f_i - 2f_{i+1} + f_{i+2})^2 = f^{\mathrm{T}} Q f \tag{6.79}$$

where

$$Q = C_n^{\mathrm{T}} C_n \tag{6.80}$$

and

$$C_n = \begin{bmatrix} 0 & 0 & 0 & \cdots & 0 & 0 & 0 & 0 \\ 1 & -2 & 1 & 0 & 0 & & & \\ 0 & 1 & -2 & 1 & 0 & \cdots & & \vdots \\ 0 & 0 & \ddots & \ddots & \ddots & & & \\ & & & \ddots & \ddots & 0 & 0 & 0 \\ \vdots & \vdots & & & 1 & -2 & 1 & 0 \\ & & & & 0 & 1 & -2 & 1 \\ 0 & 0 & 0 & \cdots & 0 & 0 & 0 & 0 \end{bmatrix} \overbrace{}^{n} \tag{6.81}$$

We now define the total cost function as a linear combination of the two cost functions:

$$J = J_1 + \alpha J_2 \tag{6.82}$$

where $\alpha \geq 0$. This function measures both the smoothness and the goodness of the fit. By changing the value of the parameter α, the relative importance of the smoothness and goodness of fit can be adjusted. The vector f that minimizes J will be the best fit for a given α and is obtained by setting the gradient of J with respect to f equal to zero:

$$\frac{\partial J_1}{\partial f} = 2(f - \hat{f})$$

$$\frac{\partial J_2}{\partial f} = 2Qf \tag{6.83}$$

$$\frac{\partial J}{\partial f} = 2(f - \hat{f}) + 2\alpha Qf = 0$$

The solution is

$$f = (I_n + \alpha Q)^{-1}\hat{f} \tag{6.84}$$

where I_n is an $n \times n$ identity matrix.

6.6.2.2 Two-Dimensional Smoothing Algorithm

In the 2-D case, there is a function of two variables, x and y, that we are trying to approximate. The values of this function at evenly spaced points in the $x-y$ plane are stored in the $n \times m$ matrix $\hat{f}$. Let f be the $n \times m$ matrix that is the smooth approximation of $\hat{f}$ that we are trying to find. In this case, J_1 is a measure of the distance between $\hat{f}$ and f and is given by

$$J_1 = \sum_{i=1}^{n}\sum_{j=1}^{m}\left(\hat{f}_{ij} - f_{ij}\right)^2 = \|\hat{f} - f\|_F^2 = \mathrm{Tr}\left[(\hat{f} - f)(\hat{f} - f)^{\mathrm{T}}\right] \tag{6.85}$$

where F stands for the Fibonacci norm (Equation 3.153). The second cost function J_2, which is a measure of the smoothness of f, is given as

$$J_2 = \sum_{i=1}^{n-2}\sum_{j=1}^{m}\left[f_{ij} - 2f_{(i+1)j} + f_{(i+2)j}\right]^2 + \sum_{i=1}^{n}\sum_{j=1}^{m-2}\left[f_{ij} - 2f_{i(j+1)} + f_{i(j+2)}\right]^2 \tag{6.86}$$

The first term in Equation 6.86 is a measure of smoothness with respect to the variable x and the second term with respect to the variable y. Equation 6.86 can be written as

$$
\begin{aligned}
J_2 &= \|C_n f\|_F^2 + \|C_m f^{\mathrm{T}}\|_F^2 = \mathrm{Tr}\left[C_n f (C_n f)^{\mathrm{T}}\right] + \mathrm{Tr}\left[C_m f^{\mathrm{T}} (C_m f^{\mathrm{T}})^{\mathrm{T}}\right] \\
&= \mathrm{Tr}\left(f^{\mathrm{T}} C_n^{\mathrm{T}} C_n f\right) + \mathrm{Tr}\left(f C_m^{\mathrm{T}} C_m f^{\mathrm{T}}\right)
\end{aligned}
\tag{6.87}
$$

where

C_n and C_m are matrices defined by Equation 6.81
Tr stands for the trace of a matrix

Since we want a fit that is both smooth and approximates the data well, we let the cost function be a linear combination of J_1 and J_2, that is, $J = J_1 + \alpha J_2$. We then try to find the minimum of J. Similar to the 1-D case, α can be used to control the relative importance of the smoothness and the goodness of the fit. To minimize J, we set the gradient of J with respect to f equal to zero:

$$
\frac{\partial J}{\partial f} = \frac{\partial J_1}{\partial f} + \alpha \frac{\partial J_2}{\partial f} = 0
\tag{6.88}
$$

since

$$
\frac{\partial J_1}{\partial f} = 2(f - \hat{f})
\tag{6.89}
$$

and

$$
\frac{\partial J_2}{\partial f} = 2\alpha f C_n^{\mathrm{T}} C_n + 2\alpha f C_m^{\mathrm{T}} C_m
\tag{6.90}
$$

Substituting Equations 6.90 and 6.89 into Equation 6.88 results in

$$
\left(\alpha C_n^{\mathrm{T}} C_n + I_n\right) f + f \alpha C_m^{\mathrm{T}} C_m = \hat{f}
\tag{6.91}
$$

where I_n is an $n \times n$ identity matrix. This equation is a special case of the Sylvester equation $(AX + XB = C)$, which can be solved by the MATLAB lyap function. Note that if either of n or m is one, then this reduces to the same equation used in the 1-D case.

6.6.2.3   Three-Dimensional Smoothing Algorithm

The 4-D smoothing algorithm can easily be derived from the 3-D case. Therefore, we first consider the 3-D case. In the 3-D case, we have a function of three variables, x, y, and z, that we are trying to approximate. The values of this function at evenly spaced points in the 3-D space are stored in an $n \times m \times l$ tensor $\hat{f}$. Let f be the

$n \times m \times l$ tensor that is a smooth approximation of $\hat{f}$. Similar to the other cases, we define J_1 as a measure of closeness of f to $\hat{f}$ to be

$$J_1 = \sum_{i=1}^{n} \sum_{j=1}^{m} \sum_{k=1}^{l} \left(\hat{f}_{ijk} - f_{ijk} \right)^2 \tag{6.92}$$

Let J_2 be a measure of the smoothness of f. Using the discrete second derivative, we have

$$J_2 = \sum_{i=1}^{n-2} \sum_{j=1}^{m} \sum_{k=1}^{l} \left[f_{ijk} - 2f_{(i+1)jk} + f_{(i+2)jk} \right]^2 + \sum_{i=1}^{n} \sum_{j=1}^{m-2} \sum_{k=1}^{l} \left[f_{ijk} - 2f_{i(j+1)k} + f_{i(j+2)k} \right]^2$$

$$+ \sum_{i=1}^{n} \sum_{j=1}^{m} \sum_{k=1}^{l-2} \left[f_{ijk} - 2f_{ij(k+1)} + f_{ij(k+2)} \right]^2 \tag{6.93}$$

The three terms in the above equation are measures of smoothness with respect to the three variables x, y, and z, respectively. Note that it is an approximation of the integral of the square of the second derivative of f along each of the three directions. Let $\otimes_n$ stand for the n-mode multiplication of a tensor with a matrix and $\langle , \rangle$ stand for tensor multiplication. The definitions of these operations are given in Appendix C. Then,

$$J_2 = \langle f \otimes_1 C_n, f \otimes_1 C_n \rangle + \langle f \otimes_2 C_m, f \otimes_2 C_m \rangle + \langle f \otimes_3 C_l, f \otimes_3 C_l \rangle \tag{6.94}$$

Similar to the 2-D case, let the overall cost function be

$$J = J_1 + \alpha J_2 \tag{6.95}$$

The minimum of J is obtained by setting the gradient of J with respect to f equal to zero. This yields

$$\frac{\partial J}{\partial f} = \frac{\partial J_1}{\partial f} + \alpha \frac{\partial J_2}{\partial f} = 0 \tag{6.96}$$

$$f - \hat{f} + \alpha f \otimes_1 \left(C_n^{\mathrm{T}} C_n \right) + \alpha f \otimes_2 \left(C_m^{\mathrm{T}} C_m \right) + \alpha f \otimes_3 \left(C_l^{\mathrm{T}} C_l \right) = 0 \tag{6.97}$$

$$f \otimes_1 \left(\alpha C_n^{\mathrm{T}} C_n \right) + f \otimes_2 \left(\alpha C_m^{\mathrm{T}} C_m \right) + f \otimes_3 \left(\alpha C_l^{\mathrm{T}} C_l + I_l \right) = \hat{f} \tag{6.98}$$

Since $\alpha C_n^{\mathrm{T}} C_n$, $\alpha C_m^{\mathrm{T}} C_m$, and $\alpha C_l^{\mathrm{T}} C_l + I_l$ are all positive definite diagonalizable matrices, we can define

$$\alpha C_n^{\mathrm{T}} C_n = R\theta R^{-1} \quad \alpha C_m^{\mathrm{T}} C_m = S\varphi S^{-1} \quad \alpha C_l^{\mathrm{T}} C_l + I_l = T\psi T^{-1} \tag{6.99}$$

where θ, φ, and ψ are diagonal matrices. Substituting Equation 6.99 into Equation 6.98 results in

$$\hat{f} = f \otimes_1 \left(R\theta R^{-1} \right) + f \otimes_2 \left(S\varphi S^{-1} \right) + f \otimes_3 \left(T\psi T^{-1} \right) \tag{6.100}$$

Multiply both sides of Equation 6.100 by the operator $\otimes_1 R^{-1} \otimes_2 S^{-1} \otimes_3 T^{-1} \hat{f} \otimes_1$

$$R^{-1} \otimes_2 S^{-1} \otimes_3 T^{-1} = f \otimes_1 \left(\theta R^{-1}\right) \otimes_2 S^{-1} \otimes_3 T^{-1} + f \otimes_2 \left(\varphi S^{-1}\right) \otimes_1 R^{-1} \otimes_3 T^{-1}$$
$$+ f \otimes_3 \left(\psi T^{-1}\right) \otimes_1 R^{-1} \otimes_2 S^{-1} \tag{6.101}$$

Define

$$\hat{F} = \hat{f} \otimes_1 R^{-1} \otimes_2 S^{-1} \otimes_3 T^{-1} \tag{6.102}$$

and

$$F = f \otimes_1 R^{-1} \otimes_2 S^{-1} \otimes_3 T^{-1} \tag{6.103}$$

Then,

$$\hat{F} = F \otimes_1 \theta + F \otimes_2 \varphi + F \otimes_3 \psi \tag{6.104}$$

Therefore,

$$\hat{F}_{ijk} = F_{ijk}\theta_{ii} + F_{ijk}\varphi_{jj} + F_{ijk}\psi_{kk} \tag{6.105}$$

$$F_{ijk} = \frac{\hat{F}_{ijk}}{\theta_{ii} + \varphi_{jj} + \psi_{kk}} \tag{6.106}$$

Once F is determined, f can be calculated from

$$f = F \otimes_1 R \otimes_2 S \otimes_3 T \tag{6.107}$$

We now extend the theory to the 4-D case. The algorithm for 4-D is as follows:

Step 1: Form and diagonalize the positive definite matrices $\alpha C_n^{\mathrm{T}} C_n$, $\alpha C_m^{\mathrm{T}} C_m$, $\alpha C_k^{\mathrm{T}} C_k$, and $\alpha C_l^{\mathrm{T}} C_l + I_l$, where I_l is the $l \times l$ identity matrix:

$$\begin{aligned} \alpha C_n^{\mathrm{T}} C_n &= R\theta R^{-1} & \alpha C_m^{\mathrm{T}} C_m &= S\varphi S^{-1} \\ \alpha C_k^{\mathrm{T}} C_k &= T\psi T^{-1} & \alpha C_l^{\mathrm{T}} C_l + I_l &= V\gamma V^{-1} \end{aligned} \tag{6.108}$$

Step 2: Compute

$$\hat{F} = \hat{f} \otimes_1 R^{-1} \otimes_2 S^{-1} \otimes_3 T^{-1} \otimes_4 V^{-1} \tag{6.109}$$

Step 3: Compute

$$F_{ijkl} = \frac{\hat{F}_{ijkl}}{\theta_{ii} + \varphi_{jj} + \psi_{kk} + \gamma_{ll}} \tag{6.110}$$

Step 4: Find f using Equation 6.107.

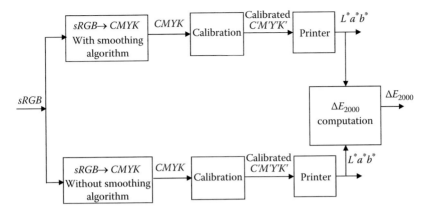

FIGURE 6.22   Image reproduction with LUTs.

6.6.3 APPLICATION TO PRINTING SYSTEMS

As an example, consider the image reproduction system shown in Figure 6.22. In this example, the input $sRGB$ image is mapped to the $CMYK$ color space by a multi-dimensional profile LUT, which is a GCR-constrained inverse of the printer (Chapter 7). The $CMYK$ pixels can be processed through 1-D calibration tone reproduction curves to generate the printer $C'M'Y'K'$. In this case, the profiling LUT is a mapping from the 3-D device-dependent $sRGB$ to the 4-D device-dependent $CMYK$ color space. Since color measurement is a noisy process, the final LUT may not be a smooth 3-D function. To produce a smooth characterization LUT, we insert a 3-D smoothing algorithm.

The results of simulations of the Lena image thorough the reproduction system of Figure 6.22 with and without a smoothing algorithm for a noisy $sRGB$ to $CMYK$ LUT are shown in Table 6.17. ΔE_{2000} is used as a measure of the performance. The lower the ΔE_{2000}, the better the color balance and image quality become. It is computed by converting the input $sRGB$ image to $L*a*b*$ and comparing the $L*a*b*$ to the printer output $L*a*b*$. The smoothing parameter α is changed from 0 to 2. The results shown in Table 6.18 correspond to the optimal choice of $\alpha = 0.5$. As can be seen, there is a 50% improvement in the 95th percentile.

The original and reproduced images with and without smoothing are shown in Figures 6.23 through 6.25. The results of the ΔE_{2000} accuracy in terms of maximum, mean, and 95th percentile ΔE_{2000} for different values of α are shown in Figure 6.26.

TABLE 6.17

Simulation Results of 3-D Profiling

	Mean ΔE_{2000}	Maximum ΔE_{2000}	95th Percentile
No noise LUT	0.465	6.954	1.664
Noise and no smoothing	2.891	12.999	6.005
Noise and smoothing	1.609	7.419	2.923

TABLE 6.18

One-Dimensional LUT

x	0	1.5	2	3	4.5	5.5	6	7	8	9
y	2.3	4.5	6.2	4.6	3.2	2.8	1.9	0.9	0.6	0

Original image   Reproduced image

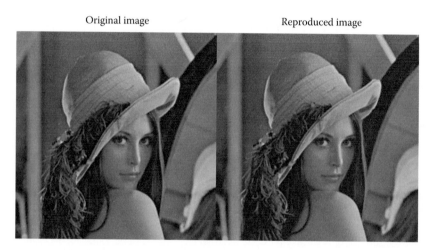

FIGURE 6.23 (See color insert following page 428.)   Original and reproduced image with noise-free profile LUT.

Original image   Reproduced image

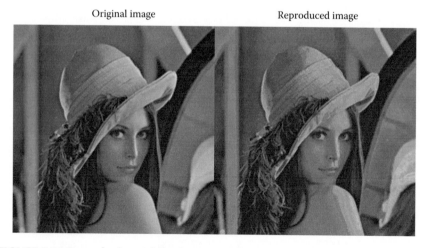

FIGURE 6.24 (See color insert following page 428.)   Original and reproduced image with noisy profile LUT.

Original image   Reproduced image

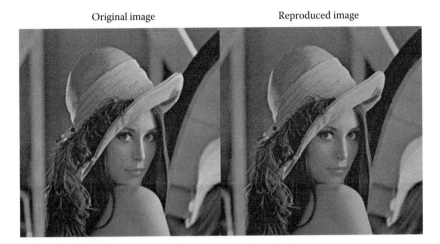

FIGURE 6.25 (See color insert following page 428.) Original and reproduced image with noisy profile LUT and smoothing algorithm.

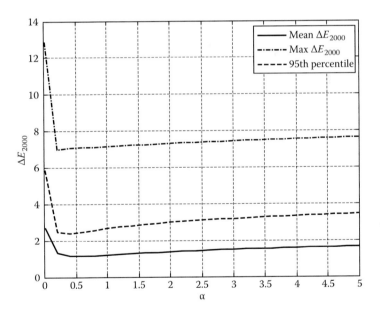

FIGURE 6.26 Maximum, mean, and 95th percentile ΔE_{2000} for different values of α.

PROBLEMS

6.1 Consider the 1-D function

$$f(x) = \ln(x+1) \quad 0 \leq x \leq 1$$

Create a 1-D LUT of size 10×2 by uniformly sampling the function from 0 to 1.
(a) Use linear interpolation to estimate the values of the function at $x_1 = 0.234$ and $x_2 = 0.568$. What is the resulting MSE?
(b) Repeat part a using cubic interpolation.

6.2 Consider the 1-D LUT shown in Table 6.18.
   (a) Use linear interpolation to estimate the value of the function at $x = 3.8$.
   (b) Repeat part a using
      (i) The Shepard technique (use L_1 norm and $\mu = 1$)
      (ii) The moving matrix with parameters $\mu = 2$ and $\varepsilon = 0$

6.3 Consider the 2-D function given by

$$z = f(x, y) = \left(x + y + xy^{1.8}\right) \ln(x + 1) \quad 0 \le x, y \le 1$$

   (a) Create a 2-D LUT of size 10×10 by uniformly sampling the function from 0 to 1 in both directions. The total number of grid points is 100.
   (b) Use bilinear interpolation to estimate the values of the function at $x = 0.25$ and $y = 0.68$. What is the resulting MSE?

6.4 Consider the 3-D function

$$g = f(x, y, z) = (xz + yz + xyz + 1) \ln(xyz + 1) \quad 0 \le x, y, z \le 1$$

   (a) Create a 3-D LUT of size $12 \times 12 \times 12$ by uniformly sampling the function from 0 to 1 in all directions. The total number of grid points is 1728.
   (b) Use trilinear interpolation to estimate the values of the function at $x = 0.75$, $y = 0.68$, and $z = 0.896$. What is the resulting MSE?
   (c) Repeat part b for the point with coordinates $x = 0.75$, $y = 0.75$, and $z = 1$. What is the resulting MSE?

6.5 Consider a *CMY* printer characterized through the 3-D LUT given in Table 6.19.
   (a) Use trilinear interpolation to find the $L*a*b*$ values corresponding to the following *CMY* patches:

   Patch 1: $CMY1 = [20 \quad 20 \quad 20]$
   Patch 2: $CMY2 = [40 \quad 40 \quad 40]$
   Patch 3: $CMY3 = [60 \quad 60 \quad 60]$
   Patch 4: $CMY4 = [80 \quad 80 \quad 80]$

   (b) Is this printer gray balanced?

TABLE 6.19

Three-Dimensional LUT

C	M	Y	L*	a*	b*
0	0	0	100.00	−0.04	0.00
0	0	85	97.61	−4.88	35.21
0	0	170	96.02	−7.53	69.72
0	0	255	94.96	−8.59	100.00
0	85	0	82.98	29.86	−7.28
0	85	85	81.49	23.82	24.18
0	85	170	80.94	17.72	87.70
0	85	255	80.51	19.95	53.95
0	170	0	69.10	55.79	−11.76
0	170	85	68.90	48.72	16.33
0	170	170	68.86	44.21	41.85
0	170	255	68.80	41.16	69.85
0	255	0	53.82	86.18	−14.25
0	255	85	54.58	78.25	9.65
0	255	170	55.07	73.16	29.49
0	255	255	55.16	70.02	49.03
85	0	0	83.70	−13.64	−20.44
85	0	85	81.13	−20.69	12.47
85	0	170	79.84	−24.61	43.54
85	0	255	78.66	−26.07	77.41
85	85	0	68.50	15.59	−25.42
85	85	85	67.73	7.81	5.02
85	85	170	67.82	2.65	32.37
85	85	255	66.72	1.31	61.35
85	170	0	56.67	40.60	−28.83
85	170	85	57.40	31.75	−0.64
85	170	170	57.65	27.53	23.35
85	170	255	56.89	25.47	47.26
85	255	0	42.76	70.00	−30.95
85	255	85	44.53	62.14	−6.25
85	255	170	45.45	57.65	13.12
85	255	255	45.60	54.93	30.82
170	0	0	66.30	−29.96	−41.98
170	0	85	65.11	−39.47	−9.95
170	0	170	64.37	−44.19	18.92
170	0	255	63.69	−46.42	49.90
170	85	0	54.01	−1.44	−43.99
170	85	85	54.92	−11.56	−14.10
170	85	170	54.96	−16.24	11.53
170	85	255	53.84	−18.73	37.31
170	170	0	44.03	23.42	−45.74
170	170	85	45.67	13.91	−17.8

(*continued*)

TABLE 6.19 (continued)
Three-Dimensional LUT

C	M	Y	L*	a*	b*
170	170	170	45.60	9.06	5.44
170	170	255	45.80	6.20	26.89
170	255	0	32.07	53.10	−47.14
170	255	85	34.28	44.77	−22.91
170	255	170	35.64	40.48	−3.57
170	255	255	36.49	37.66	13.50
255	0	0	54.56	−38.18	−55.13
255	0	85	54.79	−51.25	−24.05
255	0	170	54.76	−57.79	4.20
255	0	255	54.54	−61.72	34.77
255	85	0	44.41	−12.50	−55.51
255	85	85	45.39	−24.51	−27.21
255	85	170	45.94	−30.78	−2.37
255	85	255	46.10	−34.92	23.72
255	170	0	35.67	11.46	−56.40
255	170	85	37.20	0.42	−30.48
255	170	170	38.40	−5.71	−7.77
255	170	255	39.12	−10.12	14.22
255	255	0	24.98	42.17	−57.67
255	255	85	26.99	33.01	−35.15
255	255	170	28.80	27.68	−16.30
255	255	255	30.04	23.55	1.02

6.6 Repeat Problem 6.3 using
   (a) Tetrahedral interpolation
   (b) Shepard interpolation
   (c) Moving matrix

6.7 Show that the Shepard interpolation algorithm given by

$$\hat{y} = \begin{cases} \dfrac{\sum_{i=1}^{N} y_i d_i^{-\mu}}{\sum_{i=1}^{N} d_i^{-\mu}} & \text{if } d_i \neq 0 \quad \text{for all } i \\ y_j & \text{if } d_j = 0 \quad \text{for some } j \end{cases}$$

will result in a differentiable function if parameter $\mu > 1$.

6.8 Assume that we have a $CMY \rightarrow L*a*b*$ printer characterization function with Jacobian J at the nominal value $CMY = [120\ 67\ 145]$ given by

$$J = \frac{\partial Lab}{\partial CMY} = \begin{bmatrix} -0.1586 & -0.1223 & -0.0075 \\ -0.1267 & 0.3146 & -0.0299 \\ -0.1292 & -0.0480 & 0.4026 \end{bmatrix}$$

Find the range of parameter μ that guarantees convergence of the ICI algorithm.

6.9 Consider a scanner characterization color space transformation from a calibrated scanner *RGB* to a device-independent color space $L^*a^*b^*$ given by the 4-level (64 grid points) LUT shown in Table 6.20.

(a) Using trilinear interpolation, find the $L^*a^*b^*$ values corresponding to the following *RGB* values:

$$RGB1 = [23 \quad 23 \quad 23]$$
$$RGB2 = [123 \quad 123 \quad 50]$$

TABLE 6.20
***RGB* to $L^*a^*b^*$LUT**

R	G	B	L*	a*	b*
0	0	0	0	0	0
0	0	85	5.9	23.13	−48.17
0	0	170	19.59	39.51	−79.7
0	0	255	32.22	53.53	−108
0	85	0	30.61	−33.73	37.35
0	85	85	32.12	−22.79	−6.41
0	85	170	36.69	0.86	−51.53
0	85	255	43.76	24.87	−88.83
0	170	0	60.52	−55.37	61.32
0	170	85	61.09	−50.86	33.38
0	170	170	63	−37.41	−10.52
0	170	255	66.43	−17.56	−52.17
0	255	0	87.68	−75.03	83.09
0	255	85	87.99	−72.52	64.58
0	255	170	89.06	−64.42	26.71
0	255	255	91.05	−50.69	−14.25
85	0	0	15.2	37.22	23.69
85	0	85	18.36	40.28	−27.27
85	0	170	26.25	48.33	−68.43
85	0	255	36.23	58.92	−101.2
85	85	0	34.87	−5.74	42.49
85	85	85	36.15	0	0
85	85	170	40.11	14.81	−45.85
85	85	255	46.47	32.93	−84.28
85	170	0	62.21	−41.88	63.39
85	170	85	62.76	−38.18	35.82
85	170	170	64.59	−26.91	−7.98
85	170	255	67.9	−9.68	−49.76

(*continued*)

TABLE 6.20 (continued)
RGB to L*a*b*LUT

R	G	B	L*	a*	b*
85	255	0	88.61	−67.23	84.23
85	255	85	88.92	−64.93	65.85
85	255	170	89.97	−57.46	28.1
85	255	255	91.92	−44.7	−12.85
170	0	0	35.22	61.11	49.62
170	0	85	36.48	62.31	2.06
170	0	170	40.4	66.13	−44.77
170	0	255	46.7	72.46	−83.55
170	85	0	45.76	34.03	54.99
170	85	85	46.64	36.13	16.31
170	85	170	49.47	42.53	−30.43
170	85	255	54.32	52.39	−71.13
170	170	0	67.51	−9.43	69.75
170	170	85	67.99	−7.17	43.39
170	170	170	69.61	0	0
170	170	255	72.56	11.78	−42.12
170	255	0	91.69	−44.69	87.97
170	255	85	91.98	−42.89	70.02
170	255	170	92.97	−37.03	32.66
170	255	255	94.82	−26.77	−8.23
255	0	0	53.4	82.8	67.51
255	0	85	54.1	83.46	28.96
255	0	170	56.39	85.66	−18.59
255	0	255	60.42	89.61	−60.66
255	85	0	59.8	65.61	70.2
255	85	85	60.38	66.56	36.76
255	85	170	62.33	69.67	−9.63
255	85	255	65.81	75.07	−52.08
255	170	0	76.14	27.14	79.82
255	170	85	76.54	28.41	55.45
255	170	170	77.88	32.55	12.96
255	170	255	80.35	39.79	−29.43
255	255	0	97.15	−12.78	94.51
255	255	85	97.42	−11.53	77.34
255	255	170	98.31	−7.41	40.71
255	255	255	100	0	0

(b) Using the moving-matrix technique, find the *RGB* values corresponding to $L*a*b* = [60 \quad 82 \quad -62]$.

6.10 (a) Using the *RGB* to $L*a*b*$ LUT of Problem 6.9, estimate the Jacobian of the transformation at the nominal value $RGB = [120 \quad 67 \quad 145]$ using numerical differentiation and trilinear interpolation. The Jacobian is defined by

$$J_k = \frac{\partial L^*a^*b^*(k)}{\partial RGB(k)} = \begin{bmatrix} \frac{\partial L^*(k)}{\partial R(k)} & \frac{\partial L^*(k)}{\partial G(k)} & \frac{\partial L^*(k)}{\partial B(k)} \\ \frac{\partial a^*(k)}{\partial R(k)} & \frac{\partial a^*(k)}{\partial G(k)} & \frac{\partial a^*(k)}{\partial B(k)} \\ \frac{\partial b^*(k)}{\partial R(k)} & \frac{\partial b^*(k)}{\partial G(k)} & \frac{\partial b^*(k)}{\partial B(k)} \end{bmatrix}$$

(b) Find the range of parameter μ that guarantees convergence of the ICI algorithm that can be used to find the inverse LUT.

REFERENCES

1. H.R. Kang, *Color Technology for Electronic Imaging Devices*, SPIE Press, Bellingham, WA, 1996.
2. G. Sharma, *Digital Color Imaging Handbook*, CRC Press, Boca Raton, FL, 2003.
3. J.Z. Chang, Sequential linear interpolation of multidimensional functions, *IEEE Transactions on Image Processing*, 6, 1231–1245, Sept. 1997.
4. D. Shepard, A two-dimensional interpolation function for irregularly-spaced data, *Proceedings of the ACM National Conference*, pp. 517–524, New York, 1968.
5. R. Balasubramanian and M.S. Maltz, Refinement of printer transformation using weighted regression, *Proceedings of SPIE*, 2658, 334–340, 1996.
6. M. Xia, E. Saber, G. Sharma, and M. Tekalp, End-to-end color printer calibration by total least squares regression, *IEEE Transactions on Image Processing*, 5, 700–716, 1999.
7. R. Balasubramanian, The use of spectral regression in modeling halftone color printers, *IST/OSA Annual Conference, Optic/Imaging in the Information Age*, pp. 372–375, Rochester, Oct. 1996.
8. D. Viassolo and L.K. Mestha, A practical algorithm for the inversion of an experimental input-output color map for color correction, *Journal of Optical Engineering*, 42(3), 625, March 2003.
9. E.K. Chong and S.H. Zak, *An Introduction to Optimization*, John Wiley & Sons Inc., New York, 1996.
10. R.E. Groff, D.E. Koditschek, and P.P. Khargonekar, Piecewise linear homeomorphisms: The scalar case, *IEEE International Conference on Neural Networks*, 3, 259–264, Jul. 2000.
11. S.A. Dianat, L.K. Mestha, and A. Matthew, Dynamic optimization algorithm for generating inverse printer map with reduced measurements, *2006 IEEE International Conference on Acoustics, Speech and Signal Processing*, Toulouse, France, May 14–19, 2006.
12. Kelvin E.S., Geoffrey J.W., and Edward J.G., Reference input/output medium metric RGB color encodings (RIM/ROMM RGB), *PICS 2000 Conference*, Portland, OR, March 26–29, 2000.
13. L.K. Mestha and S. Dianat, TRC smoothing algorithm to improve image contours in 1-D color controls, US Patent 7,397,581, Jul. 8, 2008.
14. L.K. Mestha, R.E. Viturro, Y.R. Wang, and S.A. Dianat, Gray balance control loop for digital color printing systems, *NIP21: Proceedings of the IS&T's International Conference on Digital Printing Technology*, pp. 499–504, Baltimore, MD, Sept. 18–23, 2005.
15. P.K. Gurram, S.A. Dianat, L.K. Mestha, and R. Bala, Comparison of 1-D, 2-D and 3-D printer calibration algorithms with printer drift, *NIP21: Proceedings of the IS&T's International Conference on Digital Printing Technology*, pp. 505–510, Baltimore, MD, Sept. 18–23, 2005.
16. G. Sharma, R. Bala, J.R.N. Van de Capelle, M. Malz, and L.K. Mestha, Two-dimensional calibration architectures for color devices, US Patent 7,355,752, Apr. 8, 2008.

7 Three-Dimensional Control of Color Management Systems

7.1 INTRODUCTION

A color management architecture for a digital imaging system provides a means for processing digital images such that the colors produced on the output devices are a reasonable representation of the colors in the input image file. An image file may be created using a computer or may be captured by other imaging devices such as a scanner or a digital camera. The International Color Consortium (ICC) was set up to provide a standard paradigm for managing color with image capture, display, and rendering devices [1]. The algorithms described in this chapter use theoretical tools from previous chapters to produce multidimensional color transforms for managing color between devices. Numerous techniques for device characterization, gray component replacement (GCR), and constrained inverse and gamut mapping are described in detail. Some of these techniques use advanced control-based approaches. A key output from this chapter is the well-tuned custom multidimensional lookup table (LUT). These tables can be used to color-manage input images so that when the resulting images are printed on output devices, they exhibit increased color quality; better in terms of accuracy and consistency as compared to prints from nonoptimal color management systems.

7.2 IMAGE PATH ARCHITECTURE

The control of color is a system-wide problem. The color strategy followed by print shops must use standards that provide a link between color transforms and numerous devices so that the colors represented in different device-specific color spaces can be transformed to a space local to the rendering device. The preferred industry standard used by many production print shops for managing color is based on standards set by the ICC. The ICC specification is a vendor-neutral cross-platform color management standard created by a group of industry color experts to improve color workflow and is implemented in an ICC profile format. The ICC specification divides color devices into three broad categories: input devices (e.g., RGB scanners, RGB cameras, CDs, electronic documents), display devices (e.g., RGB monitors and RGB projectors), and output devices (e.g., printers). For each device class, a series of base transforms are included in the ICC profile that perform the conversion between different color spaces. A *source profile*, a three-to-three (RGB-to-*L*a*b**) or a four-to-three transform (CMYK-to- *L*a*b**), is assigned to each image file that defines a mathematical transform between the numbers in the input file (i.e., the amount of red, green, and blue light seen by the source camera) and a reference color space. This reference color space, specified in XYZ or *L*a*b**, describes the color as perceived by the human eye. These *L*a*b** or *XYZ* values are passed to the *destination profile*, which defines yet another mathematical transformation (e.g., a three-to-four transform used for a CMYK printer) between the amounts of the colorants from reference color space and the destination color space that the device would need to use to render the color. In general, these mathematical transformations are embodied as multidimensional LUTs. Using this kind of source, reference, and destination architecture, the source and destination devices from different vendors can be used in a system to produce the color with good results that is most suitable for a particular application.

The destination profiles have an additional complexity because they must also handle requests for colors the destination device cannot make. This situation is called the gamut-mapping problem and it is normally handled inside the destination profile. In the ICC profile architecture a reference color space in *L*a*b** or XYZ called the profile connection space (PCS) [1], is used for standard observers, measurement geometry, and illuminants. Also, there can be many different rendering intents (colorimetric, perceptual, saturation, etc.) included in the ICC.1:2004-10 specifications because of the need for tuning the profiles to specific applications. Device Link profiles are another special kind of ICC profiles that provide a dedicated transformation from one input device color space (e.g., CMYK) to another output device color space (CMYK) without going through the PCS. Thus the ICC profile format is made generic to the extent that each application vendor can foster their own proprietary technology to differentiate their art from their competitors. However, the basis for creating a device link profile or profiles with different rendering intents has always been an ordinary ICC profile such as the colorimetric profile. Hence, in this chapter we focus our algorithms on the generation of an accurate colorimetric profile, in particular, a destination profile.

Figure 7.1 shows a footprint of the destination ICC profile used with various components to translate color information from the PCS to the device-specific CMYK space. A color lookup table (CLUT) or a 3DLUT is a major component of the destination ICC profile that is a GCR-constrained, gamut-mapped LUT from RGB triplets to device CMYK space (Figure 7.2). Although the destination profile is simply an XYZ or *L*a*b** (input of Figure 7.1) to CMYK transform (output of Figure 7.1), in order to obtain the biggest possible color gamut with minimum amount of toner and improved color rendition accuracy, the profile is divided into several components as shown in Figure 7.1: the XYZ or *L*a*b** to RGB matrix, a tone reproduction curve (TRC), a 3DLUT, and a second TRC. This type of division will help the process of creating a destination profile. See Section 7.8 for detailed steps involved in the creation of a colorimetrically accurate destination profile.

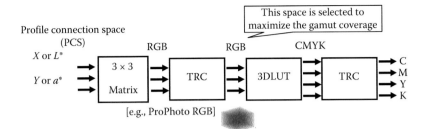

FIGURE 7.1   LUT B2A table converting PCS to device CMYK space in the ICC profile.

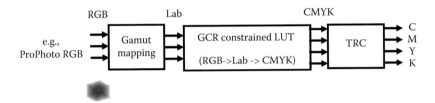

FIGURE 7.2   LUT B2A table converting PCS to device CMYK space in the ICC profile (*Note:* RGB is an intermediate color space).

7.3   PROFILING—A COMPLEX SYSTEM PROBLEM

While multidimensional profiles offer opportunity to improve color quality of output devices, the complexity involved in the creation of these profiles is high. In general, the requirements for high quality rendering are very rigid. The simplest requirements in device owners terms are "get the color right, every time, everywhere, and all the time." In technical terms, the phrase "every time, everywhere and all the time" implies achieving device process stability under all the operating environments, capturing variation in temperature, humidity, media. However, "getting the color right" on multiple devices is infeasible because different devices have different color gamuts and it is physically impossible for a device to produce a color outside its reproducible gamut. As seen in the discussion below, even with a stable device, and the best profiles, it is not easy to "get the color right" so that the color differences are maintained within the perceptual limits.

7.3.1   TIGHT COLOR RENDITION REQUIREMENTS

Very accurate colorimetrically matched reproduction is desired for high quality rendering. Colors have to be within perceptual limits in all regions of the color space that the printer can produce. That means the desired perceptual color difference for in-gamut colors between the input $L^*a^*b^*$ values to the rendered $L^*a^*b^*$ values of the pixels must be less than or equal to 1 (ΔE_{ab}^*, ΔE_{2000}, etc.) unit. This is extremely difficult to achieve due to (i) 8 bit quantization limits of the imaging system, (ii) system errors like halftone noise, development noise, etc., (iii) sensor errors, and (iv) profiling algorithm errors. In spite of improvements with respect to (i), (ii), and (iii), the profiling algorithm errors can make it impossible to achieve required rendition accuracy since the profile algorithms use a many-to-one inversion. To build a multidimensional LUT, one must find the right CMYK formulation for each color in the table, and this requires rules to choose from the many possible CMYK formulations (called gray component replacement [GCR]) and inversion methods.

7.3.2 Gamut Limitation

A device must be able to produce good color from input image files that contain colors outside the nominal device gamut (Figure 7.3). Since many devices do not have full gamut coverage, out of gamut colors are mapped to the destination device gamut that they are able to produce. For example, some colors outside a destination device gamut need to be mapped to the inside of the gamut or on the boundary as they otherwise cannot be reproduced on the output device and would simply be clipped. For instance, printing a mostly saturated blue color, as displayed on a monitor, on paper using a typical CMYK printer will likely fail. The paper blue may not be as saturated as the one displayed on the monitor. Conversely, the bright cyan of an inkjet printer may not be easily presented on an average computer monitor. Each multidimensional transform must utilize various methods to achieve the desired color and also give experienced users the opportunity to control the gamut-mapping behavior.

7.3.3 Smoothness

Reproduction of color sweeps that cross the gamut boundary must be smooth. Otherwise, output images can have contours. Generally, these image quality defects are related to gamut mapping in particular regions of the color space (e.g., dark brown region). Contours can occur due to a many-to-one mapping at the intersection between mapped colors and in-gamut colors near the gamut boundary.

FIGURE 7.3   Vector plot of out-of-gamut colors shown with respect to reproducible gamut of a printer.

7.3.4   ICC WORKFLOW

The multidimensional profile LUTs embedded in ICC profile have finite nodes at which device values are available. Input image files with color pixels not on the nodes will require interpolation. The ICC path provides simple linear interpolation with limited resolution for processing pixels at high speed. When device behavior is nonlinear as for the gamut of Figure 7.4, linear interpolation provided by ICC image path may not provide sufficient accuracy.

7.3.5   ENGINE CONDITIONS

A stable engine condition is essential to repeatedly produce the color in the original input file for the same media, toner, halftone, etc. Also, if the overall gamut of the engine varies, then the colors initially matched to the original print are no longer valid. As gamut boundaries move with time, static gamut mapping is no longer optimal, leading to unpredictable results; even custom profiles with static gamut mapping will not reproduce optimal output quality.

The first step in the profile creation stage is device characterization. Nowadays, characterization is performed by printing and measuring a set of color targets. The next step involves computing the GCR-constrained inverse and gamut mapping the out-of-gamut colors to the characterized boundary. All of these steps are too demanding for high-end printing since even a small numerical error can lead to a noticeable change in output color and a degradation in the image quality. Therefore, in the following section we first describe the underlying theory and algorithmic details of various methods applicable to characterizing a four-color print engine. We show the adaptation process using recursive least-squares (RLS) identification algorithms.

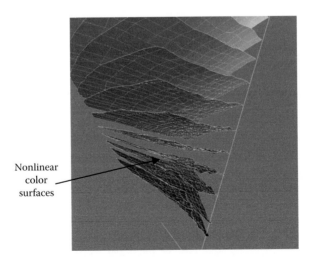

FIGURE 7.4 Nonlinear color surfaces (CYK $L*a*b*$ gamut with color surfaces for CY planes for fixed K and $M = 0$).

7.4   CHARACTERIZATION OF COLOR SYSTEMS

Conventionally, there are two different approaches to color modeling: (1) empirical or interpolation-based approaches, which essentially treat the device as a black box with inputs and outputs and (2) analytical or first principle approaches which attempt to characterize the device color response using the fewest number of measurements to arrive at analytical functions that have a physical connection to the process. Both kinds of approaches are capable of predicting the color response of the device for a variety of input images.

The empirical or interpolation-based methods are generally measurement intensive and require the use of a large set of experimentally generated data between inputs and outputs [2–6]. These models may contain only nonparametric LUTs, or parameterized analytical functions that fit the data. The function passes through the data points approximately. Thus, although we cannot use an empirical model to explain a system, we can use such a model to predict the behavior where data does not exist. One can understand that measurement data is crucial for an empirical model. Data is used to assess the structure of the functional form and then estimate its parameters. In this section, we present several empirical models with approaches to capture both time-zero and time-varying effects. Accurate first principle models (so-called "white box" models) are not available for all kinds of imaging devices. The complexity of models, errors in capturing the actual physical process in the presence of device drift over time, light scattering effects, and many other uncertainties associated with the physical device itself make it impossible to perfectly model the device over a reasonable period of time. In Chapter 10, a more elaborate, parameterized nonlinear spectral model of the printing system incorporating reasonable abstractions of the process is described. This type of model can help us to inject meaningful time-varying effects into the system.

7.4.1   LEAST-SQUARES ESTIMATION

We will consider two parts to the empirical or interpolation-based approach to printer modeling. The first part is to obtain the model at time t_0, which is typically done (at the printer manufacturer) "in the factory" but could also be performed "in the field" if an in-line color device is available [7–10]. This approach may require measurements of anywhere from one to thousands of color patches. A mathematical model is generated and initial parameters are selected from this data using data fitting techniques such as the least-squares method. Once an optimal parameterized model is selected, the second part involves an adaptation process. The adaptation process [11] is implemented using measurements from the sensor to adjust or tune the model parameters over time. It accounts for variations in printer state over time $t_0 + t$. Using the historical data and the new measurements, an adaptive algorithm will estimate the model parameters more accurately, and the predicted colors from the adapted model can better represent the real system output at that time. In automatic control literature, this kind of empirical modeling of a dynamic system is called system identification.

7.4.1.1   A Linear in the Parameters Model

A simple "linear in the parameters" model with affine, linear, quadratic, or cubic terms is introduced below. For a scalar case, if y is the output data from the sensor, and u is the input, then an input–output model can be written as

$$y = \theta_1 f_1(u) + \theta_2 f_2(u) + \cdots + \theta_n f_n(u) \tag{7.1}$$

where
y is the output (sensor) data
u is the input (e.g., CMYK values for a printer) data
θ_i is the parameter that needs fitting or identification
$f_i(u)$ is the known form of the model (e.g., affine, linear, quadratic, etc.)

In vector and matrix form,

$$y = A\theta \tag{7.2}$$

where the matrix A is named the regression matrix and it contains the form of the model and θ is the matrix containing the parameters of the model.

For example, when modeling a CMYK to $L^*a^*b^*$ input–output data, the sample output test color data is represented in $L^*a^*b^*$ and the corresponding input value is represented in CMYK. We can then write parameters of the linear affine model (Equation 7.2) as follows:

$$y = y_0 = [y_1 \quad y_2 \quad y_3]; \quad A = A_0 = [1 \quad u_1 \quad u_2 \quad u_3 \quad u_4];$$

$$\theta = \theta_0 = \begin{bmatrix} b_1 & b_2 & b_3 \\ M_{11} & M_{21} & M_{31} \\ M_{12} & M_{22} & M_{32} \\ M_{13} & M_{23} & M_{33} \\ M_{14} & M_{24} & M_{34} \end{bmatrix}; \quad \text{and} \quad y_0 = A_0 \theta_0 \tag{7.3}$$

where
$u_1 = c$
$u_2 = M$
$u_3 = Y$
$u_4 = K$
$y_1 = L^*$
$y_2 = a^*$
$y_3 = b^*$
b_i, M_{ij} are constants

Note, we used the subscript "0" to distinguish the variables between initial data and the updates coming at future time during the adaptation process. If there are N colors in the training set, then y is a matrix of size $N \times 3$.

A_0 is a matrix with N number of rows. For a linear affine model (as shown) A_0 will be of size $N \times 5$. Since the A_0 matrix is nonsquare and, therefore, noninvertible, the parameters contained in θ_0 cannot be solved using simple linear algebra or matrix methods. First we form a residue equation,

$$r = y_0 - A_0\theta_0 \tag{7.4}$$

and then form the sum of the squares of the residues,

$$S = rr^{\mathrm{T}} = (y_0 - A_0\theta_0)(y_0 - A_0\theta_0)^{\mathrm{T}} \tag{7.5}$$

Now minimize S by differentiating Equation 7.5 with respect to θ_0 and set the resulting equation to zero and solve for θ_0. This process yields the standard least-squares solution,

$$\theta_0 = \left(A_0^{\mathrm{T}}A_0\right)^{-1}A_0^{\mathrm{T}}y_0 \tag{7.6}$$

The term $\left(A_0^{\mathrm{T}}A_0\right)^{-1}A_0^{\mathrm{T}}$ is called the pseudo-inverse of matrix A_0. Equation 7.6 is used to obtain an initial guess of the parameter matrix using the measured $L*a*b*$ values for test colors whose CMYK values are included in the A_0 matrix as $u_1 = C$; $u_2 = M$; $u_3 = Y$; $u_4 = K$.

Using this technique, other parameter models can also be built easily. For example, the partial quadratic model can be modeled as

$$A_0 = \begin{bmatrix} 1 & u_1 & u_2 & u_3 & u_4 & u_1^2 & u_2^2 & u_3^2 & u_4^2 \end{bmatrix} \tag{7.7}$$

This is yet another form suitable for linear regression. Note, A_0 is a matrix of size $N \times 9$ (as shown in Equation 7.7) when N number of test colors are used.

7.4.1.2 Recursive Least-Squares Estimation Algorithm

The RLS algorithm is an adaptive learning method that can be used to refine the parameters of the model and perform system identification with the data collected with *in situ* sensors for a time-varying print engine. It can be used as a onetime estimation algorithm to improve the numerical accuracy of a least-squares model. Once the sensor data become available, it allows the system to update the parameters to reduce the error between the outputs of the linear model and the actual printer to an acceptable level. A schematic of the adaptation process is shown in Figure 7.5 for a test color. Most adaptive algorithms are of the following form:

$$x_{k+1} = x_k + \alpha e \tag{7.8}$$

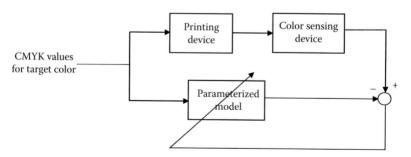

FIGURE 7.5 System block diagram showing adaptation process used to tune the parameters of the linear model.

where

x_{k+1} = new estimate

x_k = old estimate

α = weight factor

$$e = \begin{bmatrix} \text{measured} \\ \text{values} \end{bmatrix} - \begin{bmatrix} \text{model} \\ \text{output} \end{bmatrix}$$

The initial (estimated) parameters in the θ_0 matrix are obtained from Equation 7.6. In the adaptation context, we refer to θ_k as an old estimate, and θ_{k+1} as the new estimate. When the new data arrive, let a^T represent the regression matrix for the next set of data while $\bar{y}$ is the corresponding output. This additional data can be incorporated using the original equation (Equation 7.2) to obtain the overdetermined linear equation,

$$y = A\theta \tag{7.9}$$

This equation is similar in structure to Equation 7.2, with

$$A = \begin{bmatrix} A_0 \\ a^T \end{bmatrix}; \quad y = \begin{bmatrix} y_0 \\ \bar{y} \end{bmatrix} \tag{7.10}$$

The updated estimates, θ_{k+1}, are built by using all the available input and output data points as in Equation 7.11.

$$\theta_{k+1} = \left(A^T A\right)^{-1} A^T y = \left[\begin{bmatrix} A_0 \\ a^T \end{bmatrix}^T \begin{bmatrix} A_0 \\ a^T \end{bmatrix} \right]^{-1} \begin{bmatrix} A_0 \\ a^T \end{bmatrix} \begin{bmatrix} y_0 \\ \bar{y} \end{bmatrix} \tag{7.11}$$

As expected, the equation for the new estimates again has similar structure as the initial estimate Equation 7.6. The previously calculated inverse from Equation 7.6 is in fact used in the new Equation 7.11. However, we will need to calculate the inverse again as new data arrive. This means that unnecessary computation is required to solve Equation 7.11. A more efficient way of calculating the new solution utilizes Equation 7.6, which is shown next.

Few intermediate steps are necessary before the new estimates are determined. From simple matrix manipulation of Equation 7.10, we can write

$$A^T A = A_0^T A_0 + aa^T; \quad A^T y = A_0^T y_0 + a\bar{y} \tag{7.12}$$

$$\text{Also let } P = \left(A^T A\right)^{-1} \quad \text{and} \quad P_0 = \left(A_0^T A_0\right)^{-1} \tag{7.13}$$

Further manipulation is done as follows to reach a simple form for the new θ matrix. Rewriting Equation 7.11, we have

$$\theta_{k+1} = \left(A^T A\right)^{-1} A^T y = \left(A_0^T A_0 + aa^T\right)^{-1}\left(A_0^T y_0 + a\bar{y}\right) \tag{7.14}$$

Matrix inversion lemma (Equation 5.64) states that for any matrices A, B, and C of appropriate size

$$(A + BC)^{-1} = A^{-1} - A^{-1}B\left(I + CA^{-1}B\right)^{-1}CA^{-1} \tag{7.15}$$

Let $A_0^T A_0 = A$, $B = a$, and $C = a^T$. Then substituting into Equation 7.15, we get

$$\left(A_0^T A_0 + aa^T\right)^{-1} = \left(A_0^T A_0\right)^{-1} - \left(A_0^T A_0\right)^{-1} a\left(I + a^T\left(A_0^T A_0\right)^{-1} a\right)^{-1} a^T\left(A_0^T A_0\right)^{-1} \quad (7.16)$$

Let us define

$$K = \frac{P_0 a}{1 + a^T P_0 a} \quad (7.17)$$

Substituting P_0, recognizing $a^T P_0 a$ is a scalar and then substituting K Equation 7.16 becomes

$$\left(A_0^T A_0 + aa^T\right)^{-1} P = \left[I - Ka^T\right] P_0 \quad (7.18)$$

Substituting Equation 7.18 into Equation 7.14, we get

$$\begin{aligned}
\theta_{k+1} &= \left[I - Ka^T\right] P_0 \left(A_0^T y_0 + a\bar{y}\right) \\
&= P_0 A_0^T y_0 + P_0 a\bar{y} - Ka^T P_0 A_0^T y_0 - Ka^T P_0 a\bar{y}
\end{aligned} \quad (7.19)$$

But, from Equation 7.6, $\theta_0 = P_0 A_0^T y_0$ and from Equation 7.17, $P_0 a = K[1 + a^T P_0 a]$. Substituting these relationships in Equation 7.19, yields following expression for the new θ matrix. (*Note*: for convenience, we removed $k+1$ from the notation.)

$$\theta = \theta_0 + K\left(\bar{y} - a^T\theta_0\right) \quad (7.20)$$

where, again, the weight matrix K and the matrix P are given by

$$K = \frac{P_0 a}{1 + a^T P_0 a} \quad \text{and} \quad P = \left[I - Ka^T\right] P_0 \quad (7.21)$$

Notice that the new equation for θ (Equation 7.20) is very similar to the general adaptive algorithm of Equation 7.8. The matrix P_0 is updated in Equation 7.21 for each iteration. To show the iteration steps, let us replace the subscripts "0" with "K" to represent kth update and use "$K + 1$" to denote the new updates. Equations 7.20 and 7.21 then yield the following form for the old estimate and the new estimate.

$$\theta_{k+1} = \theta_k + K_{k+1}\left(y_{k+1} - a_{k+1}^T\theta_k\right) \quad (7.22)$$

$$K_{k+1} = \frac{P_k a_{k+1}}{1 + a_{k+1}^T P_k a_{k+1}} \quad \text{and} \quad P_{k+1} = \left[I - K_{k+1}a_{k+1}^T\right] P_k \quad (7.23)$$

Note: $\bar{y}$ in Equation 7.20 is replaced with y in Equation 7.22 to represent the measured values.

Simplifying Equation 7.23, it can be shown

$$K_{k+1} = P_{k+1} a_{k+1} \quad (7.24)$$

Substituting Equation 7.24 into Equation 7.22, the following recursive estimation formula emerges

$$\theta_{k+1} = \theta_k + P_{k+1}a_{k+1}\left(y_{k+1} - a_{k+1}^T\theta_k\right) \qquad (7.25)$$

where P_{k+1} is

$$P_{k+1} = P_k - \frac{P_k a_{k+1} a_{k+1}^T P_k}{1 + a_{k+1}^T P_k a_{k+1}} \qquad (7.26)$$

Equations 7.25 and 7.26 achieve the adaptive estimation by using the previous estimate and the error between the outputs to converge to a new estimate. The matrix, A_0, only enters into the recursive equation in the first guess for θ_0 (Equation 7.6). Hence, the RLS solution is more robust than the standard least-squares solution (Equation 7.6). In practice, this recursive estimation can be initiated by simply setting P_0 to a diagonal matrix, and by letting θ_0 be the best first guess. Convergence to the desired θ is obtained when the error function is driven to zero or to a level below the desired threshold.

7.4.1.3   Piecewise Linear Models

In many practical imaging devices, including printers, the linear model with the RLS algorithm cannot accurately fit one uniform function to the entire input–output data set since the functional relationship between input–output data may be locally nonlinear or may be only partially linear. The piecewise linear approach to modeling nonlinearity of many engineering systems is a well-known concept. Hence generating multiple piecewise linear (or curvilinear due to quadratic, cubic terms) models at various input points and combining them via interpolation can lead to much more efficient models than global linear models, since such models can represent the data more accurately.

To fully exploit the power of piecewise linear models, the color space is partitioned into a number of regions or clusters. The number of regions, number of training colors in each region, and the methodology required for partitioning the multidimensional color space depend on the device under study. K-means algorithm (discussed in Section 2.4.2.5) based on vector quantization—a classical quantization technique— widely used in the literature for a variety of classification, compression, and signal partitioning applications [12,13]. It has been found suitable for classifying the multidimensional color space into smaller regions or clusters. It is related to self-organizing maps, where a large set of input color nodes are divided into groups having approximately the same number of points closest to them. Linear models are generated for each region, which is identified by its centroid.

Key steps involved in constructing the initial multiple piecewise linear models based on clusters are (1) obtain training samples via experiments or historical database, (2) cluster the input or output colors (e.g., $L^*a^*b^*$ values or CMYK values) to obtain partitioned colors and their centroids, and (3) obtain parameters of the linear model for each cluster using Equation 7.3. When the adaptation is required, iteratively execute the RLS algorithm using Equation 7.25 for each of the clusters separately. A suitable cluster assignment process is required to assign each color sample to the right cluster. A simple way of doing this is by comparing the Euclidean distance of the sample output color (assuming output colors were used to create the clusters) with each centroid, and then choosing the cluster whose centroid has the shortest distance to the sample output color. After making the right cluster assignment, run the RLS algorithm. The process for generating and adapting piecewise linear models over time is illustrated in Examples 7.1 and 7.2.

Example 7.1

Use K-means (Linde–Buzo–Gray [LBG]) algorithm [12] to divide the 10^4 input–output printer modeling data into 10 clusters. Describe the critical steps in the algorithm.

SOLUTION

The method of classifying the color input–output data is described first. Let the training vectors, a mapping of the CMYK values to $L*a*b*$ be denoted as a database

$$U = [u_1 \quad u_2 \quad \dots \quad u_N] \in R^{n \times N} \longrightarrow Y = [y_1 \quad y_2 \quad \dots \quad y_N] \in R^{l \times N} \quad (7.27)$$

where

$u_1, u_2, \dots, u_N$ are vectors containing input CMYK values for each color containing four values denoted by n

$y_1, y_2, \dots, y_N$ are vectors containing output $L*a*b*$ values for corresponding colors containing three values denoted by l

In this example, $N = 10,000$ for 10^4 characterization samples, which is a predetermined number based on the number of training samples that need to be classified. Generally, as the gamut becomes larger and more nonlinear, N will become larger. N will be smaller for a calibrated printer. N could be 4^4, 8^4, or 16^4 and so on, depending on the training set.

Using the K-means algorithm, described below, the training database is partitioned into K clusters, U_k for $k = 1, 2, 3, \dots, K$ as follows:

$$U_k = \left[u_{1_1} \quad u_{2_2} \quad \dots \quad u_{N_k}\right]_k \in R^{n \times N_k} \longrightarrow Y_k$$
$$= \left[C_k \quad y_{1_1} \quad y_{2_2} \quad \dots \quad y_{N_k}\right]_k \in R^{l \times N_k} \quad (7.28)$$

where $[u_1 \quad u_2 \quad \dots \quad u_N]_k$ are the vector elements containing the N_k input CMYK values for the kth cell. These clusters will be generated by the K-means algorithm. After assigning the colors to the clusters, vectors $u_1, u_2, \dots$ of Equation 7.27 do not necessarily correspond to the $u_1, u_2, \dots$ of Equation 7.28 because of the reordering of the training sample data that occur during the clustering process. Hence, we used the notation, $u_{1_1}, u_{2_2}, \dots, u_{N_k}$ to describe new input vectors, and $[y_{1_1}, y_{2_2}, \dots, y_{N_k}]_k$ are the vector elements formed with $L*a*b*$ values, each having three elements. The number of elements could be higher when a reflectance spectral database is used for clustering (not considered in this example). Vector C_k is the centroid of the $L*a*b*$ values in a kth cluster. The relationship between K and N is as follows:

$$N = \sum_{k=1}^{K} N_k \quad (7.29)$$

Figure 7.6 shows a flowchart illustrating the method of determining the centroids. These centroids will become the centers of the respective clusters of

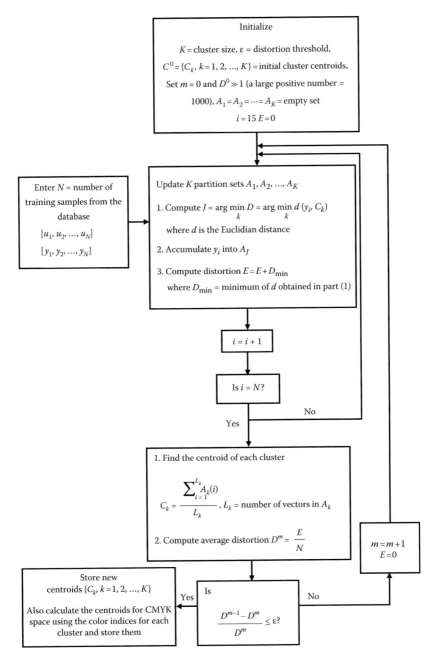

FIGURE 7.6 Flowchart of the centroid and cluster assignment in $L^*a^*b^*$ space.
Note: The training database vectors $y_1, y_2, \ldots, y_N$ are used for centroid and cluster assignments in $L^*a^*b^*$ space. If the same is required in *CMYK* space, then use training database vectors $u_1, u_2, \ldots, u_N$.

the partitioned database. Various initialization values are entered, including ε, K, m, D^0, i, and E. ε is a distortion threshold, which indicates the maximum allowable distortion, as defined by the criterion associated with desired system performance. K is the number of clusters into which the database is to be partitioned. The larger the parameter K is, the better the results obtained from the model (see Figures 7.9–7.11), but it will increase the processing time. m, D^0, i, and E are values used in the algorithm. Specifically, m and i are simply iteration counters, and may be initially set at 0 and 1, respectively. D^0 is an initial distortion setting, and is set at an arbitrary large positive number, such as 1000. E is a distortion value, which is initially set at 0.

Empty sets A_1, A_2, ..., A_K are established a *priori*. These are the clusters of the database, which will be filled with initial values and then updated until certain algorithm criteria are met, as described below. Initial cluster centroids C^0 are set equal to C_k, where $k = 1, 2, ..., K$. Thus, one centroid C_k is assigned to each empty set. The centroids C_k may be arbitrary, or may be set using a "best guess" based on previous experience. For each training sample y_i, expressed as a color vector, the Euclidean distance d to each cluster centroid C_k is determined, and from the Euclidean distances d. The cluster A_J having the minimum Euclidean distance is identified as follows:

$$J = \arg\min_k d = \arg\min_k d(y_i, C_k) \qquad (7.30)$$

Then, y_i is accumulated into A_J. Next, the distortion E is determined by

$$E = E + d_{min} \qquad (7.31)$$

where d_{min} is the minimum distortion value d obtained in earlier steps. Accumulation of the other training sample y_i into the appropriate cluster A_J continues until all training samples have been collected into the appropriate clusters. When $i = N$, that is, when all training samples have been accumulated, an updated cluster centroid C_k is determined for each cluster A_1, A_2, ..., A_K. This determination may be performed according to the following equation:

$$C_k = \frac{\sum_{i=1}^{L_k} A_k(i)}{L_k} \qquad (7.32)$$

where L_k is the number of vectors in cluster A_k. The average distortion D^m is obtained by

$$D^m = \frac{E}{N} \qquad (7.33)$$

Then, it is determined whether distortion is within the distortion threshold ε. This determination may be made by determining whether the following relation is satisfied:

$$\frac{D^{m-1} - D^m}{D^m} \leq \varepsilon \qquad (7.34)$$

If Equation 7.34 is not satisfied, the process sets $E = 0$ and $m = m + 1$, and repeats the previous steps as shown in the flowchart, beginning this time with the updated cluster centroids C_k. When Equation 7.34 is satisfied, the process stores the new centroids $C_1, C_2, \ldots, C_K$. These centroids are the "final" centroids that will be used in the clusters for the adaptation process. After the centroids are computed, a final step of accumulating the training samples into the appropriate clusters will be performed, as shown in Figure 7.7. These centroids—in $L*a*b*$ space—will be used during the adaptation process. But, when using the models, the corresponding cluster centroids in $CMYK$ space is required. This can be computed using Equation 7.32 in $CMYK$ space using color indices from the final clusters or running the flowchart of Figure 7.6 in $CMYK$ space using training database vectors $u_1, u_2, \ldots, u_N$.

The cluster centroids C_k from Figure 7.6 are used in the initialization step of Figure 7.7. However, cluster centroids obtained from some other approach can also be used here. For example, the centroids may be "critical colors," for example, a critical color in the yellow region of the gamut (colors used for controlling the printing device) or colors from a document obtained by processing the pixels in the document where an accurate model is desired. Again, begin, create an empty set, $A_1, A_2, \ldots, A_K$ and set i initially at 1. Using N number of training samples, y_i is accumulated into the appropriate cluster, that is, the cluster having the centroid with the shortest Euclidean distance from the training sample, similar to Figure 7.6. Repeat this process until all training samples are accumulated and stored into

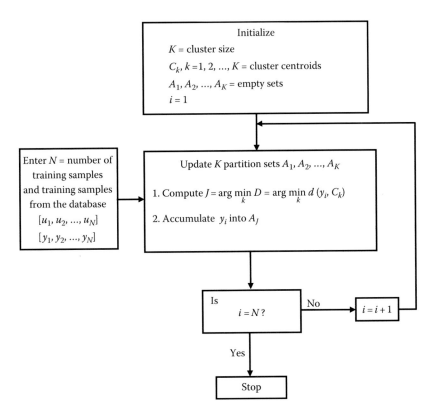

FIGURE 7.7 Flowchart of accumulating training samples into clusters in $L*a*b*$ space.

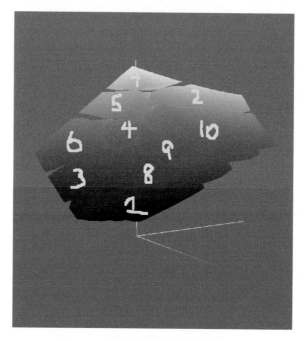

FIGURE 7.8   Ten cluster $L^*a^*b^*$ gamut (clustering was done in $L^*a^*b^*$ space).

clusters $A_1, A_2, \ldots, A_K$. Figure 7.8 illustrates pictorially the clustered color gamuts in $L^*a^*b^*$ space. Clusters obtained in CMYK space is not shown.

Example 7.2

Using RLS algorithm and training samples from Example 7.1, (a) obtain a piecewise linear affine model and (b) piecewise linear quadratic and cubic model for different clusters. Show the model accuracy as a function of the number of clusters. Describe the key steps in assigning clusters and the parameter adaptation process for the piecewise linear model.

Solution

The accuracy error with respect to the number of clusters of the piecewise linear affine model with 13 parameters in each cluster is shown graphically in Figures 7.9 through 7.11. These data were purposely produced for an uncalibrated printer. When the simulation reached 10 clusters, cluster #6 had a minimum of 550 colors and cluster #8 had a maximum of 2577 colors. As the number of training colors assigned to a particular cluster reduces, the parameter estimation becomes less accurate. The number of modeling parameters (i.e., number of elements in the θ matrix, Equation 7.2) is another factor that requires optimization, since a higher number of parameters may model the noise by overfitting the data. As might be expected, the curves in Figure 7.9 show that the accuracy error (both mean and 95%) almost decreases monotonically with an increase in the number of clusters. Thus the trend is clearly in the right direction. In general, if there are more training

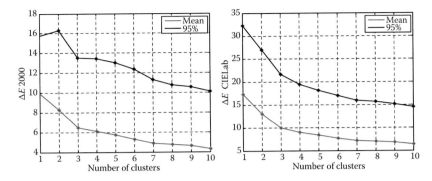

FIGURE 7.9 Error between experimental data and a piecewise linear affine model as a function of the number of clusters for test colors.

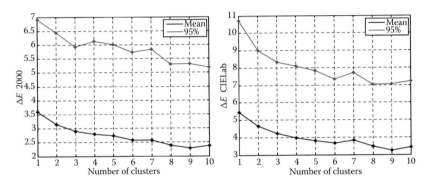

FIGURE 7.10 Error between experimental data and a piecewise linear quadratic and cubic model as a function of the number of clusters (computed for up to 10 clusters) for test colors for printer A with 10^4 characterization data.

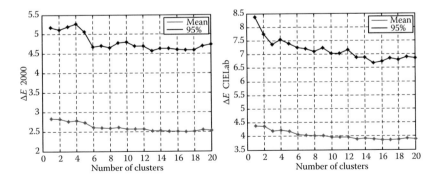

FIGURE 7.11 Error between experimental data and a piecewise linear quadratic and cubic model as a function of the number of clusters (computed for up to 20 clusters) for test colors for printer B with 16^4 characterization data.

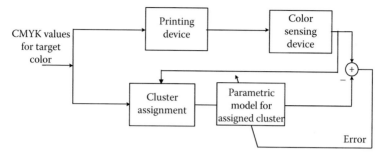

FIGURE 7.12 System block diagram showing the adaptation process with cluster assignment used to tune the parameters of the piecewise linear model.

samples available then it will be easy to increase the number of clusters, which in turn can provide improvements to the model accuracy. Although increasing number of clusters improves the accuracy, it also increases the number of models and hence the total number of parameters required for modeling the entire color space. Therefore, a trade-off between the number of model parameters, and the number of clusters is necessary.

Figure 7.12 shows a block diagram of the adaptation process. The same block diagram can be used for predicting the $L^*a^*b^*$ values of test colors when their CMYK values are known. Cluster centroid and parameter matrix θ for each of the clusters are used during this process. The prediction process is described by the following steps:

1. Receive the CMYK values for the prediction set.
2. Calculate the Euclidean distance from the current color sample to each of the cluster centroids in CMYK space.
3. Determine which of the Euclidean distances in step #2 is the shortest. This is done in the cluster assignment block, which then becomes the cluster of choice for the color whose CMYK values are to be estimated using the model.
4. Calculate the $L^*a^*b^*$ values using the updated parameters of the model corresponding to the cluster in step #3.

The above approach was used in our simulations in Figures 7.9 and 7.10 for a group of test colors that is separate from the training set. Now, during the adaptation process, we need to run following steps at a minimum:

1. Receive the $L^*a^*b^*$ values from the sensor (from the customer images or obtained by scheduling patches).
2. Calculate the Euclidean distances from the current color sample to each cluster centroid in $L^*a^*b^*$ space. *Note:* we do not use the CMYK space as in the prediction process.
3. Determine which of the Euclidean distances in step #2 is the shortest. This is done in the cluster assignment block of Figure 7.12. This becomes the cluster of choice for the color sample.
4. Run the RLS algorithm, Equations 7.25 and 7.26, to update the θ matrix corresponding to the cluster of choice based on the cluster having the centroid with the shortest Euclidean distance.

Thus, parameter updates are done adaptively based on sensor measurements to capture the current state of the device, while at other times models can be used to estimate the $L^*a^*b^*$ values using the recently updated parameters of the updated profile LUTs, for specific input CMYK nodes, whose measurements cannot be obtained easily because the printer may be scheduled for production jobs.

7.4.2 PRINCIPAL COMPONENT ANALYSIS-BASED MODEL

Though the piecewise linear models based on clustering can produce better models, the implementation may suffer from a large body of parameters to track, as well as significant complexity associated with the computation; both time and computing resources. A much better way to model the input–output data empirically is by using principal component analysis (PCA) (which is also known by many other names, such as, Karhunen–Loeve decomposition, Hotelling transform, singular value decomposition, empirical orthogonal function decomposition, etc.; Section 3.10). The PCA approach extracts key information from the data for modeling and throws away the rest. It decomposes data into linear combinations of mutually orthogonal and basis vectors of unit length (extracted from a training data set) in such a way that in that basis, the second order statistics (covariance matrix) of the data is diagonalized. This diagonal matrix is the matrix of singular values, all of them nonnegative by definition. When sorted from highest singular value to the lowest, the basis vectors become known as the most significant or principal components. As a result of this operation on the input–output data, it gives less redundancy and as good a representation of the system as possible with a few principal components.

Many researchers [14–18] have attempted to decompose the spectra in terms of principal basis vectors and use the basis vectors to model the imaging device or use the PCA to predict the reflectance measurements on multiple substrates (media) while characterizing PCAs on a reference substrate [17]. This method is widely used in other fields of engineering [19,20] and may receive more use in color systems as the demand for high performance systems increase. In PCA-based modeling, although the representation can be made with any multivariate input–output data samples [21], we show the approach when the sensed output is the reflectance spectra. A standard spectral sensor is used for measurement. Gretag spectrophotometer, which outputs 36 spectral reflectance values, evenly spaced at 10 nm over the visible spectrum over the spectrum 380–730 nm, a XRite spectrophotometer [10], and an in-line spectrophotometer [7], which has 31 outputs evenly spaced at 10 nm over the wavelength of 400–700 nm. In PCA, it is essential that the spectral training data is mutually correlated. If they are independent, PCA does not help. Also, output data with a mixture of different wavelengths (i.e., with different vector lengths) are not considered for PCA.

7.4.2.1 PCA-Based Model in Spectral Space

In this section, we first show a one-dimensional (1-D) spectral PCA-based model for a static printing system. A simple online estimation method is introduced later for a 1-D PCA to update the parameters during the adaptation process. A more advanced two-dimensional (2-D) PCA method can be found in Ref. [22] which uses a significantly smaller number of basis vectors, reduced computation and measurements for continuous adaptation. We show below how to model the printer using a 1-D PCA approach.

For N number of input colors, given a stream of output spectral training data, $R_1, R_2, \ldots, R_N$, where each R_i is a vector of length n (e.g., $n = 31$ when there are 31 reflectance values evenly spaced over the spectrum of 400–700 nm), then each of these spectra can be modeled as a random sample from a mixture model in terms of K basis functions where $K < n$ as

$$R(\lambda) \cong R_0(\lambda) + \sum_{j=1}^{K} W_j \psi_j(\lambda) \qquad (7.35)$$

where $\psi_j(\lambda)$ is the jth basis function derived from PCA analysis. $R_0(\lambda)$ is the sample mean, which is computed from the output spectral data, and W_j is the jth weight parameter.

$$R_0(\lambda) = \frac{1}{N} \sum_{i=1}^{N} R_i(\lambda) \qquad (7.36)$$

We can represent, in matrix form, the terms inside the summation of Equation 7.35 as

$$r = [\psi_1 \ \psi_2 \ \cdots \ \psi_K] \begin{bmatrix} W_1 \\ W_2 \\ \vdots \\ W_K \end{bmatrix} \qquad (7.37)$$

Equation 7.37 can be written in matrix form for each color as

$$r = BW \qquad (7.38)$$

where r is the zero-mean reflectance spectra of size $n \times 1$ of the color. If R is the full reflectance spectra of the color, then the approximated spectral data is given by $R \cong R_0 + r$ which represents the full reflectivity vector for that color at specified wavelengths.

The matrix, $B = [\psi_1 \ \psi_2 \ \cdots \ \psi_K]$, is the mixture matrix of size $n \times K$, whose columns contain the basis vectors with elements at the wavelength intervals used in the output spectral data. The vector, $W = [W_1 \ W_2 \ \cdots \ W_K]^T$, is the parameter vector of size $K \times 1$ containing the scalar weights for the color.

The basis vectors are obtained as follows. First compute the covariance matrix formed by the output spectral data

$$\Sigma = \frac{1}{N} \sum_{i=1}^{N} [R_i - R_0][R_i - R_0]^T \qquad (7.39)$$

Perform the singular value decomposition (Section 3.10) on the covariance matrix to get n number of basis functions.

$$\text{SVD}\left(\sum\right) = \sum = \Psi \Pi^2 \Psi^{\text{T}} = \sum_{i=1}^{n} \Pi_i^2 \psi_i \psi_i^{\text{T}} \tag{7.40}$$

The vectors, ψ_i, in Equation 7.40, correspond to the ith basis function. They are pairwise orthogonal and orthonormal just as the sine and cosine functions in a Fourier decomposition of the composite function. A linear weighted combination of all these basis functions represents the complete spectra. That is

$$R(\lambda) = R_0(\lambda) + \sum_{j=1}^{n} W_j \psi_j(\lambda) \tag{7.41}$$

Note that $\Psi^{\text{T}} \Psi = I =$ identity matrix. Also, $|\Pi_1| > |\Pi_2| > \cdots > |\Pi_n|$ are rank ordered singular values contained in the matrix, Π

$$\Pi = \begin{bmatrix} \Pi_1 & 0 & 0 & . & 0 \\ 0 & \Pi_2 & 0 & . & 0 \\ 0 & 0 & \Pi_3 & . & 0 \\ . & . & . & . & . \\ 0 & 0 & 0 & . & \Pi_n \end{bmatrix} \tag{7.42}$$

The parameters, W_j with $j = 1, 2, \ldots, n$, are obtained in various ways. One straightforward way is by using the orthogonality property of the basis vectors emanating from Equation 7.40 [23]. The PCA method assures that the principal directions determined by the spectral PCA stage are orthogonal to each other. Since the basis vectors are also orthonormal, weights are computed from the following dot product.

$$W_j = r^{\text{T}} \psi_j \tag{7.43}$$

If there are only few, $K < n$, dominant eigenvalues, then only few, K, basis vectors are needed to estimate the spectra as in Equation 7.35. In some situations, it is difficult to know precisely how many dominant eigenvalues (hence eigen or basis vectors) are required for a good approximation, while the rest are ignored as noise. In general, if the color gamut of the device is large, as in a 6 or 14 color press, whose gamuts are typically larger than four-color press, then more basis vectors are required for modeling the device. In Figure 7.13, the first four basis vectors are plotted as a function of wavelength. In Figure 7.14, an example of reconstructed spectra (Equation 7.35) is shown when up to nine basis vectors are used. The actual spectra from the training set along with the ΔE 2000 and ΔE CIE Lab numbers are shown in the same figure to compare with the approximated spectra. In Figure 7.15,

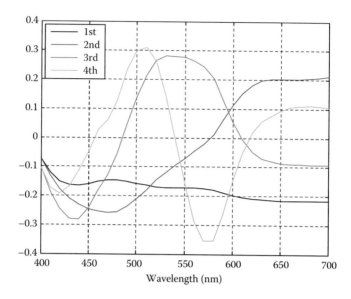

FIGURE 7.13 First four basis vectors as a function of wavelength.

the ΔE numbers are plotted as a function of the number of basis vectors to decide how many basis vectors, K, required for modeling the parameters, W_j, using Equation 7.43 for $j = 1, 2, \ldots, K$ basis vectors and for $i = 1, 2, \ldots, N$ colors. For convenience, the color subscript i is not shown in zero-mean reflectance r and W_j in the above equations. As shown in Table 7.1, for this printer, over 99% of the spectral energy is captured in the first six basis vectors.

Once the basis vectors are selected, the weights W_j are modeled in terms of input variables, CMYK, using the training samples, preferably by spanning a large portion of the operating space of the device. An adaptation estimation algorithm is sometimes more valuable for modeling the parameters.

$$W_j = f_j(C, M, Y, K) \tag{7.44}$$

The function $f_j(C, M, Y, K)$ is modeled as

$$
\begin{aligned}
f_j(C, M, Y, K) = {} & b_j + \alpha_{1j}C + \alpha_{2j}M + \alpha_{3j}Y + \alpha_{4j}K + \alpha_{5j}CM + \alpha_{6j}CY \\
& + \alpha_{7j}CK + \alpha_{8j}MY + \alpha_{9j}MK + \alpha_{10j}YK + \alpha_{11j}C^2 \\
& + \alpha_{12j}M^2 + \alpha_{13j}Y^2 + \alpha_{14j}K^2 + \cdots
\end{aligned}
\tag{7.45}
$$

For simplicity, Equation 7.45 is written in matrix form for a simple linear affine model containing five parameters for each weight, and K number of such set

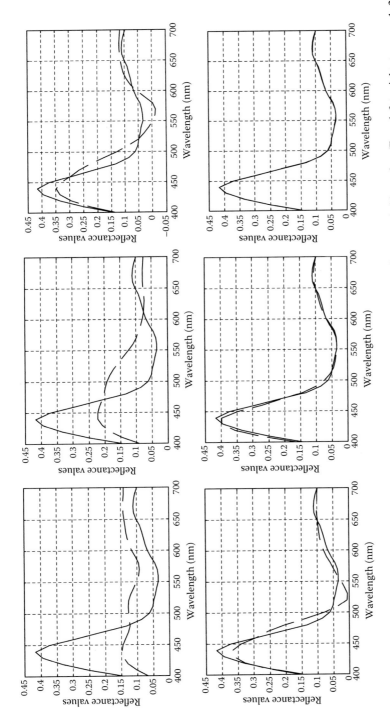

FIGURE 7.14 Reconstruction of the spectra using a sequentially larger number of basis vectors for a sample blue color. (From left to right: top row: 1, 2, and 3 basis vectors; bottom row: 4, 5, and 6 basis vectors; solid curve represents actual spectra and dashed curve represents reconstructed spectra).

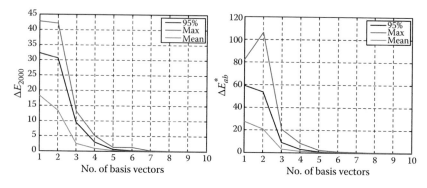

FIGURE 7.15 ΔE_{2000} and ΔE_{ab}^{*} as function of the number of basis vectors for training colors.

TABLE 7.1

ΔE Numbers Are Shown to Indicate the Approximation Accuracy

Number of Basis Vectors	ΔE_{2000}	ΔE_{ab}^{*}
1	23.95	60.92
2	21.02	62.28
3	3.41	8.82
4	2.69	5.91
5	0.35	1.09
6	0.05	0.14

$$[W_1 \quad W_2 \quad .. \quad W_K] = [1 \quad C \quad M \quad Y \quad K] \begin{bmatrix} b_1 & b_2 & . & . & b_K \\ \alpha_{11} & \alpha_{12} & . & . & \alpha_{1K} \\ \alpha_{21} & \alpha_{22} & . & . & \alpha_{2K} \\ \alpha_{31} & \alpha_{32} & . & . & \alpha_{3K} \\ \alpha_{41} & \alpha_{43} & . & . & \alpha_{4K} \end{bmatrix} \tag{7.46}$$

$$W^{\mathrm{T}} = A_0 \theta_0$$

Where $W^{\mathrm{T}} = [W_1 \ W_2 .. \ W_K]$ is a row vector containing K number of weights for the ith color. When N number of colors are used in the training set, W^{T} will be a matrix of size $N \times K$. A_0 contains N number of rows, and $CMYK$ values with a suitable structure of the model (linear affine, quadratic, cubic, etc.), and θ_0 contains the new parameter matrix with K number of columns. Equation 7.46 is very similar to Equation 7.3. Hence, the least-square solution to the parameter matrix is given by

$$\theta_0 = \left(A_0^{\mathrm{T}} A_0\right)^{-1} A_0^{\mathrm{T}} W^{\mathrm{T}} \tag{7.47}$$

The parameter matrix, θ_0, is a function of *CMYK*. When it is used in combination with the 1-D spectral PCA vectors, we have the complete model of the printer.

The training process is summarized in following key steps:

1. Receive spectral vectors, $R_1, R_2, \ldots, R_N$, of the N training set.
2. Determine the sample mean, R_0, using Equation 7.36.
3. Form covariance matrix (Equation 7.39), perform singular value decomposition (Equation 7.40), and determine K number of basis vectors ψ_j with $j = 1, 2, \ldots, K$.
4. Determine zero-mean reflectance spectra for each training spectra, $r_i = R_i - R_0$, $i = 1, 2, \ldots, N$.
5. Determine weights W_j using Equation 7.43 for $j = 1, 2, \ldots, K$ basis vectors and for $i = 1, 2, \ldots, N$ colors. Use the zero-mean spectral r from step 4. For convenience, the subscript i is not shown in r.
6. Solve the least-square solution for the parameter matrix, θ_0, using Equation 7.47, by grouping *CMYK* values corresponding to each training sample in A_0 for the appropriate model structure and forming the weight matrix, W^T from step 5.

To use the parameter vector for predicting the spectra for any new *CMYK* values, at minimum, run following key process steps:

1. Receive the *CMYK* values at which spectral reflectance are to be calculated using the model.
2. Arrange input color *CMYK* values in the matrix A_0 for the structure chosen during training step 6. Color numbers are omitted in the matrix A_0 for simplicity.
3. Determine matrix $W^T = [W_1 \ W_2 \ldots W_K]$ for all prediction colors (or W_j for $j = 1, 2, \ldots, K$ for each color) from Equation 7.46 since the parameter matrix θ_0 is known from the training process of step 6 above.
4. Obtain predicted reflectance spectra from Equation 7.41 using weight vector from prediction process step 3 and basis vectors from the training process step 3.

7.4.2.2 PCA-Based Modeling for Adaptive Estimation

For performing continuous adaptation as fresh samples become available, or to improve the accuracy of parameters instead of the least squares, an RLS algorithm (Equation 7.25) can be used with slight modifications as in Equation 7.48.

$$\theta_{k+1} = \theta_k + P_{k+1} a_{k+1} \left(W_{k+1}^T - a_{k+1}^T \theta_k \right) \tag{7.48}$$

where the equation for P_{k+1} is shown in Equation 7.26. Adaptive estimation is initiated by setting P_0 to identity and initial θ_0 to the output of the least-squares equation (Equation 7.47).

A general warning is required while using PCA method for modeling color systems. In order to gain confidence in the use of experimental data, bootstrapping is often needed. In bootstrapping, a random subset (1/2 in this case) of the data should be used for the PCA to determine PCA vectors. The results of the PCA vectors are then applied to the rest of the data set to determine how good a model with PCA can be for the whole set. This process should be repeated multiple times, by choosing each time a different random subset for PCA analysis and the rest of the data to fit with those PCA vectors. If all of the data come from one distribution, the expected amount of residual energy (actual variation–modeled variation) is insensitive to the subset chosen for analysis using PCA. Problem 7.5 illustrates the bootstrapping method for an example dataset.

Example 7.3

In a *CMYK* printing system, input *CMYK* values are sampled uniformly between 0 and 255 to create 10^4 input–output characterization data set. The spectral data for each patch is measured at 31 wavelengths from 400 to 700 nm at 10 nm intervals. Use spectral PCA mathematical techniques to analyze the characterization data and fit a model. Using PCA vectors find parameters for a least squares and RLS model. Plot the model accuracy for test colors as a function of the number of basis vectors. How many basis vectors are needed to obtain a reasonably accurate model?

SOLUTION

We will set $P_0 = I$ (Identity matrix) in the RLS Equations 7.25 and 7.26. The θ matrix is of size 23×10 (when 10 basis vectors were used) with mean propagation. All these calculations were done with absolute $L*a*b*$ values. Accuracy for approx. 2000 prediction colors is shown in Table 7.2 for different number of basis vectors. RLS algorithm is found to be more accurate than a simple least-squares regression. Improvement to the prediction accuracy saturates when the number of basis vectors becomes equal to five for both algorithms.

TABLE 7.2
Model Accuracy Shown with Respect to Spectral PCA Basis Vectors

Models	No. of Basis Vectors	ΔE_{2000}			ΔE^*_{ab}		
		Mean	95%	Max	Mean	95%	Max
Least squares	1	20.52	35.50	51.99	31.60	67.34	87.04
	2	16.06	32.83	45.31	24.87	60.81	77.89
	3	7.74	19.27	41.97	13.46	27.85	56.66
	4	7.68	19.25	42.21	13.46	29.16	57.02
	5	7.66	19.24	42.02	13.53	29.43	56.72
	10	7.65	19.28	41.79	13.52	29.25	57.73
	20	7.65	19.28	41.79	13.52	29.25	57.81
RLS	1	17.90	34.99	47.36	27.80	66.30	87.40
	2	13.19	32.80	45.50	20.46	60.16	84.97
	3	4.89	10.54	25.81	7.74	17.94	41.23
	4	4.63	10.41	25.63	7.65	17.79	46.45
	5	4.59	10.27	25.37	7.75	19.85	55.22
	10	4.58	10.20	25.60	7.71	19.79	57.44
	20	4.58	10.20	25.61	7.71	19.78	57.52

7.4.2.3 Log-PCA Model (Log-PCA)

The Log-PCA is sometimes used instead of the linear PCA described in Section 7.4.2.1. The approach is similar to linear PCA, except that the reflectance spectra are replaced by its natural log

$$R(\lambda) \rightarrow -\log(R(\lambda)) \qquad (7.49)$$

A summary of the estimation algorithm for Log-PCA is shown in Figure 7.16a and b.

Example 7.4

Use Example 7.3 input–output characterization data and apply log-PCA model to show the improvements to modeling accuracy.

SOLUTION

Table 7.3 shows the results for two different least-squares models. Both algorithms give similar accuracy when compared to Example 7.3 in which the spectral data were not converted with log-PCA.

7.4.2.4 Piecewise Linear PCA Model

The main purpose of the PCA approach is to model the printer data set from a few principal components and a few weights (Equation 7.43), since a few principal components can capture the essential characteristics of the printer. To improve the accuracy of the printer model, clustered linearized PCA models can be generated since the clustering approach can model the nonlinearities well. It requires a reasonably large

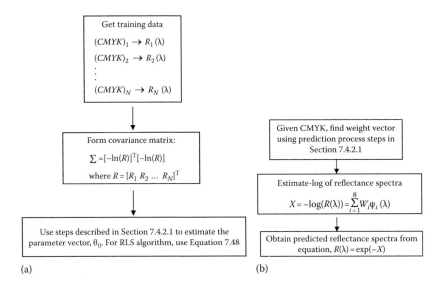

(a)　　　　　　　　　　　　　　　　　　(b)

FIGURE 7.16 (a) Parameter training algorithm shown for log-PCA model. (b) Spectral prediction algorithm using model parameters and log-PCA.

TABLE 7.3

Model Accuracy Shown with Respect to Log-PCA Basis Vectors

Models	No. of Basis Vectors	ΔE_{2000} Mean	95%	Max	ΔE_{ab}^* Mean	95%	Max
Least squares	1	19.21	36.94	40.25	32.43	74.68	90.97
	2	14.43	36.93	44.02	23.23	61.45	74.60
	3	5.92	11.00	14.93	7.66	15.06	27.48
	4	5.33	10.68	14.79	7.75	14.86	26.72
	5	4.95	10.21	14.82	7.38	14.49	26.82
	10	4.96	10.15	14.90	7.36	14.38	26.89
	20	4.96	10.15	14.90	7.36	14.38	26.89
RLS	1	19.21	36.94	40.25	32.43	74.68	90.97
	2	14.43	36.93	44.02	23.23	61.45	74.60
	3	5.92	11.00	14.93	7.66	15.06	27.48
	4	5.33	10.68	14.79	7.75	14.86	26.72
	5	4.95	10.21	14.82	7.38	14.49	26.82
	10	4.96	10.15	14.90	7.36	14.38	26.89
	20	4.96	10.15	14.90	7.36	14.38	26.89

number of training samples to adequately capture PCA vectors within each cluster. If $[y_{1_1} y_{2_2} \cdots y_{N_k}]_k$ are the vector elements of the spectral sensor at specified wavelengths, each having some specified number of elements (say 31 elements when the data is captured between 400 and 700 nm at 10 nm interval) and vector C_k is the centroid of the spectral values in a kth cluster, then K-means algorithm is one of the candidates available for us to cluster the data in spectral space. After clustering, conceptually the rest of the modeling procedure is very similar to the one described in the previous sections. Hence, we have not repeated the entire process again. Example 7.5 illustrates the performance improvement gained with this approach for a nonlinear printer.

Example 7.5

A clustered PCA-based model is to be constructed for a 10^4 spectral characterization data. Use the K-means algorithm to cluster the color data in spectral space. Obtain spectral PCA vectors for each cluster. Use five or six PCA basis vectors. Model each cluster with RLS algorithm. Show the performance of your model for a test data as a function of the number of clusters.

SOLUTION

Results are summarized in Table 7.4. Cluster assignments are done in the output (i.e., $L^*a^*b^*$) space. Clearly this model is more accurate when compared to Example 7.4.

7.4.2.5   Yule–Nielson Corrected PCA Model

Yule–Nielson correction factor can sometimes be introduced to the measured spectral data from the printer before applying the PCA. This factor is somewhat related

TABLE 7.4

Accuracy of Clustered Spectral PCA Model (ΔE_{2000} and ΔE_{ab}^* Space)

Models	No. of Basis Vectors	ΔE_{2000}			ΔE_{ab}^*		
		Mean	95%	Max	Mean	95%	Max
RLS	1	4.98	10.94	53.00	7.98	23.30	162.73
	2	4.11	9.51	41.81	7.29	20.33	102.39
	3	3.76	7.32	31.23	6.51	16.46	100.60
	4	3.40	7.82	32.47	5.67	14.46	92.96
	5	3.14	7.25	30.88	5.13	12.64	91.39
	6	3.07	7.12	53.74	5.02	12.49	137.04
	7	3.01	7.10	50.57	4.89	12.23	127.96
	8	2.92	6.47	20.89	4.73	11.11	57.62
	9	2.74	6.68	19.95	4.23	10.58	49.31
	10	2.72	6.49	22.61	4.23	10.79	39.65

to the physical effects of light scattering and mechanical dot growth [24,52,56]. The approach is similar to normal PCA, except that the reflectance spectra are raised to the power of $(1/m)$, where m is determined by trial and error to achieve the best prediction.

$$R(\lambda) \rightarrow R(\lambda)^{1/m} \tag{7.50}$$

Values of m are somewhat arbitrary and could be set anywhere between 1 (for a glossy substrate) and 2 (for a perfect diffuser). It can also exceed 2.0.

As clearly seen from previous discussions, successful printer characterization based purely on experimental data and its piecewise linear representation may need a large number of measurements. Consequently, models based on the fundamental physical process are much more preferred, if available, since the nonlinearities of the physical process can be captured in the analytical functions. Often physics-based approaches would require fewer measurements. The following sections illustrate various color models studied by many researchers.

7.4.3 NEUGEBAUER MODEL

One well-known spectral modeling technique is the Neugebauer color mixing model [25–64] commonly used to model the output color in digital printing processes. The Neugebauer model determines the spectral reflectance function in terms of the weighted sum of spectral reflectance functions obtained from one, two, and more combinations of available colorants and a substrate on which the colorants are formed. The resulting colorant combinations are referred to as the Neugebauer primaries. In a three-color system, for example, having cyan, magenta, yellow (CMY) colorants, there are 8 ($2^3 = 8$) Neugebauer primaries. In a four-color system, for example, having cyan, magenta, yellow, black (*CMYK*) colorants, there will be 16 ($2^4 = 16$) Neugebauer primaries. In an N color system, there will be 2^N known spectral reflectance functions. In mathematical form, the vector corrected

Neugebauer model for a *CMYK* printer with corrections to the penetration and scattering of light onto paper [52–56] is given by the following equation

$$R(\lambda) = \left[\sum_{i=1}^{K} W_i R_i^{\frac{1}{m}}(\lambda) \right]^m \tag{7.51}$$

where
   $R(\lambda)$ is the average output spectral reflectance (estimated)
   W_i is the Demichel weighting of the *i*th Neugebauer primary
   $R_i(\lambda)$ is the reflectance spectra of the *i*th Neugebauer primary which is the *i*th basis vector, and a free parameter
   m is the Yule–Nielson correction factor

The penetration represented by the m factor was expressed as a function of wide band reflectance. More recently, the m factor has been expressed in terms of narrow-band spectral curves and is allowed to vary over a wide range to obtain a best fit with the measurement [47,48]. The Neugebauer model, Equation 7.51, assumes that the reflectance of a spatial area is the additive combination of the reflectances of the primary colors and their overprints. K is equal to the number of Neugebauer primaries, which is 16 for a four-color *CMYK* printer. They are the measured reflectance spectra of the corresponding primary samples with 100% area coverage on paper white:

$$R_i(\lambda) \Rightarrow \text{Reflectance Spectra of } \{W, C, M, Y, K, CM, CY, MY, MK, YK, CK, CMY,$$
$$CMK, MYK, CYK, CMYK\}$$

Weights, W_i, can be tuned from spectral reflectance measurements using standard least square or RLS algorithms as described in earlier sections. The weights can be further represented as individual fractional area coverages as in Demichel's equation [56–58] and then these fractional area coverages can be tuned from spectral measurements. Alternatively, fixing the weights, W_i, and tuning of Neugebauer primaries with least squares or weighted least squares using training spectral reflectance measurements are other basic ideas being attempted for *CMYK* printers [30].

7.4.3.1 Parameterized Model for Neugebauer Weights

For convenience, let us denote the Yule–Nielson corrected *i*th reflectance values at a given wavelength by the vector, $r_i = R_i^{1/m}$, where $i = 1, 2, \ldots, K$. With the corresponding new notation for the estimated spectra, $r = R^{1/m}$, the Neugebauer equation (Equation 7.51) can be represented in vector form as

$$r = BW \tag{7.52}$$

The matrix, $B = [r_1 \; r_2 \; \ldots \; r_K]$, is the matrix of size $n \times K$ (e.g., $n = 31$) when there are 31 reflectance values, and vector $W = [W_1 \; W_2 \ldots W_K]^{\mathrm{T}}$ is the Demichel weight vector.

With a few matrix operations, we can express the weights as a function of the primaries and the r vector as follows:

$$W^{\mathrm{T}} = r^{\mathrm{T}}B\left(B^{\mathrm{T}}B\right)^{-1} \tag{7.53}$$

The weight vector W can now be modeled in terms of a new parameter matrix θ_0 (with linear affine/quadratic/cubic model parameters) and input variables, $CMYK$, using the training samples as in Equations 7.44 through 7.47. The new parameter matrix θ_0 can be estimated using the least square Equation 7.47 or RLS Equation 7.48.

Example 7.6

For a four-color $CMYK$ printer, develop a spectral Neugebauer model (Equation 7.51) with Demichel weights for 16 Neugebauer primaries. Estimate the weights using training samples with least-squares algorithms and Yule–Nielson factor (m) as a free parameter.

    a. Plot the accuracy of the model as a function of m.
    b. For the best m, plot the model accuracy as a function of the number of uniformly sampled data sets (e.g., 3^4, 4^4, 5^4, 6^4, etc.).

SOLUTION

When Yule–Nielson factor m is equal to 1, there is no light scattering in the paper, and $m = 2$ corresponds to Lambertian or perfectly diffused scattering in the paper [31]. For an experimental printer considered in this simulation, we used Bayer's dithering technique to produce a halftone patch. As the halftone screen changes, the optimal m could be different. A trial and error approach can be adopted to come up with a best value for m. Figure 7.17 shows the model accuracy as a function of parameter m and Figure 7.18 shows the model accuracy as a function of the number of uniformly sampled data sets for best value of m from Figure 7.17.

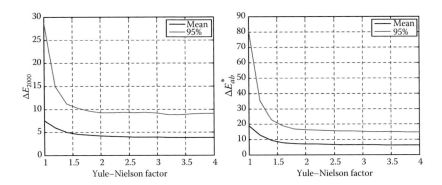

FIGURE 7.17 The accuracy of the Neugebauer model for IT8 colors plotted as a function of the Yule–Nielson factor. Demichel weights are modeled with respect to input $CMYK$ values using least-squares algorithm with linear affine and quadratic models.

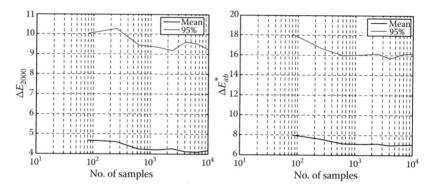

FIGURE 7.18 The accuracy of Neugebauer model plotted as a function of the number of characterization samples.

Evidently, the accuracy of Neugebauer models cannot be improved beyond a certain number based on increased training samples. Since nonlinearities of the physical process are captured in the function, the model gives reasonably good prediction within a few hundred training samples (Table 7.5).

TABLE 7.5

Accuracy of Neugebauer Models (ΔE_{2000} and ΔE_{ab}^* Space)

No. of Samples	ΔE_{2000}			Max Color		
	Mean	95%	Max	L*	a*	b*
81	4.66	10.03	15.14	18.76	−5.30	28.86
256	4.60	10.27	16.39	9.29	17.24	0.23
625	4.24	9.45	17.51	7.87	18.33	1.82
1296	4.19	9.35	17.86	6.98	18.34	2.30
2401	4.25	9.19	17.79	7.33	18.44	2.36
4096	4.11	9.61	18.03	4.60	16.62	3.14
6561	4.10	9.51	18.07	4.98	17.07	2.96
10000	4.17	9.24	18.14	5.28	17.50	2.82

No. of Samples	ΔE_{ab}^*			Max Color		
	Mean	95%	Max	L*	a*	b*
81	7.98	18.10	32.40	65.74	−3.19	41.31
256	7.59	16.72	31.15	53.58	68.69	66.94
625	7.11	15.99	32.02	52.15	67.95	67.91
1296	7.07	16.03	33.33	51.77	68.32	69.21
2401	7.12	16.15	33.06	51.92	68.22	68.94
4096	6.96	15.68	38.30	50.91	69.73	74.10
6561	6.99	16.10	36.93	50.83	69.32	72.76
10000	7.06	16.22	36.71	51.19	69.36	72.54

7.4.3.2 Dot Area Coverages and Neugebauer Weights

For binary printers, multiple colors are achieved by varying dot area coverages of the primary colors. For the case where the dot locations for the colorants are statistically independent, Demichel's equations [56,57] can be used to express the Demichel weights as the primary dot area coverages. These steps are not always necessary unless there is a requirement for further tuning of the Neugebauer weights to improve the prediction accuracy of the model, and a need for reduction in number of training samples due to cost reasons.

For random mixing, since the separations are printed independently and the relative position of dots is random, the Demichel weights corresponding to the primaries can be obtained by a probabilistic model as

$$
\begin{aligned}
W_1 &= (1-c)(1-m)(1-y)(1-k) & W_9 &= (1-c)my(1-k) \\
W_2 &= c(1-m)(1-y)(1-k) & W_{10} &= (1-c)m(1-y)k \\
W_3 &= (1-c)m(1-y)(1-k) & W_{11} &= (1-c)(1-m)yk \\
W_4 &= (1-c)(1-m)y(1-k) & W_{12} &= cmy(1-k) \\
W_5 &= (1-c)(1-m)(1-y)k & W_{13} &= cm(1-y)k \\
W_6 &= cm(1-y)(1-k) & W_{14} &= c(1-m)yk \\
W_7 &= c(1-m)y(1-k) & W_{15} &= (1-c)myk \\
W_8 &= c(1-m)(1-y)k & W_{16} &= cmyk
\end{aligned}
\tag{7.54}
$$

where c, m, y, and k are the actual fractional areas covered by cyan, magenta, yellow, and black colorant toner dots, respectively. These areas are functions of the input digital counts (control values) C, M, Y, and K, which are integers between 0 and 255. The mappings from the input digital counts C, M, Y, and K to the fractional area coverage values c, m, y, and k are called dot growth functions or dot area functions (see Example 7.7). Thus, for a fixed λ, Equation 7.51 (with Equation 7.54) represents a fourth order polynomial in c, m, y, and k. Figure 7.19 shows a schematic input–output diagram of the printer model with the weighted Neugebauer Equation 7.51, the Demichel mixing Equation 7.54, and the dot growth functions $C \to c$, $M \to m$, $Y \to y$, and $K \to k$.

Another commonly used halftone configuration is the dot-on-dot screen, where the c, m, y, and k dots are placed at the same screen angle and phase as illustrated in Figure 7.20 for an ideal dot pattern with no noise in a four-colorant system. In this system, the colorants are drawn with a decreasing area coverage [29,43]. For this screen design, if p_i (with $i=1, 2, 3, 4$) represent the printer colorants of increasing dot area coverage, and a_i (with $i=1, 2, 3, 4$) represent the corresponding dot area coverages, then the Neugebauer Equation 7.51 contains five primaries (i.e., $K=5$). The five primaries consist of the corresponding measured reflectance spectra and are denoted by $R_i(\lambda) \in \{R_{p_1 p_2 p_3 p_4}(\lambda), R_{p_2 p_3 p_4}(\lambda), R_{p_3 p_4}(\lambda), R_{p_4}(\lambda), R_w(\lambda)\}$. The weights corresponding to the primaries are expressed as

$$
W_i \in \{a_1, (a_2 - a_1), (a_3 - a_2), (a_4 - a_3), 1 - a_4\}. \tag{7.55}
$$

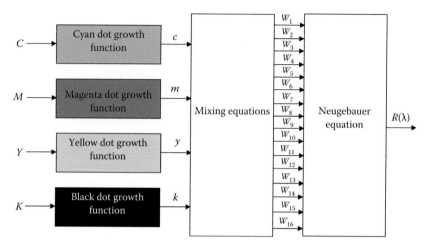

FIGURE 7.19 Block diagram of the Neugebauer model for a color printer.

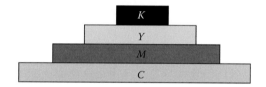

FIGURE 7.20 Dot-on-dot halftone screen example.

The ideal dot-on-dot mixing model does not consider noise and misregistration effects. Therefore, a combination of dot-on-dot and random mixing model is proposed [31] to improve the model accuracy by giving relative weights to the reflectance predicted by the dot-on-dot model and the random mixing model. The predicted spectral reflectance in the combined model is given by

$$R(\lambda) = (1 - \alpha)R_d(\lambda) + \alpha R_r(\lambda) \qquad (7.56)$$

where
   $R_d(\lambda)$ is the spectral reflectance that is predicted by the dot-on-dot model using Equation 7.55
   $R_r(\lambda)$ is the spectral reflectance predicted by the random mixing model using Equation 7.54
   α is a weighting parameter between 0 and 1 that is used to relatively weight the two models

The interested reader may refer to the original work on this subject in Ref. [31].

While the basic form of the mixing equations is similar to Equation 7.51, the weights W_i are modeled differently based on the way the dots are positioned. In order to use the Neugebauer model to model the printer accurately, we need to estimate the dot area coverages from the digital counts $CMYK$ as well as the Yule–Nielson correction factor m. Use these values in Equation 7.51 along with the measured Neugebauer primaries (i.e., reflectance spectra of the primaries, $R_i(\lambda)$, $i = 1, 2, \ldots, 16$ for four-color $CMYK$ printer) to predict the spectral reflectance for any $CMYK$ input. The dot area coverages may be determined from spectral reflectance samples using a parameter fitting or any of the estimation algorithms (least square, RLS, total least square [43], robust estimation [44], genetic algorithm, neural networks, etc.). As an example, we show how to model the dot area coverages for a random halftone screen with the least-squares algorithm and the RLS algorithm.

7.4.3.3   Estimation of Dot Area Coverages Using Least Squares

An approach of relating the dot area coverage to the corresponding digital counts, $CMYK$, is shown for one of the separations (e.g., cyan) and the same method is carried out for other colorants to find their respective dot area coverages. The relationship between the dot area coverage and digital counts is called dot growth function, which can be obtained by solving an optimization problem in (a) spectral space or (b) $L*a*b*$ space.

7.4.3.3.1   Optimization in Spectral Space

Consider a set of cyan reflectance measurements $R_{Ci}(\lambda)$ corresponding to input digital values $C = i$, $M = Y = K = 0$. To estimate the dot area coverage for cyan in a least squared error sense in the spectral reflectance space, the following metric is minimized

$$J = \sum_{k=1}^{N} \left\{ w(\lambda)[R_{Ci}(\lambda_k)]^{\frac{1}{m}} - \left[\hat{R}_{Ci}(\lambda_k)\right]^{\frac{1}{m}} \right\}^2 \qquad (7.57)$$

In Equation 7.57, $\hat{R}_{Ci}(\lambda)$ is the estimated reflectance obtained using the Neugebauer model and is given by the following equation

$$\hat{R}_{Ci}(\lambda) = c_i P_C(\lambda) + (1 - c_i) P_W(\lambda) \qquad (7.58)$$

where $P_C(\lambda)$ and $P_W(\lambda)$ are the reflectance spectra of cyan primary and paper white, respectively. To improve the modeling accuracy, a wavelength dependent weight $w(\lambda)$ is often included to emphasize the errors in the regions of the visible spectrum to which the human visual system is most sensitive. Spectral weights can be selected as $w(\lambda) = \max(\bar{x}(\lambda), \bar{y}(\lambda), \bar{z}(\lambda))$, where $\bar{x}(\lambda)$, $\bar{y}(\lambda)$, $\bar{z}(\lambda)$ are tristimulus functions (i.e., color matching functions: see Appendix A). Other weighting functions are explored in Ref. [31].

To minimize the error metric in Equation 7.57, the derivative of J with respect to c_i is set equal to zero. For $w(\lambda) = 1$, this gives

$$\frac{\partial J}{\partial c_i} = \frac{\partial}{\partial c_i} \sum_{k=1}^{N} \left\{ [R_{Ci}(\lambda_k)]^{\frac{1}{m}} - [c_i P_C(\lambda_k) + (1 - c_i) P_W(\lambda_k)]^{\frac{1}{m}} \right\}^2 = 0 \qquad (7.59)$$

Solving Equation 7.59 for the unknown c_i, we obtain

$$c_i = \frac{\sum_{k=1}^{N}\left\{[P_C(\lambda_k)]^{1/m}-[P_W(\lambda_k)]^{1/m}\right\}\left\{[R_{Ci}(\lambda_k)]^{1/m}-[P_W(\lambda_k)]^{1/m}\right\}}{\sum_{k=1}^{N}\left\{[P_C(\lambda_k)]^{1/m}-[P_W(\lambda_k)]^{1/m}\right\}^2} \quad (7.60a)$$

$$i = 0, 1, 2, \ldots, 255$$

The above process is repeated for other three colorants m, y, and k. A dot growth function (curve) can be easily obtained by using single separation step wedges at discrete digital counts (see Example 7.7) and then plotting *CMYK* values against respective digital counts *CMYK*. For case $w(\lambda) \neq 1$ Equation 7.60a is modified by multiplying each reflectance spectra by $w^m(\lambda)$. The modified equation is given by

$$c_i = \frac{\sum_{k=1}^{N}\left\{[w^m(\lambda_k)P_C(\lambda_k)]^{1/m}-[w^m(\lambda_k)P_W(\lambda_k)]^{1/m}\right\}\left\{[w^m(\lambda_k)R_{Ci}(\lambda_k)]^{1/m}-[w^m(\lambda_k)P_W(\lambda_k)]^{1/m}\right\}}{\sum_{k=1}^{N}\left\{[w^m(\lambda_k)P_C(\lambda_k)]^{1/m}-[w^m(\lambda_k)P_W(\lambda_k)]^{1/m}\right\}^2}$$

$$(7.60b)$$

7.4.3.3.2 Optimization in L*a*b* Color Space

In this approach, the error minimization is carried out in the $L^*a^*b^*$ space to assign visually meaningful weights. Unlike in the previous approach, since the transformation from spectral domain to color space $L^*a^*b^*$ is nonlinear; a closed-form least square solution is intractable and thus a simple numerical iterative solution is preferred.

Assuming that the dot growth functions are separation independent, as in the previous section, a set of N cyan, magenta, yellow, black, and neutral patches are printed and their $L^*a^*b^*$ values are measured. Let ΔE_C, ΔE_M, ΔE_Y, ΔE_k and ΔE_g be the CIELAB error between the measured $L^*a^*b^*$ values and their corresponding predicted values for cyan, magenta, yellow, and neutral patches from the Neugebauer model with the appropriate mixing equation. The overall error metric, ΔE, is defined as the weighted sum of the above CIELab errors. Thus,

$$\Delta E(i) = \alpha_c \Delta E_c(i) + \alpha_m \Delta E_m(i) + \alpha_y \Delta E_y(i) + \alpha_k \Delta E_k(i) + \alpha_g \Delta E_g(i) \quad (7.61)$$

where the weights α_c, α_m, α_y, α_k, and α_g are chosen based on region of the color space where more accurate reproduction is desirable (e.g., $\alpha_c = \alpha_m = \alpha_y = \alpha_k = 0.15$, and $\alpha_g = 0.4$ for more accurate prediction along the neutral axis rather than along individual colorant axes). The above metric is minimized with respect to the dot areas $c(i)$, $m(i)$, $y(i)$, and $k(i)$.

Example 7.7

A four-color *CMYK* printer has 16 usage basis colors for combinations of two states (0% and 100% area coverages) whose measured spectral curves are shown in Table 7.6 at 10 nm interval between 400 and 700 nm. The relationship between weights of the Neugebauer primaries to the dot area coverages (c, m, y, k) is modeled by Demichel dot model. Obtain the dot growth functions showing the relationship between the digital input C, M, Y, K and the corresponding dot area coverages c, m, y, k ($0 \leq c, m, y, k \leq 1$) for measured spectral samples shown in Table 7.7 using

    a. The least square error minimization (Equation 7.60a)
    b. The weighted least square error minimization (Equation 7.60b)

SOLUTION

Figure 7.21 shows the dot growth functions constructed for each separation using the least-squares error minimization formula, Equation 7.60, and synthesized spectral curves from Table 7.7. To complete the function, dot areas for intermediate digital counts are obtained through linear interpolation. To predict the spectral reflectance for any new color, we need to know their digital counts and follow the Neugebauer calculations shown in the block diagram of Figure 7.19. Given the digital counts ($CMYK$), we use the individual dot growth functions to find their corresponding dot area coverages. All the dot area coverages are substituted in the weight Equation 7.54. Then the spectra are calculated using basis vectors from Table 7.6, these weights, and Equation 7.51.

The usual ideas of error minimization (Equation 7.57) led to a closed form optimal solution for determining the dot area coverages (i.e., CMYK values) corresponding to their digital counts (i.e., CMYK values). However, this operation is done with one important assumption: each dot area coverage value for a given separation is independent of other three separations. To further improve the dot area coverage solution, samples with mixed colors have to be used while calculating the dot growth functions. To accommodate the use of mixed color samples, the error minimization steps should be performed with a mixture model so that the factors involved with multiple separations are also taken into consideration.

7.4.3.4   Cellular Neugebauer Model (Lab-NB)

The basic Neugebauer model, Equation 7.51, used basis vectors that are combinations of the spectral curves for primary colors with 0% or 100% area coverages and their over prints. These area coverages form the grid points of a cube when CMY separations are involved. Intermediate area coverages, such as 50%, were not included. Figure 7.22 shows the structure of the cube for a cellularized Neugebauer model with grid points also having 50% area coverage combinations. For these grid points, the Neugebauer primaries will increase from $2^3 = 8$ to $3^3 = 27$. Weights for the cellular model are obtained using Demichel weights, Equation 7.54 for random screens, or Equation 7.55 for dot-on-dot screen or using various least-squares regressions. A much finer division of the cube can be obtained by using a larger number of grid points, which presumably can lead to improved accuracy.

Other physics-based models considered in printing systems are Clapper–Yule [38,55,65] model, which introduces correction for errors by modeling the internal scattering, the ink transmissions, and the surface reflectance.

TABLE 7.6

Reflectance Spectra of Neugebauer Primaries

Primaries	Reflectance Spectra between 400 and 700 nm at 10 nm Interval																														
White	0.46	0.74	0.93	1.02	1.03	1.01	0.97	0.95	0.93	0.92	0.01	0.90	0.90	0.89	0.89	0.89	0.89	0.89	0.89	0.89	0.90	0.90	0.90	0.90	0.90	0.91	0.91	0.91	0.90	0.91	0.91
C	0.21	0.37	0.50	0.61	0.68	0.72	0.73	0.73	0.71	0.67	0.63	0.56	0.49	0.40	0.31	0.23	0.16	0.11	0.08	0.07	0.06	0.06	0.06	0.06	0.06	0.07	0.07	0.07	0.07	0.07	0.07
M	0.22	0.35	0.44	0.48	0.47	0.38	0.27	0.18	0.11	0.07	0.05	0.05	0.05	0.05	0.06	0.06	0.08	0.16	0.32	0.50	0.64	0.74	0.79	0.82	0.83	0.84	0.84	0.84	0.85	0.85	0.85
Y	0.05	0.06	0.06	0.06	0.06	0.07	0.08	0.09	0.15	0.30	0.47	0.63	0.76	0.81	0.83	0.84	0.84	0.84	0.85	0.85	0.85	0.85	0.86	0.86	0.86	0.86	0.86	0.86	0.86	0.86	0.87
K	0.03	0.03	0.03	0.03	0.03	0.03	0.03	0.03	0.03	0.03	0.03	0.03	0.03	0.03	0.03	0.03	0.03	0.03	0.03	0.03	0.03	0.03	0.03	0.03	0.03	0.03	0.03	0.03	0.03	0.03	0.03
CM	0.14	0.25	0.33	0.39	0.42	0.37	0.29	0.21	0.13	0.09	0.06	0.05	0.04	0.04	0.03	0.03	0.04	0.06	0.07	0.07	0.08	0.09	0.09	0.10	0.10	0.11	0.11	0.11	0.10	0.10	0.10
CY	0.03	0.04	0.05	0.05	0.06	0.08	0.09	0.09	0.14	0.26	0.36	0.43	0.44	0.39	0.32	0.23	0.16	0.11	0.08	0.07	0.06	0.06	0.06	0.07	0.07	0.07	0.07	0.07	0.07	0.07	0.06
CK	0.02	0.02	0.02	0.02	0.02	0.02	0.02	0.02	0.02	0.02	0.02	0.02	0.02	0.02	0.02	0.02	0.02	0.02	0.02	0.02	0.02	0.02	0.02	0.02	0.02	0.02	0.02	0.02	0.02	0.02	0.02
MY	0.06	0.07	0.08	0.08	0.08	0.07	0.07	0.05	0.04	0.04	0.04	0.04	0.04	0.04	0.05	0.08	0.16	0.32	0.49	0.62	0.72	0.77	0.80	0.81	0.81	0.82	0.82	0.82	0.82	0.83	0.83
MK	0.02	0.02	0.02	0.02	0.02	0.02	0.02	0.02	0.02	0.02	0.02	0.02	0.02	0.02	0.02	0.02	0.02	0.02	0.02	0.02	0.02	0.02	0.02	0.02	0.02	0.02	0.02	0.02	0.02	0.02	0.02
YK	0.02	0.02	0.02	0.03	0.03	0.03	0.04	0.04	0.04	0.04	0.04	0.05	0.05	0.04	0.04	0.03	0.03	0.03	0.03	0.03	0.03	0.04	0.04	0.05	0.05	0.05	0.06	0.06	0.06	0.06	0.05
CMY	0.02	0.03	0.03	0.04	0.04	0.04	0.04	0.04	0.04	0.04	0.04	0.05	0.05	0.04	0.04	0.03	0.03	0.03	0.03	0.03	0.03	0.04	0.04	0.05	0.05	0.05	0.06	0.06	0.06	0.06	0.05
CMK	0.01	0.01	0.01	0.01	0.01	0.01	0.01	0.01	0.01	0.01	0.01	0.01	0.01	0.01	0.01	0.01	0.01	0.01	0.01	0.01	0.01	0.01	0.01	0.01	0.01	0.01	0.01	0.01	0.01	0.01	0.01
CYK	0.01	0.01	0.01	0.01	0.01	0.01	0.01	0.01	0.01	0.01	0.01	0.01	0.01	0.01	0.01	0.01	0.01	0.01	0.01	0.01	0.01	0.01	0.01	0.01	0.01	0.01	0.01	0.01	0.01	0.01	0.01
MYK	0.01	0.01	0.01	0.01	0.01	0.01	0.01	0.01	0.01	0.01	0.01	0.01	0.01	0.01	0.01	0.01	0.01	0.01	0.01	0.01	0.01	0.01	0.01	0.02	0.02	0.02	0.02	0.02	0.02	0.02	0.02
CMYK	0.01	0.01	0.01	0.01	0.01	0.01	0.01	0.01	0.01	0.01	0.01	0.01	0.01	0.01	0.01	0.01	0.01	0.01	0.01	0.01	0.01	0.01	0.01	0.01	0.01	0.01	0.01	0.01	0.01	0.01	0.01

TABLE 7.7
Reflectance Spectra for Obtaining Dot Growth Function

Reflectance Spectra between 400 and 700 nm at 10 nm Interval

Digital Count	Reflectance Spectra between 400 and 700 nm at 10 nm Interval																											
C																												
28	0.44	0.72	0.91	0.99	1.01	0.99	0.95	0.93	0.92	0.90	0.89	0.88	0.87	0.86	0.86	0.85	0.84	0.83	0.83	0.83	0.83	0.83	0.83	0.83	0.83	0.84	0.84	0.84
57	0.42	0.69	0.88	0.96	0.99	0.97	0.94	0.92	0.90	0.89	0.87	0.86	0.84	0.83	0.81	0.80	0.79	0.78	0.77	0.77	0.77	0.77	0.77	0.77	0.77	0.77	0.77	0.78
85	0.41	0.68	0.86	0.95	0.97	0.96	0.93	0.91	0.90	0.88	0.86	0.85	0.83	0.81	0.79	0.77	0.76	0.75	0.74	0.74	0.74	0.74	0.74	0.74	0.74	0.74	0.74	0.74
113	0.36	0.60	0.77	0.86	0.90	0.90	0.88	0.87	0.85	0.83	0.80	0.78	0.74	0.71	0.67	0.63	0.59	0.57	0.55	0.54	0.54	0.54	0.54	0.54	0.54	0.54	0.54	0.55
142	0.33	0.55	0.72	0.81	0.86	0.87	0.85	0.84	0.82	0.80	0.77	0.73	0.69	0.64	0.59	0.54	0.50	0.47	0.45	0.44	0.43	0.43	0.43	0.43	0.43	0.43	0.43	0.44
170	0.28	0.48	0.63	0.73	0.79	0.81	0.80	0.79	0.78	0.75	0.71	0.66	0.61	0.55	0.48	0.42	0.36	0.32	0.30	0.28	0.28	0.28	0.28	0.28	0.28	0.27	0.28	0.28
198	0.27	0.46	0.61	0.71	0.77	0.79	0.79	0.78	0.76	0.74	0.70	0.65	0.59	0.52	0.45	0.39	0.33	0.28	0.26	0.26	0.25	0.24	0.24	0.24	0.24	0.24	0.24	0.24
227	0.24	0.41	0.55	0.66	0.72	0.76	0.76	0.75	0.73	0.70	0.66	0.60	0.53	0.46	0.38	0.30	0.23	0.19	0.16	0.16	0.14	0.14	0.14	0.14	0.14	0.13	0.14	0.14
255	0.21	0.37	0.50	0.61	0.68	0.73	0.73	0.73	0.71	0.67	0.63	0.56	0.49	0.40	0.31	0.23	0.16	0.11	0.08	0.07	0.06	0.06	0.06	0.06	0.06	0.06	0.06	0.06
M																												
28	0.44	0.72	0.90	0.98	0.99	0.97	0.92	0.89	0.87	0.86	0.85	0.84	0.83	0.83	0.83	0.83	0.83	0.83	0.84	0.85	0.86	0.87	0.88	0.88	0.88	0.89	0.89	0.90
57	0.42	0.69	0.87	0.94	0.95	0.92	0.87	0.84	0.81	0.79	0.77	0.77	0.76	0.76	0.76	0.76	0.77	0.77	0.79	0.81	0.84	0.84	0.86	0.88	0.88	0.89	0.89	0.89
85	0.41	0.68	0.85	0.92	0.93	0.90	0.84	0.81	0.78	0.76	0.75	0.74	0.73	0.73	0.73	0.73	0.74	0.76	0.79	0.83	0.83	0.85	0.87	0.87	0.88	0.89	0.89	0.89
113	0.36	0.59	0.75	0.81	0.76	0.69	0.64	0.60	0.59	0.56	0.54	0.53	0.53	0.53	0.53	0.53	0.55	0.59	0.67	0.74	0.74	0.80	0.82	0.84	0.85	0.86	0.87	0.87
142	0.33	0.54	0.68	0.74	0.68	0.59	0.54	0.48	0.45	0.43	0.42	0.42	0.42	0.42	0.42	0.42	0.45	0.50	0.59	0.69	0.69	0.76	0.79	0.82	0.84	0.85	0.86	0.86
170	0.29	0.47	0.59	0.64	0.59	0.48	0.43	0.39	0.36	0.33	0.29	0.27	0.27	0.26	0.27	0.30	0.36	0.45	0.49	0.61	0.61	0.71	0.76	0.79	0.82	0.83	0.83	0.84
198	0.28	0.45	0.56	0.61	0.56	0.44	0.39	0.36	0.33	0.29	0.25	0.24	0.23	0.23	0.23	0.26	0.33	0.38	0.46	0.59	0.59	0.70	0.76	0.78	0.80	0.82	0.84	0.84
227	0.24	0.40	0.50	0.54	0.53	0.45	0.34	0.29	0.19	0.15	0.13	0.12	0.12	0.12	0.12	0.16	0.23	0.23	0.38	0.54	0.54	0.67	0.74	0.76	0.79	0.81	0.81	0.83
255	0.22	0.35	0.44	0.48	0.47	0.38	0.27	0.18	0.11	0.07	0.05	0.05	0.05	0.05	0.05	0.06	0.08	0.16	0.32	0.50	0.50	0.64	0.70	0.74	0.80	0.80	0.79	0.82
Y																												
28	0.43	0.69	0.87	0.95	0.96	0.94	0.91	0.89	0.88	0.88	0.89	0.89	0.89	0.89	0.89	0.89	0.89	0.89	0.89	0.89	0.89	0.89	0.89	0.89	0.89	0.89	0.90	0.90
57	0.40	0.64	0.80	0.87	0.88	0.87	0.84	0.85	0.84	0.82	0.87	0.87	0.88	0.88	0.89	0.89	0.89	0.89	0.89	0.89	0.89	0.89	0.89	0.88	0.89	0.89	0.89	0.90
85	0.38	0.61	0.77	0.83	0.86	0.84	0.81	0.79	0.80	0.81	0.83	0.85	0.88	0.88	0.88	0.88	0.89	0.89	0.89	0.89	0.89	0.89	0.89	0.88	0.89	0.89	0.89	0.90
113	0.29	0.46	0.57	0.61	0.62	0.61	0.60	0.59	0.61	0.68	0.74	0.80	0.85	0.87	0.87	0.87	0.88	0.88	0.88	0.88	0.88	0.88	0.88	0.87	0.88	0.88	0.88	0.89
142	0.24	0.37	0.45	0.49	0.50	0.49	0.48	0.52	0.48	0.60	0.69	0.77	0.83	0.86	0.86	0.87	0.87	0.87	0.87	0.87	0.86	0.87	0.87	0.86	0.87	0.87	0.88	0.88
170	0.16	0.25	0.30	0.32	0.32	0.33	0.33	0.37	0.33	0.48	0.60	0.72	0.81	0.84	0.85	0.85	0.86	0.86	0.86	0.86	0.86	0.86	0.86	0.85	0.86	0.86	0.87	0.87
198	0.15	0.22	0.26	0.28	0.28	0.29	0.29	0.34	0.29	0.45	0.58	0.70	0.80	0.84	0.85	0.85	0.86	0.86	0.86	0.86	0.86	0.86	0.86	0.84	0.86	0.86	0.87	0.87

(continued)

TABLE 7.7 (continued)

Reflectance Spectra for Obtaining Dot Growth Function

Digital Count	Reflectance Spectra between 400 and 700 nm at 10 nm Interval																										
227	0.09	0.13	0.15	0.15	0.16	0.16	0.17	0.18	0.23	0.37	0.52	0.67	0.78	0.82	0.84	0.84	0.85	0.85	0.85	0.85	0.85	0.85	0.86	0.86	0.86	0.86	0.86
255	0.05	0.06	0.06	0.06	0.06	0.07	0.08	0.09	0.15	0.30	0.47	0.63	0.76	0.81	0.83	0.84	0.84	0.84	0.84	0.85	0.85	0.85	0.85	0.85	0.86	0.86	0.86
K																											
28	0.42	0.69	0.87	0.94	0.95	0.93	0.90	0.88	0.86	0.85	0.84	0.84	0.83	0.83	0.83	0.83	0.82	0.83	0.83	0.83	0.83	0.83	0.83	0.83	0.83	0.83	0.84
57	0.39	0.63	0.79	0.86	0.88	0.86	0.82	0.81	0.79	0.78	0.77	0.77	0.76	0.76	0.76	0.76	0.76	0.76	0.76	0.76	0.76	0.76	0.76	0.76	0.77	0.77	0.77
85	0.37	0.61	0.76	0.82	0.84	0.82	0.79	0.77	0.76	0.75	0.74	0.73	0.73	0.73	0.72	0.72	0.72	0.72	0.73	0.73	0.73	0.73	0.73	0.73	0.73	0.73	0.73
113	0.27	0.43	0.54	0.59	0.60	0.58	0.56	0.55	0.54	0.53	0.53	0.52	0.52	0.52	0.52	0.52	0.52	0.52	0.52	0.52	0.52	0.52	0.52	0.52	0.52	0.52	0.52
142	0.21	0.34	0.42	0.46	0.46	0.46	0.44	0.43	0.42	0.42	0.41	0.41	0.41	0.41	0.40	0.40	0.40	0.40	0.40	0.41	0.41	0.41	0.41	0.41	0.41	0.41	0.41
170	0.14	0.21	0.26	0.28	0.28	0.28	0.27	0.26	0.26	0.25	0.25	0.25	0.25	0.25	0.25	0.25	0.25	0.25	0.25	0.25	0.25	0.25	0.25	0.25	0.25	0.25	0.25
198	0.12	0.18	0.22	0.24	0.24	0.23	0.23	0.22	0.22	0.22	0.21	0.21	0.21	0.21	0.21	0.21	0.21	0.21	0.21	0.21	0.21	0.21	0.21	0.21	0.21	0.21	0.21
227	0.07	0.09	0.11	0.12	0.12	0.12	0.11	0.11	0.11	0.11	0.11	0.11	0.11	0.11	0.11	0.11	0.11	0.11	0.11	0.11	0.11	0.11	0.11	0.11	0.11	0.11	0.11
255	0.03	0.03	0.03	0.03	0.03	0.03	0.03	0.03	0.03	0.03	0.03	0.03	0.03	0.03	0.03	0.03	0.03	0.03	0.03	0.03	0.03	0.03	0.03	0.03	0.03	0.03	0.03
W																											
0	0.46	0.74	0.93	1.02	1.03	1.01	0.97	0.95	0.93	0.92	0.91	0.90	0.90	0.89	0.89	0.89	0.89	0.89	0.89	0.89	0.89	0.89	0.90	0.90	0.90	0.90	0.90

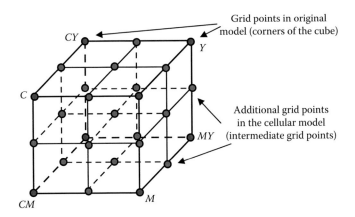

FIGURE 7.21   Dot growth functions plotted as a function of the digital count.

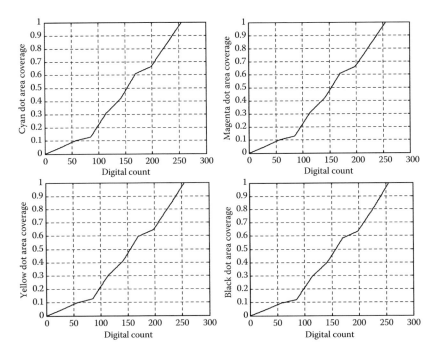

FIGURE 7.22   A CMY color space is divided into subcubes.

The Hoffman–Schmelzer model [66] considers the interaction between the light, ink, and the paper. It has been used for the packaging industry with some heuristic corrections [67]. Kubelka–Munk and Beer–Bouguer models are other models to be considered. Derivation of these models can be found in many publications, Refs. [38,68,69]. These models are, however, not widely used for halftone printing.

7.4.4   DEVICE DRIFT MODEL

A systematic drift in color, which may occur due to changes in humidity and temperature, can be modeled by sampling a few colors and then updating the printer model over time. In this section, we examine a method to predict color drift in digital printers using measurements from a sensor. Two methods are discussed: (1) channel independent (scalar) autoregressive (AR) model and (2) channel dependent vector autoregressive (VAR) model. The scalar AR model can be used for predicting single channel drift, that is, drift in the print density on the photoconductor or on the paper, lightness on the paper, and chroma shift or hue shift. The VAR model can predict the color drift specified in terms of reflectance spectra or the $L*a*b*$ on the paper.

7.4.4.1   Autoregressive (AR) Model Applied to Printer Drift Prediction

Let time t_0 be the time at which the printer forward model (map) is initially constructed from measurements. It is easy to build a printer drift model if the measurements for all the patches used for constructing the forward model are available. If performed during run time, this would require too many measurements, which increase cost and decrease productivity. Hence a more preferred method would be to use a few measurements to sample the color drift and to update the entire model. Thus a drifted printer model, $P(t)$, is built from the initial forward printer model, $P(t_0)$, at time $t_0 + t$ based on few measurements of printed color patches. Printed colors are selected in critical regions of the color space to maximize the sensitivity to color drift.

We describe the general AR model for signal prediction and explain how it can be applied to our specific problem of printer drift prediction. Consider the measurable output $y(n)$ from the printing system. The function $y(n)$ could be $L*$ or $a*$ or $b*$ or chroma or hue or print density for a single color patch. The output of the system is given by $y(n)$ at time index n. It is assumed that N data samples are available. We use an AR model of order P of a stationary zero-mean process $y(n)$ of the form

$$y(n) = -\sum_{i=1}^{P} a_i y(n - i) + e(n) \tag{7.62}$$

where $\{a_i\}_{i=1}^{P}$ are the AR parameters and $e(n)$ is the zero-mean white noise process error with variance σ^2. In this model, the predicted output is given by

$$\hat{y}(n) = -\sum_{i=1}^{P} a_i y(n - i) \tag{7.63}$$

The error between the measured output and the predicted output is given by

$$e(n) = y(n) - \hat{y}(n) \tag{7.64}$$

To determine the coefficients of the AR model, the mean-squared error (MSE) between the predicted output and the measured output is minimized. From Equations 7.63 and 7.64, the error between the measured output and the predicted output is given by

$$e(n) = y(n) + \sum_{i=1}^{P} a_i y(n-i) \tag{7.65}$$

The MSE can be expressed as the experted value of the square of the error

$$E\left[e^2(n)\right] = E\left[\left\{y(n) + \sum_{i=1}^{P} a_i y(n-i)\right\}^2\right] \tag{7.66}$$

Equation 7.66 can be expanded as

$$E\left[e^2(n)\right] = E\left[\left\{y(n) + \sum_{i=1}^{P} a_i y(n-i)\right\}\left\{y(n) + \sum_{i=1}^{P} a_i y(n-i)\right\}\right] \tag{7.67}$$

Expanding the above equation results in

$$E\left[e^2(n)\right] = E\left[y^2(n)\right] + E\left[\sum_{i=1}^{P} a_i y(n-i)y(n)\right] + E\left[y(n)\sum_{i=1}^{P} a_i y(n-i)\right]$$

$$+ E\left[\sum_{i=1}^{P} a_i y(n-i)\sum_{j=1}^{P} a_j y(n-j)\right] \tag{7.68}$$

Interchanging the expectation and summation, we have

$$E\left[e^2(n)\right] = E\left[y^2(n)\right] + \sum_{i=1}^{P} a_i E[y(n-i)y(n)] + \sum_{i=1}^{P} a_i E[y(n)y(n-i)]$$

$$+ \sum_{i=1}^{P}\sum_{j=1}^{P} a_i a_j E[y(n-i)]y(n-j)] \tag{7.69}$$

Performing the expectation, we get

$$E\left[e^2(n)\right] = r_{yy}(0) + \sum_{i=1}^{P} a_i r_{yy}(i) + \sum_{i=1}^{P} a_i r_{yy}(-i) + \sum_{i=1}^{P}\sum_{j=1}^{P} a_i a_j r_{yy}(j-i) \tag{7.70}$$

Therefore,

$$E\left[e^2(n)\right] = r_{yy}(0) + 2\sum_{i=1}^{P} a_i r_{yy}(i) + \sum_{i=1}^{P}\sum_{j=1}^{P} a_i a_j r_{yy}(j-i) \tag{7.71}$$

where $r_{yy}(i)$ is the autocorrelation of the output process, $y(n)$, at ith lag. As the signal $y(n)$ is real, autocorrelation function r_{yy} is symmetric ($r_{yy}(i) = r_{yy}(-i)$). We minimize the above expression to obtain the optimum values for a_i, $i = 1, 2, \ldots, P$. Applying the orthogonality principle, we obtain the well-known Yule–Walker equations

$$\sum_{j=1}^{P} r_{yy}(j - i)a_j = -r_{yy}(i) \quad i = 1, 2, \ldots, P \tag{7.72}$$

The above equations can be written in matrix form as

$$
\begin{bmatrix}
r_{yy}(0) & r_{yy}(1) & r_{yy}(2) & \cdots & r_{yy}(P-1) \\
r_{yy}(-1) & r_{yy}(0) & r_{yy}(1) & & r_{yy}(P-2) \\
\vdots & & & \ddots & \\
r_{yy}(2-P) & & & & \\
r_{yy}(1-P) & & \cdots & & r_{yy}(0)
\end{bmatrix}
\begin{bmatrix}
a_1 \\
a_2 \\
a_3 \\
\vdots \\
a_P
\end{bmatrix}
= -
\begin{bmatrix}
r_{yy}(1) \\
r_{yy}(2) \\
r_{yy}(3) \\
\vdots \\
r_{yy}(P)
\end{bmatrix}
\tag{7.73}
$$

Since the output of the system $y(n)$ is real, its autocorrelation is symmetric, that is, $r_{yy}(i) = r_{yy}(-i)$, so Equation 7.73 can be rewritten as

$$
\begin{bmatrix}
r_{yy}(0) & r_{yy}(1) & r_{yy}(2) & \cdots & r_{yy}(P-1) \\
r_{yy}(1) & r_{yy}(0) & r_{yy}(1) & & r_{yy}(P-2) \\
\vdots & & & \ddots & \\
r_{yy}(P-2) & & & & \\
r_{yy}(P-1) & & \cdots & & r_{yy}(0)
\end{bmatrix}
\begin{bmatrix}
a_1 \\
a_2 \\
a_3 \\
\vdots \\
a_P
\end{bmatrix}
= -
\begin{bmatrix}
r_{yy}(1) \\
r_{yy}(2) \\
r_{yy}(3) \\
\vdots \\
r_{yy}(P)
\end{bmatrix}
\tag{7.74}
$$

And the variance of the error function $e(n)$, MSE is given by

$$\sigma_e^2 = r_{yy}(0) + r_{yy}(1)a_1 + r_{yy}(2)a_2 + \cdots + r_{yy}(P)a_P \tag{7.75}$$

Combining the above two equations we obtain

$$
\begin{bmatrix}
r_{yy}(0) & r_{yy}(1) & r_{yy}(2) & \cdots & r_{yy}(P-1) & r_{yy}(P) \\
r_{yy}(1) & r_{yy}(0) & r_{yy}(1) & \cdots & r_{yy}(P-2) & r_{yy}(P-1) \\
r_{yy}(2) & r_{yy}(1) & r_{yy}(0) & \cdots & r_{yy}(P-3) & r_{yy}(P-2) \\
\vdots & & & \ddots & & \\
r_{yy}(P-1) & & & & \ddots & r_{yy}(1) \\
r_{yy}(P) & & \cdots & & \cdots & r_{yy}(0)
\end{bmatrix}
\begin{bmatrix}
1 \\
a_1 \\
a_2 \\
\vdots \\
a_{P-1} \\
a_P
\end{bmatrix}
=
\begin{bmatrix}
\sigma_e^2 \\
0 \\
0 \\
\vdots \\
0 \\
0
\end{bmatrix}
\tag{7.76}
$$

The AR coefficients and the variance of error are determined by solving the above system of linear equations using least-squares regression equations for a set of N number of measured data samples. The autocorrelation and variance of the data can be found using MATLAB functions.

Now we will discuss how this model can be applied to predict the printer drift. Say we have the outputs of the system in $L^*a^*b^*$ and density, d, measured using a color sensor at time index n corresponding to time t. The initial time t_0 is assumed to be zero, with corresponding time index $n = 0$. We predict the output in each channel (say cyan), independent of the other channels (i.e., magenta, yellow, and black), using the above model (Equation 7.74) with $y(n) = L^*(n) - L^*(0)$ for predicting L^*, $y(n) = a^*(n) - a^*(0)$ for predicting a^*, $y(n) = b^*(n) - b^*(0)$ for predicting b^*, $y(n) = d(n) - d(0)$ for predicting print density represented by the symbol d. After estimating the AR coefficients for the three channels using Equation 7.74 or Equation 7.76, we can predict the new values $\hat{L}^*(n)$, $\hat{a}^*(n)$, $\hat{b}^*(n)$, $\hat{d}(n)$ from the P previous values of the outputs using Equation 7.62. That is

$$\hat{y}(n) = c - \sum_{i=1}^{P} a_i y(n - i) \qquad (7.77)$$

where c is the initial value of L^* for predicting L^* and the corresponding initial values for predicting a^*, or b^*, and density d. This approach ignores the interaction between $L^*a^*b^*$ and d treating them independent of each other. The model order P can be determined using statistical techniques including the minimization of an order selection criterion [70] (not discussed in this book).

7.4.4.2 Vector Autoregressive Model Applied to Printer Drift Prediction

In the vector AR model, we consider the dependence of one channel on the other channels while predicting the output of the printer. So in the vector AR model, the measured output process is a 3×1 vector random process defined by

$$y(n) = \begin{bmatrix} L^*(n) - L^*(0) \\ a^*(n) - a^*(0) \\ b^*(n) - b^*(0) \end{bmatrix} \qquad (7.78)$$

The predicted output in terms of the P (previously) measured output values is given by the following equation:

$$\hat{y}(n) = -\sum_{i=1}^{P} A_i y(n - i) \qquad (7.79)$$

The error signal between measured and predicted outputs is given by

$$e(n) = y(n) - \hat{y}(n) = y(n) + \sum_{i=1}^{P} A_i y(n - i) \qquad (7.80)$$

where A_i for $i = 1, \ldots, P$ are 3×3 matrices which define the VAR matrix coefficients. The prediction error is assumed to be zero-mean white noise process with unknown covariance matrix Σ. The error is minimized in the least squares sense (similar to the scalar case). The MSE is given by

$$E\left[e^2(n)\right] = E\left[e(n)e^{\mathrm{T}}(n)\right]$$

$$E\left[y(n) + \sum_{i=1}^{P} A_i y(n-i)\right]\left[y^{\mathrm{T}}(n) + \sum_{i=1}^{P} y^{\mathrm{T}}(n-i)A_i^{\mathrm{T}}\right]$$

$$= R_{yy}(0) + \sum_{i=1}^{P} A_i R_{yy}^{\mathrm{T}}(i) + \sum_{i=1}^{P} R_{yy}(i)A_i^{\mathrm{T}} + \sum_{i=1}^{P}\sum_{i=1}^{P} A_i R_{yy}(j-i)A_j^{\mathrm{T}} \qquad (7.81)$$

where $R_{yy}(i)$ is the 3×3 correlation matrix of the output of the system at lag i and is given by

$$R_{yy}(i) = \begin{bmatrix} r_{LL}(i) & r_{La}(i) & r_{Lb}(i) \\ r_{aL}(i) & r_{aa}(i) & r_{ab}(i) \\ r_{bL}(i) & r_{ba}(i) & r_{bb}(i) \end{bmatrix} \qquad (7.82)$$

Since $y(n)$ is real, $R_{yy}(-i) = R_{yy}^{\mathrm{T}}(i)$.

The diagonal elements of the positive definite matrix $R_{yy}(i)$ are the autocorrelation of the three components of the color vector $L^*a^*b^*$ and the off diagonal elements are measure of correlation between the three coordinates of the $L^*a^*b^*$ color vector. We now optimize the cost function given by Equation 7.81 with respect to matrix A_i. The result is similar to the Yule–Walker equations for the scalar case and is given by

$$\begin{bmatrix} R_{yy}(0) & R_{yy}(1) & \cdots & R_{yy}(P) \\ R_{yy}(1) & R_{yy}(0) & \cdots & R_{yy}(P-1) \\ \vdots & \vdots & & \vdots \\ R_{yy}(P) & R_{yy}(P-1) & & R_{yy}(0) \end{bmatrix}\begin{bmatrix} A_0 \\ A_1 \\ \vdots \\ A_P \end{bmatrix} = \begin{bmatrix} \Sigma \\ 0 \\ \vdots \\ 0 \end{bmatrix} \qquad (7.83)$$

where
A_0 is a 3×3 matrix with $A_0(i, j) = 1$
0 is a 3×3 matrix of zero elements
Σ is the 3×3 covariance matrix of the prediction error signal $e(n)$

Σ is the covariance matrix of the prediction error function and is given by

$$\Sigma = R_{yy}(0) + R_{yy}(1)A_1^{\mathrm{T}} + R_{yy}(2)A_2^{\mathrm{T}} + \cdots + R_{yy}(P)A_P^{\mathrm{T}} \qquad (7.84)$$

After estimating the VAR matrix coefficients $\{A_i\}_{i=1}^{P}$, we can predict the new values of $\hat{L}^*(n)$, $\hat{a}^*(n)$, and $\hat{b}^*(n)$ from P previous values of the output using the following equation:

$$\hat{y}(n) = c - \sum_{i=1}^{P} A_i y(n-i) \qquad (7.85)$$

where $c = [L^*(0) \quad a^*(0) \quad b^*(0)]^{\mathrm{T}}$.

Comments about the experimental validation of AR drift models are shown next.

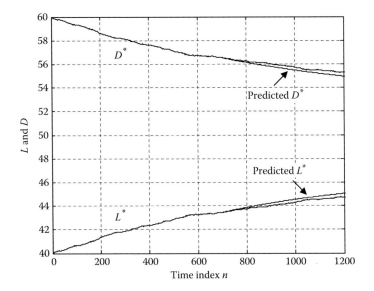

FIGURE 7.23 Measured and predicted L on a drifted print engine.

We used two sets of printer drift data. One data set has 33 color patches on the neutral or gray axis with measurements at 1200 different time samples. The second data set has 3131 colors in the printer gamut with measurements at 10 different time samples. A large number of temporal training data set is used for building scalar or VAR drift models. The data were divided into two parts, one part is used to build the drift model and the second part is used for prediction. An AR model of order 1, 2, and 3 is constructed using the first 700 samples of the data and the rest of the data is used to verify the model. The measured and predicted values of L^* and $D^* = 100 - L^*$ for a gray patch are shown in Figure 7.23.

To test the vector AR model, the second data set is used. Since the data record length is small, we used interpolation to increase the data set record length. Again similar to the scalar case, the data were divided into an estimation and a prediction part. The techniques are quantified by using $\Delta E_{ab}^* = \|\mathrm{Lab}_{\text{measured}} - \mathrm{Lab}_{\text{estimated}}\|$ between the actual measured drift data and predicted drift data for the predicted population. The prediction error statistics, average ΔE_{ab}^*, maximum ΔE_{ab}^*, and 95th percentile for vector regressive models of order 1, 2, and 3 are tabulated in Table 7.8.

TABLE 7.8
Error Statistics for VAR Model

VAR Model Order	$\Delta E_{ab\ \text{mean}}^*$	$\Delta E_{ab\ \text{max}}^*$	$\Delta E_{ab\ 95\%}^*$
$P = 1$	3.4	12.8	6.5
$P = 2$	2.8	10.9	5.4
$P = 3$	2.7	7.7	4.7

The AR and VAR models can predict drift data accurately if the measured data record length is large enough for estimation of the parameters of the models (Equations 7.76 and 7.83). The accuracy of the model depends on the data length as well as sensor noise. If the noise model of the color sensor is available, it can be removed from the drift measurements of the printer, which can result in a more accurate drift prediction.

7.5   GCR SELECTION AND INVERSION

In a four-color *CMYK* printer, as *K* increases, the total gamut volume created by mixing *CMY* with *K* decreases (Figure 7.24). Gray component replacement (GCR) provides a method to substitute black (*K*) for *CMY* mixtures in rendering a given color, which results in an extension of the darker region of the gamut by changing lightness

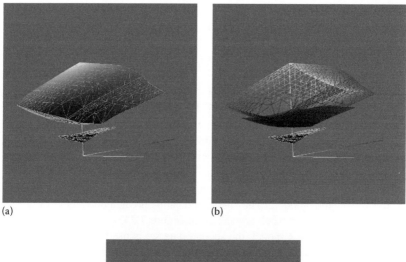

(a)   (b)

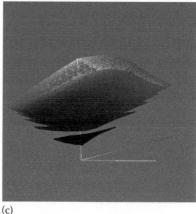

(c)

FIGURE 7.24   (a) *CMY* gamut with $K = 0$ (top), and $K = 255$ (bottom). (b) *CMY* gamut with $K = 0$ (wire), $K = 153$ (solid; middle), and $K = 255$ (bottom). (c) *CMY* gamut with $K = 0$, 51, 102, 153, 204, 255.

(or darkness) as compared to printers without K separation. It helps to reproduce shadows, gray areas, and muted tones in images. Other benefits of introducing K are toner savings and improved stability. A potential disadvantage in using high K levels throughout the gamut is the dirty/grainy appearance that can arise in flesh tones, sky tones, and other important colors. Smoothness and gamut coverage must also be taken into account while rendering pixels with black. GCR is a critical element in the inversion process since it introduces redundant solutions. Thus, GCR optimization requires a delicate trade-off among these competing requirements, and depends strongly on the physics of the printer. Generally, print vendors fine-tune the addition of black intelligently either by using complex algorithms or by using carefully designed experiments. Experiments are often done with many iterations to get the right amount of K. Once the tuning is done, the GCR will be included as a part of the multidimensional LUT. We will further elaborate on this topic later.

Printer characterization gives an approximation to the forward map (from device-specific *CMYK* space into the device-independent CIELab or spectral space). An accurate multidimensional profile LUT is intended to provide the inverse of the printer (or inverse of the characterized printer map) for every input color node described as *L*a*b** which is a transformation of the node from CIELab into a *CMYK* color space. Numerous techniques are available in the literature [71–80] that use variety of methods to characterize the printer and then compute inverse. For pixels not on the nodes, during real-time image processing, numerous interpolation schemes are used. Among these, tetrahedral interpolation is the most commonly used method inside the ICC profile. For any given node, in control terminology, the four-color forward printer map can be considered as a four-input three-output plant with four input variables (*CMYK*) used to control three outputs (*L*a*b**). Since the forward map is the representation of a printer with an over-actuated system, there exist many possible combinations of *CMYK* values that produce a given *L*a*b**. To avoid this kind of degeneracy and determine a unique combination for four separations, GCR constraints are required. These constraints are included in the inverse map (*L*a*b** to *CMYK* LUT).

The inverse map can be thought of as cascade of two LUTs: (1) a three-to-three mapping LUT, CIELab to *CMY* and then (2) a three-to-four mapping LUT, *CMY* to *CMYK*. In such a LUT architecture, conceptually an *RGB* node will be transformed first to three-color space, CIELab, then to another three-color space, *CMY*, finally followed by a transformation to four-color space, *CMYK*, prior to rendering. This is described next.

7.5.1   A Simple GCR Function

A simple GCR function, *CMY* to *CMYK* transformation, is shown in Figure 7.25 [71]. The GCR function is described in terms of two components: (1) black addition and (2) under color removal (UCR). As the figure indicates, we first subtract $0.5x^2$ from all three-component values of *CMY* and add black colorant per Equation 7.86.

$$K = x^2 \qquad (7.86)$$

where $x = \min(C, M, Y)$.

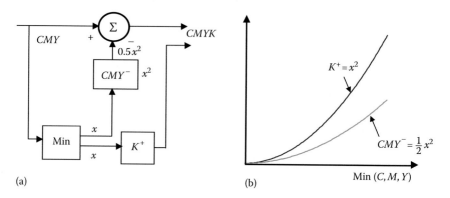

(a)   (b)

FIGURE 7.25 A simple GCR/UCR function (a) block diagram view (b) plot of GCR function with respect to min (C, M, Y).

For example, $CMY = [0.5 \quad 0.3 \quad 0.4]$ color patch (CMY values in 0 to 1 range) is equivalent to $CMYK = [0.455 \quad 0.255 \quad 0.355 \quad 0.09]$ patch using the above GCR/UCR algorithm. Soon it will become clear that this method has the disadvantage of not producing sufficiently optimized colors for the entire color gamut since inherently the practice of starting with *CMY* separations and subsequently adding black leads to loss of gamut. However, in Ref. [81] a max-gamut GCR is developed which starts with *CMY* separations and still preserves the total gamut volume. Figure 7.26 shows the comparison between the two gamuts; one gamut produced

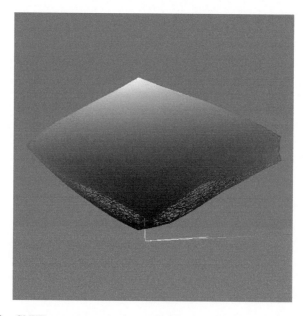

FIGURE 7.26 *CMYK* gamut compared to a *CMY* to $L^*a^*b^*$ printer gamut with the simple *CMY* to *CMYK* GCR (wire: A *CMYK* to $L^*a^*b^*$ gamut of a four-color printer; solid: a gamut of a *CMY* to $L^*a^*b^*$ printer with GCR/UCR of Figure 7.25).

by a four-color *CMYK* printer (e.g., a calibrated *CMYK* to $L^*a^*b^*$ printer map), the other gamut produced by a three-color *CMY* printer (e.g., a *CMY* to $L^*a^*b^*$ printer map including the GCR) using the GCR function of Figure 7.25.

Once the GCR function is selected using the *CMYK* to $L^*a^*b^*$ printer model, an augmented *CMY* to $L^*a^*b^*$ printer map can be generated by including the GCR function in the system, to produce a new forward map, *CMY* to $L^*a^*b^*$. Since this is a deterministic forward map; it gives unique colors for a unique combination of *CMY* and does not require any constraints during inversion. Inversion of this kind of unconstrained three-input to three-output forward map is reasonably easy and will be described next.

7.5.2 INVERSION OF A THREE-TO-THREE FORWARD MAP

The inversion mapping process introduced in the previous section is illustrated in Figure 7.27 starting from the intermediate RGB space. The *CMY* values are converted to *CMYK* using the GCR/UCR algorithms (e.g., Equation 7.86). In this section let us consider that the GCR/UCR functions are embedded inside the augmented forward printer map, Q. The inverse of this map, Q shown symbolically as Q^{-1}: $L^*a^*b^* \rightarrow CMY$, where input $L^*a^*b^*$ points are on a three-dimensional (3-D) grid of size $N \times N \times N$. This grid has a dynamic range $0 \leq L^* \leq 100$, $-127 \leq a^* \leq 128$, and $-127 \leq b^* \leq 128$.

To generate the ICC workflow, we start with a uniformly or nonuniformly sampled *RGB* color LUT in the range 0–255. This is an intermediate color space in the final ICC destination profile LUT (see Figures 7.1 and 7.2). Next, we apply a suitable transformation in the forward direction from *RGB* to $L^*a^*b^*$ and map the out-of-gamut colors in $L^*a^*b^*$ to a suitable node on the device boundary (or inside) of the printer model. Now, to express these *RGB* nodes (in the reverse direction) in the PCS, a conversion from *RGB* to $L^*a^*b^*$ or *XYZ* is required which is purely an inverse transformation starting from each of the *RGB* color nodes of the uniformly or nonuniformly sampled *RGB* grid. Since *RGB* space is not device independent, any of the standard *RGB*s (e.g., *genRGB*, *ROMMRGB*, etc.) are potential candidates for use in the multidimensional profile LUTs. Figure 7.28 shows a 3-D view of the nodes when constructed with *sRGB*, *ROMMRGB*, and *genRGB* color spaces. Figure 7.2

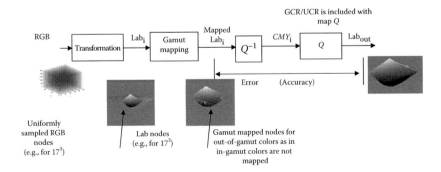

FIGURE 7.27   Block diagram illustrating a forward and inverse printer process with GCR in the augmented printer map, Q.

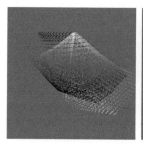

FIGURE 7.28   Three-dimensional view of node colors when super imposed on a printer gamut (left: *sRGB* to *L*a*b**), (middle: *ROMMRGB* to *L*a*b**), and (right: ProPhoto *RGB* to *L*a*b**).

shows the image path in the forward direction when we start from the intermediate *RGB* space. The reverse path to PCS space can be found in Figure 7.1. The overall profile accuracy depends on the choice of this intermediate color space since the spacing between the nodes in this color space has different effects in different regions of the printer gamut. This in turn can affect the interpolation accuracy of pixels that are not located on the nodes since all interpolation operations inside the ICC profile workflow use linear interpolation such as the tetrahedral interpolation to process images at high speed. Thus, a correspondence in the reverse direction (i.e., from nodes in the *RGB* color space to *L*a*b*/XYZ* values in the PCS) and a correspondence in the forward direction from the *RGB* nodes to *L*a*b** nodes are generated just before constructing the printer inverse (i.e., *L*a*b** to *CMY*).

7.5.2.1   Inverse by Working on the Printer Model

Different numerical algorithms can be used to compute the printer inverse map starting from the printer model [71–88]. They include Shepard's algorithm [89], the moving matrix (MM) algorithm [90], iteratively clustered interpolation (ICI) algorithm [91], 3-D root finding algorithm [92], conjugate gradient algorithm [93]. Various other methods described in Refs. [71–88] and [94] can be used in the inversion. Comparison of the inversion performance of some of these methods is described in Section 6.4.6. Although round trip accuracy (Section 7.7) is a good quantitative metric for comparison between various known inversion methods, execution time, computational complexity (i.e., number of complex operations such as multiplications, matrix inversions performed, etc.) also play an important role in the choice of the inversion approach. For example, (1) Shepard's method is easy to implement but relatively less accurate than others, (2) the MM method has accuracy better than Shepard's but because of complex operations involved in it, it has high execution time and the method often fails for nodes near the boundary, and (3) the ICI algorithm provides by far the best results in terms of accuracy, execution time, and complexity. Although ICI algorithm is iterative, its use for a nondeterministic system is difficult. Hence a more general control-based approach is developed and is described in the next section. It is more suitable for accurately inverting color maps with three or more separations by either working on the printer forward map or by directly iterating on the printer with color sensors.

7.5.2.2 Control-Based Inversion

We describe the control-based inversion for a linear three-input three-output forward map or a printer. In this approach, a feedback control algorithm can be used at each node to accurately convert the in-gamut $L*a*b*$ nodes to CMY color space. Since GCR is already embedded inside the augmented printer model (CMY to $L*a*b*$), we do not have to constrain the control algorithm when compared to the control approaches applied directly on a nonunique, $CMYK$ to $L*a*b*$ printer. The algorithm iterates on the printer model (or the printer) for each in-gamut node by using the $L*a*b*$ node as target. A block diagram shown in Figure 7.29 illustrates how inversion operations are executed at a given $L*a*b*$ node.

We show the use of linear multiple-input multiple-output (MIMO) state-feedback controllers [95] to update the final CMY values that will further refine the inversion errors. The problem lies in the design of stable controllers for each of the node colors, which is done by representing printer characterization data in a suitable form such as the state variable form.

Considering the printer input–output characteristic as linear (which is generally true at the nominal CMY values, see Figure 7.30), we first develop a state space model for the CMY to $L*a*b*$ printer model (or the printer). After that, we design the feedback controller for this system. In Section 8.6, a derivation of the linear state space model is shown at the nominal CMY values for a node color whose target $L*a*b*$ values are given. This approach requires the use of nominal CMY values for each node because of the use of linear state space form. However, these values can come from any of the less accurate inversion algorithms described above (e.g., output of the ICI algorithm). Thus, for an individual node color, the system with the integrator in Figure 7.29 can be expressed in state space form as

$$x(k + 1) = Ax(k) + Bu(k) \qquad (7.87)$$

where

$x(k)$ represents the $L*a*b*$ values from the printer model obtained at iteration k
A is an identity matrix

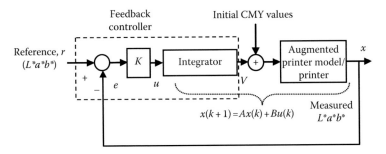

FIGURE 7.29 Closed-loop control algorithm with a gain matrix and an integrator as the controller.

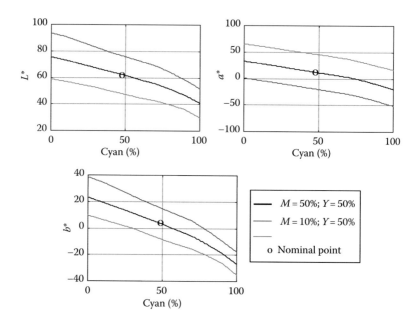

FIGURE 7.30 Diagram representing $L^*a^*b^*$ values when C separation is varied at constant M and Y.

 B is the Jacobian matrix computed around the initial CMY value (or measured using techniques outlined in Ref. [95])

 $u(k)$ is the control law applied to the input of the printer

The Jacobian matrix B is different for each node color and computed as follows:

$$B = \begin{bmatrix} \dfrac{\partial L^*}{\partial C} & \dfrac{\partial L^*}{\partial M} & \dfrac{\partial L^*}{\partial Y} \\ \dfrac{\partial a^*}{\partial C} & \dfrac{\partial a^*}{\partial M} & \dfrac{\partial a^*}{\partial Y} \\ \dfrac{\partial b^*}{\partial C} & \dfrac{\partial b^*}{\partial M} & \dfrac{\partial b^*}{\partial Y} \end{bmatrix} \tag{7.88}$$

The control law is designed using MIMO state-feedback controllers (based on linear quadratic regulator [LQR] or pole-placement as shown in Chapter 6 or any other multivariable digital control techniques [96–100]). Thus, $u(k) = -Ke(k)$, where $e(k)$ is the error between the target $L^*a^*b^*$ and the model $L^*a^*b^*$ at iteration k. In pole-placement design (Chapter 6), the gain matrix K is derived based on the pole values specified such that the closed-loop system model shown in Figure 7.29 is stable. This is achieved by assigning pole values within the range $[0,1)$ in the Z-domain. For the loop to be stable, the eigenvalues of the closed-loop system should remain inside the unit circle of the Z-domain even when the printer drifts. Hence, the pole values

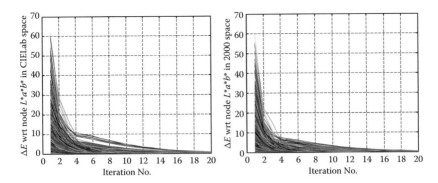

FIGURE 7.31 Convergence plot for 5192 in-gamut nodes using nominal values calculated from MM algorithm.

TABLE 7.9

Accuracy Statistics for In-Gamut Nodes before the Start and End of Iterations

	ΔE_{2000}			ΔE_{ab}^*		
	Mean	**95%**	**Max**	**Mean**	**95%**	**Max**
Start of iteration	1.86	6.37	60.18	2.67	9.18	62.41
End of iteration	0.00	0.00	0.58	0.01	0.00	0.77

have to be carefully tuned to avoid oscillations during iteration (or instability) and potential actuator saturation for near-boundary colors.

Figure 7.31 shows the convergence plot for all the in-gamut nodes when the nominal *CMY* values are calculated using the MM algorithm mentioned in Section 6.3.2. Nominal *CMY* values can also be calculated using other algorithms (e.g., ICI from Section 6.4.3). This plot clearly indicates that the convergence error (shown in Table 7.9 for both ΔE_{ab}^* and ΔE_{2000}) is near zero for all the nodes at the end of the iterations, which is a significant improvement when compared with other inversion algorithms used in the industry.

During iterations, the ΔE errors shown in Figure 7.31 for node colors at the gamut boundary, may go higher in a more recent iteration than the previous one, which can result in the choice of wrong inverse (i.e., *CMY*). That is, node colors near the boundary may reach saturation (i.e., their *CMY* values may go to 0 or 255). This can lead to limit cycles during control iterations, meaning ΔE descent with respect to iterations can oscillate. Since iterations contain the history of oscillations, a unique best actuator algorithm shown in the diagrams of Figure 7.32 can be used to select proper *CMY* values that give the best inverse for the node colors at the gamut boundary. Assume that there are N_0 colors that will be adjusted by the algorithm, which will iterate N_t times and all iteration history is stored. The best *CMY* selection algorithm starts from the second block in Figure 7.32a where the algorithm picks the

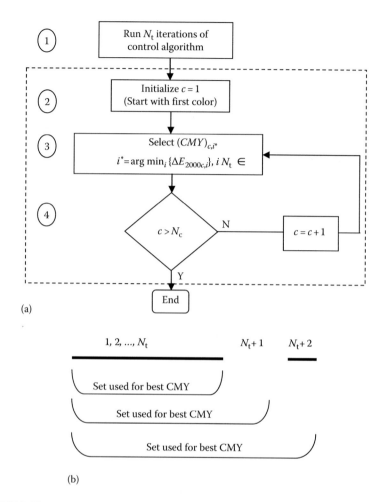

(a)

(b)

FIGURE 7.32 (a) Best *CMY* selection flow diagram for node colors shown for N_t iterations and N_c colors. (b) Best CMY selection progression for control-based algorithm shown for N_t iterations and beyond.

first color for the analysis. The third block seeks *CMY* values for each color c that generated the minimum ΔE value across all N_t iterations. The selected *CMY* values are stored. The fourth block checks whether all colors considered in the algorithm have already been analyzed. If so, the selection algorithm stops and the best *CMY* values selected will be used for the respective node colors. Otherwise, the algorithm continues the process for the next color in the list. Thus the best actuator algorithm selects the best *CMY* that leads to minimum ΔE convergence by assessing ΔE results across all the iterations. An example is shown in Problem 7.7.

Linear state-feedback design also has other shortcomings such as not reaching zero steady state error for colors near the boundary. These, and many other control system-related shortcomings are overcome by using a multiplicity of gain matrices per node color as opposed to single gain matrix.

7.5.2.3 Inverse by Iterating Directly on the Printer

If we can directly iterate on the printer for any node color using control algorithms, the historical methodology of building a printer model and then inverting that model for developing multidimensional LUTs may not be necessary. The iteration process contains the following steps.

   a. Find nominal *CMY* values for the node colors of interest using ICI or other inversion algorithms on a coarse printer model. A coarse printer model is a less accurate representation of the printer. The corresponding *CMYK* values are calculated for these nominal *CMY* values, using the *CMY* to *CMYK* GCR function.

   b. Create a test image containing the color patches of the determined *CMYK* values and print. The $L^*a^*b^*$ values of the color patches on the printed test image are measured.

   c. Using the controller shown in Figure 7.29 (e.g., gain matrix and the integrator), process the error between the measured $L^*a^*b^*$ values and the target $L^*a^*b^*$ values. The gain matrix can be found using pole-placement or LQR methods *a priori* at the nominal *CMY* values or using any of the control approaches shown in Refs. [96–100].

   d. Continue with steps a through c for a few iteration cycles.

The best actuator selection method described in the previous subsection may be required for node colors near the gamut boundary. Improvement of accuracy in performance for boundary colors can also be achieved by scheduling multiplicity of gain matrices for every node color using MIMO model-predictive-control methods [100].

Example 7.8

Generate a GCR constrained ICC destination profile with a 3-to-3 control-based inversion using the GCR function of Figure 7.25 for a test printer. Evaluate the quantitative round trip accuracy and gamut volume. Compare the effective volume with respect to the gamut volume obtained from the characterization data. Use a suitable gamut-mapping strategy described in Section 7.6. Evaluate the visual response of the ICC profile for a test image.

SOLUTION

We constructed a 33^3 ICC profile using the steps shown in Section 7.8 and the GCR of Figure 7.25. The steps involved in evaluating the ICC profiles are covered in detail in Section 7.7.

   The *CMYK* response of the ICC profile for *RGB* neutral sweep is shown in Figure 7.33. Clearly, most of the neutrals are produced using *CMY* separations and no black. This is good for high quality rendering. The rapid rise of magenta and yellow separations between $R=G=B=0$ to about 50 digital counts is caused by mapping out-of-gamut colors to the printer gamut. This could cause image quality defects, such as contours for dark colors if left unattended. Also a sharp transition in the cyan separation for dark colors can cause problems. In Figure 7.34, an L^* linearity plot is shown. Clearly, we see problems with dark colors (once again) that is a problem at

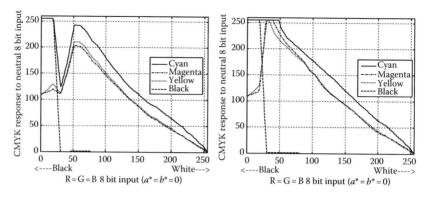

FIGURE 7.33 GCR response to neutral *RGB* sweep (left: with MM inversion; right: control-based inversion).

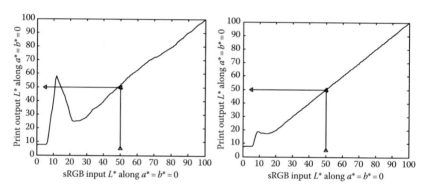

FIGURE 7.34 Input L^* vs. output L^* response (left: with MM inversion; right: control-based inversion).

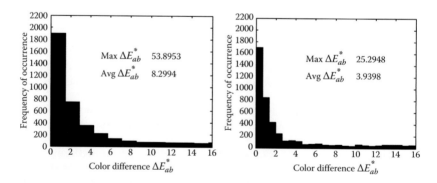

FIGURE 7.35 Histogram of roundtrip accuracy shown in terms of ΔE_{ab}^* (left: with MM inversion; right: control-based inversion).

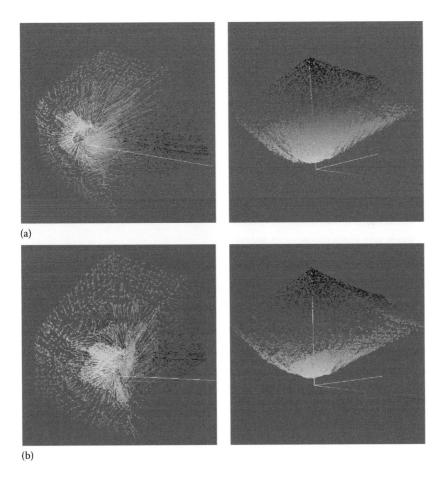

(a)

(b)

FIGURE 7.36 (a) Vector plots between colors produced with A2B1 (forward) and colors produced with combined maps (B2A1 (inverse) and A2B1 (forward)) with MM-based inversion (left: view from the bottom, right: view from the side). (b) Vector plots between colors produced with A2B1 (forward) and colors produced with combined maps (B2A1 (inverse) and A2B1 (forward)) with control-based inversion (left: view from the bottom, right: view from the side).

the bottom of the gamut for pixels outside of it. A ΔE_{ab}^* histogram for in-gamut round trip colors is shown in Figure 7.35. Most round trip colors are not on the grid. The profile inaccuracy is largely due to the inaccuracy caused by interpolation. Although nodes have converged to nearly zero ΔE, most inaccuracies are near the region below the *CMY* gamut (see Figure 7.36). These errors can be associated to GCR methods used in this process. Also, the effective gamut utilization with this kind of GCR was found to be around 84%, which means a loss of 16% printable volume. This is largely associated with the multidimensional profile LUT with the 3-to-3 inversion and GCR. Figures 7.37 through 7.40 show the visual effects of inversion algorithms on various images as compared to the original image. Clearly, all other things being equal, objectionable color contents (shifts, contours, blocking, etc.) can be associated to inversion algorithms.

FIGURE 7.37 (See color insert following page 428.) Lena image rendered with ICC profiles from Example 7.8 (top: original; bottom: simulated with MM inversion; middle: simulated with control-based inversion).

7.5.3 Brief Review of GCR Methods

The GCR methods available in the literature can be divided into two categories: (a) transformations with fixed GCRs and (b) transformations with flexible GCRs. Fixed GCRs are most commonly used in multidimensional LUTs. Flexible GCRs allow the user to have control over the gray component across all levels of GCRs.

In Ref. [71], several methods for determining the black (K) component are reviewed. One method is black addition in which black (K) is calculated as a function of a scaled inverse of L^*. In another method, black (K) is calculated as a function of the minimum value of the other color components, such as C, M, and Y for the *CMY* color space. In a third method, a three-input four-output transform, subject to certain constraints, is used to calculate the black component. The constraints placed on the transform include a requirement for the sum of the color component values at a node to be less than a threshold. For example, in

FIGURE 7.38 (See color insert following page 428.) Hairs image rendered with ICC profiles from Example 7.8 (left: original; middle: simulated with MM inversion; right: simulated with control-based inversion).

FIGURE 7.39 (See color insert following page 428.) Image showing the simulation of Daffodil plant (left: original image; top: MM inversion; bottom: control-based inversion).

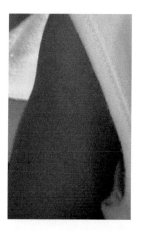 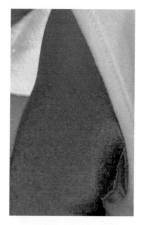 

FIGURE 7.40 (See color insert following page 428.)   Image showing the rendering of color sweeps with ICC profile from Example 7.8 (left: original image; middle: MM inversion; right: control-based inversion).

CMYK color space, $C + M + Y + K$ would be constrained to be less than a threshold. A second constraint is to maintain *K* between certain minimum and maximum values. In Refs. [82,83], another UCR/GCR strategy is proposed in which the optimization is done to reduce moiré. In this method, the UCR/GCR strategy is used to characterize the moiré as a function of the color components and to select the optimized output color components when the moiré function is minimized. In another method, Ref. [87], a flexible method for estimating the black component comprises (1) determining maximum black component, (2) adjusting the black component amounts based on chroma, and (3) determining the other color components. But this method has a disadvantage of not producing sufficiently optimized colors for the entire color gamut.

Various other GCR control strategies are shown in [84–86]. In the Pareto-optimal approach [101,102] ICC-based profile LUTs are built to convert images to *CMYK*, and consistency (i.e., a colorimetric accuracy match) is introduced through carefully selected optimization parameters that attempt to use unique *CMYK* solutions during the inversion process and yet preserve arbitrary GCR flexibility within a printer's reproducible limits. Although the Pareto-optimal approach, one of the more flexible GCR methods, provides good control of multiple GCR schemes with full *CMYK* gamut utilization and a controlled substitution of black into a single methodology, it can lead to loss of accuracy during inversion. Other inversion methods are shown in Ref. [103] to generate ICC profile, and in Ref. [104] generate LUTs to minimize metameric effects.

7.5.4 GCR Constrained 4-to-3 Inverse

The *CMY* gamut (referred to as the gamut with *K* not equal to zero and with CMY to CMYK GCR) is generally limited by the type of GCR function used as compared to the full *CMYK* gamut. Due to this limitation, the designer should be cognizant of three regions in the $L^*a^*b^*$ color space for color reproduction: (1) the out-of-gamut region, (2) the *CMY* gamut, and (3) the region outside the *CMY* gamut but within the *CMYK* gamut. When the available gamut is limited by the *CMY* gamut, the out-of-gamut colors have to be mapped to the boundary of the *CMY* gamut as opposed to the *CMYK* gamut. This makes the interactions between gamut mapping and the GCRs even more complex. Hence a preferred approach would be to completely eliminate the loss of gamut due to GCRs, that is, to use the same gamut volume for minimum or maximum black. A 4-to-3 control-based, constrained inversion generated by (a) iterating on the printer model or (b) by directly iterating on the printer can offer full *CMYK* gamut for any GCR curve and high accuracy inversion to node colors inside the gamut.

7.5.4.1 A 4-to-3 Control-Based Inversion

The 4-to-3 control-based inversion is applicable to a four-color printer. As in previous method (Figure 7.29) the node colors in the device-independent color space, $L^*a^*b^*$ are generated by applying suitable transformations on a uniformly/nonuniformly sampled *RGB* color grid (e.g., *genRGB*). Figure 7.41 shows the block diagram of the 4-to-3 control-based inversion, which is clearly very similar to Figure 7.29. The key differences between 3-to-3 and 4-to-3 inversion are

a. "Initial (or nominal) *CMYK* values" contain the GCR strategy for in-gamut node colors, whereas, in the 3-to-3 inversion, a GCR strategy is defined in the *CMY* to *CMYK* transformation.
b. No augmented printer or printer model is used. This allows the use of full *CMYK* gamut when creating the inverse.
c. Printer/printer model Jacobian (Equation 7.89) and gain matrices contain the sensitivity of output color to black separation. In addition vectors $\{x, u, V\}$ and matrices $\{A, B, C\}$ in the state variable model and the feedback contain terms associated with black separation.

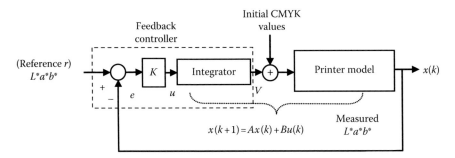

FIGURE 7.41 Closed-loop control algorithm with a gain matrix and the integrator as controller for a 4-to-3 control-based inversion.

$$B = \begin{bmatrix} \dfrac{\partial L^*}{\partial C} & \dfrac{\partial L^*}{\partial M} & \dfrac{\partial L^*}{\partial Y} & \dfrac{\partial L^*}{\partial K} \\[2mm] \dfrac{\partial a^*}{\partial C} & \dfrac{\partial a^*}{\partial M} & \dfrac{\partial a^*}{\partial Y} & \dfrac{\partial a^*}{\partial K} \\[2mm] \dfrac{\partial b^*}{\partial C} & \dfrac{\partial b^*}{\partial M} & \dfrac{\partial b^*}{\partial Y} & \dfrac{\partial b^*}{\partial K} \end{bmatrix} \tag{7.89}$$

A final multidimensional profile LUT (i.e., ICC profile) is generated based on the final $L^*a^*b^*$ to $CMYK$ LUT, which is the sum of the initial (or nominal) LUT containing the GCR and the newly found correction $\Delta CMYK$ (vector V in Figure 7.41) through iterations on the $CMYK$ printer or the $CMYK$ printer model.

The inversion approach uses a 3-to-4 control-based algorithm with a MIMO gain matrix. The Jacobian matrix at the nominal $CMYK$ values for each node color is used to compute the gain matrix. The pole-placement algorithm, place() [105,106] (see Section 5.2.3) or LQR methods (see Section 5.3.2) are candidate algorithms used for computing the gain matrix. Furthermore, best actuator algorithm (similar to the one described in Figure 7.32) would be required for node colors near the boundary to select the best $CMYK$ values during the iteration run. Gain scheduling algorithms based on model-predictive control technology [100] are other methods suitable for use in the control algorithm.

7.5.4.2   K-Restricted GCR

The function used to generate the black (K) component can be a parametric multidimensional function known as the K-function or K-restricted GCR function. The K-function produces the first (initial) estimated values for the black component based on the $L^*a^*b^*$ values of the node colors. The initial estimated values for black (K) are then used to produce values for other separations, C, M, and Y, in the initial $CMYK$ LUT of Figure 7.42.

The multidimensional K-function may be defined by the following equation:

$$K = \frac{1}{U^{\eta} e^{-\alpha \left(a^{*2} + b^{*2}\right)/L^{*2}} + 1} - \frac{1}{2} U^{\eta} \tag{7.90}$$

where

$$U = 2 \left(\frac{1}{e^{-\beta \left(\frac{L^*}{L_0 - L^*}\right)^2} + 1} - 0.5 \right) \tag{7.91}$$

and L_0, η, β, and α are predetermined values, selected based on the color gamut. Figure 7.42 shows two graphs of the function generated by Equations 7.90 and 7.91. The graph shown on the left side is the L^* response (vertical axis) with respect to K values (horizontal axis). The graph on the right contains K values (vertical axis) as a

FIGURE 7.42 A graph of K-restricted multidimensional GCR function generated with $\alpha = 0.01$, $\beta = 1.5$, and $\eta = 0.5$. (a) L^* with respect to K digital count (left), (b) color gamut (center), and (c) K digital count with respect to a^* and b^*.

function of a^* or b^* (horizontal axis) and using $a^* = 0$ (against b^* as the horizontal axis) or $b^* = 0$ (against a^* as the horizontal axis) for different values of L^*. The color gamut with L^* (vertical axis) and a^* (horizontal axis) is shown in the center sandwiched between these graphs. These curves show that the K values are weakly dependent on either the a^* or the b^* values. L_0, η, β, and α are parameters of the multidimensional K-function which satisfy following properties:

1. $K \rightarrow 1$ when $L^* \rightarrow 0$ so that maximum value of K is used at the dark end of the color space.
2. The function is symmetrical with respect to a^* and b^*, and K increases when the color is away from the neutral zone.
3. $K \rightarrow 0$ when $L^* \rightarrow L_0$. Since L_0 is selected to be around 100 (top of the gamut), only a small amount of black will be used near the white. Parameter L_0 affects the middle part of the curve near the neutral region (shown by double arrow), which is the place where most colors require adjustment to obtain pleasing appearance on the paper. Figure 7.43 shows K as a function of L^* for $a^* = 0$ and $b^* = 0$ for three different values of L_0 with $\alpha = 0.01$, $\beta = 1.5$, and $\eta = 0.5$.

For a digital production printer gamut, typical values used for α, β, and η are $\alpha = 0.01$, $\beta = 1.5$, and $\eta = 0.5$. A large value of α gives a strong dependence of K on a^* and b^*, while a small value of α means that K is mostly determined by luminosity. The parameter η determines the shape of the K-function curve at small L^* values (see left side of the graph in Figure 7.42). Larger value of η means the

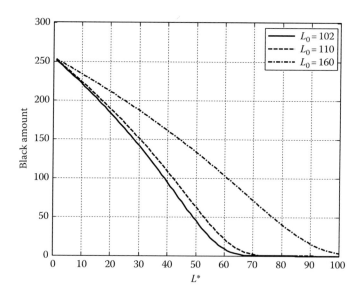

FIGURE 7.43   A neutral response of K-restricted GCR functions with $\alpha = 0.01$, $\beta = 1.5$, and $\eta = 0.5$ shown for three different values of L_0.

K-function curve is more rounded at small $L*$. The parameter β determines the shape of the K-function in the middle range of $L*$. To set the parameters of Equations 7.90 and 7.91, the information at the boundary of the color gamut can be used since there is unique correspondence between $L*a*b*$ to CMYK at that location.

Because there are four parameters $(\alpha, \beta, \eta, L_0)$ to be optimized and the search space is huge, the simultaneous perturbation stochastic approximation (SPSA) method [107] can be used to determine a local region where the parameters could be trained using boundary values.

Complete procedure that is used to create the K-restricted (GCR constrained) $L*a*b*$ to CMYK LUT is shown below. The following steps are described for one in-gamut node color:

a. Create a uniformly/nonuniformly sampled high density CMYK to $L*a*b*$ LUT. This data can be obtained from the printer model (e.g., a 16^4 CMYK to $L*a*b*$ LUT). These data can also be obtained by printing a set of patches and measuring $L*a*b*$ values with a sensor.

b. Calculate the K value for the node color ($L*a*b*$ data) using the K-function (Equations 7.90 and 7.91).

c. Find all ΔE_{2000} values between node $L*a*b*$ value and the $L*a*b*$ values from the characterization LUT of step (a). Rank order the colors with respect to ΔE_{2000} values and pick all colors with $\Delta E_{2000} < 6$ (or any reasonable threshold which is set depending on the desired accuracy). There could be multiple CMYK values for any given node color. Record their corresponding CMYK values.

d. Compare K values between steps (b) and (c) for each of the colors picked in step (c). Pick few colors that are closest to K value in the K-function.

e. Compute the Jacobian and gain matrices for each of the colors picked in step (d). Run a 4-to-3 iterative control algorithm of Figure 7.41 on a printer model. If experimental characterization input–output data is available, then use numerical interpolation techniques described in Chapter 6 in place of the printer model.

f. Compare the ΔE_{2000} numbers and select CMYK values with the smallest ΔE_{2000} numbers. This gives the K-restricted GCR constrained CMYK value for the node color.

g. Repeat steps (b) through (f) until the CMYK values are found for all the in-gamut node colors.

h. Use the CMYK values from step (g) to complete the ICC profile. If the ICC profile has undesirable kinks in CMYK response between neighboring nodes, then another preconditioning step is required, which would involve applying multidimensional filtering techniques [108] (Chapter 6) to soften the kinks. Filtering may make the in-gamut CMYK values inaccurate, but these inaccuracies can be minimized by running the steps described in Section 7.5.5 using the filtered K-restricted profile as the source.

Example 7.9 uses these steps to create a K-constrained ICC profile.

Example 7.9

Construct a K-restricted GCR for a uniformly sampled printer input–output data using the analytic functions of Equations 7.90 and 7.91 and a 4-to-3 control-based inversion approach. Show the advantages with respect to gamut and image quality (IQ).

Solution

Figure 7.44 shows the CMYK response curves (also called corner plots) for a first inverted LUT produced with a 4-to-3, K-constrained control-based algorithm. These plots are drawn from white ($C=M=Y=K=0$) to the dark black ($C=0$, $M=0$, $Y=0$, $K=255$), dark red ($C=0$, $M=Y=K=255$), dark green ($C=255$, $M=0$, $Y=K=255$) and dark blue ($C=M=255$, $Y=0$, $K=255$) corners of the gamut (see Section 7.7.2 to understand more about the corner plots). Values 0–100 on the x-axis represent points starting from white with 100 on the corner. The x-axis is extended to reach out-of-gamut colors until it touches the sides of the cube formed by the device-independent space. Solid curves represent CMYK response after one post filtering step. Clearly, the CMYK response shown by dashed curves for both in-gamut colors and out-of-gamut colors is not smooth. This is because the starting CMYK values obtained in step (c) contain local minima (depending on the resolution of the CMYK grid). This LUT produces contours in some images (see the "neck" region shown in Figure 7.45) due to nonsmooth CMYK formulations caused by the inversion algorithm. Also, the K response shown by dashed dot curve in Figure 7.44 (d,f,h) is calculated from Equations 7.90 and 7.91. The inversion algorithm tries to follow the general trend. Except for a nonsmooth CMYK response, the profile LUT is, in general, in the right direction.

To avoid the formulation jumps (i.e., non-smooth CMYK response), we apply the preconditioning steps mentioned in step (h) above and generate the second inverted LUT. We filter the LUT by applying the multidimensional smoothing algorithm and refine the accuracy by applying a 4-to-3 control-based inversion on the printer model using post-filtered CMYK values as the starting points for the inversion. After performing this step a few times, we obtain a final inverted LUT to produce smooth CMYK response (see response curves in Figure 7.46 shown just before and after the sixth preconditioning step). In Figure 7.46, the differences between solid and dashed curves are negligible as compared with similar figures in Figure 7.44. Solid curves represent the responses before final filtering. The dashed curves are smooth. They are obtained after the final filtering step. Figures 7.47 and 7.48 show the image quality improvement due to preconditioning steps. Table 7.10 shows the round trip accuracy at the end of each preconditioning step. It is to be noted that each precondition step includes filtering and 4-to-3 control-based inversion (Figure 7.41) carried out using post-filtered CMYK values as the initial CMYK LUT in the inversion process. Filtering is applied to all nodes (i.e., in-gamut and out-of-gamut nodes) in the multidimensional profile. Without the control-based inversion on the post-filtered CMYK values, there is significant loss in the round trip accuracy. In this example, with limited node size (33^3), we see an effective gamut utilization of greater than 94%.

The CMYK formulations between adjacent nodes have to be smooth. As shown in this example, multidimensional filtering (Section 6.6) combined with additional iterative control procedure on filtered data may be necessary to soften the kinks in CMYK formulations. Alternatively, step (c) can be started with previously inverted LUTs using methods such as the MM.

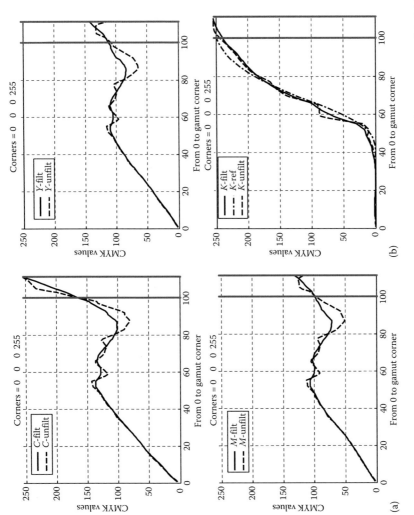

FIGURE 7.44 Inversion response curves for *K*-restricted GCR (without preconditioning steps). (a) C and M response toward dark black corner ('0' on the *x*-axis corresponds to white (C=M=Y=K=0)). (b) Y and K response toward black corner ('0' on the *x*-axis corresponds to white (C=M=Y=K=0)).

(continued)

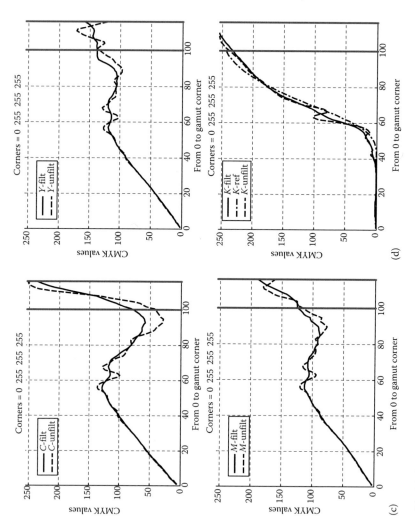

FIGURE 7.44 (continued) (c) C and M response toward dark red corner ('0' on the x-axis corresponds to white (C=M=Y=K=0)). (d) Y and K response toward dark red corner ('0' on the x-axis corresponds to white (C=M=Y=K=0)).

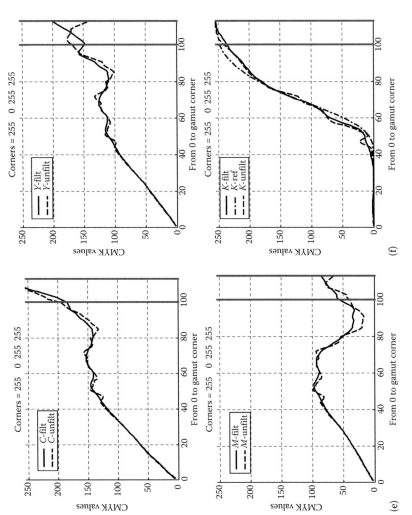

FIGURE 7.44 (continued) (e) C and M response toward dark green corner ('0' on the x-axis corresponds to white (C = M = Y = K = 0)). (f) Y and K response toward dark green corner ('0' on the x-axis corresponds to white (C = M = Y = K = 0)).

(continued)

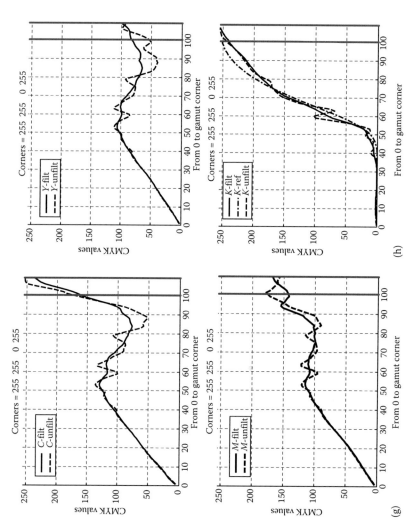

FIGURE 7.44 (continued) (g) C and M response toward dark blue corner ('0' on the x-axis corresponds to white ($C = M = Y = K = 0$)). (h) Y and K response toward dark blue corner ('0' on the x-axis corresponds to white ($C = M = Y = K = 0$)).

FIGURE 7.45 (See color insert following page 428.)   Hair image rendered with K-restricted ICC profile whose inversion response is shown in Figure 7.44. (left: original; middle: simulated using ICC profile from Example 7.8 with 3-to-3 control-based inversion; right: simulated with K-restricted 4-to-3 control-based inversion without preconditioning [step (h)]).

The above procedure may take more computing to complete all the steps, especially due to large time spent in step (c). Hence, a faster way to build the profiles is to run the procedure in two groups. The first group can be used to build the initial *CMYK* LUT and the gain matrix in the factory on a nominal print engine for all the node colors (includes in-gamut and out-of-gamut node colors). The second group can be used for running the iterative control algorithm of Figure 7.41 on a drifted print engine in the field. These initial *CMYK* and gain matrices can be carried as source/starting LUTs for the control algorithm. For this kind of architecture, it is important to note that the initial *CMYK* and gain matrix values need to be found for colors that the printer cannot render in its nominal state but are required to cover the drifted printer gamut. This can be done by extrapolating the printer model for the nominal state or through extrapolation of the initial *CMYK* values for the in-gamut node colors produced on a printer under nominal state.

Apart from the speed advantage, there are other benefits in carrying the GCR-constrained initial *CMYK* and gain matrix values to the field. We can incorporate different fixed GCR/UCR approaches as starting LUTs as future upgrades since each of these LUTs can be fine tuned intelligently using carefully designed experiments in the factory. For example, a maximum/medium/minimum black solution can be easily created by adjusting the parameters of the GCR equations and still achieve full *CMYK* gamut utilization for images by maximizing the accuracy for all the in-gamut node colors. These and other extensions to multiple GCR/UCR combinations can be comprehended within the single initial LUT structure.

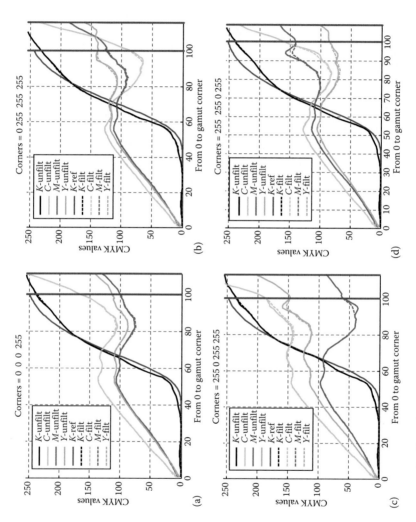

FIGURE 7.46 Response curves for K-restricted GCR (after preconditioning iterations) to different gamut corners from white ($C = M = Y = K = 0)$).

FIGURE 7.47 (See color insert following page 428.) Image showing improvements to contours with preconditioning steps for *K*-restricted GCR (top left: original image; top right: processed image with ICC profile from *K*-restricted GCR after one step [contours are visible]; bottom: same after preconditioning steps [contours are removed]).

7.5.4.3 Tricolor GCR

In some printers, the overall color printer gamut can be represented as a composite of the gamut subclasses, wherein each gamut subclass is comprised of a subset of printer color separations [109]. Nodes are assigned to one of the gamut subclasses for efficiently calculating the *CMYK* color separation values using one of the inversion techniques (MM, ICI, or control-based). A three-gamut class method classifies the colors into $CYK \rightarrow L^*a^*b^*$, $CMK \rightarrow L^*a^*b^*$, and $MYK \rightarrow L^*a^*b^*$ sub-gamuts that cover the whole printer gamut. This approach reduces the dimensionality of the four-color process to three-color groups. Use of partitioned gamuts provides the following advantages:

1. Improved node color accuracy with one of the separations always held to zero.
2. Improved toner usage for high area coverage printing by identifying the most toner efficient *CMYK* values for in-gamut colors.

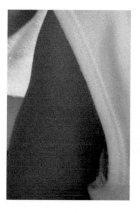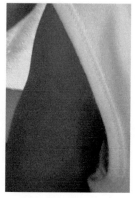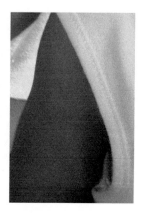

FIGURE 7.48 (See color insert following page 428.) Image showing the rendering of color sweeps (left: original image; middle: processed image with ICC profile from K-restricted GCR after one step [contours are visible]; right: same after preconditioning steps [contours are removed]).

TABLE 7.10

Round Trip Error with Respect to Number of Iterations Used in the Preconditioning Method with Filtering Step Included in the K-Restricted GCR Algorithm

Preconditioning Steps	ΔE_{2000}			ΔE_{ab}^*		
	Mean	95%	Max	Mean	95%	Max
None	0.75	2.47	12.67	1.35	3.54	23.82
1	1.08	3.82	6.18	1.90	5.35	27.63
2	1.13	4.06	6.52	1.96	5.72	29.20
3	1.16	4.29	6.79	2.01	5.96	29.54
4	1.18	4.39	6.98	2.04	6.06	29.68
5	1.19	4.49	7.23	2.06	6.14	29.77
6	1.21	4.53	7.31	2.08	6.21	29.82

3. Removes the nonuniqueness (degeneracy) problem due to the use of a 3-to-3 inversion by selecting appropriate sub-gamuts.
4. Provides more room for the controller on the actuators when the control-based inversion is used.
5. Allows for a unique solution for each pixel once a gamut class has been chosen. Hence, in-gamut color sweeps can be rendered without contours.

Since one of the toners, C, M, or Y, is always equal to zero for any given color, the toner usage is usually smaller than that of other GCRs.

Figure 7.49 shows the CMK, MYK, and CYK gamuts along the chromatic axes when viewed from the top of the gamut. In a typical four-color printer, there exists

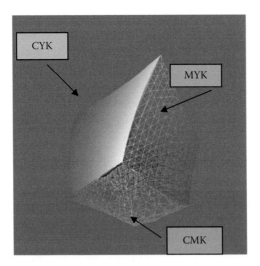

FIGURE 7.49   Top view with *CMK*, *MYK*, and *CYK* gamuts.

yet another zero black tricolor (CMY) gamut class. It overlaps with other classes that include the *K* separation. While using this GCR, the node colors have to be carefully assigned to one of the CMK, CYK, or MYK gamuts and the overlapping CMY gamut should be avoided. Otherwise, it can lead to severe formulation jumps and discontinuities unless some form of iterative blending is introduced. The overlap region with the CMK, CYK, and MYK gamuts are generally nonexistent except for numerical precision. Whenever there is an overlap with a CMY gamut class, this strategy has two components: (1) minimum black strategy and (2) maximum black strategy. For maximum black strategy, use black everywhere (i.e., no CMY gamut) including neutrals (Figure 7.50). Minimum black strategy would not work for images due to discontinuities or abrupt changes in separations, as seen in the *CMYK* response for the colors on the neutral axis from $C = M = Y = K = 0$ to $K = 255$ corner $(C = M = Y = 0$ and $K = 255)$ (Figure 7.51). Whereas, for rendering spot colors, a minimum black strategy with CMY gamut class can be used without worrying about discontinuities.

A smooth GCR can be obtained for a minimum black strategy by using the steps of Example 7.9 (multidimensional filtering and control-based iterations multiple times) of a four tricolor gamut class GCR [109,110] (i.e., CMY, CMK, MYK, and CYK). The blended GCR response is shown in Figure 7.52.

7.5.5   GCR Retrieval from Historical Profiles

In this workflow, particularly in the prepress and/or in digital front end (DFEs), the documents are designed using various layout tools and their color appearance is fine tuned by typically proofing on a workgroup digital printer or the press itself. When the prints are made, it is expected that the appearance on the destination printer follows the proof. If it does not follow the proof, then adjustments are made in many

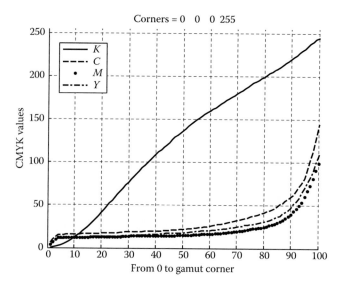

FIGURE 7.50 Example of a *CMYK* response to colors on neutral axis for three gamut class GCR ('0' on the *x*-axis corresponds to white (C = M = Y = K = 0)).

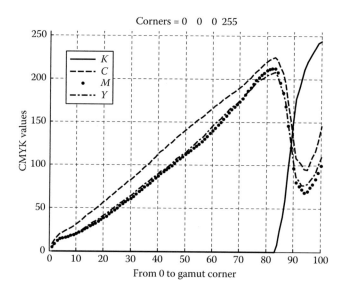

FIGURE 7.51 *CMYK* response to colors on neutral axis for four gamut class GCR with *CMY* gamut as the fourth gamut class ('0' on the *x*-axis corresponds to white (C = M = Y = K = 0)).

places, including the color management profile LUTs. One of the key adjustments is the selection of GCR methods. The GCR methods fine-tune the use of *CMYK* separations for improving the appearance. In particular, some of the key colors (e.g., black in flesh tones and sky tones) need less black. Sometimes, a maximum

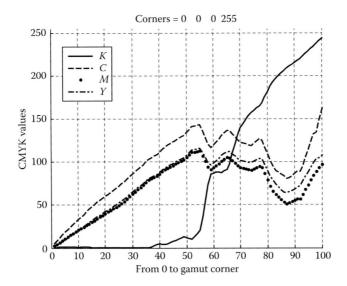

FIGURE 7.52 *CMYK* response to colors on neutral axis for four gamut class GCR after two iterations of preconditioning steps ('0' on the x-axis corresponds to white (C = M = Y = K = 0)).

K GCR is preferred over either a medium K GCR, or a low K GCR, etc. These adjustments are stored as standard profiles (e.g., ICC profiles). When a print job has already been produced, and a future reprint may be expected, then the color in the future reprint has to be retrieved from the original electronic documents with as much original GCR design as possible while making sure that the embedded or the associated graphics are intact.

Below we show a method to reproduce images by retrieving GCRs from different sources. Direct use of a GCR-embedded profile as an initial *CMYK* LUT may produce contours and other undesirable image artifacts since the profile LUT may not represent the inverse of the printer in which the GCRs have to be retrieved. Hence, in this section we show a preconditioning/processing method that is applied to a profile. Using this approach, the GCRs can be reproduced with reasonable precision as compared with the desired GCRs, resulting in expected image quality appearance in documents without contours or other undesirable image artifacts, which otherwise would have appeared.

The preconditioning method includes the following steps:

1. Extract the *CMYK* values for all the node colors (L*a*b* values) from the GCR contained input profile.
2. Run the 4-to-3 control iterations for each node color using the appropriate printer model or printer using the GCR contained in the input profile. This step involves using an appropriate controller in the loop (as in Figure 7.41) based on the Jacobian matrix extracted from the printer model at the *CMYK* values of interest. After convergence (i.e., when the errors are near zero), record new *CMYK* values corresponding to each node colors.

3. Filter the *CMYK* values generated in step 2 using multidimensional filtering algorithms [108], if required.
4. Rerun steps 2 and 3, if required, in a computer program. After convergence, store new *CMYK* values corresponding to each node color in the starting LUTs for use in the run-time algorithm.

7.5.6 *K*-SUPPRESSION METHODS

Now, during the updating process in the field where the press would operate, due to print engine drifts, this 4-to-3 control approach can inject black toner in regions where it is not needed (e.g., scum dots in neutral flesh tones or excessive black in the flesh tones) even after designing good GCRs with no black in the initial *CMYK* LUT. Figure 7.53 illustrates a quantitative assessment of the problem where scum dots for neutral flesh tones or excessive black in flesh tones, sky tones and other important tone scales can appear dirty/grainy or nonuniform. Such an appearance would be unacceptable for high quality color rendition. This is particularly enhanced at the neutrals because of the high degree of degeneracy, that is, the number of *CMYK* combinations that can produce the same color is high near neutrals but low near the boundary of the gamut. There is only one unique *CMYK* solution on the surface of the gamut.

The gain matrix shown in Figure 7.41 (for each node color) is designed with LQR to minimize a selected quadratic objective function over the iteration length. This will provide suppression of the scum dots.

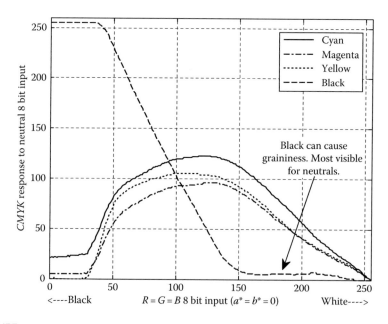

FIGURE 7.53   *CMYK* separation without the use of a *K*-suppression algorithm for neutrals.

Construction of the gain matrix: We use the LQR for designing the gain matrix of the feedback controller that is used to compute the *CMYK* values at the profile nodes for the updated printer model. The linear quadratic controller (Section 5.3.2) minimizes a selected quadratic objective function for a node color over the iteration length, N, which is shown below

$$J = \frac{1}{2} \sum_{k=0}^{N-1} x^{\mathrm{T}}(k)Qx(k) + u^{\mathrm{T}}(k)Ru(k) \qquad (7.92)$$

where $x(k)$ is the state vector containing $L^*a^*b^*$ values and $u(k)$ is the actuator vector for four-color system. The state space formulation of Figure 7.41 is used for obtaining the gain matrix with LQR design.

Since our problem is focused on suppressing black, we set the values in the Q and R matrices as follows:

$$Q = \mathrm{diag}[q_1 \quad q_2 \quad q_3] \qquad (7.93a)$$

$$R = \mathrm{diag}[r_1 \quad r_2 \quad r_3 \quad \alpha] \qquad (7.93b)$$

The Q matrix is 3×3 with very small fixed values for the elements (e.g., $q_1 = q_2 = q_3 = 1 \times 10^{-3}$). Varying the scale value, r, can allow the K to be suppressed (or not). Generally, when the user finds excessive black in neutrals, they can change the values of the parameter r. The R matrix contains α as the weight which is used to suppress black.

The gain matrix equation is obtained by using the procedure described in Chapter 5. We state the final equations below.

Gain matrix:

$$K(k) = R^{-1}B^{\mathrm{T}}P(k+1)\left[I + BR^{-1}B^{\mathrm{T}}P(k+1)\right]^{-1}A \qquad (7.94a)$$

Recursive equation to compute P(k):

$$P(k) = A^{\mathrm{T}}P(k+1)A - A^{\mathrm{T}}P(k+1)BR^{-1}\left(I + B^{\mathrm{T}}P(k+1)BR^{-1}\right)$$
$$\times B^{\mathrm{T}}P(k+1)A + Q \qquad (7.94b)$$

Boundary condition:

$$P(N) = 0 \qquad (7.94c)$$

It turns out that the state space model for each node color has an A matrix which is equal to an identity matrix (i.e., $A = \mathrm{diag}[1 \ 1 \ 1]$). Note that k refers to iteration number in the above equation. Figure 7.54 illustrates the neutral response without the scum dots. Thus, by using the LQR design, we are able to automatically provide low gain for black dots in the regions where black is not desirable. The new gain matrix uses the weight profile to emphasize the removal of black dots via the appropriate values in R matrix.

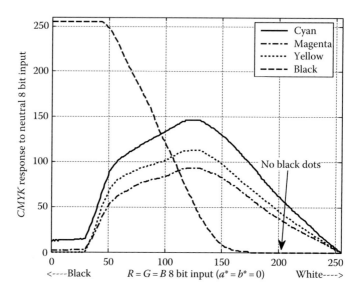

FIGURE 7.54   *CMYK* separation with the use of a *K*-suppression algorithm.

7.6   GAMUT-MAPPING METHODS

The printer gamut is generally limited when compared to the gamut of the source digital image due to the physical limitations of the printer's colorants. Colors present in the source gamut but not in the output gamut are said to be "out-of-gamut" and must be accounted for before the digital image can be printed. That is, areas in the output document where these colors are found cannot be simply left blank in the final printed document. The typical computer monitor has a wider gamut than the typical color printer (see Figure 7.55). For example, bright secondary colors (such as

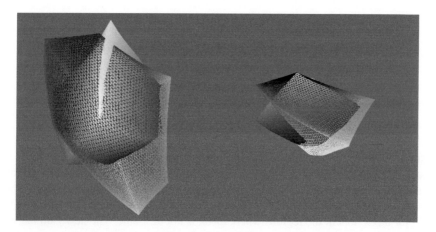

FIGURE 7.55 (See color insert following page 428.)   Comparison between display gamut and a typical color printer gamut (wire: sRGB display gamut; solid: printer gamut; left: top view; right: side view).

lime green) that appear on the display monitor cannot be printed using subtractive process colors.

The out-of-gamut colors are converted to printable colors through a transformation called gamut mapping or gamut compression. In other words, gamut mapping is a means of tone scale modification that attempts to preserve the original appearance of a color image captured with a wider gamut device on a smaller gamut device. Otherwise, these colors cannot be reproduced on the printer and would simply be clipped. It is a key feature used in every color reproduction device.

There are at least two gamuts involved in mapping colors: (a) source/image gamut (i.e., gamut of the image whose colors need mapping) and (b) destination gamut (i.e., the gamut of the device in which the image will be reproduced). Particularly when displays are involved, as is normally the case when soft proofing images, there will be three gamuts involved: (1) display/monitor gamut, (2) source/image gamut, and (3) printer gamut. For instance, producing a mostly saturated blue color as displayed on a monitor as a printed output on paper (using a typical *CMYK* printer) will likely fail. The blue color printed on paper with *CMYK* toners/inks may not be as saturated as the one seen on a monitor. Conversely, the bright cyan of an inkjet printer may not be easily presented on an average computer monitor. In addition to this problem, a gamut-mapping algorithm that produces pleasing results for one image may not work well for another. This is true for pictorial images as well as business graphics and computer-generated images. The color management system can utilize various gamut-mapping methods to achieve the desired results and give experienced users control of the gamut-mapped outcome.

There are several techniques used for gamut mapping [111–133]. There is no unique gamut-mapping method that satisfies all requirements for image reproduction, such as pleasing color, contrast, lightness, chroma, hue, and the like across all imaging devices. Some gamut-mapping algorithms offer feature enhancements in one region of the gamut and others are more favorable elsewhere in the gamut. As such, device designers generally compromise regarding the gamut-mapping functions they wish to employ in their respective color management systems.

Techniques used for dealing with out-of-gamut colors include gamut clipping and gamut compression. In gamut clipping, all out-of-gamut colors are mapped to a color on the gamut "surface" in some way that minimizes the degradation of the resultant output, while in-gamut colors are left unaltered. A common form of clipping involves a ray-based approach, wherein a ray is drawn from a desired out-of-gamut color to a point on a neutral axis. The location or point where the ray penetrates the gamut surface is the gamut-mapped color. Such a strategy is implemented to preserve hue through the gamut-mapping operation. In gamut compression, both in-gamut and the out-of-gamut colors are altered in order to map the entire range of image colors to the printer gamut. For efficient computational processing of images in practical systems, the gamut-mapping operation is incorporated into the nodes of a 3-D LUT.

7.6.1 GAMUT MAPPING WITH RAY-BASED CONTROL MODEL

All gamut-mapping methods need to determine accurately whether the node color is inside or outside the destination gamut. A color that is wrongly considered as outside the gamut may be handled incorrectly by the gamut-mapping algorithm, which can lead to unacceptable artifacts (e.g., contours) and color errors when images are rendered through those multidimensional LUTs. Also, for spot colors, if the out-of-gamut colors are wrongly assigned as "in-gamut", then the consequence of this decision might lead to mapping the colors to a wrong region of the destination gamut. In such a case, the reproduction of the mapped color can be "way off" from the original. Such mistakes can easily be avoided by not mapping the wrongly assigned in-gamut color. We present below a control-based method to decide whether node colors are inside or outside a printable gamut.

Consider a ray (modeled by a line), expressed in $L*a*b*$ space by Equation 7.95 with start and end points given by x_0 and x_c, where x_0 and x_c have coordinates $\{L_0 \quad a_0 \quad b_0\}$ and $\{L_c \quad a_c \quad b_c\}$, respectively:

$$x = x_0 + mi \tag{7.95}$$

where

$$x = \begin{bmatrix} L \\ a \\ b \end{bmatrix}, \quad x_0 = \begin{bmatrix} L_0 \\ a_0 \\ b_0 \end{bmatrix}, \quad m = \frac{1}{N}\begin{bmatrix} L_c - L_0 \\ a_c - a_0 \\ b_c - b_0 \end{bmatrix} \tag{7.96}$$

x_c is the node color whose in–out determination has to be made. x_0 is inside the gamut (e.g., a centroid of the gamut). In Equation 7.95, i is the index that is incremented from 0 to a convenient integer until the node color x_c. When $i=0$, the line will start at $x=x_0$. When $i=N$, the line will end on the node color, where N is equal to the number of points along the line. Figure 7.56 shows schematically a chroma plane and a line starting from point x_0 to the node color x_c. For $i=0$ to N, there will be a total of $N+1$ colors whose values are given by x along the ray between x_0 and x_c (including extreme points). Now, 4-to-3 control iterations, as in Figure 7.41, are carried out using x value as the target for $i=0$ to N. A GCR LUT can

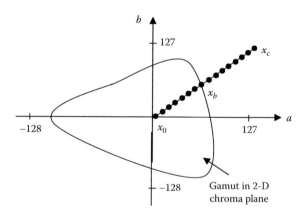

FIGURE 7.56 A line between start and end points.

be used to extract the nominal *CMYK* values for a given *x* value. The Euclidean norm (ΔE_{ab}^*) is computed for each iteration. Tables 7.11 and 7.12 show an example of the ΔE_{ab}^* convergence numbers for 101 colors, each with 10 iterations. If the ΔE_{ab}^* numbers are close to zero (or close to a predefined threshold value, e.g., 0.05) after steady state is reached, then those colors are considered inside the printer gamut. If the ΔE_{ab}^* numbers are larger than the threshold value, then we can say that those colors are out-of-gamut. In Table 7.12, color numbers 77–100 are out-of-gamut. Color numbers 0–76 are considered as being inside the gamut.

TABLE 7.11

Example Illustrating the Convergence of a Ray between $x_0 = [L_0 = 50, a_0 = 0, b_0 = 0]$ and $x_c = [L_c = 80, a_c = 50, b_c = 30]$ for Colors between 0 and 50

Color#	L^*	a^*	b^*	1.00	2.00	3.00	4.00	5.00	6.00	7.00	8.00	9.00	10.00	C	M	Y	K
								Iteration#							Best CMYK		
0	50.00	0.00	0.00	8.22	2.77	0.94	0.30	0.09	0.03	0.01	0.00	0.00	0.00	87.86	70.74	70.26	98.23
1	50.30	0.50	0.30	8.17	2.79	0.94	0.30	0.10	0.04	0.01	0.01	0.00	0.00	86.40	72.45	70.67	97.30
2	50.60	1.00	0.60	7.97	2.74	0.88	0.29	0.11	0.04	0.01	0.01	0.00	0.00	86.21	75.18	71.86	95.88
3	50.90	1.50	0.90	8.49	2.93	0.94	0.33	0.12	0.04	0.02	0.01	0.00	0.00	84.60	76.62	72.17	94.94
4	51.20	2.00	1.20	7.99	2.70	0.89	0.33	0.13	0.05	0.02	0.01	0.00	0.00	84.93	79.85	73.70	93.29
5	51.50	2.50	1.50	7.94	2.72	0.94	0.35	0.13	0.04	0.02	0.01	0.00	0.00	83.59	81.40	74.18	92.27
6	51.80	3.00	1.80	8.06	2.69	0.95	0.35	0.13	0.05	0.02	0.01	0.00	0.00	83.53	83.71	75.79	90.57
7	52.10	3.50	2.10	8.04	2.80	1.02	0.37	0.15	0.07	0.03	0.01	0.01	0.00	82.56	85.22	76.62	89.15
8	52.40	4.00	2.40	7.85	2.75	1.02	0.43	0.20	0.09	0.04	0.02	0.01	0.00	82.34	87.32	78.14	87.17
9	52.70	4.50	2.70	8.08	2.86	1.12	0.49	0.22	0.11	0.05	0.02	0.01	0.01	81.75	89.12	79.34	85.37
10	53.00	5.00	3.00	7.67	2.74	1.19	0.55	0.26	0.13	0.07	0.03	0.02	0.01	82.43	91.73	81.68	82.63
11	53.30	5.50	3.30	8.20	2.97	1.29	0.60	0.29	0.15	0.07	0.04	0.02	0.01	80.97	92.71	82.04	81.36
12	53.60	6.00	3.60	7.77	3.00	1.36	0.64	0.32	0.16	0.08	0.04	0.02	0.01	81.21	94.93	83.75	78.81
13	53.90	6.50	3.90	8.21	3.26	1.46	0.70	0.35	0.18	0.09	0.05	0.02	0.01	80.25	96.27	84.57	77.09
14	54.20	7.00	4.20	7.66	3.18	1.49	0.73	0.38	0.19	0.09	0.04	0.02	0.01	80.03	98.05	85.94	74.98
15	54.50	7.50	4.50	8.20	3.43	1.59	0.80	0.39	0.18	0.08	0.04	0.02	0.01	78.59	98.88	86.43	73.79
16	54.80	8.00	4.80	7.90	3.48	1.65	0.80	0.36	0.16	0.07	0.03	0.02	0.01	78.06	100.30	87.51	72.00
17	55.10	8.50	5.10	7.68	3.49	1.73	0.79	0.36	0.17	0.08	0.04	0.02	0.01	77.35	101.59	88.33	70.29
18	55.40	9.00	5.40	7.39	3.49	1.67	0.75	0.37	0.18	0.09	0.05	0.02	0.01	77.53	103.29	89.70	67.88
19	55.70	9.50	5.70	7.63	3.60	1.66	0.75	0.36	0.18	0.09	0.05	0.02	0.01	75.82	103.94	89.89	66.82
20	56.00	10.00	6.00	7.46	3.65	1.64	0.79	0.38	0.19	0.09	0.05	0.02	0.01	76.52	106.15	91.63	63.92
21	56.30	10.50	6.30	7.61	3.66	1.66	0.80	0.39	0.19	0.09	0.05	0.02	0.01	75.39	107.26	92.18	62.38
22	56.60	11.00	6.60	7.18	3.37	1.55	0.73	0.35	0.17	0.08	0.03	0.01	0.01	75.09	108.90	93.10	60.26
23	56.90	11.50	6.90	7.40	3.39	1.54	0.71	0.31	0.12	0.05	0.02	0.01	0.00	74.04	110.08	93.64	58.93
24	57.20	12.00	7.20	7.13	3.29	1.51	0.64	0.25	0.10	0.04	0.01	0.01	0.00	73.74	111.73	94.61	57.14
25	57.50	12.50	7.50	7.33	3.31	1.48	0.57	0.23	0.09	0.04	0.01	0.01	0.00	72.72	112.84	95.15	55.74
26	57.80	13.00	7.80	7.05	3.09	1.26	0.49	0.19	0.07	0.03	0.01	0.00	0.00	72.56	114.43	96.15	53.79
27	58.10	13.50	8.10	6.60	2.94	1.15	0.45	0.18	0.07	0.03	0.01	0.00	0.00	72.09	115.83	96.97	52.04
28	58.40	14.00	8.40	6.82	2.99	1.16	0.45	0.17	0.07	0.03	0.01	0.00	0.00	71.06	116.89	97.41	50.65
29	58.70	14.50	8.70	7.00	2.98	1.15	0.44	0.17	0.06	0.03	0.01	0.00	0.00	70.12	118.01	97.91	49.20
30	59.00	15.00	9.00	6.22	2.63	1.01	0.40	0.16	0.07	0.03	0.01	0.00	0.00	70.30	119.75	99.02	46.91
31	59.30	15.50	9.30	6.21	2.51	0.99	0.40	0.16	0.07	0.03	0.01	0.00	0.00	69.38	120.76	99.51	45.39
32	59.60	16.00	9.60	6.46	2.53	1.02	0.41	0.16	0.06	0.03	0.01	0.00	0.00	68.64	121.90	100.13	43.76
33	59.90	16.50	9.90	6.59	2.61	1.04	0.42	0.16	0.06	0.03	0.01	0.00	0.00	67.48	122.81	100.52	42.45

(continued)

TABLE 7.11 (continued)

Example Illustrating the Convergence of a Ray between $x_0 = [L_0 = 50, a_0 = 0,$ $b_0 = 0]$ and $x_c = [L_c = 80, a_c = 50, b_c = 30]$ for Colors between 0 and 50

Color#	L^*	a^*	b^*	1.00	2.00	3.00	4.00	5.00	6.00	7.00	8.00	9.00	10.00	C	M	Y	K
								Iteration#							Best CMYK		
34	60.20	17.00	10.20	5.85	2.34	0.95	0.38	0.15	0.06	0.02	0.01	0.00	0.00	66.70	123.93	101.10	40.85
35	60.50	17.50	10.50	5.74	2.35	0.95	0.38	0.15	0.06	0.02	0.01	0.00	0.00	65.85	125.02	101.57	39.28
36	60.80	18.00	10.80	5.49	2.28	0.93	0.38	0.15	0.06	0.02	0.01	0.00	0.00	65.84	126.46	102.43	37.08
37	61.10	18.50	11.10	5.74	2.39	0.98	0.40	0.16	0.06	0.03	0.01	0.00	0.00	65.38	127.61	103.05	35.26
38	61.40	19.00	11.40	5.23	2.17	0.89	0.36	0.14	0.06	0.02	0.01	0.00	0.00	64.83	128.72	103.63	33.52
39	61.70	19.50	11.70	5.50	2.28	0.92	0.37	0.15	0.06	0.02	0.01	0.00	0.00	63.52	129.45	103.81	32.39
40	62.00	20.00	12.00	4.92	2.02	0.82	0.33	0.13	0.06	0.03	0.02	0.01	0.00	64.24	131.15	105.14	29.57
41	62.30	20.50	12.30	5.52	2.26	0.90	0.38	0.19	0.09	0.05	0.02	0.01	0.01	62.80	131.84	105.26	28.25
42	62.60	21.00	12.60	5.14	2.13	0.89	0.42	0.20	0.09	0.05	0.02	0.01	0.01	61.87	132.76	105.70	26.50
43	62.90	21.50	12.90	5.25	2.19	0.98	0.46	0.22	0.11	0.05	0.03	0.01	0.01	60.75	133.57	106.04	24.88
44	63.20	22.00	13.20	4.37	1.89	0.94	0.47	0.24	0.12	0.06	0.03	0.02	0.01	60.73	135.18	107.09	22.10
45	63.50	22.50	13.50	4.96	2.16	1.06	0.52	0.26	0.13	0.07	0.03	0.02	0.01	59.25	136.04	107.22	20.90
46	63.80	23.00	13.80	4.78	2.22	1.08	0.53	0.26	0.13	0.07	0.03	0.02	0.01	58.42	137.28	107.77	19.00
47	64.10	23.50	14.10	4.44	2.17	1.07	0.52	0.26	0.13	0.07	0.04	0.02	0.01	57.62	138.44	108.23	17.05
48	64.40	24.00	14.40	4.15	2.03	0.99	0.49	0.25	0.13	0.06	0.03	0.02	0.01	56.46	139.40	108.49	15.49
49	64.70	24.50	14.70	4.76	2.31	1.12	0.55	0.27	0.11	0.04	0.02	0.01	0.00	55.10	140.25	108.63	14.46
50	65.00	25.00	15.00	4.16	2.08	1.05	0.43	0.17	0.06	0.02	0.01	0.00	0.00	54.81	141.69	109.43	12.70

The *CMYK* values achieved during convergence are also shown in last four columns corresponding to each row. They are the best *CMYK* values for the chosen GCR used to determine the nominal *CMYK* that give the minimum ΔE_{ab}^* error. In other words, these values represent the best *CMYK* values for minimum ΔE_{ab}^* (for both in & out-of-gamut colors). From this table, it is also easy to find the mapped *CMYK* values for the node color, x_c, when the node color is projected along the ray in the ray-based gamut mapping. As is the case in gamut clipping method—one of the ray-based approaches the gamut-mapped color will be on the surface when this approach is used. This is obtained when the ray penetrates the surface. It is equal to 77th color in Table 7.12. The *CMYK* values of the nearest out-of-gamut node give the gamut-mapped *CMYK* values for the node color x_c when the mapping is done along the ray. In gamut compression method—another ray-based approach—the gamut-mapped colors are moved inward along the ray to the nearest node with a scaling associated with the local region of the gamut. A ray-based control model can also be used to perform an accurate linear mapping for the compression algorithms. Thus, in Table 7.12 (see marked boxes), the node color, x_c (1) is out-of-gamut by 15.79 ΔE_{ab}^*, (2) can be gamut mapped with minimum ΔE_{ab}^* with a gamut-mapped *CMYK* value of $C = 0$, $M = 169.76$, $Y = 110.61$, $K = 0$, (3) can be gamut mapped along the ray on the gamut surface with a gamut-mapped *CMYK* value of $C = 0$, $M = 150.98$, $Y = 103.28$, $K = 0$, and (4) can be mapped inward along the ray as in compression methods (e.g., say for example, x_c can be mapped to color# 72 with a gamut-mapped *CMYK* value of $C = 7.94$, $M = 148.99$, $Y = 103.93$, $K = 2.51$).

TABLE 7.12

Example Illustrating the Convergence of a Ray between $x_0 = [L_0 = 50, a_0 = 0, b_0 = 0]$ and $x_c = [L_c = 80, a_c = 50, b_c = 30]$ for Colors between 51 and 100

Color#	L^*	a^*	b^*	Iteration# 1.00	2.00	3.00	4.00	5.00	6.00	7.00	8.00	9.00	10.00	Best CMYK C	M	Y	K
51	65.30	25.50	15.30	4.54	2.23	1.00	0.39	0.15	0.06	0.02	0.01	0.00	0.00	53.66	142.62	109.70	11.61
52	65.60	26.00	15.60	4.14	2.02	0.85	0.34	0.13	0.05	0.02	0.01	0.00	0.00	51.71	143.12	109.48	11.13
53	65.90	26.50	15.90	4.33	2.03	0.83	0.34	0.14	0.06	0.03	0.01	0.01	0.00	50.16	143.82	109.50	10.35
54	66.20	27.00	16.20	4.12	1.94	0.78	0.31	0.13	0.05	0.02	0.01	0.00	0.00	48.97	144.70	109.73	9.31
55	66.50	27.50	16.50	4.32	1.91	0.79	0.34	0.15	0.07	0.03	0.01	0.01	0.00	47.38	145.31	109.81	8.44
56	66.80	28.00	16.80	4.03	1.74	0.79	0.36	0.17	0.08	0.04	0.02	0.01	0.00	44.94	145.63	109.48	8.10
57	67.10	28.50	17.10	3.95	1.73	0.79	0.36	0.17	0.08	0.04	0.02	0.01	0.00	44.15	146.57	110.03	6.62
58	67.40	29.00	17.40	3.87	1.78	0.83	0.40	0.19	0.09	0.04	0.02	0.01	0.00	42.12	146.96	109.90	5.97
59	67.70	29.50	17.70	4.28	2.05	0.98	0.47	0.23	0.11	0.05	0.03	0.01	0.01	39.91	147.25	109.66	5.47
60	68.00	30.00	18.00	3.55	1.67	0.77	0.36	0.17	0.08	0.04	0.02	0.01	0.00	37.91	147.63	109.53	4.82
61	68.30	30.50	18.30	3.76	1.79	0.84	0.40	0.18	0.09	0.04	0.02	0.01	0.00	35.26	147.76	109.05	4.62
62	68.60	31.00	18.60	3.72	1.75	0.82	0.38	0.17	0.08	0.04	0.02	0.01	0.00	33.47	148.24	108.98	3.82
63	68.90	31.50	18.90	4.12	1.95	0.92	0.43	0.20	0.09	0.04	0.02	0.01	0.00	31.57	148.67	108.86	3.10
64	69.20	32.00	19.20	3.44	1.57	0.72	0.33	0.15	0.07	0.03	0.02	0.01	0.00	28.35	148.58	108.16	3.25
65	69.50	32.50	19.50	3.80	1.78	0.83	0.39	0.18	0.09	0.04	0.02	0.01	0.00	25.46	148.58	107.62	3.15
66	69.80	33.00	19.80	3.92	1.81	0.85	0.40	0.19	0.09	0.04	0.02	0.01	0.00	23.25	148.85	107.39	2.60
67	70.10	33.50	20.10	4.20	1.92	0.88	0.41	0.19	0.09	0.04	0.02	0.01	0.00	21.01	149.11	107.14	2.07
68	70.40	34.00	20.40	3.81	1.76	0.81	0.38	0.18	0.08	0.04	0.02	0.01	0.00	17.64	148.89	106.34	2.33
69	70.70	34.50	20.70	4.29	2.00	0.93	0.41	0.15	0.06	0.02	0.01	0.00	0.00	14.75	148.83	105.61	2.43
70	71.00	35.00	21.00	3.63	1.53	0.57	0.22	0.09	0.03	0.01	0.01	0.00	0.00	11.89	148.58	104.61	2.97
71	71.30	35.50	21.30	4.27	1.77	0.67	0.26	0.10	0.04	0.02	0.01	0.00	0.00	10.80	149.24	104.80	1.97

(continued)

TABLE 7.12 (continued)

Example Illustrating the Convergence of a Ray between $x_0 = [L_0 = 50, a_0 = 0, b_0 = 0]$ and $x_c = [L_c = 80, a_c = 50, b_c = 30]$ for Colors between 51 and 100

Color#	L^*	a^*	b^*	1.00	2.00	3.00	4.00	5.00	6.00	7.00	8.00	9.00	10.00	C	M	Y	K
								Iteration #							Best CMYK		
72	71.60	36.00	21.60	3.70	1.38	0.52	0.20	0.08	0.03	0.01	0.00	0.00	0.00	7.94	148.99	103.93	2.51
73	71.90	36.50	21.90	4.28	1.61	0.62	0.24	0.09	0.04	0.01	0.01	0.00	0.00	5.95	149.18	103.61	2.31
74	72.20	37.00	22.20	2.98	1.20	0.49	0.20	0.08	0.04	0.02	0.01	0.00	0.00	4.32	149.55	103.46	1.80
75	72.50	37.50	22.50	3.61	1.44	0.58	0.23	0.10	0.05	0.02	0.01	0.01	0.00	3.22	150.18	103.59	0.71
76	72.80	38.00	22.80	2.71	1.06	0.41	0.19	0.10	0.05	0.03	0.01	0.01	0.00	0.44	150.10	102.96	1.09
77	73.10	38.50	23.10	2.87	1.23	0.73	0.52	0.41	0.34	0.33	0.32	0.32	0.32	0.00	150.98	103.28	0.00
78	73.40	39.00	23.40	2.59	1.33	1.03	0.95	0.92	0.91	0.91	0.91	0.91	0.91	0.00	151.80	103.58	0.00
79	73.70	39.50	23.70	3.22	1.83	1.60	1.54	1.52	1.51	1.51	1.50	1.50	1.50	0.00	152.62	103.90	0.00
80	74.00	40.00	24.00	3.38	2.31	2.14	2.12	2.11	2.11	2.10	2.10	2.10	2.10	0.00	153.41	104.17	0.00
81	74.30	40.50	24.30	4.16	3.02	2.79	2.74	2.72	2.71	2.70	2.70	2.69	2.69	0.00	154.25	104.53	0.00
82	74.60	41.00	24.60	4.46	3.66	3.43	3.36	3.32	3.31	3.30	3.29	3.28	3.28	0.00	155.07	104.89	0.00
83	74.90	41.50	24.90	5.08	4.27	4.03	3.96	3.92	3.90	3.89	3.88	3.87	3.87	0.00	155.90	105.25	0.00
84	75.20	42.00	25.20	5.34	4.70	4.56	4.53	4.51	4.50	4.49	4.48	4.47	4.47	0.00	156.65	105.49	0.00

85	75.50	42.50	25.50	6.17	5.42	5.22	5.15	5.12	5.10	5.08	5.07	5.06	5.06	0.00	157.52	105.89	0.00
86	75.80	43.00	25.80	6.81	6.03	5.82	5.76	5.72	5.69	5.67	5.66	5.65	5.65	0.00	158.34	106.24	0.00
87	76.10	43.50	26.10	7.59	6.64	6.42	6.35	6.32	6.30	6.28	6.26	6.26	6.25	0.00	159.20	106.52	0.00
88	76.40	44.00	26.40	7.83	7.10	6.96	6.92	6.90	6.88	6.87	6.86	6.85	6.85	0.00	159.89	106.78	0.00
89	76.70	44.50	26.70	8.84	7.86	7.63	7.56	7.52	7.49	7.47	7.45	7.44	7.43	0.00	160.84	107.20	0.00
90	77.00	45.00	27.00	9.25	8.32	8.15	8.12	8.10	8.09	8.07	8.06	8.05	8.05	0.00	161.57	107.37	0.00
91	77.30	45.50	27.30	10.08	9.04	8.81	8.75	8.71	8.68	8.66	8.65	8.64	8.63	0.00	162.47	107.79	0.00
92	77.60	46.00	27.60	10.51	9.55	9.37	9.33	9.30	9.28	9.26	9.25	9.24	9.23	0.00	163.20	108.02	0.00
93	77.90	46.50	27.90	11.35	10.27	10.03	9.96	9.91	9.88	9.85	9.84	9.82	9.82	0.00	164.11	108.45	0.00
94	78.20	47.00	28.20	11.95	10.83	10.61	10.55	10.51	10.48	10.46	10.44	10.43	10.42	0.00	164.91	108.68	0.00
95	78.50	47.50	28.50	12.79	11.55	11.27	11.17	11.12	11.08	11.05	11.03	11.02	11.01	0.00	165.84	109.12	0.00
96	78.80	48.00	28.80	13.11	12.09	11.85	11.76	11.70	11.66	11.63	11.61	11.60	11.59	0.00	166.47	109.45	0.00
97	79.10	48.50	29.10	14.02	12.72	12.45	12.37	12.32	12.28	12.25	12.23	12.22	12.21	0.00	167.46	109.66	0.00
98	79.40	49.00	29.40	14.55	13.37	13.08	12.98	12.91	12.86	12.83	12.80	12.79	12.78	0.00	168.20	110.11	0.00
99	79.70	49.50	29.70	15.35	14.02	13.70	13.59	13.52	13.47	13.43	13.40	13.38	13.37	0.00	169.10	110.44	0.00
100	80.00	50.00	30.00	15.79	14.54	14.27	14.17	14.11	14.07	14.02	14.00	13.98	13.97	0.00	169.76	110.61	0.00

Thus, this control-based iteration of the ray model not only gives an accurate determination of whether or not the colors are located inside or outside the gamut surface, but also performs accurate gamut mapping along the ray axis. Additional constraints can be included during the iterations while performing error minimization (i.e., ΔE_{2000} etc.) using ray-based control model to provide variations to the components such as lightness, L^*, chroma, C, and hue, H.

Next we show how to apply the ray-based control model to perform few well-known gamut-mapping strategies.

7.6.2 CENTROID CLIPPING

In centroid gamut mapping, the node colors with a constant hue are mapped toward a focal point inside the gamut [131]. In our illustration, we use the method that maps to the gamut surface along the ray toward the gamut centroid (x_0)—the focal point. Though this type of gamut mapping reduces chroma, it is straightforward to use with ray-based control model to map colors accurately.

Figure 7.57 shows rays pointing from the sample node colors to the centroid ($L^* = 60.59$, $a^* = 6.62$, $b^* = 13.11$), which are overlaid on top of the actual printer gamut. Figure 7.58 shows the ΔE_{ab}^* convergence values at the end of 10 iterations for six out-of-gamut node colors shown in Table 7.13. Each ray was divided into 101 points. Clearly, the ΔE_{ab}^* convergence value during iteration deviates to greater than zero at the gamut intersection as shown by arrows. Mapped *CMYK* values, either for minimum ΔE_{ab}^* mapping (also called nearest point clipping) or for centroid clipping to the gamut surface, can be easily obtained using the procedure outlined in Section 7.6.1.

FIGURE 7.57   Rays pointing to the focal point in centroid gamut mapping.

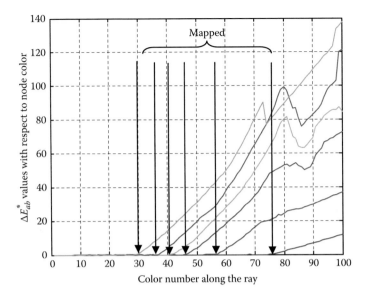

FIGURE 7.58 Converged ΔE_{ab}^* values shown for six node colors.

TABLE 7.13
Out-of-Gamut Color Considered in Figure 7.58

Out of Gamut Node Colors	L^*	a^*	b^*
1	0.00	0.00	0.00
2	−0.12	7.68	−54.12
3	−0.36	23.59	−91.50
4	−0.72	46.12	−122.55
5	−1.19	68.22	−150.08
6	−1.75	88.88	−175.31

7.6.3 SOFT GAMUT MAPPING WITH RAY-BASED CONTROL MODEL

A ray-based control model can be used to implement accurate gamut compression algorithms, one of which is the soft gamut mapping illustrated below.

Let x_b be the nearest out-of-gamut color along the ray between the in-gamut color, x_0, and the out-of-gamut node color, x_c. For this discussion, x_c could be selected as the color on the boundary of the source color space. This approach is shown graphically in Figure 7.59 with a dashed curve. All the out-of-gamut colors between x_0 and x_c are mapped to the region between x_0 and x_b nonlinearly inside the gamut surface using a smooth curve.

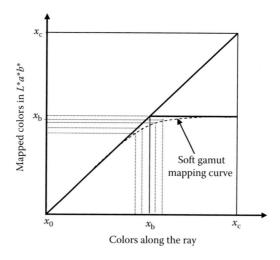

FIGURE 7.59   Soft gamut mapping.

A soft gamut-mapping curve can be constructed using a piecewise linear compression function shown below followed by a 1-D smoothing algorithm described in Section 6.5.3.1.

$$x = (1 - \beta)\left[x_0 + \frac{x_b - x_0}{N}i\right] + \beta\left[x_0 + \frac{x_c - x_0}{N}i\right] \quad \text{for } 0 \le i \le i_b \qquad (7.97)$$

$$x = (1 - \beta)\left[x_0 + \frac{x_b - x_0}{N}i\right] + \beta x_b \quad \text{for } i_b \le i \le N \qquad (7.98)$$

where $0 \le \beta \le 1$ is the compression parameter, i is the index incremented between 0 to a convenient integer value until x_c. Figure 7.60 shows L^*, a^*, b^* curves with respect to color indices i for three different values of β.

When $\beta = 0$, the interval $\{x_0 \text{ to } x_c\}$ is mapped linearly into $\{x_0 \text{ to } x_b\}$. A ray-based control model is run for $\beta = 0$ and $i = 0$ (representing color x_0) to $i = N$ (representing color x_c which is now mapped to x_b) to obtain the *CMYK* values corresponding to the colors along the ray. If the out-of-gamut node color happens to fall anywhere between x_b and x_c then it is mapped according to the rules described below. With $\beta = 0$, all colors including the in-gamut colors are moved inward except for the node color x_c which is clipped to the boundary. As β increases to a value greater than 0, the in-gamut colors are compressed less. For $\beta = 1$, all in-gamut colors are passed through the mapping without any compression. All colors greater than x_b are mapped to x_b. Smoother compression can be achieved using the 1-D filtering of the function shown in Equations 7.97 and 7.98. In Figure 7.60, we used the 1-D filtering algorithm described in Section 6.6.2 with α set equal to 100,000. Both β and α contribute to the overall smoothness of the soft gamut-mapping curves and can be customized differently to each region of the gamut.

A conventional gamut-mapping approach is described below that preserves the lightness and hue angle.

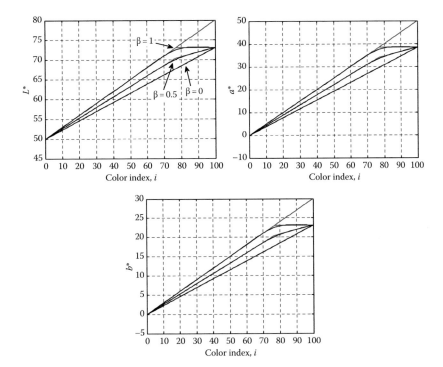

FIGURE 7.60 Soft gamut-mapping curves for L^* (top left), a^* (top right), and b^* (bottom center) for $\beta = 0, 0.5, 1$ shown with and without smoothing.

7.6.4 GAMUT MAPPING FOR CONSTANT LIGHTNESS AND HUE

An algorithm that preserves the lightness and hue angle of the node *Lab* color coordinates is shown below. For convenience, the "*"s are removed from the color coordinates, $L^*a^*b^*$. If the original color in the *CIELab* color space is denoted by $[L\ \ a\ \ b]$ and its corresponding mapped color by $[L'\ \ a'\ \ b']$, then, to preserve the lightness and hue relationship, the following constraints must be satisfied:

$$L' = L \quad \text{and} \quad \tan^{-1}\left(\frac{b'}{a'}\right) = \tan^{-1}\left(\frac{b}{a}\right) \tag{7.99}$$

Graphically, as illustrated in Figure 7.61, this can be shown on a chroma plane at constant L^* by drawing a ray from the desired color to the centroid of the gamut.

First we need to determine whether a given input color is inside or outside the printer gamut. This is achieved by the following two steps:

1. Transform the color gamut into spherical coordinates where the center of the sphere is at the centroid of the gamut, approximately equal to $Lab = [50\ 0\ 0]$. In these new color coordinates, each node color is specified by its spherical coordinates r, α, and θ given by the equations,

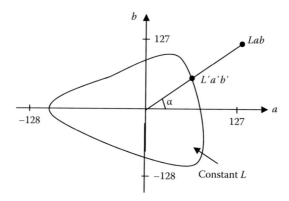

FIGURE 7.61   Gamut mapping for constant lightness and hue.

$$r = \sqrt{(L - 50)^2 + a^2 + b^2}$$

$$\alpha = \tan^{-1}\left(\frac{b}{a}\right) \quad \text{and} \quad \theta = \tan^{-1}\left(\frac{L - 50}{\sqrt{a^2 + b^2}}\right) \qquad (7.100)$$

In these equations, r is the distance from the gamut centroid to the given color point, α is the hue angle with the dynamic range from $0°$ to $360°$ and θ is the angle in the constant α plane from $0°$ to $180°$.

2. Find the intersection of the gamut with a straight line connecting the input color point having coordinates $[r_0 \quad \alpha_0 \quad \theta_0]$ to the centroid of the gamut. This is achieved by searching the gamut boundary points with coordinates that fall into the range $\alpha_0 \pm \Delta\alpha$ and $\theta_0 \pm \Delta\theta$. The average of these boundary points is denoted by $[r_{ave} \, \alpha_{ave} \, \theta_{ave}]$. Now the desired color point is inside the gamut if $r_0 < r_{ave}$, otherwise it is outside the printer gamut.

Once we determine that the color is outside the printer gamut, we map it to the nearest point on the gamut boundary that best satisfies the conditions shown in Equation 7.99.

7.6.5   MERIT-BASED GAMUT MAPPING

There is no unique gamut-mapping method that can satisfy requirements like pleasing color, contrast, lightness, chroma, hue, etc. More than 90 gamut-mapping algorithms [111] can be used for mapping colors to a suitable region of the printer gamut. Some offer feature enhancements in one region of the gamut. Others do the same in different regions of the gamut. As such, device designers generally compromise in the gamut-mapping functions they wish to employ in their respective color management systems. Merit-based feedback system used for gamut mapping offers a way to automatically select the gamut-mapping algorithms from the library of algorithms to optimize a merit function [134]. In this method, all node colors are clustered in different regions of interest within the color space and each cluster is associated with at least one

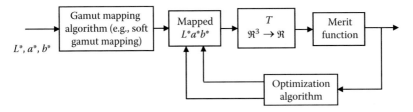

FIGURE 7.62   Merit-based gamut mapping with closed-loop optimization.

candidate gamut-mapping function. A merit function is constructed based on a high-level gamut-mapping strategy with each cluster. At least one merit function is associated with each cluster. Gamut-mapping parameters are iteratively varied (or tuned) to optimize their values. The merit-based gamut mapping uses a closed-loop optimization technique (Figure 7.62) to iteratively tune various mapping parameters associated with the gamut-mapping algorithm. A number of different candidate algorithms are iterated for each defined cluster within the color space. Once optimal mapping parameters are obtained for each candidate, a best (that minimizes the merit function) is picked for that cluster. The optimized gamut-mapping functions thus obtained for adjacent clusters are then blended together using multidimensional smoothing algorithms or other methods described in Ref. [134] to generate a composite smooth function that collectively exploits the local advantages of each cluster. The entire process is done automatically in a computer program.

For illustration of this technique we can choose the soft gamut-mapping approach, introduced in Section 7.6.3, with two gamut-mapping parameters, β and α (or x_0 in the centroid clipping algorithm). Other parameters may be used depending on the gamut-mapping algorithm. In Figure 7.62, a transformation block "T" maps the out-of-gamut $L^*a^*b^*$ values to a 1-D function for each color point. This could be, for example, ΔE_{2000} formula, which is to say that a ΔE_{2000} mapping is being emulated. The ΔE_{2000} function calculates distances between the out-of-gamut $L^*a^*b^*$ values and the mapped $L^*a^*b^*$ values for each node color of the current cluster. The merit function is selected based on what is required by the device designer during optimization. A mean-squared error function can be used as the merit function. In this case, the merit function determines the mean-squared error of the values calculated using the ΔE_{2000} function for each node color belonging to a particular cluster that needs to be mapped. The output of the merit function is a single value which represents the numerical merit of the mapping parameters. An optimization block manipulates the gamut-mapping parameters during successive iterations to generate the best, or optimum, merit function value. The optimization algorithm could be a brute-force approach, that selects an exhaustive set of combinations of β and α within their limits. A brute-force approach in many instances does not take much time to generate the final results given present processor speeds. When the optimum merit function value is reached (or exceeded) the iteration loop ends for the cluster being processed. Note that the above calculation can be done for more than one gamut-mapping algorithm for each cluster, and the one with the best merit function output can be chosen.

Clusters can be formed by running the vector quantization algorithm (Section 2.4.7) for the $L^*a^*b^*$ nodes or can be formed in or around the vicinity of the corners, edges, and faces. In the entire color space, with this method, it is not impractical to imagine more than 50 clusters being processed intelligently by the computer with 90 different gamut-mapping algorithms.

7.6.6 BLACK POINT COMPENSATION

RGB images often contain colors that are darker than the darkest color a printer can make. Minimum color error can be obtained by mapping the out-of-gamut colors to the darkest color the printer can make. However, in that case, all details in the shadow regions may be lost. They often contain information the viewer knows should be there (the folds of a dark coat, the strands of dark hair) and their absence can be very disturbing. Therefore, an L^* mapping technique for black point compensation similar to the one shown in Figure 7.60 is often used for *RGB* images. It retains the details for dark colors, though at a reduced contrast, and at the cost of lightening some of the darkest colors (see Figure 7.63). For implementing the black point compensation (i.e., input L^* to output L^* map along the neutral axis)

FIGURE 7.63 (See color insert following page 428.)   Example of black point compensation (top left: original, top right: without BPC, bottom: with BPC).

using soft gamut mapping with ray-based control method, we define x_c equal to zero (e.g., 0,0,0) x_0 along the neutral axis (e.g., 50,0,0) and fine tune the parameters α and β. After that, each node color in the inverted LUT is updated with the new input L^* to output L^* map (see Problem 7.12).

7.7 EVALUATION OF PROFILES

7.7.1 GAMUT UTILIZATION AND ROUND TRIP ACCURACY

Color gamut and its utilization depend on various factors. Colorants of toner (or ink), process physics, type of GCRs, inversion methods, and node spacings are some of the factors that contribute to the gamut of colors that a printer can produce. The reproducible gamut is generally smaller than the available process gamut.

The printing industry has access to numerous commercially available profiling packages. It is not impossible to find that colors that can be produced using one profile may be impossible with another. Hence, it is important to assess the gamut utilization for a given profile from an awareness and competitive point of view. Otherwise, the printer may not reproduce all the colors in an *RGB* image even though the printer's *CMYK* gamut is capable of producing those colors.

One way to quantify the gamut utilization is through computational procedures by using A2B and B2A tags of the ICC Profile. A2B tag represents the printer model and B2A tag contains the gamut-mapped inverse LUT. Using them in sequence for a set of in-gamut test colors described in $L^*a^*b^*$ space can sometimes give a good indication of the quality of inversion. By computing ΔE (ΔE_{ab}^*, ΔE_{2000} or any other ΔE) values for a set of in-gamut test colors between their input $L^*a^*b^*$ values for the B2A tag and the output $L^*a^*b^*$ values from the A2B tag, we can quantify the round trip accuracy. At the same time if the test colors are sufficiently dense and well distributed, we can also generate the gamut volume. Ideally, the inversion round trip error for a colorimetric profile must be zero and the round trip gamut volume should be close to 100% when compared to the native *CMYK* gamut. If the gamut volume is not 100%, then one should produce a vector plot between the two sets of $L^*a^*b^*$ values to identify the regions of colors (e.g., chromatic greens/reds) where the deficiencies are present.

The gamut volume and round trip accuracy is computed as follows. Create a large number of in-gamut test colors in *CMYK* space (e.g., 5000). Obtain their estimated $L^*a^*b^*$ values for each of the test colors using the A2B tag. Now use these estimated $L^*a^*b^*$ values as inputs to the B2A tag. The output of B2A tag is then fed as inputs to the A2B tag to produce second set of $L^*a^*b^*$ values. The reproducible color gamut of the printer represented in the printer model is the volume covered by the second set of colors. The gamut volume can be obtained using the methods described in Ref. [135]. The roundtrip accuracy is equal to the ΔE between first and the second set of $L^*a^*b^*$ values.

Another way to quantify the gamut utilization is by using the printer instead of the A2B tag of the ICC profile. First a set of $L^*a^*b^*$ values are measured using a sensor, such as an in-line or offline spectrophotometer for a set of *CMYK* test colors.

Measurement guidelines, such as same paper, same white point, same stock backing, averaging with multiple sets, etc., have to be strictly followed when a spectrophotometer is used. Measurement procedures adopted during the printer characterization should match those during the profile validation stage. Otherwise, validation will not be accurate. For determining round trip accuracy, use the measured $L^*a^*b^*$ values as inputs to the B2A tag. Create patches for the output of B2A tag and print and measure. If this procedure is difficult to implement, then create patches using the measured $L^*a^*b^*$ values and print them through the ICC profile and measure the printed patches. This procedure will eventually result in a second set of $L^*a^*b^*$ values. Compare the gamuts between the second and the first set of $L^*a^*b^*$ values. Also obtain the round trip accuracy statistics. A vector plot of these $L^*a^*b^*$ values will give additional information about the rendering accuracy, which is particularly useful to accessing the spot roundtrip inaccuracy in localized regions of the color space.

7.7.2 GAMUT CORNER PLOTS AND NEUTRAL RESPONSE

Gamut corner plots can give additional measurement of the quality of inversion and gamut-mapping algorithms. They can be used to identify unsmooth color gradients and help ensure smoothness in *CMYK* formulations in critical regions of the transformation.

A generic *CMYK* printer gamut has at least 14 distinct corners, 24 edges, and 12 faces. The corner *CMYK* and corresponding $L^*a^*b^*$ values for a representative printer are shown in Table 7.14. Figure 7.64a and b show these corners when viewed from the top and bottom, respectively, and Figure 7.64c shows all 14 lines emanating through the corners until they touch the sides of the $L^*a^*b^*$ cube. Figures 7.65(a to d) and 7.66

TABLE 7.14

CMYK and L*a*b* Values for a Typical EP Production Printer Gamut Corners

Corners	C	M	Y	K	L*	a*	b*
Cyan	255	0	0	0	50.05	−29.98	−57.79
Magenta	0	255	0	0	50.83	84.90	−6.47
Yellow	0	0	255	0	95.21	−7.63	111.12
White	0	0	0	0	100.00	0.00	−0.01
Red	0	255	255	0	51.93	72.34	47.80
Green	255	0	255	0	50.34	−63.37	35.09
Blue	255	255	0	0	20.89	43.68	−51.07
Dark cyan	255	0	0	255	17.51	−10.08	−21.40
Dark magenta	0	255	0	255	17.34	24.71	−0.70
Dark yellow	0	0	255	255	27.78	−3.11	27.80
Dark black	0	0	0	255	12.64	1.11	1.00
Dark blue	255	255	0	255	14.70	21.17	−24.02
Dark green	255	0	255	255	27.01	−33.17	16.33
Dark red	0	255	255	255	23.41	30.65	17.67

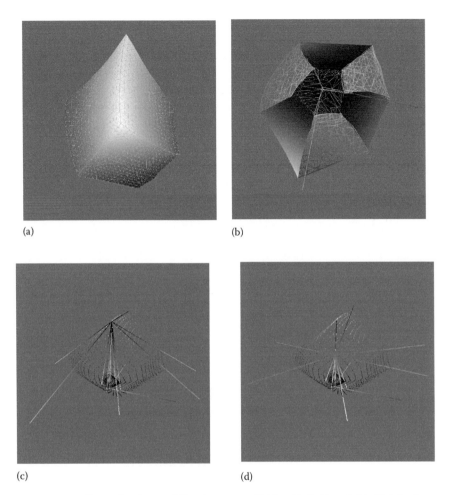

(a) (b)

(c) (d)

FIGURE 7.64 (See color insert following page 428.) (a) (left) and (b) (right): Gamut views from top and from bottom, respectively; (c) lines to gamut corners (cyan, magenta, yellow, white, and blue corners from 100,0,0 point); (d) lines to gamut corners (cyan, magenta, yellow, white, and blue corners from 50,0,0 point).

show the *CMYK* response (left) and round trip accuracy (right) when a line stimulus is used from $L^* = 100$, $a^* = b^* = 0$ point to 14 different gamut corners. Similar plots are shown from $L^* = 50$, $a^* = b^* = 0$ in Figure 7.65(e–i). The x-axis contains 101 discrete points along the line between the starting point ($L^* = 50$, $a^* = b^* = 0$) to the corner. The corner is shown with a thick vertical line in each of these plots. The x-axis, representing the points on the gamut corner axis, is extended until it touches the sides of the $L^*a^*b^*$ cube. The corner point shown is always at 100. The space between 0 and 100 units is the in-gamut axis and the space between the red vertical lines (at 100) to the end is the out-of-gamut region. The line can be expressed in $L^*a^*b^*$ space by the line equation with a start point $\{L_0\ a_0\ b_0\}$ and corner point $\{L_c\ a_c\ b_c\}$ shown by Equations 7.95 and 7.96. For the purpose of drawing corner plots, in Equation 7.95, when $i = 0$ the line will start at $x = x_0$. When $i = N$, the line will intersect the gamut corner point $\{L_c\ a_c\ b_c\}$. Note that when the line stimulus is given between $\{100,0,0\}$ and the dark corner $\{0,0,0\}$, the corner plot shows the neutral response of the inversion algorithm (Figure 7.66). Clearly, the round trip accuracy corner plots from (50,0,0) show lower accuracy numbers as compared with similar corner plots from (100,0,0) point.

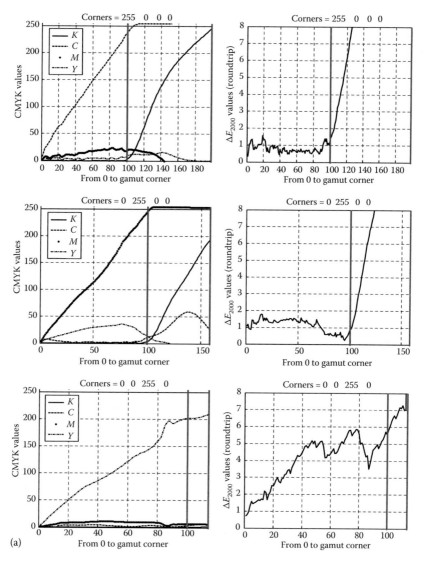

FIGURE 7.65 (a) *CMYK* response (left) and corresponding round trip accuracy (right) to cyan, magenta, yellow corners from 100,0,0 point ('0' on *x*-axis corresponds to 100,0,0 point).

We deliberately chose the profiles with comparatively higher round trip accuracy for near-boundary nodes to highlight the usefulness of these plots in identifying ΔE_{2000} errors. Clearly, the corner plots provide useful indication of the round trip accuracy in critical regions. They also provide a visual indication of the *CMYK* response to continuous $L^*a^*b^*$ stimuli along the line. If the *CMYK* response does not show a smooth continuous function with respect to x-axis, then we can expect

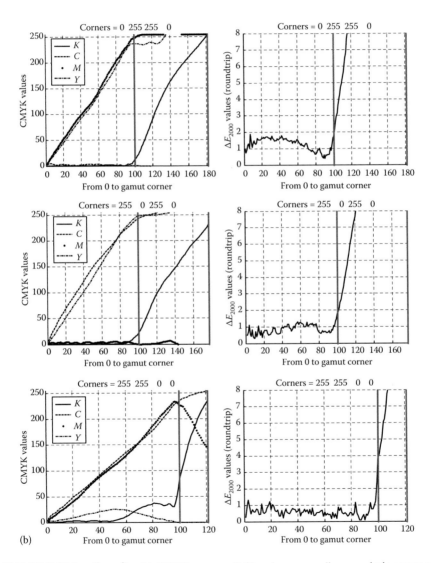

FIGURE 7.65 (continued) (b) *CMYK* response (left) and corresponding round trip accuracy (right) to red, green, blue corners from 100,0,0 point ('0' on x-axis coresponds to 100,0,0 point).

(continued)

undesirable contours when rendering color sweeps (e.g., tints). In Figure 7.65c and d, examples of sharp transitions in the round trip accuracy for dark magenta corner [0 255 0 255] and dark blue corner [255 255 0 255] can be seen, which can lead to contours. In such cases, smoothing methods should be applied to smooth out the transitions. Sometimes, these transitions are caused by sparse nodes in the $L*a*b*$ to $CMYK$ LUT. But, highly dense LUTs can also be built to reduce such jumps in $CMYK$ formulations and improve roundtrip accuracy. But, higher LUT size

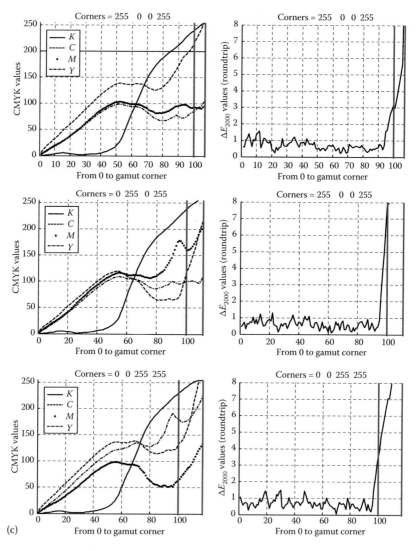

(c)

FIGURE 7.65 (continued)   (c) CMYK response (left) and corresponding round trip accuracy (right) to dark cyan, dark magenta, dark yellow corners from 100,0,0 print ('0' on x-axis corresponds to 100,0,0 point).

requires more memory, more time for creating profiles, and, hence, may not be suitable for general use.

It is also critical to note that the *CMYK* response of out-of-gamut colors to the line stimuli should also be smooth. If discontinuities occur for the out-of-gamut colors, then different mapping strategies may have to be used to preserve the

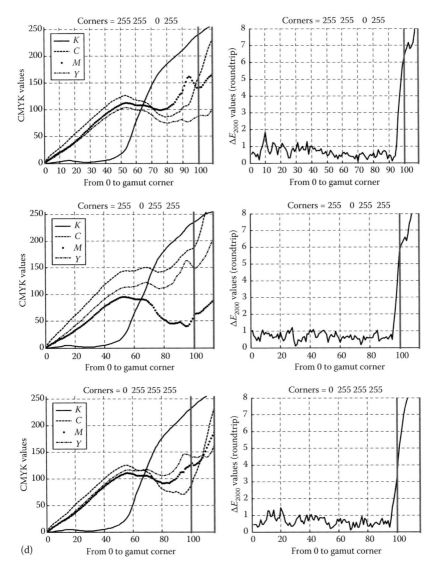

(d)

FIGURE 7.65 (continued) (d) CMYK response (left) and corresponding round trip accuracy (right) to dark blue, dark green, and dark red corners from 100,0,0 point ('0' on x-axis corresponds to 100,0,0 point).

(*continued*)

smoothness. Otherwise, visual artifacts can be expected from the final profile. Thus, the corner plots typically probe deep into the local regions of a color gamut and can be a good indicator of the local inversion and mapping problems.

7.7.3 VISUAL EVALUATION OF PROFILES

No single profile can please everyone. Visual evaluation is largely influenced by the observer characteristics, viewing conditions, and preferences to a certain color and

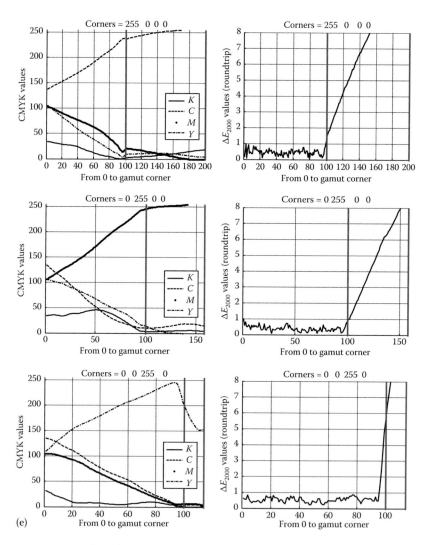

FIGURE 7.65 (continued) (e) CMYK resoponse (left) and corresponding round trip accuracy (right) to cyan, magenta, and yellow corners from 50,0,0 point ('0' on x-axis corresponds to 50,0,0 point).

its appearance with the image content. Often, rating experiments are carried out with weights applied to certain image quality attributes, and attribute tables are populated by comparing images to proofs.

Ultimate color rendering performance on a variety of print media depends on how well the profiles perform for different paper stocks, halftone screens, and print

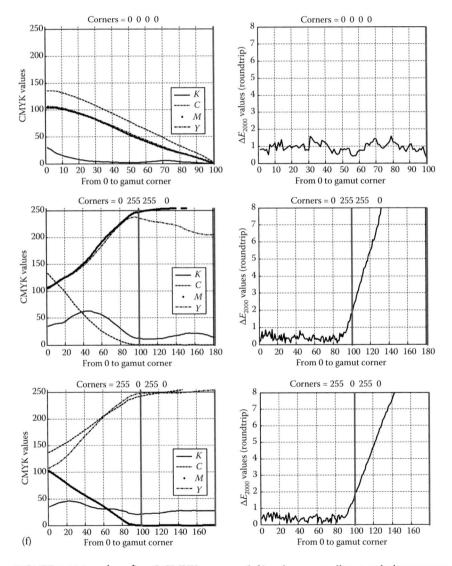

FIGURE 7.65 (continued) (f) CMYK response (left) and corresponding round trip accuracy (right) to white, red, green corners from 50,0,0 point ('0' on x-axis corresponds to 50,0,0 point).

(*continued*)

conditions. Many of the profile-induced image quality defects can be spotted easily by performing soft proofing on a calibrated monitor. For example, by using Adobe PhotoShop software or MATLAB image processing tool box, the blocking in hair can be easily observed, which is a good example of not having proper black point compensation.

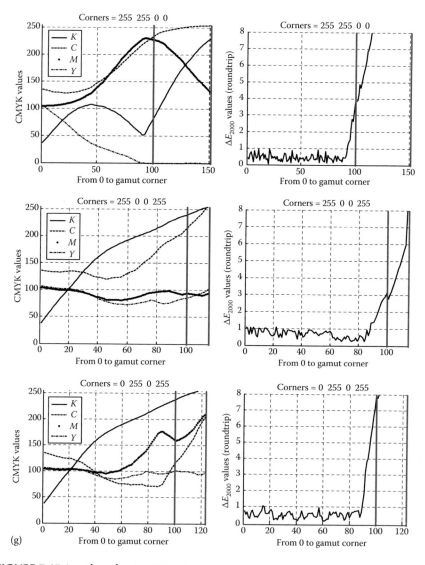

FIGURE 7.65 (continued)   (g) CMYK response (left) and corresponding round trip accuracy (right) to blue, dark cyan, dark yellow corners from 50,0,0 point ('0' on x-axis corresponds to 50,0,0 point).

If the in-gamut colors in the original image do not match the colors in the rendered image, then the errors can be associated to round trip accuracy with the actual printing device. This could be due to inversion errors or the errors in the printer model, if a printer model is used as a surrogate characterization of the printer.

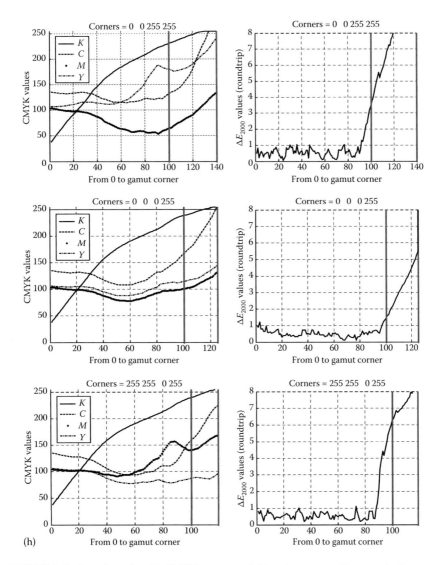

FIGURE 7.65 (continued) (h) CMYK response (left) and corresponding round trip accuracy (right) to dark yellow, dark black, and dark blue corners from 50,0,0 point ('0' on x-axis corresponds to 50,0,0 point).

(*continued*)

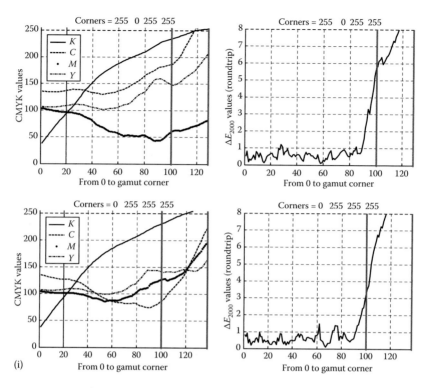

FIGURE 7.65 (continued) (i) CMYK response (left) and corresponding round trip accuracy (right) to dark green, dark red corners from 50,0,0 point ('0' on *x*-axis corresponds to 50,0,0 point).

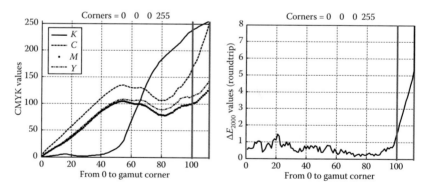

FIGURE 7.66 *CMYK* response and round trip accuracy to neutral axis—gamut corner plot from 100,0,0 point to dark black corner. ('0' on *x*-axis corresponds to 100,0,0 point).

Or else, the error may be only due to the inversion process, if iterations are carried out directly on the printer and the measurement process and system noise are assumed to be well within the reasonable limits when compared to the total round trip accuracy of the actual printing device. Improving accuracy can give rise to

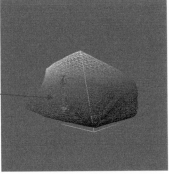

FIGURE 7.67 (See color insert following page 428.) Example of in-gamut color sweeps—susceptible to contours (see Figure 7.39 for performance with MM algorithm vs. the control-based inversion).

reduced contours, balanced neutrals, improved highlights, improved image details and can be an enabler for photo rendering if the halftone screens and printer resolution is up to the desired quality. Emulation with respect to industry color standards like GRACoL, ISO, Japan Color or other source color spaces becomes easy provided the colorants have enough gamut coverage.

Often, the in-gamut colors are shifted deliberately to offset bias in the printing system or to render more preferred outcome. For example, sky colors are generally centered on a 280 degree hue angle and may be purposely detuned with a hue shift for preference matching during the profile creation stage. Any intentionally induced color shifts may not give a good match to the original, but has the effect to render more preferred color. Nonsmooth color transitions can cause contours. In Figure 7.39, grass contours occurred due to an inaccuracy in the profiles created by the MM algorithm. These colors are inside the printer gamut (see Figure 7.67). Also, contouring can occur when color sweeps intersect the gamut boundary. For example, when color sweeps occur from dark out-of gamut color to light in-gamut color as in human arm (Figure 7.68) or sensitive regions, such as in skin tones, we may see gamut-mapping issues like many-to-one mapping. Inaccuracies in the inversion process can also lead to nonideal behavior in gamut-mapped colors as seen in Figure 7.38 and in the corner plots (roundtrip accuracy) shown in Figure 7.65, which is largely due to unpredictability associated with near-boundary colors.

Since GCR plays an important role in high quality color reproduction using toner or inks, when the tone scales appear dirty or grainy or nonuniform with black toner, then tone scale suppression may be required. In addition to this, if the *CMYK* formulations are not smooth between nodes of the profile LUT, we can expect to see contours. Hence the task of achieving smoothness without loosing roundtrip accuracy is a real challenge to high quality reproduction with multidimensional LUTs.

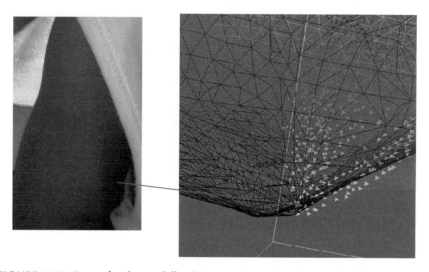

FIGURE 7.68 (See color insert following page 428.)   Example of out-of-gamut color sweeps—susceptible to contours (see Figure 7.40 for performance with MM algorithm vs. the control-based inversion).

7.8   AN EXAMPLE SHOWING HOW TO BUILD MULTIDIMENSIONAL INVERSE LUT

The ICC profile architecture assumes a reference color space in $L^*a^*b^*$ or XYZ values (called PCS) for standard observers, measurement geometry, and number of illuminants. For this example to build an ICC profile for a printing device with media relative colorimetric rendering intent, let us restrict the colorimetric options to a 2 degree observer, and a D50 illuminant source with the PCS defined by $L^*a^*b^*$ values. We adopt relative-to-paper colorimetry to incorporate a standard reference media to reduce paper dependency. That is, the paper white will have the PCS values of {100, 0, 0}. All reflectance measurements are done with a specified measurement geometry (0/45 or 45/0). The ICC profiles with other rendering intents (e.g., perceptual, saturation etc.) can also be easily built by incorporating proper tone scale adjustments, compression, and other compromises to the basic colorimetric profile.

A 3DLUT is a major component of the destination ICC profile which is a GCR-constrained, gamut-mapped transformation LUT from RGB triplets to device $CMYK$ space. Hence we first show the detailed steps involved in building the 3DLUT starting from a printer model. We use the footprint shown in Figure 7.1 to describe the steps. For convenience, we use a low resolution LUT of size 6^3 to show the numerical values. The performance is evaluated for a high resolution 33^3 LUT using the same printer model.

Step 1: Choose a working color space for RGB triplets. There are numerous choices for this (Adobe RGB, Apple RGB, CIE RGB, ColorMatch RGB, NTSC RGB, Wide

Gamut *RGB*, sRGB, *ROMMRGB*, etc.). Gamuts produced by the chosen color space should be compared to the gamut of the printer in Lab space prior to selecting the final working color space so that all colors that are reproducible by the printer gamut are covered by the working color space. To limit our options, we choose ProPhoto *RGB* as the color space. This choice was made because the ProPhoto *RGB* space gives reasonable gamut coverage (see Figure 7.28).

To convert the *RGB* color space to *Lab*, we first determine the matrix for converting *RGB* to *XYZ*, then use the *XYZ* to *Lab* equation (see Appendix A). Given the chromaticity coordinates of the *RGB* system (x_r, y_r), (x_g, y_g), and (x_b, y_b) and its reference white (X_W, Y_W, Z_W), the 3×3 conversion matrix, *M*, for converting *RGB* to *XYZ* is given by

$$[X \quad Y \quad Z] = [R \quad G \quad B]M \tag{7.101}$$

where, $M = \begin{bmatrix} S_r X_r & S_r Y_r & S_r Z_r \\ S_g X_g & S_g Y_g & S_g Z_g \\ S_b X_b & S_b Y_b & S_b Z_b \end{bmatrix}$,

$X_i = x_i/y_i$; $Y_i = 1$; $Z_i = (1 - x_i - y_i)/y_i$ with index $i = r, g, b$ and,

$$[S_r \quad S_g \quad S_b] = [X_W \quad Y_W \quad Z_W] \begin{bmatrix} X_r & Y_r & Z_r \\ X_g & Y_g & Z_g \\ X_b & Y_b & Z_b \end{bmatrix}^{-1} \tag{7.102}$$

The *XYZ* values in Equation 7.101 are in the nominal range (0.0–1.0). For converting *XYZ* to *RGB* the *M* matrix is inverted (i.e., M^{-1}). The *RGB* values must be linear (i.e., Equation 7.101 uses $\gamma = 1$, see also Equations 7.105 and 7.106 for use with $\gamma \neq 1$, which gives non-linear relationship) and in the nominal range (0.0–1.0). They also have to be scaled. If the values are in the range (0, 255), then divide the *RGB* values by 255.0.

It is important to note here that the reference whites are consistent with ICC specifications. For example, if the sRGB color space is used as the working space, then direct use of sRGB gives a mismatch between ICC profiles defined relative to D50 reference white since sRGB is defined relative to D65 light source. Before they can be used in a D50 environment, they should be adopted to a D50 reference white from their native reference white. Chromatic adaptation algorithms can be used to convert the source color (sRGB with D65 reference white) into the destination color (sRGB with D50 reference white) by a linear transformation (Bradford, *XYZ* scaling, and Von Kries are popular transforms), which is dependent on the source reference white and the destination reference white.

For a 2-degree observer and a D50 illuminant [136], $X_W = 0.96422$, $Y_W = 1.0$, $Z_W = 0.82521$. From Appendix A (Table A.2), chromaticity coordinates for a ProPhoto *RGB* system are given by

$$\begin{aligned} x_r &= 0.7347; \quad y_r = 0.2653; \\ x_g &= 0.1596; \quad y_g = 0.8404; \\ x_b &= 0.0366; \quad y_b = 0.0001; \end{aligned} \tag{7.103}$$

Substituting these parameters in Equation 7.102, the conversion matrix, M, is obtained as

$$M = \begin{bmatrix} 0.7977 & 0.2880 & 0 \\ 0.1352 & 0.7119 & 0 \\ 0.0314 & 0.0001 & 0.8252 \end{bmatrix} \qquad (7.104)$$

Given an RGB color such that its components are in the nominal range [0.0, 1.0] and its gamma is γ and the RGB system is not sRGB:

$$[X \quad Y \quad Z] = [r \quad g \quad b]M \qquad (7.105)$$

where

$$r = R^{\gamma}; \quad g = G^{\gamma}; \quad b = B^{\gamma} \qquad (7.106)$$

For sRGB, r, g, b values can be found from the following equations:

$$r = \begin{cases} R/12.92 & R \leq 0.04045 \\ ((R+0.055)/1.055)^{2.4} & R > 0.04045 \end{cases}$$

$$g = \begin{cases} G/12.92 & G \leq 0.04045 \\ ((G+0.055)/1.055)^{2.4} & G > 0.04045 \end{cases} \qquad (7.107)$$

$$b = \begin{cases} B/12.92 & B \leq 0.04045 \\ ((B+0.055)/1.055)^{2.4} & B > 0.04045 \end{cases}$$

Now, given the working space definition for RGB color space, we know the transformation matrix. Using a proper choice of γ (e.g., $\gamma = 1.8$ for ProPhoto RGB) we can determine the XYZ values and hence $L*a*b*$ values.

Step 2: Form a uniformly sampled RGB LUT (e.g., 17^3). Convert the RGB colors to XYZ and then to $L*a*b*$ using Equation 7.105 and the equations shown in Appendix A for a 2-degree observer and D50 illuminant.

A low resolution LUT of size 6^3 was formed to show numerical values by varying the first channel R least rapidly and varying third channel B most rapidly between 0 and 255 with a uniformly sampled RGB grid. In this example, the color index 1 corresponds to first grid point with $R = 0 = G = B$ and index 216 corresponds to the last grid point $R = G = B = 255$ with increments in between, according to the uniform sampling approach described above.

Step 3: Separate $L*a*b*$ nodes into in-gamut and out-of-gamut colors.

After creating the RGB grid points, the data is converted to $L*a*b*$ and then separated into two groups: in-gamut and out-of-gamut colors. We used the ray-based control model described in Section 7.6.1 to separate the colors into two groups. These colors are the nodes in L*a*b* space at which the device CMYK values are found that populate the multidimensional CLUT of the ICC profile. The RGB and the corresponding L*a*b* node values are shown in Table 7.15 for in-gamut colors. The out-of-gamut nodes are shown in Tables 7.16 and 7.17.

Step 4: Perform gamut mapping for out-of-gamut colors.

In this example, we used a simple centroid clipping (Section 7.6.2, Figure 7.57) with ray-based control model. The gamut-mapped L*a*b* and their corresponding CMYK values are shown in Tables 7.16 and 7.17. Clearly, all of the out-of-gamut nodes are mapped to the gamut boundary.

Step 5: Perform constrained GCR inversion for in-gamut colors.

We used the K-restricted GCR method during the inversion process. Table 7.15 contains the CMYK values corresponding to each in-gamut node. Now rearrange the in-gamut and out-of-gamut RGB to CMYK LUTs to form a single LUT with color index incremented from 1 to 216. To create an ICC profile, this LUT can be used to fill the CLUT inside the ICC structure.

TABLE 7.15
RGB → L*a*b* and L*a*b* → CMYK LUT for In-Gamut Node Colors

Color Index	RGB Values			Node $L^*a^*b^*$ (Out-of-Gamut Colors)			GCR Constrained CMYK			
	R	G	B	L^*	a^*	b^*	C	M	Y	K
44	51	51	51	28.16	0.00	0.00	111	80	85	182
45	51	51	102	28.17	4.98	−39.26	207	142	12	92
50	51	102	51	46.00	−56.96	30.76	221	9	246	21
51	51	102	102	46.00	−52.85	−8.51	227	14	145	17
80	102	51	51	36.86	48.31	15.00	6	218	105	110
81	102	51	102	36.87	50.72	−24.27	101	216	4	25
86	102	102	51	50.94	−2.24	39.27	107	88	209	55
87	102	102	102	50.94	0.00	0.00	137	105	107	28
88	102	102	153	50.94	3.31	−31.78	155	102	27	11
122	153	102	51	57.24	39.55	50.12	21	160	206	18
123	153	102	102	57.24	41.01	10.86	39	154	89	12
124	153	102	153	57.24	43.19	−20.93	61	151	4	1
128	153	153	51	69.38	−3.46	71.05	62	46	229	20
129	153	153	102	69.38	−2.07	31.79	80	67	141	5
130	153	153	153	69.38	0.00	0.00	96	72	73	2
170	204	204	51	85.46	−4.38	98.79	20	14	249	14
171	204	204	102	85.46	−3.40	59.52	30	30	162	4
172	204	204	153	85.46	−1.92	27.73	38	33	94	2
173	204	204	204	85.46	0.00	0.00	51	32	33	3

TABLE 7.16

RGB → L*a*b* → CMYK **LUT for Out-of-Gamut Node Colors (between Color Index 1–108)**

Color Index	RGB Values			Node $L^*a^*b^*$ (Out-of-gamut Colors)			Gamut Mapped $L^*a^*b^*$			Gamut Mapped CMYK			
	R	G	B	L^*	a^*	b^*	L^*	a^*	b^*	C	M	Y	K
1	0	0	0	0.00	0.00	0.00	10.60	1.16	2.29	143	112	120	255
2	0	0	51	0.00	6.97	−48.55	14.85	6.88	−33.44	255	183	0	158
3	0	0	102	0.01	24.27	−87.80	20.92	18.18	−52.99	255	210	0	13
4	0	0	153	0.03	48.36	−119.56	34.25	24.78	−44.60	193	200	0	0
5	0	0	204	0.05	70.40	−147.26	40.62	27.67	−39.81	166	187	0	0
6	0	0	255	0.08	90.29	−172.28	44.26	29.21	−36.94	152	179	0	0
7	0	51	0	23.43	−70.88	40.40	36.81	−42.98	30.58	209	1	255	113
8	0	51	51	23.44	−63.97	−8.15	31.98	−47.73	−3.26	255	0	175	119
9	0	51	102	23.44	−49.53	−47.42	33.84	−33.81	−30.47	255	11	67	110
10	0	51	153	23.45	−32.73	−79.19	35.71	−19.75	−48.73	255	56	9	62
11	0	51	204	23.45	−15.57	−106.91	39.98	−5.69	−53.50	220	107	0	0
12	0	51	255	23.46	1.26	−131.96	45.37	4.42	−46.37	186	116	0	0
13	0	102	0	43.77	−107.74	75.47	52.27	−49.99	43.98	202	5	255	15
14	0	102	51	43.77	−104.48	26.91	50.00	−63.38	21.80	233	0	224	0
15	0	102	102	43.77	−96.96	−12.36	49.75	−60.19	−3.32	233	0	161	0
16	0	102	153	43.78	−86.79	−44.14	49.75	−53.63	−23.82	238	0	106	0
17	0	102	204	43.78	−75.07	−71.87	49.92	−45.26	−40.85	243	0	59	0
18	0	102	255	43.78	−62.53	−96.92	50.26	−35.91	−54.56	244	0	14	0
19	0	153	0	60.23	−137.41	103.85	60.46	−45.95	46.23	183	0	238	0
20	0	153	51	60.23	−135.39	55.29	60.46	−47.34	29.14	186	0	206	0
21	0	153	102	60.24	−130.54	16.02	60.45	−47.56	14.26	189	0	167	0
22	0	153	153	60.24	−123.66	−15.76	60.45	−46.79	1.27	191	0	130	0
23	0	153	204	60.24	−115.29	−43.49	60.44	−45.19	−10.94	193	0	102	0
24	0	153	255	60.24	−105.88	−68.55	60.44	−42.32	−22.41	197	0	75	0
25	0	204	0	74.60	−163.30	128.62	64.52	−40.96	45.45	167	0	222	0
26	0	204	51	74.60	−161.86	80.06	64.66	−42.24	32.53	170	0	199	0
27	0	204	102	74.60	−158.37	40.79	64.80	−42.88	21.41	173	0	172	0
28	0	204	153	74.60	−153.29	9.00	64.87	−42.15	11.86	174	0	146	0
29	0	204	204	74.60	−146.96	−18.73	65.01	−41.76	3.08	175	0	119	0
30	0	204	255	74.60	−139.63	−43.79	65.08	−40.18	−5.10	175	0	101	0
31	0	255	0	87.58	−186.69	150.99	66.94	−38.81	45.51	158	0	216	0
32	0	255	51	87.58	−185.59	102.43	67.07	−39.51	34.55	160	0	194	0
33	0	255	102	87.58	−182.90	63.16	67.21	−39.81	25.37	163	0	173	0
34	0	255	153	87.58	−178.93	31.38	67.34	−39.77	17.68	165	0	153	0
35	0	255	204	87.58	−173.91	3.65	67.48	−39.41	10.70	165	0	132	0
36	0	255	255	87.58	−168.01	−21.41	67.61	−38.78	4.13	165	0	114	0
37	51	0	0	13.17	52.99	22.70	25.02	41.39	20.30	0	255	178	163
38	51	0	51	13.17	55.28	−25.85	23.60	44.58	−17.28	112	255	0	112
39	51	0	102	13.18	60.75	−65.11	30.01	41.53	−37.34	173	231	0	0
40	51	0	153	13.19	68.42	−96.88	38.79	35.05	−37.48	157	202	0	0
41	51	0	204	13.20	77.63	−124.59	43.77	31.83	−35.77	147	183	0	0

TABLE 7.16 (continued)

RGB → L*a*b* → CMYK LUT for Out-of-Gamut Node Colors (between Color Index 1–108)

Color Index	RGB Values			Node L*a*b* (Out-of-gamut Colors)			Gamut Mapped L*a*b*			Gamut Mapped CMYK			
	R	G	B	L*	a*	b*	L*	a*	b*	C	M	Y	K
42	51	0	255	13.22	87.85	−149.62	46.62	30.58	−34.89	139	172	0	0
43	51	51	0	28.16	−2.08	48.56	35.46	−0.12	40.58	45	51	255	180
46	51	51	153	28.17	12.03	−71.04	36.28	10.68	−50.01	208	156	0	0
47	51	51	204	28.18	20.58	−98.77	44.23	13.67	−43.39	177	147	0	0
48	51	51	255	28.19	30.17	−123.82	47.96	15.80	−40.29	163	141	0	0
49	51	102	0	46.00	−58.67	79.31	52.35	−30.27	50.52	163	23	255	53
52	51	102	153	46.01	−46.94	−40.30	46.59	−44.80	−38.16	255	5	69	13
53	51	102	204	46.01	−39.65	−68.02	48.13	−32.94	−56.26	255	10	11	1
54	51	102	255	46.01	−31.33	−93.08	52.14	−15.39	−48.48	201	67	0	0
55	51	153	0	61.63	−101.34	106.26	61.07	−43.04	55.96	177	0	255	0
56	51	153	51	61.63	−99.97	57.70	61.11	−46.14	35.18	183	0	214	0
57	51	153	102	61.63	−96.65	18.43	61.14	−47.08	15.88	186	0	170	0
58	51	153	153	61.63	−91.82	−13.35	61.15	−45.55	−0.91	188	0	120	0
59	51	153	204	61.64	−85.76	−41.08	61.16	−43.27	−16.15	192	0	89	0
60	51	153	255	61.64	−78.74	−66.14	61.15	−39.05	−29.29	195	0	52	0
61	51	204	0	75.59	−135.49	130.33	65.40	−38.86	50.62	161	0	229	0
62	51	204	51	75.59	−134.39	81.77	65.62	−40.62	36.11	165	0	203	0
63	51	204	102	75.59	−131.68	42.51	65.77	−41.09	23.25	168	0	172	0
64	51	204	153	75.60	−127.71	10.72	65.92	−41.06	12.26	171	0	143	0
65	51	204	204	75.60	−122.67	−17.01	66.00	−39.92	2.27	170	0	114	0
66	51	204	255	75.60	−116.76	−42.07	66.07	−38.41	−7.03	171	0	95	0
67	51	255	0	88.34	−164.40	152.31	67.67	−36.99	48.61	154	0	219	0
68	51	255	51	88.34	−163.49	103.75	67.95	−38.46	37.13	157	0	198	0
69	51	255	102	88.34	−161.24	64.48	68.09	−38.70	26.98	159	0	174	0
70	51	255	153	88.34	−157.92	32.70	68.23	−38.63	18.50	161	0	152	0
71	51	255	204	88.34	−153.69	4.97	68.37	−38.26	10.83	162	0	129	0
72	51	255	255	88.35	−148.67	−20.09	68.51	−37.64	3.65	162	0	111	0
73	102	0	0	28.21	80.31	48.64	37.76	58.57	38.16	0	255	215	72
74	102	0	51	28.21	81.32	0.08	35.66	64.14	3.08	8	255	38	78
75	102	0	102	28.21	83.79	−39.19	37.93	60.64	−23.50	95	232	0	0
76	102	0	153	28.22	87.44	−70.97	44.41	47.03	−28.93	103	195	0	0
77	102	0	204	28.22	92.08	−98.69	47.97	39.95	−30.49	108	176	0	0
78	102	0	255	28.23	97.55	−123.74	50.24	35.72	−30.68	109	159	0	0
79	102	51	0	36.86	47.33	63.56	42.44	37.77	51.70	1	188	255	103
82	102	51	153	36.87	54.26	−56.05	44.94	38.06	−32.54	123	184	0	0
83	102	51	204	36.87	58.77	−83.78	49.45	31.13	−32.43	123	158	0	0
84	102	51	255	36.88	64.10	−108.83	51.82	27.89	−32.01	121	150	0	0
85	102	102	0	50.94	−3.15	87.83	54.17	0.12	62.80	56	52	255	107
89	102	102	204	50.95	7.54	−59.51	53.70	7.28	−38.82	154	104	0	0
90	102	102	255	50.95	12.55	−84.57	55.72	9.62	−36.22	143	105	0	0

(continued)

TABLE 7.16 (continued)

RGB → L*a*b* → CMYK LUT for Out-of-Gamut Node Colors (between Color Index 1–108)

Color Index	RGB Values			Node $L^*a^*b^*$ (Out-of-gamut Colors)			Gamut Mapped $L^*a^*b^*$			Gamut Mapped CMYK			
	R	G	B	L^*	a^*	b^*	L^*	a^*	b^*	C	M	Y	K
91	102	153	0	64.90	−49.22	111.89	62.86	−22.69	64.97	124	18	255	30
92	102	153	51	64.90	−48.39	63.34	64.19	−39.31	55.05	165	0	244	0
93	102	153	102	64.90	−46.35	24.07	64.60	−42.64	23.30	173	0	177	0
94	102	153	153	64.90	−43.32	−7.72	64.64	−40.33	−6.47	176	0	99	0
95	102	153	204	64.90	−39.45	−35.45	64.45	−34.61	−30.35	180	0	38	0
96	102	153	255	64.91	−34.84	−60.51	63.83	−24.48	−42.10	173	16	0	0
97	102	204	0	77.98	−88.92	134.45	67.98	−33.98	64.68	149	0	252	0
98	102	204	51	77.98	−88.18	85.89	68.51	−36.51	46.22	151	0	212	0
99	102	204	102	77.98	−86.34	46.62	68.85	−37.54	29.03	156	0	176	0
100	102	204	153	77.98	−83.61	14.83	69.11	−37.59	13.95	159	0	136	0
101	102	204	204	77.98	−80.11	−12.90	69.12	−35.88	0.37	158	0	102	0
102	102	204	255	77.98	−75.93	−37.96	69.12	−33.83	−11.91	160	0	74	0
103	102	255	0	90.19	−123.32	155.51	69.77	−33.66	57.25	143	0	229	0
104	102	255	51	90.19	−122.65	106.95	70.22	−35.39	43.61	146	0	204	0
105	102	255	102	90.19	−121.01	67.68	70.51	−36.13	31.39	150	0	176	0
106	102	255	153	90.20	−118.56	35.89	70.66	−35.94	20.86	153	0	151	0
107	102	255	204	90.20	−115.41	8.16	70.81	−35.48	11.40	153	0	121	0
108	102	255	255	90.20	−111.63	−16.90	70.81	−34.18	2.76	152	0	102	0

TABLE 7.17

RGB → L*a*b* → CMYK LUT for Out-of-Gamut Node Colors (between Color Index 109–216)

Color Index	RGB Values			Node $L^*a^*b^*$ (Out-of-Gamut Colors)			Gamut Mapped $L^*a^*b^*$			Gamut Mapped CMYK			
	R	G	B	L^*	a^*	b^*	L^*	a^*	b^*	C	M	Y	K
109	153	0	0	40.39	102.43	69.63	47.36	69.38	50.13	0	251	221	0
110	153	0	51	40.39	103.05	21.07	46.15	75.57	18.80	0	247	95	0
111	153	0	102	40.39	104.59	−18.20	46.55	74.71	−8.65	26	233	0	0
112	153	0	153	40.39	106.87	−49.98	50.29	57.75	−19.06	60	204	0	0
113	153	0	204	40.39	109.83	−77.71	52.52	47.90	−23.22	70	177	0	0
114	153	0	255	40.40	113.37	−102.76	53.83	42.38	−25.71	78	156	0	0
115	153	51	0	46.20	80.07	79.65	49.87	61.34	62.68	0	224	255	9
116	153	51	51	46.20	80.68	31.09	48.07	71.05	28.75	0	237	128	0
117	153	51	102	46.20	82.19	−8.18	47.28	76.52	−6.58	14	234	0	0
118	153	51	153	46.20	84.45	−39.96	51.74	54.48	−19.53	61	197	0	0
119	153	51	204	46.20	87.36	−67.69	53.90	44.16	−24.46	73	159	0	0

TABLE 7.17 (continued)

RGB → L*a*b* → CMYK LUT for Out-of-Gamut Node Colors (between Color Index 109–216)

Color Index	RGB Values			Node $L^*a^*b^*$ (Out-of-Gamut Colors)			Gamut Mapped $L^*a^*b^*$			Gamut Mapped CMYK			
	R	G	B	L^*	a^*	b^*	L^*	a^*	b^*	C	M	Y	K
120	153	51	255	46.21	90.85	−92.75	55.27	37.79	−26.06	83	148	0	0
121	153	102	0	57.24	38.96	98.68	58.33	28.45	70.87	14	121	255	50
125	153	102	204	57.24	46.00	−48.66	58.47	31.63	−26.11	86	130	0	0
126	153	102	255	57.24	49.38	−73.72	59.02	26.72	−27.70	96	123	0	0
127	153	153	0	69.38	−4.02	119.61	65.91	0.18	77.55	45	43	255	61
131	153	153	204	69.38	2.68	−27.73	69.38	2.68	−27.73	105	66	0	0
132	153	153	255	69.38	5.90	−52.79	66.22	6.16	−29.07	111	82	0	0
133	153	204	0	81.36	−44.12	140.28	71.08	−19.00	77.33	94	8	255	23
134	153	204	51	81.36	−43.59	91.72	74.72	−27.52	66.56	115	0	231	0
135	153	204	102	81.36	−42.29	52.45	75.75	−29.08	41.83	117	0	178	0
136	153	204	153	81.36	−40.34	20.66	76.48	−29.30	18.89	124	0	118	0
137	153	204	204	81.36	−37.81	−7.07	76.28	−26.92	−2.13	126	0	74	0
138	153	204	255	81.36	−34.77	−32.13	75.76	−23.59	−19.92	129	0	29	0
139	153	255	0	92.87	−80.50	160.12	73.35	−27.79	71.18	120	0	247	0
140	153	255	51	92.87	−80.01	111.56	74.15	−29.76	54.46	120	0	210	0
141	153	255	102	92.87	−78.79	72.29	74.64	−30.53	38.86	124	0	177	0
142	153	255	153	92.87	−76.97	40.51	75.12	−30.99	25.44	129	0	144	0
143	153	255	204	92.87	−74.61	12.77	75.28	−30.34	12.96	131	0	110	0
144	153	255	255	92.87	−71.76	−12.29	75.28	−29.04	1.56	132	0	87	0
145	204	0	0	51.01	121.73	87.95	56.52	55.54	44.92	0	206	189	0
146	204	0	51	51.01	122.17	39.39	56.28	58.62	24.94	0	196	113	0
147	204	0	102	51.01	123.26	0.12	55.95	63.19	6.81	0	202	66	0
148	204	0	153	51.01	124.89	−31.67	55.71	66.94	−9.72	0	199	3	0
149	204	0	204	51.02	127.01	−59.39	56.67	55.98	−16.62	37	181	0	0
150	204	0	255	51.02	129.57	−84.45	57.34	48.42	−20.06	54	159	0	0
151	204	51	0	55.28	105.21	95.31	58.15	51.97	50.92	0	197	207	0
152	204	51	51	55.28	105.65	46.76	58.02	54.65	29.43	0	190	127	0
153	204	51	102	55.28	106.73	7.49	57.81	59.18	10.16	0	190	76	0
154	204	51	153	55.29	108.35	−24.30	57.62	63.59	−7.84	0	190	14	0
155	204	51	204	55.29	110.45	−52.03	58.18	53.86	−16.53	35	171	0	0
156	204	51	255	55.29	112.99	−77.08	58.63	45.98	−20.26	54	152	0	0
157	204	102	0	64.13	71.74	110.56	62.63	44.07	69.15	0	158	237	0
158	204	102	51	64.13	72.17	62.00	62.70	45.62	42.20	0	176	173	0
159	204	102	102	64.13	73.23	22.73	62.82	48.58	19.17	0	155	96	0
160	204	102	153	64.13	74.81	−9.05	62.98	52.65	−1.85	0	155	35	0
161	204	102	204	64.13	76.87	−36.78	62.68	48.07	−16.33	31	149	0	0
162	204	102	255	64.13	79.36	−61.84	62.19	39.35	−20.62	56	132	0	0
163	204	153	0	74.61	33.38	128.64	69.21	23.08	84.16	6	98	255	18
164	204	153	51	74.61	33.79	80.08	72.23	29.17	68.69	0	111	216	0
165	204	153	102	74.61	34.82	40.81	72.58	30.73	36.79	0	115	134	0

(continued)

TABLE 7.17 (continued)

$RGB \rightarrow L^*a^*b^* \rightarrow CMYK$ LUT for Out-of-Gamut Node Colors (between Color Index 109–216)

Color Index	R	G	B	L^*	a^*	b^*	L^*	a^*	b^*	C	M	Y	K
	RGB Values			Node $L^*a^*b^*$ (Out-of-Gamut Colors)			Gamut Mapped $L^*a^*b^*$			Gamut Mapped CMYK			
166	204	153	153	74.61	36.35	9.02	73.00	32.93	9.49	0	105	59	0
167	204	153	204	74.61	38.35	−18.71	72.51	33.59	−13.93	28	105	0	0
168	204	153	255	74.62	40.76	−43.77	68.73	26.42	−19.88	60	103	0	0
169	204	204	0	85.46	−4.78	147.35	74.65	0.18	88.95	29	35	255	35
174	204	204	255	85.47	2.32	−25.06	80.62	3.16	−17.62	68	40	0	0
175	204	255	0	96.19	−40.93	165.84	77.86	−16.44	87.19	73	3	255	15
176	204	255	51	96.19	−40.55	117.28	81.42	−20.97	74.05	82	0	221	0
177	204	255	102	96.19	−39.61	78.01	82.84	−22.27	53.68	83	0	178	0
178	204	255	153	96.19	−38.19	46.23	83.38	−22.06	34.31	85	0	124	0
179	204	255	204	96.19	−36.36	18.49	83.73	−21.32	16.61	89	0	91	0
180	204	255	255	96.19	−34.13	−6.57	83.38	−19.46	0.52	94	0	54	0
181	255	0	0	60.61	139.17	104.50	60.60	49.70	42.81	0	187	177	0
182	255	0	51	60.61	139.50	55.94	60.60	51.14	27.46	0	176	119	0
183	255	0	102	60.61	140.34	16.67	60.60	53.42	14.36	0	166	85	0
184	255	0	153	60.61	141.59	−15.12	60.60	55.89	2.81	0	175	50	0
185	255	0	204	60.61	143.22	−42.85	60.60	58.53	−8.15	0	174	15	0
186	255	0	255	60.62	145.21	−67.90	60.60	54.43	−14.84	24	160	0	0
187	255	51	0	63.94	126.25	110.25	61.73	47.29	46.14	0	180	184	0
188	255	51	51	63.94	126.58	61.69	61.77	48.61	30.11	0	173	128	0
189	255	51	102	63.95	127.41	22.42	61.82	50.71	16.51	0	158	90	0
190	255	51	153	63.95	128.66	−9.37	61.85	52.39	4.68	0	158	54	0
191	255	51	204	63.95	130.28	−37.10	61.94	56.09	−6.97	0	161	20	0
192	255	51	255	63.95	132.25	−62.15	61.84	53.10	−14.74	21	157	0	0
193	255	102	0	71.20	98.54	122.77	64.68	42.01	55.33	0	156	205	0
194	255	102	51	71.20	98.87	74.21	64.79	43.06	37.24	0	160	154	0
195	255	102	102	71.21	99.69	34.94	64.95	44.78	22.06	0	150	103	0
196	255	102	153	71.21	100.92	3.15	64.89	44.81	9.08	0	136	65	0
197	255	102	204	71.21	102.52	−24.58	65.37	49.77	−3.85	0	148	28	0
198	255	102	255	71.21	104.46	−49.64	65.21	49.18	−14.18	17	145	0	0
199	255	153	0	80.29	64.70	138.42	69.65	33.34	70.75	0	122	225	0
200	255	153	51	80.29	65.02	89.86	69.85	34.07	49.18	0	127	176	0
201	255	153	102	80.29	65.82	50.59	70.15	35.33	31.29	0	129	123	0
202	255	153	153	80.29	67.02	18.81	70.25	36.22	15.90	0	117	81	0
203	255	153	204	80.29	68.59	−8.92	70.74	38.54	1.76	0	114	39	0
204	255	153	255	80.29	70.49	−33.98	71.23	41.11	−12.32	11	117	0	0
205	255	204	0	90.06	29.31	155.28	77.25	19.44	93.44	0	85	255	0
206	255	204	51	90.06	29.63	106.72	77.84	20.08	67.88	0	89	203	0
207	255	204	102	90.07	30.41	67.46	78.28	20.89	45.72	0	89	142	0
208	255	204	153	90.07	31.58	35.67	79.16	22.34	27.32	0	93	101	0
209	255	204	204	90.07	33.10	7.94	79.75	23.83	9.75	0	84	50	0

TABLE 7.17 (continued)
RGB → *L*a*b** → *CMYK* LUT for Out-of-Gamut Node Colors
(between Color Index 109–216)

Color Index	RGB Values			Node $L^*a^*b^*$ (Out-of-Gamut Colors)			Gamut Mapped $L^*a^*b^*$			Gamut Mapped CMYK			
	R	G	B	L^*	a^*	b^*	L^*	a^*	b^*	C	M	Y	K
210	255	204	255	90.07	34.95	−17.12	81.67	26.88	−8.51	0	83	0	0
211	255	255	0	100.00	−5.47	172.41	81.48	0.22	97.54	14	27	255	20
212	255	255	51	100.00	−5.16	123.85	90.93	−2.45	98.38	0	16	231	0
213	255	255	102	100.00	−4.41	84.58	91.72	−2.09	69.57	0	22	171	0
214	255	255	153	100.00	−3.27	52.79	93.50	−1.64	46.24	0	20	111	0
215	255	255	204	100.00	−1.79	25.06	96.26	−0.99	23.93	0	13	64	0
216	255	255	255	100.00	0.00	0.00	100.00	0.00	0.00	1	1	0	0

PROBLEMS

7.1 Using the RLS algorithm, obtain a global linear model for a *CMY* gray color [$C = 127 = M = Y$, $K = 0$]. The solution should include some details of the experimental procedure.

7.2 Using the RLS algorithm, obtain a local linear model for a *CMY* gray color [$C = 127 = M = Y$, $K = 0$]. The solution should include some details of the experimental procedure.

7.3 Using the RLS algorithm, obtain a quadratic and cubic model for the printer. Calculate the modeling errors with respect to the virtual printer model shown in Chapter 10. Is a quadratic and cubic model a good choice for this virtual printer?

7.4 In a printing system *CMYK* values are sampled between 0 and 255 in a uniform grid to obtain up to 10^4 input–output characterization data. Patches are printed and measured by the spectrophotometer. The spectra are measured at 31 wavelengths from 400 to 700 nm at 10 nm intervals. Use PCA mathematical techniques to analyze the set of multivariate input–output characterization data. Show the ΔE (ΔE^*_{ab}, ΔE_{2000}) error plot with respect to number of principal components. Determine the parameter matrix using least squares. Plot the model accuracy for the test colors as a function of the number of basis vectors. How many basis vectors are needed to obtain a reasonably accurate model?

7.5 Use bootstrapping techniques to gain confidence in the PCA-based modeling.

7.6 A clustered PCA-based model is to be constructed for a 10^4 spectral characterization data. Use the *K*-means algorithm to cluster the color data in spectral space. Obtain PCA vectors for each cluster. Model each cluster with the RLS algorithm. Plot the performance of your model for test data as a function of the number of clusters.

7.7   Simulate a node color convergence near the gamut boundary using a pole-placement algorithm. Show ΔE_{ab}^{*} convergence errors with and without the best actuator algorithm. Use the Neugebauer model to represent the printer.

7.8   Simulate a node color convergence using a pole placement algorithm and a 4 to 3 control-based inversion. Use the Neugebauer model to represent the printer. Choose any suitable GCR as nominal *CMYK* values for the node color. Show ΔE_{ab}^{*} convergence errors with and without the best actuator algorithm.

7.9   Extend Problem 7.8 for additional in-gamut nodes (say, seven gray targets with L^* of 25–90, in steps of 10 with $a^* = b^* = 0$) along the neutral axis. If the nodes are not in-gamut change the L^* values so that they are inside the printer's gamut. Change nominal *CMYK* values and simulate the converge performance. Do you still see near zero convergence error? Restrict the K separation using K-restricted function of Equation 7.90 and simulate converge performance with 3-to-3 control loop. What do you see?

7.10   Run a 4 to 3 control loop for the nodes in Problem 7.9. Is the convergence error near zero at steady state? Plot actual K separation values along the neutral axis from 4 to 3 control loop and superimpose the values on those obtained from Equation 7.90. What do you see? Explain.

7.11   Produce a ray-based control model for mapping an out-of-gamut spot color described with, $x_c = [L_c = 80, a_c = 50, b_c = 30]$, and the gamut centroid, $x_0 = [L_0 = 50, a_0 = 0, b_0 = 0]$. Repeat this for another ray drawn toward the same gamut centroid for another suitable out-of-gamut node color.

7.12   Black point compensation algorithm is used to retain shadow details in images since many images contain colors that are darker than the darkest color a printer can make. Using the algorithm described in Section 7.6.3 or a nonlinear input L^* to output L^* mapping technique, design a black point compensation method that can retain shadow details. Simulate the effects of parameters on shadow details.

7.13   Repeat the profiling example shown in Section 7.8 to build a 17^3 destination ICC profile using MATLAB. Obtain Gamut corner plots and round trip accuracy (overall and to corner points). Simulate the performance of some test images.

REFERENCES

1. International Color Consortium Specification, ICC. 1:2004-10 (Profile version 4.2.0.0), Image technology colour management—Architecture, profile format, and data structure.
2. P. Heuberger, B. Ninness, T. Oliveira e Silva, P. Van den Hof, and B. Wahlberg, Modeling and identification with orthogonal basis functions, *36th IEEE CDC Precon-ference Workshop# 7*, San Diego, CA, Dec. 1997.
3. L. Ljung, *System Identification—Theory for the User*, 2nd edn., PTR Prentice Hall, Upper Saddle River, NJ, 1999.
4. S.R. Schmidt and R.G. Launsby, *Understanding Industrial Designed Experiments*, AIR Academy Press & Associates, Colorado Springs, CO, ISBN 1-880156-03-2, 2005.

5. J. Hardeberg and F. Schmitt, Color printer characterization using a computational geometry approach, in *IS&T and SID's 5th Color Imaging Conference: Color Science, Systems and Applications*, Scottsdale, AZ, 1997.

6. P.C. Hung, Colorimetric calibration in electronic imaging devices using a look-up table model and interpolations, *Electronic Imaging*, 2, 53–61, 1993.

7. L.K. Mestha, F.F. Hubble II, T.L. Love, G. Skinner, K.J. Mihalyov, D.A. Robbins, and D.M. Diehl, Low cost LED based spectrophotometer, in *Proceedings of International Congress of Imaging Science (ICIS'06)*, Rochester, NY, May 7–12, 2006.

8. K.D. Vincent, Colorimeter and calibration system, US Patent 5,272,518, Dec. 21, 1993; Other US Patents in Color Sensor 7,262,853; 4,917,500; 5,137,364; 5,107,332; 5,150,174; 5,272,518; 5,303,037; 5,377,000; 5,671,059; 5,838,451; 5,844,680; 5,963,333; 6,020,583; 6,075,595; 6,147,761; 6,157,454; 6,538,770; 6,556,300; 6,567,170; 6,574,425; 6,603,551.

9. W.D. Jung, R.W. Jung, and A.R. Loudermilk, Apparatus and method for measuring color, US Patent 7,400,404, Jul. 15, 2008; Other US Patents 7,397,541; 7,397,562.

10. S.H. Peterson and M.A. Cargill, Color measurement instrument, US Patent 7,262,853, Aug. 28, 2007.

11. L.K. Mestha and O.Y. Ramirez, On-line model prediction and calibration system for a dynamically varying color reproduction device, US Patent 6,809,837, Oct. 26, 2004; US Patent 5,818,960 (Kodak); 5,612,902 (Apple).

12. L.K. Mestha, S.A. Dianat, F.G. Polo, and G.W. Skinner, Reference database and method for determining spectra using measurements from an LED color sensor, and method of generating a reference database, US Patent 7,383,261, Jun. 3, 2007.

13. T. Jaaskelainen, J. Parkkinen, and S. Toyooka, Vector-subspace model for color representation, *Journal of the Optical Society of America A*, 7, 725–730, Apr. 1990.

14. J.P.S. Parkkinen, J. Hallikainen, and T. Jaaskelainen, Characteristic spectra of Munsell colors, *Journal of the Optical Society of America A*, 6, 318–322, Feb. 1989.

15. L.K. Mestha and S.A. Dianat, Systems and methods for determining spectra using dynamic Karhunen–Loeve algorithms with measurements from LED color sensor, US Patent 6,584,435, Jun. 24, 2003.

16. L.K. Mestha, Y.R. Wang, S.A. Dianat, D.E. Koditschek, E. Jackson, and T.E. Thieret, Coordinitization of tone reproduction curve in terms of basis functions, US Patent 5,749,020, May 5, 1997.

17. M. Shaw, G. Sharma, R. Bala, and E.N. Dalal, Color printer characterization adjustment for different substrates, *Color Research & Application*, 28(6), 454–467, Oct. 2003.

18. K. Ioka, T. Wada, and K. Ishii, Method and apparatus for calculating image correction data and projection system, US Patent 7,129,456, Oct. 31, 2006; Other US Patent 7,347,525.

19. A. Hyvarinen, J. Karhunen, and E. Oja, *Independent Component Analysis*, Chapter 6, Wiley-Interscience, New York, ISBN 0-474-40540-X, 2001.

20. I.T. Jolliffe, *Principal Component Analysis*, Springer, New York, ISBN 0-387-95442-2, 2001.

21. G.N. Ali, A.K. Mikkilineni, E.J. Delp, J.P. Allebach, P.-J. Chiang, and G.T. Chiu, Application of principal component analysis and Gaussian mixture models to printer identification, and Printer identification based on texture features, *NIP20: Proceedings of the IS&T's International Conference on Digital Printing Technologies*, pp. 301–305, Salt Lake City, UT, Oct. 31–Nov. 5, 2004.

22. S.A. Schweid and L.K. Mestha, An image output color management system and method, US Patent Application, Attorney Docket 20070653, Dec. 2007.

23. B. Noble and J.W. Daniel, *Applied Linear Algebra*, Prentice-Hall, Englewood Cliffs, NJ, ISBN 0-13-041260-0, 1968.

24. R.S. Berns, L.A. Taplin, and T.Z. Liang, Spectral color reproduction with six color output, US Patent 6,698,860, Mar. 2, 2004.

25. H.E.J. Neugebauer, Die theoretischen Grundlagen des Mehrfarbenbuchdrucks, *Zeitschrift fur wissenschaftliche Photographie Photophysik and Photochemie*, 36, 73–89, Apr. 1937, reprinted in [22, pp. 194–202].

26. J.A.S. Viggiano, Modeling the color of multicolored halftones, in *TAGA Proceedings*, Rochester, NY, pp. 44–62, 1990.

27. J.A.S. Viggiano, The color of halftone tints, in *TAGA Proceedings*, Rochester, NY, pp. 647–661, 1985.

28. R. Balasubramanian, Colorimetric modeling of binary color printers, in *Proceedings of the IEEE International Conference on Image Processing*, Vol. 2, Washington, DC, pp. 327–330, Nov. 1995.

29. R. Balasubramanian, A printer model for dot-on-dot halftone screens, *Proceedings of the SPIE: Color Hard Copy and Graphic Arts IV*, 2413, 356–364, 1995.

30. R. Balasubramanian, The use of spectral regression in modeling halftone color printers, *IST/OSA Annual Conference, Optics Imaging in the Information Age*, Rochester, NY, pp. 372–375, Oct. 1996.

31. R. Balasubramanian, Optimization of the spectral Neugebauer model for printer characterization, *Journal of Electronic Imaging*, 8(2), 156–166, 1999.

32. S. Bandyopadhyay and T. Paul, A new model of printer characterization, *NIP20: Proceedings of the IS&T's International Conference on Digital Printing Technologies*, pp. 425–428, Salt Lake City, UT, Oct.–Nov. 2004.

33. F.H. Imai, D.R. Wyble, and R.S. Berns, A flexibility study of spectral color reproduction, *Journal of Imaging Science and Technology*, 47, 543, 2003.

34. D.R. Wyble and R.S. Berns, A critical review of spectral models applied to binary color printing, *Color Research & Application*, 25(1), 4–19, Feb. 2000.

35. P. Emmel and R.D. Hersch, A unified model for color prediction of halftoned print, *Journal of Imaging Science and Technology*, 44(4), 351–385, 2000.

36. P. Emmel and R.D. Hersch, Modeling ink spreading for color prediction, *Journal of Imaging Science and Technology*, 46(3), 237–246, 2002.

37. S. Gustavson, The color gamut of halftone reproduction, in *Proceedings of the Fourth IS&T/SID Color Imaging Conference*, Scottsdale, AZ, Nov. 19–22, 1996.

38. P.M. Roberts, Color ink model processes for printers, US patent 6,848,768, Feb. 2005.

39. L. Yang, Spectral model of halftone on a fluorescent substrate, in *NIP20: Proceedings of the IS&T's International Conference on Digital Printing Technologies*, pp. 405–409, Salt Lake City, UT, Oct.–Nov. 2004.

40. R. Rolleston and R. Balasubramanian, Accuracy of various types of Neugebauer models, in *Proceedings of IS&T/SID Color Imaging Conference: Transforms and Portability of Color*, Vol. 2, pp. 32–37, Nov. 1993.

41. K.J. Heuberger, Z.M. Jing, and S. Persiev, Color transformations and lookup tables, in *Proceedings of the TAGA/ISCC*, Vol. 2, pp. 863–881, 1992.

42. D.W. Couwenhoven, G. Braun, K.E. Spaulding, and G.J. Woolfe, Calibrating a digital printer using a cost function, US Patent 7,245,395, Jul. 17, 2007.

43. M. Xia, E. Saber, G. Sharma, and A.M. Tekalp, End-to-end color printer calibration by total least squares regression, *IEEE Transactions on Image Processing*, 7(5), 700–716, May 1999.

44. M. Rotea, C. Lena, and D. Viassolo, Robust estimation algorithm for spectral Neugebauer models, *Proceedings of the IEEE Conference on Decision and Control*, 4, 4109–4114, Dec. 2003; M. Rotea and C. Lana, A robust estimation algorithm for printer modeling, *Proceedings of the American Control Conference*, 3, 2636–2641, Jul. 2004; C. Lana, M. Rotea, and D. Viassolo, Characterization of color printers using robust parameter estimation, in *11th Color Imaging Conference: Color Science and Engineering Systems, Technologies, Applications*, Scottsdale, AZ, pp. 224–231, Nov. 3, 2003.

45. S. Zuffi, R. Schettini, and G. Mauri, Spectral-based printer modeling and characterization, *Journal of Electronic Imaging*, 14, 023008, May 2005; S. Zuffi and R. Schettini, Innovative method for spectral-based printer characterization, *Proceedings of SPIE*, 4663, 1, Jan. 2002.

46. W.A. Rozzi, Spectral modeling of photographic printing based on dye concentration, US Patent 6,633,408, Oct. 14, 2003.

47. C. Hua and K. Huang, Advanced cellular YNSN printer model, in *Proceedings of IS&T/SID's 5th Color Imaging Conference*, Scottsdale, AZ, pp. 231–234, Nov. 1997.

48. K. Iino and R.S. Berns, Building color management modules using linear optimization I. desktop color system, *Journal of Imaging Science and Technology*, 42(1), 79–94, 1998.

49. K. Iino and R.S. Berns, Building color management modules using linear optimization II. Prepress system for offset printing, *Journal of Imaging Science and Technology*, 42(1), 79–94, 1998.

50. M. Mahy, Method and apparatus for calculating color gamuts, US Patent 5,832,109, Nov. 1997.

51. W. Rohodes, Fifty years of the Neugebauer equations, *Proceedings of SPIE*, 1184, 7–18, 1989.

52. J.A.C. Yule and W.J. Nielsen, The penetration of light into paper and its effect on halftone reproduction, in *TAGA Proceedings*, Vol. 3, pp. 65–76, May 1951.

53. F.R. Ruckdeschel and O.G. Hauser, Yule–Nielsen effect in printing: A physical analysis, *Applied Optics*, 17(21), 3376–3383, 1978.

54. Y. Liu, Spectral reflectance modification of Neugebauer equations, *TAGA Proceedings*, pp. 154–172, 1991.

55. F.R. Clapper and J.A.C. Yule, Reproduction of color with halftone images, in *Proceedings of the 7th Annual Technical Meeting TAGA*, pp. 1–14, May 1955.

56. J.S. Arney and C.D. Arney, Modeling the Yule–Nielson effect, *Journal of the Imaging Science and Technology*, 40, 233, 1996.

57. E. Demichel, in Procede, Vol. 26, pp. 17–21, 26–27, 1924.

58. I. Amidror and R.D. Hersch, Neugebauer and Demichel: Dependence and independence in *n*-screen superpositions for colour printing, *Color Research & Application*, 25(4), 267–277, 2000.

59. H.B. Archer, An inverse solution to the Neugebauer equation for the calculation of grey balance requirements, Porc. IARIGAI 16, 1981.

60. I. Pobboravsky, Methods of computing ink amounts to produce a scale of neutrals for photomechanical reproduction, *TAGA Proceedings*, pp. 10–34, 1966.

61. I. Pobboravsky and M. Pearson, Computation of dot areas required to match a colorimetrically specified color using the modified Neugebauer equations, *TAGA Proceedings*, pp. 65–77, 1972.

62. P.H. Chappuis, A modified factorization of the Neugebauer equations to compute accurate mask characteristics for neutral balance, *TAGA Proceedings*, pp. 256–280, 1977.

63. M. Mahy and P. Delabastita, *Inversion of the Neugebauer equations, Color Research & Application*, 21(6), 365–374, 1996.

64. K.V. Velde and P. Delabastita, Color separation method, US Patent 7,265,870, Sep. 2007.

65. F.R. Clapper and J.A.C. Yule, The effect of multiple internal reflections on the densities of half-tone prints on paper, *Journal of the Optical Society of America*, 43(7), 600, 1953.

66. H. Von Schmelzer, Farbrezeptberechnung, *Farbe + Lack*, 84(4), 208, 1977.

67. D Hasler, B. Zimmermann, and P. Ehbets, Efficient characterization of printing systems for the packaging industry, in *NIP20: Proceedings of the IS&T's International*

Conference on Digital Printing Technologies, pp. 357–363, Salt Lake City, UT, Oct. 31–Nov. 5, 2004.

68. P. Kubelka, New contributions to the optics of intensely light-scattering materials: Part I, *Journal of the Optical Society of America*, 38, 448–457, 1947.

69. H.R. Kang, *Color Technology for Electronic Imaging*, SPIE Optical Engineering Press, New York, ISBN 0-8194-2108-1, 1996.

70. H. Akaike, A new look at statistical model identification, *IEEE Transactions on Automatic Control*, AC-19, 716–723, 1973.

71. H.R. Kang, Printer-related color processing techniques, *Proceedings of SPIE*, 2413, 410–419, 1995.

72. H.R. Kang, Applications of color mixing models to electronic printing, *Journal of Electronic Imaging*, 3, 276–287, 1994.

73. J.M. Kasson, S.I. Nin, W. Plouffe, and J.L. Hafner, Performing color space conversions with three-dimensional linear interpolation, *Journal of Electronic Imaging*, 4(3), 226–250, 1995.

74. G. Marcu and K. Iwata, *RGB*-YMCK color conversion by application of the neural networks, in *Color Imaging Conference: Transforms and Transportability of Color, IS&T*, Springfield, VA, pp. 27–32, 1993.

75. S. Abe and G. Marcu, A neural network approach for *RGB* to YMCK color conversion, in *IEEE Region 10's Ninth Annual International Conference*, IEEE Press, Los Alamitos, CA, pp. 6–9, 1994.

76. Y. Arai, Y. Nakano, and T. Iga, A method of transformation from CIE *L*a*b** to CMY value by a three-layered neural network, in *IS&T/SID Color Imaging Conference: Transforms & Transportability of Color*, IS&T, Springfield, VA, pp. 41–44. 1993.

77. S. Tominaga, Color control using neural networks and its application, in *Proceedings of SPIE*, Vol. 2658, SPIE Press, Bellingham, WA, pp. 253–260, 1996.

78. P. Hung, Smooth colorimetric calibration technique utilizing the entire color gamut of *CMYK* printers, *Journal of Electronic Imaging*, 3(4), 415–424, 1994.

79. J.W. Birkenshaw, M. Scott-Taggart, and K.T. Tritton, The black printer, in *TAGA Proceedings*, pp. 403–429, 1986.

80. R. Bala, Device characterization, Chapter 5, in *Digital Color Imaging Handbook*, G. Sharma Ed., CRC Press, Boca Raton, FL, 2003.

81. M. Maltz, Smooth gray component replacement strategy that utilizes the full device gamut, US Patent 7,411,696, Aug. 12, 2007.

82. R. Balasubramanian and R. Eschbach, Reducing multiseparation color moire via a variable undercolor removal and gray-component replacement strategy, *Journal of Imaging Science & Technology*, 45(2), 152–160, March/April, 2001.

83. R. Balasubramanian and R. Eschbach, Design of UCR and GCR strategies to reduce moire in color printing, *IS&TPICS Conference*, pp. 390–393, 1999.

84. M.C. Mongeon, Image transformation into device dependent color printer description using 4th-order polynomial regression and object oriented programming development of image processing modules, in *Proceedings of SPIE*, Vol. 2658, SPIE Press, Billingham, WA, pp. 341–352, 1996.

85. C. Nakamura and K. Sayanagi, Gray component replacement by the Neugebauer equations, *Proceedings of SPIE*, 1184, 50–63, 1989; M. Tsukada and J. Tajima, New algorithm for UCR using direct color mapping, *Proceedings of SPIE*, 2413, 365–374, 1995.

86. H. Ogatsu, K. Murai, and S. Kita, A flexible GCR based on CIE *L*a*b**, *Proceedings of SPIE*, 2414, 123–133, 1995.

87. US Patent 5,502,579 to Kita et al. (Kita '529) and US Patent 5,636,290 to Kita et al. (Kita '290).

88. M. Mahy, Color separation method for same, US Patent 5,878,195, Mar. 2, 1999; 4,482,917; 5,425,134; 5,508,827; 5,553,199.

89. D. Shepard, A two-dimensional interpolation function for irregularly-spaced data, *Proceedings of ACM National Conference*, pp. 517–524, 1967.

90. R. Balasubramanian and M.S. Maltz, Refinement of printer transformation using weighted regression, *Proceedings of SPIE*, 2658, 334–340, 1996.

91. D.E. Viassolo, S.A. Dianat, and L.K. Mestha, A practical algorithm for the inversion of an experimental input–output color map for color correction, *Journal of Optical Engineering*, 42(3), 625–631, Mar. 2003.

92. W.A. Rozzi, Gamut-preserving color imaging, US Patent Application 20020105659, Aug. 8, 2002.

93. R.F. Poe and H.S. Gregory, Jr., Multicolorant process control, US Patent 5,857,063, Jan. 5, 1999.

94. E.K. Chong and S.H. Zak, *An Introduction to Optimization*, John Wiley & Sons Inc., New York, 1996.

95. L.K. Mestha, R.E. Viturro, Y.R. Wang, and S.A. Dianat, Gray balance control loop for digital color printing systems, *NIP20: Proceedings of the IS&T's International Conference on Digital Printing Technologies*, pp. 499–504, Baltimore, MD, Sep. 2005.

96. H.K. Khalil, Nonlinear systems, 3rd edn., Prentice Hall, ISBN 0-13-067389-7, 2002.

97. K.S. Narendra and A.M. Annaswamy, *Stable Adaptive Systems*, Dover Publications, Mineola, NY, ISBN-10-0-486-442268, May 2005.

98. F.L. Lewis, *Optimal Control*, John Wiley & Sons, Inc., New York, ISBN 0-471-81240-4, 1986.

99. K. Ogata, *Discrete-Time Control Systems*, Prentice Hall, Inc., Upper Saddle River, NJ, ISBN 9-13-034281-5, 1995.

100. A.E. Gil and L.K. Mestha, Spot color controls and method, US Patent Application 20080043264, Feb. 2008.

101. D. Littlewood and G. Subbarayan, Controlling the gray component with Pareto-optimal color space transformations, *Journal of Imaging Science and Technology*, 46(6), Nov./Dec. 2002.

102. D. Littlewood, P. Drakopoulos, and G. Subbarayan, Pareto-optimal formulations for cost versus colorimetric accuracy trade-offs in printer color management, *ACM Transactions on Graphics*, 21(2), 132–174, 2002.

103. W. Lan, A.N. Zhang, and B. Ma, Addressing the problems of color science and management in toner-based digital print, *NIP20: Proceedings of the IS&T's International Conference on Digital Printing Technologies*, pp. 342–346, Salt Lake City, UT, Oct. 31–Nov. 5, 2004.

104. S. Nuyan, Method and apparatus for controlling the spectral reflectance of a material, US Patent 6,052,194, Apr. 18, 2000.

105. Control System Toolbox for use with MATLAB, The Math Works Inc., MA, 1992.

106. J. Kautzky and N.K. Nichols. Robust eigenstructure assignment in state feedback control, Numerical Analysis Report NA/2/83, School of Mathematical Sciences, Flinders U., Bedford Park, S.A. 5042, Australia.

107. J.C. Spall, Stochastic optimization: Stochastic approximation and simulated annealing, in *Encyclopedia of Electrical and Electronics Engineering*, J.G. Webster, Ed., Wiley, New York, Vol. 20, pp. 529–542, 1999.

108. L.K. Mestha and S. Dianat, TRC smoothing algorithm to improve image contours in 1-D color controls, US Patent 7,397,581, Jul. 8, 2008 (Extensions to multidimensional smoothing described in Sections 6.6.2.2 and 6.6.2.3).

109. L.K. Mestha, A.E. Gil, and M. Hoffmann, A method for classifying a printer gamut into subgamuts for improved spot color accuracy, US Patent Application, Attorney Docket 20061087, Dec. 2007.

110. M. Maltz, S.J. Harrington, and S.A. Bennett, Blended look-up table for printing images with both pictorial and graphical elements, US Patent 5,734,802, Mar. 31, 1997.

111. J. Morovic, *Color Gamut Mapping*, John Wiley & Sons, New York, ISBN-10: 0470030321, 2007.

112. R.S. Gentile, E. Walowit, and J.P. Allebach, A comparison of techniques for color gamut mismatch compensation, *Journal of Imaging Technology*, 16(5), 176–181, 1990.

113. M.C. Stone, W.B. Cowan, and J.C. Beatty, Color gamut mapping and the printing of digital color images, *ACM Transactions on Graphics*, 7(4), 249–292, 1988.

114. J. Morovic and M.R. Lou, The fundamentals of gamut mapping: A survey, *Journal of Imaging Science and Technology*, 45(3), 283–290, 2001.

115. G.J. Braun, D.W. Couwenhoven, K.E. Spaulding, and G.J. Woolfe, Color gamut mapping using a cost function, US Patent 7,239,422, Jul. 3, 2007.

116. S.W. Marshall, E. Jackson, S.J. Harrington, and Y.R. Wang, Color gamut mapping for accurately mapping certain critical colors and corresponding transforming of nearby colors and enhancing global smoothness, US Patent 6,625,306, Sep. 23, 2003.

117. J. Morovic and M. Ronnier Luo, The fundamentals of Gamut mapping, *Color Research & Application*, UK, 2000.

118. P. Zolliker and K. Simon, Continuity of gamut mapping algorithms, *Journal of Electronic Imaging*, 15(1), 13004, Jan.–Mar. 2006.

119. J. Morovic and M.R. Luo, Calculating medium and image gamut boundaries for gamut mapping, *Color Research & Application*, 25(6), 394–401, Oct. 2000.

120. E.M. Granger, Gamut mapping for hard copy using the ATD color space, *Proceedings of SPIE*, 2414, 27, 1995.

121. T. Hoshino and R.S. Berns, Color gamut mapping techniques for color hard copy images, *Proceedings of SPIE*, 1909, 152–165, 1993.

122. K.E. Spaulding, R.N. Elison, and J.R. Sullivan, UltraColor: A new gamut mapping strategy, in *IS&T/SPIE Symposium on Electronic Imaging: Science & Technology*, San Jose, CA, Feb. 5–10, 1995.

123. P.-C. Hung and R.S. Berns, Determination of constant hue loci for a CRT gamut and their predictions using color appearance spaces, *Color Research & Application*, 20(5), Oct. 1995.

124. M.C. Stone and W.B. Cowan, Color gamut mapping and the printing of digital color images, *ACM Transactions on Graphics*, 7(4), Oct. 1988.

125. M. Stokes, Selectively pleasing color gamut mapping in a color computer graphics system, US Patent 5,611,030, Mar. 11, 1997.

126. M. Wolski, J.P. Allebach, and C.A. Bouman, Gamut mapping squeezing the most out of your color system, in *IS&T and SID's 2nd Color Imaging Conference: Color Science, Systems and Applications*, pp. 89–92, Nov. 1994.

127. E.D. Montag and M.D. Fairchild, Gamut mapping: Evaluation of chroma clipping techniques for three destination gamuts, *The Sixth Color Imaging Conference: Color Science, Systems, and Applications*, Scottsdale, AZ, pp. 57–61, 1998.

128. E.D. Montag and M.D. Fairchild, Psychophysical evaluation of gamut mapping techniques using simple rendered images and artificial gamut boundaries, *IEEE Transactions on Image Processing*, 6, 977–989, 1997.

129. G. Finlayson and S. Hordley, Improving gamut mapping color constancy, *IEEE Transactions on Image Processing*, 9(10), 1774–1783, Oct. 2000.

130. H.-S. Lee and D. Han, Implementation of real time color gamut mapping using neural network, *2005 IEEE Mid-Summer Workshop on Soft Computing in Industrial Applications*, Helsinki University of Technology, Espoo, Finland, June 28–30, 2005.

131. K.M. Braun, R. Balasubramanian, and R. Eschbach, Development and evaluation of six gamut-mapping algorithms for pictorial images, *IS&T/SID 7th Color Imaging Conference*, Scottsdale, AZ, pp. 144–148, 1999.

132. R. Kimmel, D. Shaked, M. Elad, and I. Sobel, Space-dependent color gamut mapping: A variational approach, *IEEE Transactions on Image Processing*, 14(6), 796–803, Jun. 2005.

133. P. Zolliker and K. Simon, Retaining local image information in gamut mapping algorithms, *IEEE Transactions on Image Processing*, 16(3), 664–672, Mar. 2007.

134. A.E. Gil, L.K. Mestha, and M. Maltz, Merit based gamut mapping in a color management system, US Patent Application, Attorney Docket 20071654, Jun. 2008.

135. R. Saito and H. Kotera, Extraction of image gamut surface and calculation of its volume, *NIP16: Proceedings of the IS&T's International Conference on Digital Printing Technologies*, Vancouver, BC, Canada, pp. 566–569, Oct. 2000.

136. ASTM E308–01, Standard practice for computing the colors of objects by using the CIE system, ASTM International, Aug. 10, 2001.

FIGURE 2.62 RGB LENA image.

FIGURE 2.63 Halftone LENA image.

FIGURE 3.23 Original RGB image.

FIGURE 3.24 Compressed RGB image.

FIGURE 3.25 Error RGB image.

FIGURE 6.15 View of 240 test colors used for comparing inversion algorithms.

Original image   Reproduced image

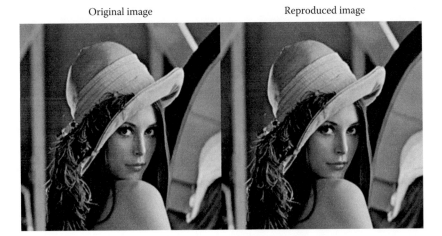

FIGURE 6.23   Original and reproduced image with noise-free profile LUT.

Original image   Reproduced image

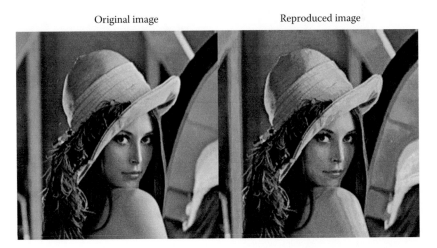

FIGURE 6.24   Original and reproduced image with noisy profile LUT.

FIGURE 6.25 Original and reproduced image with noisy profile LUT and smoothing algorithm.

FIGURE 7.37 Lena image rendered with ICC profiles from Example 7.8 (top: original; bottom: simulated with MM inversion; middle: simulated with control-based inversion).

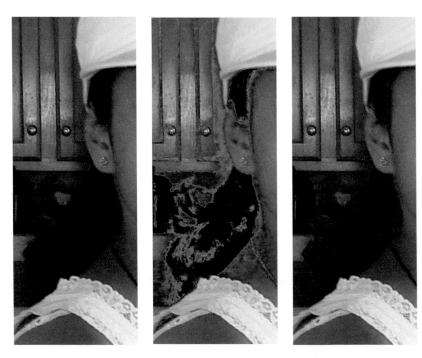

FIGURE 7.38 Hairs image rendered with ICC profiles from Example 7.8 (left: original; middle: simulated with MM inversion; right: simulated with control-based inversion).

FIGURE 7.39 Image showing the simulation of Daffodil plant (left: original image; top: MM inversion; bottom: control-based inversion).

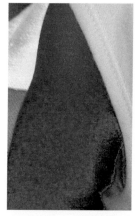

FIGURE 7.40 Image showing the rendering of color sweeps with ICC profile from Example 7.8 (left: original image; middle: MM inversion; right: control-based inversion).

FIGURE 7.45 Hair image rendered with K-restricted ICC profile whose inversion response is shown in Figure 7.44. (left: original; middle: simulated using ICC profile from Example 7.8 with 3-to-3 control-based inversion; right: simulated with K-restricted 4-to-3 control-based inversion without preconditioning [step (h)]).

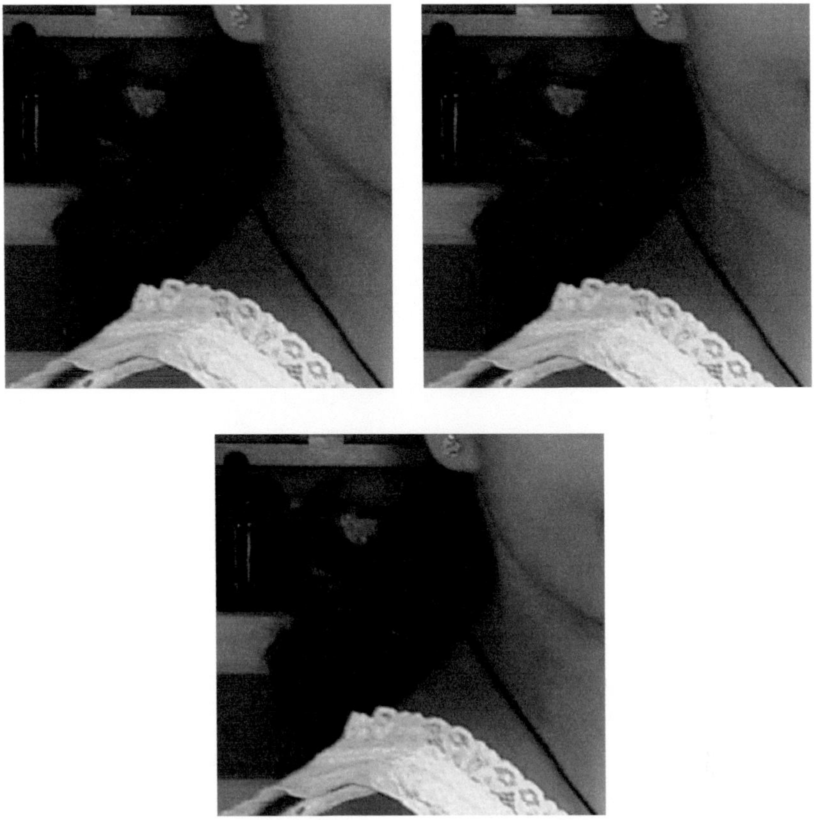

FIGURE 7.47 Image showing improvements to contours with preconditioning steps for *K*-restricted GCR (top left: original image; top right: processed image with ICC profile from *K*-restricted GCR after one step [contours are visible]; bottom: same after preconditioning steps [contours are removed]).

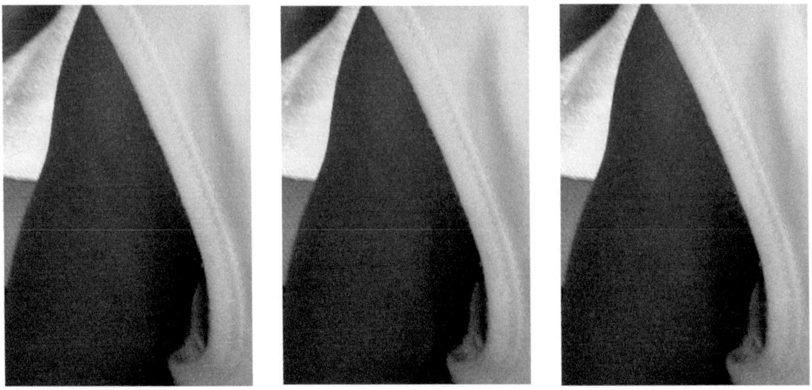

FIGURE 7.48 Image showing the rendering of color sweeps (left: original image; middle: processed image with ICC profile from *K*-restricted GCR after one step [contours are visible]; right: same after preconditioning steps [contours are removed]).

FIGURE 7.55 Comparison between display gamut and a typical color printer gamut (wire: sRGB display gamut; solid: printer gamut; left: top view; right: side view).

FIGURE 7.63 Example of black point compensation (top left: original, top right: without BPC, bottom: with BPC).

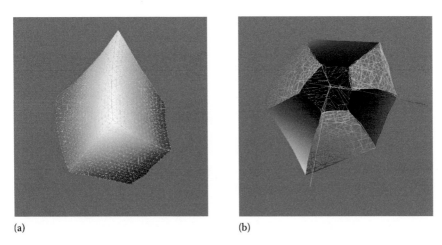

(a)   (b)

FIGURE 7.64 (a) (left) and (b) (right): Gamut views from top and from bottom, respectively.

FIGURE 7.67 Example of in-gamut color sweeps—susceptible to contours (see Figure 7.39 for performance with MM algorithm vs. the control-based inversion).

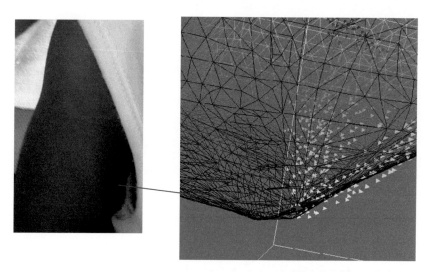

FIGURE 7.68 Example of out-of-gamut color sweeps—susceptible to contours (see Figure 7.40 for performance with MM algorithm vs. the control-based inversion).

FIGURE 8.2 Image reproduction system.

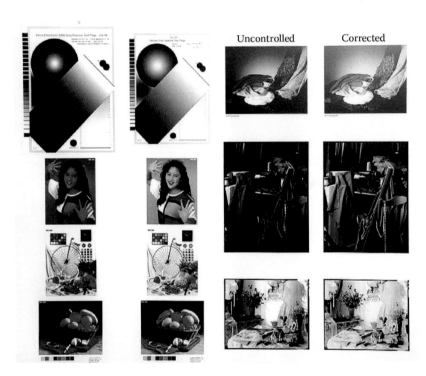

FIGURE 8.5 *CMYK* test images with and without gray-balanced print engine.

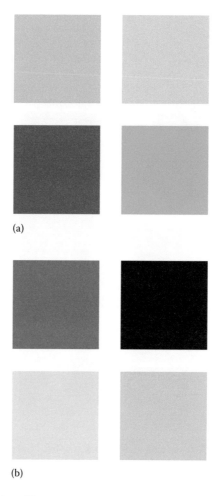

(a)

(b)

FIGURE 8.26 Examples of Pantone colors reproducible by a digital production printer on a coated stock: (a) top left: 108, top right: 1215, bottom left: 1255, and bottom right: 135; and (b) top left: 7530, top right: 7546, bottom left: cool gray 2, and bottom right: warm gray 2.

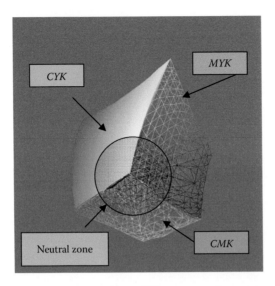

FIGURE 8.28 Top view of *CMK*, *MYK*, *CYK* gamuts and neutral zone (CMY gamut removed from the figure).

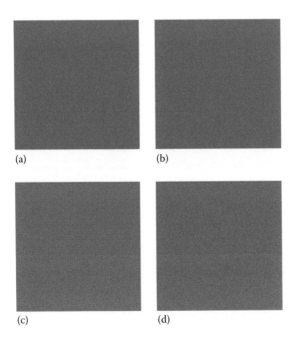

(a)

(b)

(c)

(d)

FIGURE A.4 Patches near gray with ΔE_{ab}^* color difference of 1 (a is the reference patch).

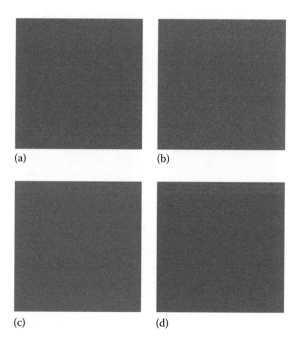

(a) (b)

(c) (d)

FIGURE A.5 Patches near gray with ΔE_{ab}^* color difference of 2 (a is the reference patch).

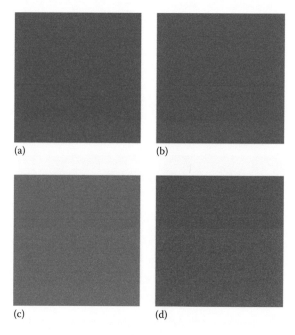

(a) (b)

(c) (d)

FIGURE A.6 Patches near gray with ΔE_{ab}^* color difference of 5 (a is the reference patch).

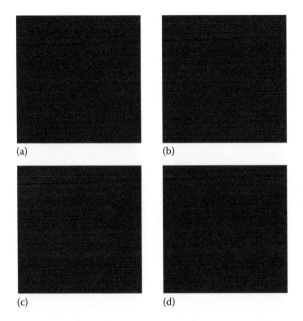

(a)   (b)

(c)   (d)

FIGURE A.7 Patches off gray with ΔE_{ab}^* color difference of 1 (a is the reference patch).

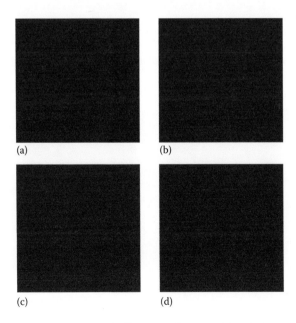

(a)   (b)

(c)   (d)

FIGURE A.8 Patches off gray with ΔE_{ab}^* color difference of 2 (a is the reference patch).

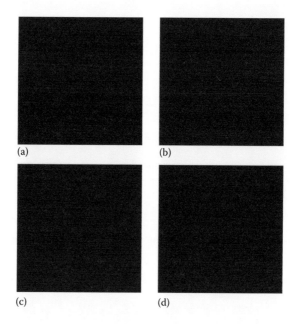

(a)   (b)

(c)   (d)

FIGURE A.9 Patches off gray with ΔE_{ab}^* color difference of 5 (a is the reference patch).

FIGURE A.10 Two color patches of Example A.5.

8 One-Dimensional, Two-Dimensional, and Spot-Color Management and Control Methods

8.1 INTRODUCTION

A key attribute of high-quality color printing is color consistency within a job, from job-to-job, and from printer-to-printer. For example, a publisher might want to print millions of copies of a book on dozens of print engines. One risk is that the different print engines may produce copies that appear different. Print engine colors drift. For high-quality consistent color rendition, the processes internal and external to the print engines have to be controlled at regular print intervals. Profiling and calibration are two important control functions normally implemented external to the print engine, that is, in the digital front end (DFE). These functions involve the printing, measuring, and processing of control patches for a particular paper and halftone screen. They are normally executed on an "as needed" basis at a lower frequency than the internal control functions.

The profiling and calibration of a conventional four-color (cyan, magenta, yellow, and black) digital printer involves (1) generating a three-dimensional (3-D) profiling lookup table (LUT) for mapping the device-independent $L^*a^*b^*$ or XYZ color space to device-dependent $CMYK$ color space based on measurement of color patches, as described in Chapter 7, and (2) constructing device tone reproduction curves (TRCs) that are one-dimensional (1-D) or two-dimensional (2-D) input $CMYK$ to output $CMYK$ maps, both in device-specific space. The 1-D/2-D $CMYK$ maps are used to account for device variability over time. Figure 8.1 shows a schematic of the image-processing path with the destination profile, $CMYK$ to $CMYK$ calibration map, and a print engine with numerous details omitted for clarity. This chapter provides detailed algorithms and methodology to generate 1-D/2-D calibration maps. We show both model-based and (more sophisticated) control-based techniques. At the end of the chapter, we extend the control-based methods to generate an accurate spot-color recipe in device-specific space ($CMYK$) to render fixed colors that may be Pantone colors, customer logo colors, colors in a customer's proprietary marked patterns, or customer-defined colors in the form of an index color table.

431

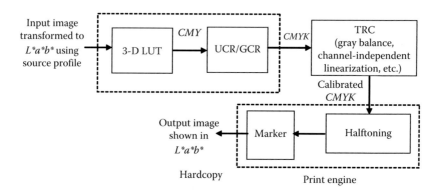

FIGURE 8.1 Image path with input and output in device-independent color space, $L*a*b*$ (UCR: under color removal; GCR: gray component replacement).

8.2 PRINCIPLES OF COLOR MANAGEMENT

Consider a digital imaging system with variety of color devices such as a digital camera, image scanner, cathode ray tube (CRT) or liquid crystal display (LCD) monitor, digital printer, and offset press. A color management system will make the color conversion and reproduction among these devices as accurate as possible. This includes profiling and calibration of different devices to provide nearly identical reproduction across them. For example, if an image is scanned by a scanner, displayed on an LCD monitor, and then printed on a digital printer, we would like the printed image to be perceived as being identical to what we see on the monitor. This is called what you see is what you get (WYSIWYG). Obviously, this is not possible across all the imaging systems due largely to physical limitations on reproducibility of each device. Our goal is to make the reproduction process as accurate as possible (within physical limits) so that output is device, media, and halftone independent. For example, consider the image reproduction system shown in Figure 8.2, where a digital camera captures an image that is displayed on an LCD monitor and then printed on a digital printer.

The image captured by the digital camera is in *RGB* color space. The camera is generally calibrated and hence it complies to some calibrated *RGB* space (e.g., *sRGB*). These *RGB* values are generally displayed for visualization on a computer

FIGURE 8.2 (See color insert following page 428.) Image reproduction system.

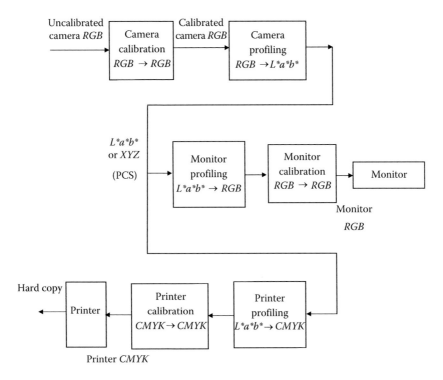

FIGURE 8.3 Color reproduction system (a simple example) (PCS: profile connection space).

monitor (e.g., an LCD monitor, CRT monitor, etc.). Typically, monitors are also profiled and calibrated. To produce a hardcopy, media, such as coated or glossy paper, the image files with camera *RGB* values are sent to the printer. As described in Chapter 7, there is a common universal color space (e.g., profile connection space (PCS) in international color consortium (ICC) architecture) that is used to communicate color values between various devices. This space is a device-independent color space (*L*a*b** or *XYZ*). Color capability of each device can be universally understood, reviewed, and compared normally in device-independent color space. By profiling and/or calibrating a device with multidimensional or 1-D LUTs, it is possible to convert the device-dependent colors into a device-independent color space. An example of this flow is shown schematically in Figure 8.3.

8.3 ONE-DIMENSIONAL GRAY-BALANCE CALIBRATION

Gray-balance calibration is a process by which 1-D LUTs are produced for the separation colors (e.g., *CMY* with $K=0$) using the measurement of output color near neutral. Grayness is an indication of how clean a process color is in a printing system as compared to a theoretical ideal. A raw print engine is normally not gray balanced. Based on appearance, neutral gray balance is often defined as the

TABLE 8.1

G7 Gray-Balance Chart for a Commercial-Coated Paper

CMY (%)			Nominal CIELab	
C	M	Y	a*	b*
0	0	0	0	−2
12.5	9	9	0	−2
25.1	18.8	18.8	0	−2
37.3	29	29	0	−2
49.8	40	40	0	−2
62.7	52.9	52.9	0	−1.5
75.3	66.3	66.3	0	−1

color of a 50% black ink tint on paper. To ensure the best chance of matching the output of devices with different colored papers or black ink, GRACoL Standards Organization [1] defines neutral gray balance in colorimetric terms with non-equal CMY as [50 C, 40 M, 40 Y] = [$a* = 0.0$, $b* = -2$] with a tolerance of ±0.5 for $a*$ and ±1.0 for $b*$. Since most people prefer a slightly bluer gray, $b*$ is set close to −2.0 instead of 0, as in neutral color. A similar definition is specified by GRACoL standards, as shown in Table 8.1 for other CMY area coverages on a commercial-coated paper, where paper is near neutral with $a* = 0.0$ and $b* = -2$. The definition of gray balance for different paper types is still under discussion by the Standards Organization. However, in Ref. [1], for gray balance, target $a*$ and $b*$ values are shown for a subset of nonstandard paper.

If we disregard the effects of paper, then a good gray has zero chroma (i.e., $a* = b* = 0.0$). When equal amounts of cyan, magenta, and yellow are printed on white paper, a well-balanced printer should produce an equivalent neutral gray. However, a brownish color rather than a neutral gray often occurs due to paper effects. As in GRACoL standards, the CMY values for one or all three of the colors may need readjustment to properly reproduce the gray scale within standards.

Rather than trying to develop a gray-balance approach for various definitions, we show a general approach in which the gray-balance calibration is performed to make the engine produce equivalent neutral gray with equal amounts of CMY colors [2,3]. For example, the equivalent neutral gray means that input digital count of $C = M = Y = 40$ should produce gray with $L^* = 100 - \frac{100}{255} \times 40 = 84.3$ and $a* = b* = 0$. The K separation does not come into this definition. To have a gray-balanced device, the input CMY digital counts have to be passed through a transformation LUTs to produce device $C'M'Y'$, which is then used to make prints. The three 1-D transformations $C' = f_c(C)$, $M' = f_c(M)$, and $Y' = f_c(Y)$ that map the input $CMY = [C\ M\ Y]^T$ to the device $CMY = [C'\ M'\ Y']^T$ for every digital count (or area coverage) are called tone reproduction curves. In practice, these three TRCs are given by three 1-D LUTs. Each LUT has 256 entries for input digital counts of 0–255. This should not to be confused with channel-wise linearization, as in the paper-based level 3 controls described in Chapter 9 and later in this chapter. Such 1-D TRCs linearize each separation independently, and, therefore, do not always yield a neutral gray balance for $C = M = Y$.

Since the TRCs (gray-balanced or 1-D channel-wise linearized) contain all the possible entries and are independent of each other, each separation can be processed on a real-time basis without interpolation after RIPping. If a color sensing instrument becomes available, then the construction/generation of the gray-balance TRCs and processing of each pixel with the updated TRCs can also be used as a means to improve the stability of neutral colors. Whereas, when real-time processing is applied with 1-D channel-wise TRCs, we can only stabilize colors along the separations (not neutral colors, such as $C = M = Y$). Hence, a system can be engineered to automatically gray balance the print engine and to automatically linearize the separations independently to overcome environmental and other disturbances that cause color variations on paper. When 1-D channel-wise linearizations are used in conjunction with a gray-balance calibration, proper separation rules have to be applied while creating and using the TRCs to avoid undesirable loop interactions.

8.4 TWO-DIMENSIONAL CALIBRATION

With 1-D gray-balanced TRC LUTs, as described in Section 8.3, we can achieve color balance to neutral colors by constantly monitoring the colors along the neutral axis (i.e., the L^* axis in the $L^*a^*b^*$ color space) and producing color-balanced TRCs. This will provide good control along a sensitive color axis. Although, this approach can make other colors reasonably accurate, it does not control colors everywhere in the color space like the 3-D color control approaches described in Chapter 7. Accordingly, one might ask the following question: Why not just do 3-D color control more frequently to stabilize the color output of engines instead of simply updating gray-balance TRCs on a more frequent basis? There are three reasons why this strategy is not preferred: (1) processing of 3-D LUTs for high-speed RIPping is not cost effective when compared to processing 1-D gray-balance TRCs, (2) memory requirements are very minor for 8-bit TRC processing (256 bytes of memory per separation TRC), and (3) the number of levels that can be corrected in a 3-D LUT is limited along each separation axis to a much smaller number (e.g., 33 in a 33^3 profile LUT) as compared to 256 levels in the 1-D TRCs, thus limiting the quality levels for rendering neutral colors. A full resolution 3-D LUT with full lookup can avoid the shortfall in levels, but can be prohibitively large (a full 3-D LUT size for a 8-bit system would be $3 \times (256)^3$ bytes $= 49.75$ M Bytes of storage), especially when customized for each media and halftone screen.

Gray balance alone can introduce undesirable hue shifts in the reproductions of sweeps from white to secondary colors, such as those colors lying along the red axis ($M = Y$ and $C = K = 0$) as well as those along the blue and green axes. On the other hand, with a 2-D calibration [4], we can achieve gray balance as well as control over colors lying on other critical axes in the color space. In Section 8.6.3, the 2-D calibration approach is discussed. In this case, the TRCs are 2-D and the LUTs representing each TRC would have $256^2 = 65,536$ entries, quite reasonable as compared to full resolution 3-D LUTs. Obviously, as we try to control more axes in the color space, the LUT size increases. The 2-D TRCs offer significant levels to control as compared to 1-D channel-wise or 1-D gray-balance TRCs and hence can provide more accurate color rendition than 1-D TRCs, but are still less accurate than 3-D LUTs [5].

8.5   ONE-DIMENSIONAL AND TWO-DIMENSIONAL PRINTER CALIBRATION USING PRINTER MODELS

In this section, we show how to construct the 1-D and 2-D printer calibration TRC LUTs if an updated printer model is given. The model can be in any of the forms described in Chapters 7 or 10 or even a simple input–output LUT obtained experimentally while characterizing the printer. We first discuss the algorithms for 1-D calibration, which involves both 1-D channel-wise linearization and 1-D gray balance to equivalent neutral targets.

8.5.1   ONE-DIMENSIONAL CHANNEL-WISE (INDEPENDENT) CALIBRATION

In channel-independent calibration, each channel is independently linearized. By linearization, we would like to make the ΔE_{ab}^* obtained by measuring color patches from paper to be linear so that the channel-linearized printer can emulate an ideal printer, which has the characteristic of linearized ΔE_{ab}^* from paper. The ΔE_{ab}^* from paper is the Euclidean norm between target color and the paper white in the device-independent color space $(L^*a^*b^*)$. The ΔE_{2000} metric is not normally used for the paper-based 1-D channel-wise calibration.

As an example, let us assume that we choose the cyan channel for linearization and printed N cyan patches $(M = Y = K = 0)$ that represent the step wedge with the cyan digital count of $d_0 = 0, d_1, d_2, \ldots, d_{N-1} = 255$, and measured the $L^*a^*b^*$ values of each cyan patch using a color sensor. Let the corresponding measured $L^*a^*b^*$ value of the ith patch be $(L^*a^*b^*)_i$. Note the first patch is the paper white which means that $(L^*a^*b^*)_0$ is the $L^*a^*b^*$ of the paper. We now compute the ΔE_{ab}^* from paper for each patch as follows:

$$\Delta E_i = \|(Lab)_i - (Lab)_0\| \quad i = 0, 1, \ldots, N - 1 \tag{8.1}$$

The function $f(.)$ is formed by normalizing the ΔE values to obtain a value of 255 for the cyan value of 255. It is defined as

$$d_i \rightarrow f(d_i) = \frac{255 \Delta E_i}{\max_i (\Delta E_i)} \tag{8.2}$$

To linearize the printer, we need ΔE_i to be a linear function of d_i. To achieve this, we find the inverse of the transformation given by Equation 8.2, which is the TRC of the cyan channel. That TRC can be represented by following equation:

$$\text{Cyan TRC: } c = f^{-1}(C) \tag{8.3}$$

We repeat the above process for all the other three channels to obtain the TRCs for the remaining magenta, yellow, and black channels by printing and measuring corresponding color patches. Instead of step wedges, pure sweeps covering 0–255 digital counts can also be used for channel-wise linearization. In such cases, the measurement aperture of the sensing process becomes more important than the gradient of the sweep. The following example shows the 1-D channel-independent linearization process for the cyan channel using patches representing step wedges.

Example 8.1

Assume that $N = 11$ Cyan patches with digital counts of 0, 25, 50, 75, 100, 125, 150, 175, 200, 225, and 255 are printed on a color digital printer. The measured $L^*a^*b^*$ values of these 11 patches and their corresponding ΔE from paper are shown in Table 8.2. Note that the first row, which corresponds to $CMYK = [0 \ 0 \ 0 \ 0]$, is the paper white. Therefore, all the ΔE values are measured with respect to this reference. In practice, the $L^*a^*b^*$ of different spots on the paper can be measured and the average of all these $L^*a^*b^*$ values used as the reference paper white to avoid errors in the measurement process. The function $f(d)$ is obtained by dividing the ΔE values by its maximum and multiplying the results by 255. The function is shown in the last column of Table 8.2. The TRC is obtained by finding the inverse of the function $f(d)$. The inverse can be simply obtained by interchanging the dependent and independent variables as shown in Table 8.3. Also see Section 9.9 for additional information about finding an inverse for different conditions. The TRC for the digital counts between the data points of Table 8.3 is obtained by curve fitting or interpolation. The function $f(d)$ and the resulting TRC for the cyan channel are shown in Figure 8.4. In this figure, the nodes represent the measured points and the solid curve is obtained with linear interpolation. A higher-order interpolation approach can be adopted based on the need.

TABLE 8.2
Measured $L^*a^*b^*$ Values of Cyan Patches

d	L*	a*	b*	ΔE_{ab}^*	f(d)
0	97.81	0.004	−0.22	0	0
25	94.24	−6.43	−9.78	12.06	42
50	90.13	−9.52	−14.96	19.15	66.68
75	86.67	−13.08	−16.07	23.37	81.37
100	82.23	−16.71	−19.90	30.15	104.97
125	77.87	−18.87	−21.34	34.64	120.57
150	74.44	−20.29	−25.32	39.85	138.71
175	70.45	−24.03	−34.33	49.89	173.68
200	67.34	−27.49	−39.25	56.63	197.14
225	63.43	−31.05	−44.05	63.77	221.99
255	57.50	−35.59	−49.97	73.26	255

TABLE 8.3
Cyan TRC Function

d	$\mathbf{TRC} = f^{-1}(d)$
0	0
42	25
66.68	50
81.37	75
104.97	100
120.57	125
138.71	150
173.68	175
197.14	200
221.99	225
255	255

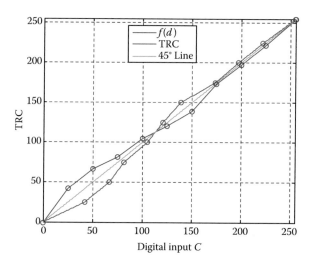

FIGURE 8.4 Channel-independent cyan TRC.

8.5.2 Gray-Balanced Calibration

Recall that gray-balance calibration is performed (i.e., gray-balance TRCs for cyan, magenta, and yellow channels are generated) to make the print engine produce equivalent neutral gray accurately with equal amounts of *CMY* colors. That is, after gray-balance calibration, an input with a digital count of

$$C = M = Y = d \quad \text{and} \quad K = 0 \tag{8.4}$$

produces gray level with

$$L^* = 100 - \frac{d}{255} \times 100, \quad a^* = b^* = 0 \qquad (8.5)$$

The TRC for the black channel is also constructed as a second step to perform channel-wise linearization. The main reason for the adaptation of the gray-balance calibration for the *CMY* channels is the fact that the human visual system is very sensitive to colors near the neutral axis. Small color differences near the neutral are more noticeable than larger color differences far from it. Once the gray balance is achieved with TRCs and the black channel is linearized, the printer will tend to make other colors containing combinations of *CMYK* lot more pleasing and closer to what is required by the press (see the images in Figure 8.5 with and without gray balance). However, the choice of gray-balance reference or target values matters a lot with respect to the neutral appearance [1]. If the gray-balance reference is altered, then the color balance will also be altered and may result in an error in color.

In system theoretic terms, the outcome of the gray-balance calibration can be thought of as a way of generating inverse maps to linearize the printer along the chosen reference/target axis. For example, when the target values are selected on the neutral axis, the printer with a gray-balanced inverse map will be linear to

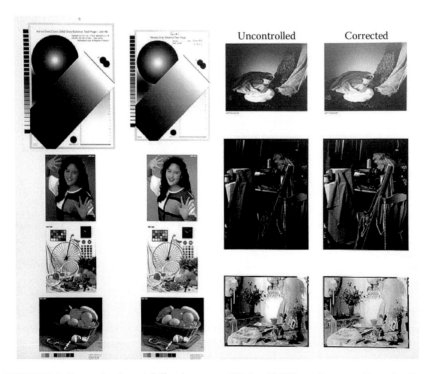

FIGURE 8.5 (See color insert following page 428.) *CMYK* test images with and without gray-balanced print engine.

neutral colors, but is not linear for other regions of the color axes in the 3-D *CMY* color space. If the reference values are off-neutral, then colors are adjusted by changing *CMY* to follow the off-neutral axis, effectively making the printer linear to the off-neutral reference axis. To perform gray-balance calibration, we take the following steps:

(1) Generate a set of *CMY* color patches in and around the neutral axis (our reference axis). Print and measure the corresponding *L*a*b** values and form a printer forward map. This can be achieved either by creating M^3 uniformly spaced *CMY* patches to form a LUT or by measuring fewer patches to fit a model and then use the model to create the desired LUT (i.e., uniformly spaced *CMY* → *L*a*b** forward map).

(2) Utilize interpolation (ICI or tetrahedral) or a regression method to compute *N CMY* values that yield a neutral measurement ($a^* = b^* = 0$) for the L^* values spanning the dynamic range of the printer ($L_1^*, L_2^*, \ldots, L_N^*$). If the *CMY* gamut coverage is such that the $a^* = b^* = 0$ axis is not within the *CMY* gamut, then create smooth off-neutral reference values. In particular, this may be required for dark regions of the color space. Let the *CMY* values corresponding to L_i^* be $CMY_i = [C_i \ M_i \ Y_i]$. Now compute

$$D_i^* = \frac{255}{100}\left(100 - L_i^*\right) \quad \text{for } i = 1, 2, \ldots, N \tag{8.6}$$

(3) Create the three TRCs using the following functions:

(a) Cyan TRC:

$$C_{\text{out}} = f_1(C_{\text{in}}), \quad \text{where } f_1 : D_i^* \to C_i \tag{8.7}$$

(b) Magenta TRC:

$$M_{\text{out}} = f_2(M_{\text{in}}), \quad \text{where } f_2 : D_i^* \to M_i \tag{8.8}$$

(c) Yellow TRC:

$$Y_{\text{out}} = f_3(Y_{\text{in}}), \quad \text{where } f_3 : D_i^* \to Y_i \tag{8.9}$$

Note that the above three TRCs are only defined for *N* input digital counts. Linear or cubic interpolation is used to obtain all the values from 0 to 255, in steps of one digital count. Also, TRCs should be smooth and monotonically increasing functions of their arguments. If required, 1-D smoothing techniques described in Section 6.6.2.1 are used to generate smooth TRCs for all three channels.

Example 8.2

Assume that a $9 \times 9 \times 9$ *CMY* to *L*a*b** characterization LUT is measured for a digital color printer. This LUT is used to compute the *CMY* values corresponding to the gray values of $a_i^* = b_i^* = 0$, $L_i^* = 15 + 10i$, $i = 1, 2, \ldots, 6$. The ICI algorithm (Section 6.4.3) is used to find the right combination of the *CMY* values. The results are shown in Table 8.4. Note that since ICI is an iterative technique, the iterations are terminated once the error or number of iterations reach a predefined threshold. Hence, exact gray with $a_i^* = b_i^* = 0$ may not be achieved. Table 8.5 shows the three gray-balance TRC functions. Figures 8.6 through 8.8 show the plots of the cyan, magenta, and yellow gray-balanced TRCs, respectively. The available seven data points on the TRCs are marked by a circle and the other values are obtained through linear interpolation (for internal points) or linear extrapolation (for external points). Note that since smooth TRCs are constructed, the final TRCs do not pass through the data points, particularly in the shadow region. This has the effect of making gray balance less accurate in those regions, but may be essential to reproduce smooth, contourless shadow colors.

TABLE 8.4
Input *CMY* Values that Generate Gray

L^*	a^*	b^*	C	M	Y
25	0.005	0.005	235.25	237.18	221.58
35	0	−0.002	203.44	200.61	187.41
45	0.003	0.004	164.89	169.45	155.1
55	−0.002	−0.003	131.58	129.78	132.97
65	0.002	0.003	97.88	104.02	104.85
75	0.0017	0.001	68.48	70.85	68.73

TABLE 8.5
Three Gray-Balance TRCs

$D^* = C_{in}, M_{in}, Y_{in}$	C_{out}	M_{out}	Y_{out}
191	235.25	237.18	221.58
166	203.44	200.61	187.41
140	164.89	169.45	155.1
115	131.58	129.78	132.97
89	97.88	104.02	104.85
64	68.48	70.85	68.73

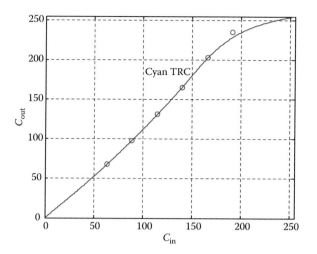

FIGURE 8.6   Cyan gray-balanced TRC.

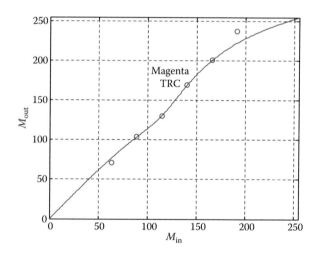

FIGURE 8.7   Magenta gray-balanced TRC.

8.5.3   Two-Dimensional Calibration

One-dimensional calibration is limited to color balancing the printer to one of the color axes or to achieve a linearized ΔE_{ab}^* response with respect to paper. Achieving both simultaneously with TRCs is not simple. A full 3-D function with input C_{in}, M_{in}, and Y_{in} to output C_{out}, M_{out}, and Y_{out} is being applied in many color management systems as an alternative to 1-D calibration. In this case, the three 3-D TRCs (or LUTs) would be

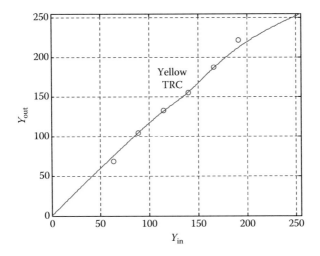

FIGURE 8.8   Yellow gray-balanced TRC.

$$C_{out} = f_1(C_{in}, M_{in}, Y_{in}) \tag{8.10}$$

$$M_{out} = f_2(C_{in}, M_{in}, Y_{in}) \tag{8.11}$$

$$Y_{out} = f_3(C_{in}, M_{in}, Y_{in}) \tag{8.12}$$

where C_{in}, M_{in}, and Y_{in} are the inputs of the 3-D map and C_{out}, M_{out}, and Y_{out} are the outputs of the 3-D map, which are the actual device *CMY* values sent to the printer. Alternatively, in 2-D calibration, three 2-D TRCs are constructed as follows:

$$C_{out} = f_1(C_{in}, M_{in} + Y_{in}) \tag{8.13}$$

$$M_{out} = f_2(M_{in}, C_{in} + Y_{in}) \tag{8.14}$$

$$Y_{out} = f_3(Y_{in}, C_{in} + M_{in}) \tag{8.15}$$

For example, the cyan TRC given by Equation 8.13 suggests that the calibrated cyan value is a function of input cyan and sum of input magenta and yellow. The domain of this function or LUT is illustrated in Figure 8.9, and the implementation of Equations 8.13 through 8.15 is represented by the block diagram of Figure 8.10.

The advantage of 2-D calibration over 1-D is that we can obtain good control over five color axes, as shown in Figure 8.9. By controlling the gray axis, we can achieve good gray balance; controlling the cyan primary axis, we can achieve channel-independent linearization. By controlling other axes such as primary to black, we can linearize the printer along those axes, achieving better control of the printer gamut. The process of linearization along these axes is similar to the channel-independent linearization.

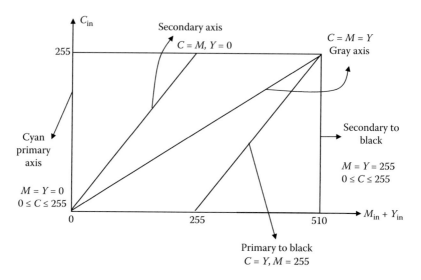

FIGURE 8.9   Two-dimensional cyan TRC with different control axes.

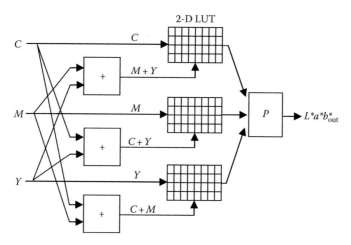

FIGURE 8.10   Two-dimensional calibration.

Example 8.3

Assume that a $13 \times 13 \times 13$ *CMY* to $L^*a^*b^*$ characterization LUT is measured for a digital color printer. The 2-D LUTs for *C*, *M*, and *Y* were filled in by controlling different axis shown in Figure 8.9. Fifteen grid points are chosen on each control axis. The scatter plot of these grid points for the 2-D cyan TRC in $(C, M+Y)$ plane is shown in Figure 8.11.

The corresponding 2-D LUT for the cyan TRC corresponding to the control grid points of Figure 8.11 is shown in Table 8.6. This is obtained using the ICI algorithm.

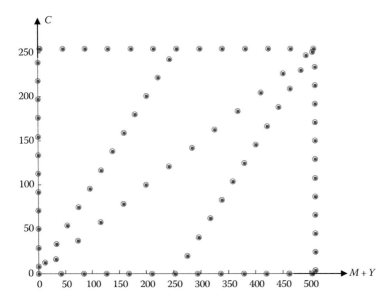

FIGURE 8.11   Scatter plot of grid control point for 2-D cyan TRC in $(C, M+Y)$ plane.

Note that the LUT includes the control grid points of Figure 8.11. The complete LUT for every point in the $(C, M+Y)$ plane is constructed using interpolation/smoothing. Similar LUTs are constructed for magenta and yellow TRCs.

To evaluate the performance of 2-D calibration against 1-D, a total of 180 in-gamut patches were generated. The patch values are processed through the calibration LUTs and the simulated printer for both 1-D and 2-D calibrations. The accuracy results are compiled in Table 8.7. Two-dimensional calibration outperforms 1-D with respect to mean and 95th percentile ΔE_{ab}^{*}.

8.6   ONE-DIMENSIONAL AND TWO-DIMENSIONAL PRINTER CALIBRATION WITH STATE-FEEDBACK METHODS

The approach of the previous section regarding the gray-balance and channel-independent calibrations is based on printer model constructed using a high-resolution LUT or an accurate printer model. In this section, we show how control approaches using state-feedback methods [2,3] can be used to design 1-D and 2-D calibration TRCs. Other, more robust, control approaches are also available [6].

Consider the gray-balance control loop with the gain-weighted MIMO (multi-input and multi-output) integral controller shown in Figure 8.12. In this figure, $x(k)=[L^{*}(k)\ a^{*}(k)\ b^{*}(k)]^{\mathrm{T}}$ are the states, as defined by the state variable model for a $CMY \rightarrow L^{*}a^{*}b^{*}$ printer described in Section 7.5.2.2, and, in this case, the states are also the output of the system. The outputs are the measured $L^{*}a^{*}b^{*}$ values from the color sensor. $V(k)=[\delta C(k)\ \delta M(k)\ \delta Y(k)]^{\mathrm{T}}$. $r=[L_{\text{target}}^{*}\ a_{\text{target}}^{*}\ b_{\text{target}}^{*}]^{\mathrm{T}}$ is the target $L^{*}a^{*}b^{*}$, and K is the 3×3 feedback gain matrix. The printer in this loop can be an (accurate) printer model or an actual printer. The sensor can be an inline or offline spectrophotometer. If an inline sensor is used, then the processes can be automated to

TABLE 8.6

Two-Dimensional LUT for the 2-D Cyan TRC

C_{in}	$M_{in} + Y_{in}$	C_{out}	C_{in}	$M_{in} + Y_{in}$	C_{out}
0	0	0	129	510	95
0	42	16	134	0	106
0	84	34	138	138	114
0	126	51	142	284	144
0	168	70	146	401	86
0	210	89	150	510	79
0	252	110	155	0	131
0	294	130	159	159	137
0	336	151	163	326	181
0	378	174	167	422	70
0	420	198	171	510	63
0	462	224	176	0	157
0	504	251	180	180	162
3	510	251	184	368	212
8	0	6	188	443	54
12	12	9	192	510	47
16	32	20	197	0	183
20	275	231	201	201	188
24	510	224	205	410	235
29	0	23	209	464	38
33	33	26	213	510	31
37	74	38	218	0	209
41	296	206	222	222	214
45	510	197	226	452	248
50	0	39	230	485	21
54	54	43	234	510	16
58	116	57	239	0	235
62	317	179	243	243	240
66	510	169	247	494	253
71	0	56	251	506	3
75	75	60	255	4	253
79	158	75	255	46	234
83	338	151	255	88	215
87	510	142	255	130	194
92	0	72	255	172	173
96	96	76	255	214	151
100	200	92	255	256	128
104	359	124	255	298	107
108	510	115	255	340	85
113	0	89	255	382	64
117	117	93	255	424	42
121	242	111	255	466	21
125	380	101	255	508	1

TABLE 8.7

Round-Trip Accuracy Numbers for 1-D and 2-D Calibrations

Calibration Type	Mean ΔE_{ab}^*	95th Percentile ΔE_{ab}^*
One-dimensional gray-balance calibration	5.13	8.23
One-dimensional channel-independent calibration	3.12	9.06
Two-dimensional calibration	1.75	5.32

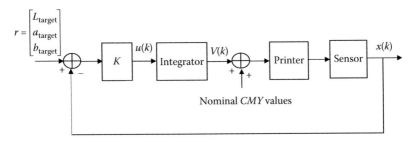

FIGURE 8.12 Gray-balance control loop.

eliminate the need for operator intervention. Otherwise, it can be a semiautomated system.

We have shown the above control loop for one color patch. The nominal *CMY* values are constructed using any of the following procedure:

(1) When a printer model is not available, use $C = M = Y = d$ and $K = 0$ where d is computed from Equation 8.16 by knowing the gray-balance target for the chosen patch (i.e., L^*, a^*, and b^* values):

$$d = \frac{(100 - L^*)*255}{100};$$

$a^* = b^* =$ gray-balance target (or 0 for equivalent neutral)   (8.16)

(2) When a printer model is available, compute *CMY* for the $L^*a^*b^*$ gray-balance target using the ICI algorithm.

To construct three TRCs, we can select *N* number of patches using the nominal *CMY* values obtained for each of the patch gray-balance targets. As a result of this, we would have *N* number of control loops running in parallel. If an actual printer is used, these loops will be iterating on the printer. If a printer model is available, then the control iterations are carried out as described in Section 7.5.2.2.

For any nominal *CMY* input, we can approximately represent the dynamics of the single-color reproduction system using a first-order finite difference equation given by

$$x(k + 1) = BV(k) + x_0 \tag{8.17}$$

where
x_0 is the $L^*a^*b^*$ vector corresponding to the nominal *CMY*
B is the Jacobian matrix given by

$$B = \begin{bmatrix} \frac{\partial L^*}{\partial C} & \frac{\partial L^*}{\partial M} & \frac{\partial L^*}{\partial Y} \\ \frac{\partial a^*}{\partial C} & \frac{\partial a^*}{\partial M} & \frac{\partial a^*}{\partial Y} \\ \frac{\partial b^*}{\partial C} & \frac{\partial b^*}{\partial M} & \frac{\partial b^*}{\partial Y} \end{bmatrix} \tag{8.18}$$

The Jacobian matrix B is computed at the nominal *CMY* using the coarse printer model. The Jacobian matrix can also be measured directly on the printer using techniques described in Ref. [7].

Since we are using an integrator

$$V(k) = V(k - 1) + u(k) \tag{8.19}$$

substituting Equation 8.19 into Equation 8.17 results in

$$x(k + 1) = BV(k) + x_0 = BV(k - 1) + Bu(k) + x_0 \tag{8.20}$$

Assuming that the Jacobian is not changing between two consecutive prints, Equation 8.17 can be written for print number k as

$$x(k) = BV(k - 1) + x_0 \tag{8.21}$$

Subtracting Equation 8.21 from Equation 8.20 results in

$$x(k + 1) = x(k) + Bu(k) + Ax(k) + Bu(k) \tag{8.22}$$

where A is a 3×3 identity matrix. Using the state feedback, $u(k) = K[r - x(k)]$, the closed-loop state equation becomes

$$x(k + 1) = (A - BK)x(k) + BKr \tag{8.23}$$

The gain matrix, K, can then be designed using pole-placement or optimal control techniques, as described in Chapter 6.

8.6.1   POLE-PLACEMENT DESIGN

In pole-placement design, the closed-loop poles are chosen to be at locations $P = [p_1 \quad p_2 \quad p_3]$, where a pole is chosen for each of the three color patches that is within the unit circle of the complex z-plane. We chose the poles to be positive real numbers to avoid oscillations and possible overshoots. Typical values are between 0.3 and 0.6. We summarize below some key process steps used to generate gray-balance TRCs using control approaches for a manual off-line sensor.

Step 1: Select a set of N uniformly spaced gray-target patches whose gray-balance targets are given. For equivalent neutral gray, the target values lie on the L^* axis $(a^* = b^* = 0)$. Let the L^* of these N gray patches be $\{d_i\}_{i=1}^{N}$, where $d_1 = 0$ and $d_N = 100$. For each target, a control loop is needed, resulting in N control loops running in parallel.

Step 2: Estimate an initial set of N nominal CMY values using the steps described above.

Step 3: Print these N patches with CMY values and measure their color (i.e., $L^*a^*b^*$) values.

Step 4: Compute ΔE color difference between measured $L^*a^*b^*$ with the target $L^*a^*b^*$. If the color difference is small, stop the iteration process, save the printer input $CMY_i = [C_i \ M_i \ Y_i]$ corresponding to the ith patch, and go to Step 5; otherwise process the error to generate $V(k)$ and the new set of N printer input CMY values, and continue with Step 3.

Step 5: Form the three functions $\{D_i, \ C_i\}_{i=1}^{N}$, $\{D_i, \ M_i\}_{i=1}^{N}$, and $\{D_i, \ Y_i\}_{i=1}^{N}$, where $D_i = 2.55 \, (100 - d_i)$. These functions are the three gray-balance TRCs corresponding to cyan, magenta, and yellow channels, respectively. A smoothing algorithm can be used to interpolate between points to generate full TRCs for all digital values from 0 to 255. If $d_1 \neq 0$ or $d_N \neq 100$, extrapolation is used to find the TRC values below d_1 and above d_N. Special highlight or shadow corrections described in Section 8.6.2 may be required prior to generating the final gray-balance TRCs.

Example 8.4

In this example, we select 7 gray targets with L^* of 25–90 in steps of 10 with $a^* = b^* = 0$ as gray-balance targets. The Neugebauer model is used as the surrogate printer and the system is simulated in MATLAB. The pole-placement design is used with three real poles, $P = [0.2 \quad 0.3 \quad 0.2]$. The nominal CMY values for the ith patch is estimated to be $CMY_{nominal} = [255 - 2.55L_i^* \quad 255 - 2.55L_i^* \quad 255 - 2.55L_i^*]$. The B matrix is estimated at the nominal CMY using numerical differentiation. The plot of $e(k)$ for the first patch as a function of print number is shown in Figure 8.13. The plot of ΔE convergence for all the seven control patches as a function of print (iteration) number k is shown in Figure 8.14. The final gray-balanced TRCs for all three separations are shown in Figures 8.15 through 8.17.

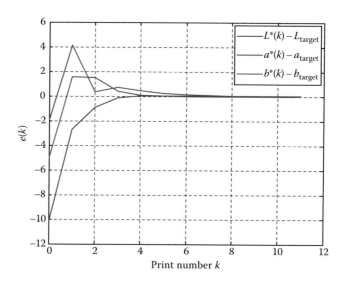

FIGURE 8.13   Error $e(k)$ with respect to iteration number, k.

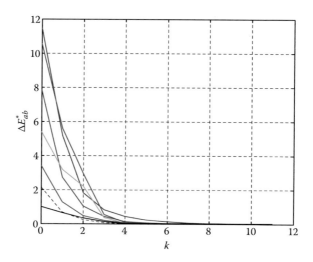

FIGURE 8.14 $\Delta E_{ab}^*(k)$ as a function of print number.

8.6.2   Highlight and Shadow Corrections

The TRC has highlight and shadow regions. A highlight is a color or shade with a very low digital value, indicating a region where very little toner/ink is deposited on the paper. Given a 0–255 digital value range, highlights typically occur between 0 and 20. A shadow is a color or shade with a very high digital value, typically above 230.

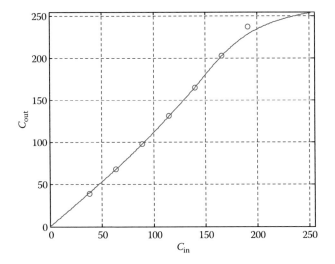

FIGURE 8.15 Cyan gray-balanced TRC for pole-placement design.

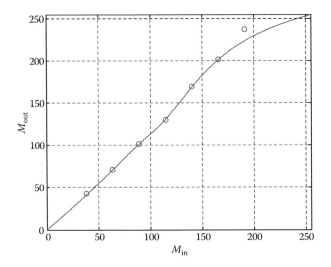

FIGURE 8.16 Magenta gray balanced TRC for pole-placement design.

Figure 8.18 is an example of a cyan TRC shown with typical highlight and shadow regions. Determining TRCs using a calibration approach with interpolation, curve fitting, or smoothing as described above works well for most of the range of digital values. However, it does not work well for highlights or shadows in high-end printing systems, particularly those of electrophotographic printers.

Calibration data for highlights is difficult to produce because the print engine is not capable of reliably depositing a small amount of toner and the sensing of the color in highlight region can be very noisy. Most marking engines can reliably

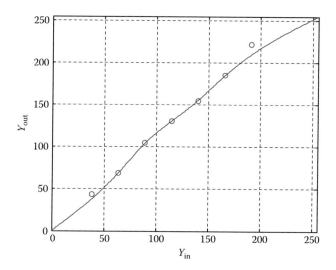

FIGURE 8.17   Yellow gray balanced TRC for pole-placement design.

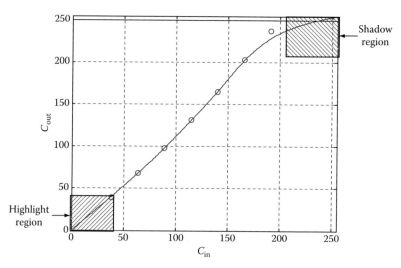

FIGURE 8.18   Cyan gray-balanced TRC with highlight (shaded lower left region) and shadow (shaded upper right region) regions.

deposit average quantities of toner, but not small quantities. Hence, particularly electrophotographic printers require overcompensation in the highlights due to lack of development. The contribution of the paper to the sensing measurements is larger in a highlight region, introducing a noise factor not as dominant in typical measurements. As such, the highlight region of most TRCs has low quality because the calibration data tend to be low quality.

The shadow regions of most TRCs also have low quality characteristics such as non-smooth development or contours. Adding more toner does not change the color much at the saturation region because it is nearly saturated. Here, full saturation is based on the physical device. Note that it is possible for a person to specify a color that is more saturated than the physical device can deliver. TRCs in the shadow region can be low quality because of the physical device, gray-balance algorithms such as smoothing algorithms, and the user specifications. If the TRCs are not smooth and well behaved in the shadow region, unwanted contours can occur in dark saturated colors. In this section, we show how to produce TRCs that work well for highlights and shadow regions.

8.6.2.1 Highlight Corrections

We illustrate highlight corrections using the highlight region of the cyan gray-balance TRC, which is shown in Figure 8.19. In this figure, active control patches near the highlight region are at cyan digital input values of $C_{in} = 13$ and $C_{in} = 30$. After gray balance, their corresponding outputs are $C_{out} = 20$ and $C_{out} = 46$. The passive control patches are at $C_{in} = 0$ and $C_{in} = 1$ digital counts. The active control patch is used during iterations to find its C_{out} value with control-based gray-balance calibration. The passive shadow control patch is not directly used during iteration. The C_{out} value for the passive shadow patch is assigned (or user selected). We can have more than one passive control patches. Nodes are included in Figure 8.19 for the four C_{in} values mentioned above. The variation in digital count appears as development on the paper for all active control patches. If nothing is done to compensate for the highlights, then, for this example, the lightest printable highlight input values are near $C_{in} = 13$. The passive control patch with the input highlight value of $C_{in} = 1$ is a value that is well within the highlight region and is not developable. In other words, a value of 1 can be

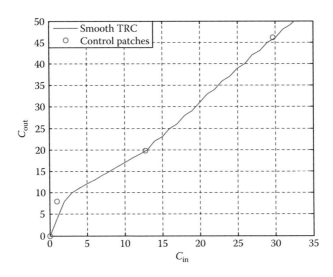

FIGURE 8.19 Highlight region of cyan gray-balanced TRC.

present in the image pixel and a print engine can try to print it, but the printed result is far from certain. $C_{in} = 0$ is another passive control patch as in this case (ideally) no toner is developed. Using the active control patches at $C_{in} = 13$ and $C_{in} = 30$, extrapolation can be used to produce a target C_{out} highlight value corresponding to the input at $C_{in} = 1$. In Figure 8.19, we used $C_{out} = 8$. We also used $C_{out} = 0$ corresponding to the $C_{in} = 0$ patch. After smooth extrapolation was done with these two passive patches ($C_{in} = 0$ and $C_{in} = 1$), $C_{in} = 0$ was mapped to 0, $C_{in} = 1$ was mapped to 4, $C_{in} = 13$ was mapped to 20, and $C_{in} = 30$ was mapped to 46. So, although we chose a $C_{out} = 8$ corresponding to $C_{in} = 1$, after extrapolation and smoothing, the actual value at $C_{in} = 1$ was mapped to a lower value. Extrapolation, such as linear or polynomial extrapolation, of three or more active control patches can also be used to produce adequate results. The algorithms produce better results when more passive and active control patches in the neighborhood of the highlight are used to define the TRCs in the extreme highlight region. If the stability of marking engines is not good for highlights, then this strategy has to be reconsidered to produce good highlights on a dynamically varying process.

8.6.2.2 Shadow Corrections

As discussed above, calibration TRCs in the shadow region can also be problematic. Control points cannot be joined with linear interpolation in the shadow region without good shadow region smoothing. When there are no active control patches in that region, then algorithms producing TRCs must extrapolate.

In Figure 8.21, we show the approach to solve shadow region problems for the cyan separation. The maximum desired saturation is 255 because that is the extreme value along the input cyan axis. We use another passive control patch at $C_{in} = 255$,

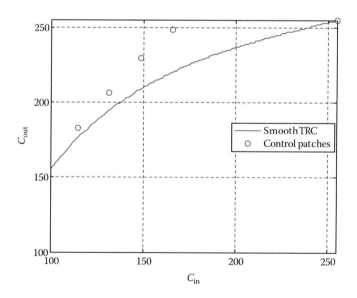

FIGURE 8.20 Cyan gray-balanced TRC with shadow region.

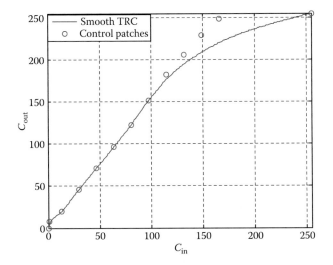

FIGURE 8.21 Cyan gray-balanced TRC with highlight and shadow corrections for C_{in}.

which is the maximum amount of toner that the marking engine can deposit. The first active shadow control patch is at $C_{in} = 166$ at which the gray-balance algorithm produces a output digital value, $C_{out} = 249$. In our illustration in Figure 8.21, we used one passive control patch at $C_{in} = 255$ whose output digital value was assigned as $C_{out} = 255$ since, for this example, the maximum desired saturation is 255. Recall that more than one passive control patch can be selected for shadow corrections. In Figure 8.21, we show a second active control patch at $C_{in} = 149$ ($C_{out} = 230$) and the third active control patch at $C_{in} = 132$ ($C_{out} = 206$), and so on. If the final TRC were to pass through all of these control patches, then pixels with C_{in} greater than or equal to 200 will be saturated to 255. This is not desirable for producing good quality images. Hence, the shadow-correction algorithm applied (post-gray-balance control iterations) should create a smooth transition to the final passive control point, $C_{in} = 255$, and in the course of such a correction, the TRC may deviate from the active control points.

The approach described above is used to provide highlight and shadow corrections to all separation TRCs. Passive control patches may have different output digital values for highlight and shadow regions between separations since each development station may have different constraints with respect to the highlight and shadow developabilities. In Figures 8.22 through 8.24, we show the TRCs with highlight and shadow compensations for magenta, yellow, and black. Note that the black TRC is produced using channel-wise linearization process, because it is processed independently during gray-balance calibration.

8.6.3 TWO-DIMENSIONAL PRINTER CALIBRATION WITH STATE-FEEDBACK METHODS

The 2-D TRCs for printer calibration can be constructed using a state feedback approach similar to 1-D calibration. The first step would be to select N_i control grid

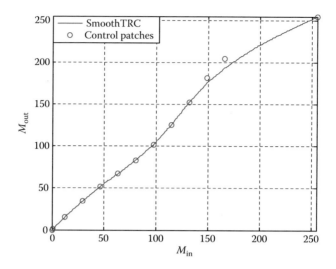

FIGURE 8.22　Magenta gray balanced TRC with highlight and shadow corrections for M_{in}.

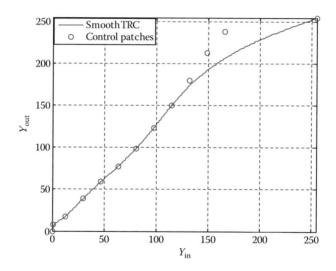

FIGURE 8.23　Yellow gray balanced TRC with highlight and shadow corrections for Y_{in}.

prints along each of the ith axis of Figure 8.9. For example, N_1 grid points are placed on the gray axis, N_2 points on the cyan primary axis, N_3 along secondary to black, and so on. Once all these grid points are selected, $N_1 \times N_2 \times N_3 \times \cdots$, control loops running in parallel are used to control all of them. The rest of the points in the plane of Figure 8.9 are computed using a 2-D interpolation/smoothing algorithm. This same process is repeated for all channels.

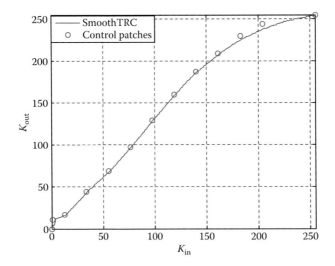

FIGURE 8.24 Black channel-wise linearized TRC with highlight and shadow corrections for K_{in}.

8.6.4 PREDICTIVE GRAY BALANCE

In the conventional gray-balance control strategy, N gray patches are used resulting in N control loops running in parallel. In principle, it is desirable to use a large number of target patches (i.e., large N) to better capture the printer nonlinearity. It is also desirable to use large-size patches or multiple patches (at different locations on the paper) for the same *CMY* input to reduce the process and sensor noise by averaging measurements over a large patch or over multiple patches with same CMY input. Due to cost constraints and competitive pressure, large N or multiple patches or oversized patches are not allowed. Use of lower number of control patches is preferred in practice to save cost. Typically, the dimension and the layout of these target N patches are designed to fit a single page to save printing and measurement time as well as material (paper and toner). For a typical printer, 10–11 patches with one being pure white to extract media white can be used to construct *CMY* TRCs and another 10–11 patches for building black K TRC. By using predictive gray-balance techniques, we can reduce the total number of control patches by a factor of two or more that require printing and measurement. Therefore, the predictive calibration method becomes especially useful for long print runs (e.g., 20 h) with single/multiple jobs where calibration pages are interspersed within/between print jobs. For a 20 h long run on a cut sheet digital press, with two test pages for calibration printed every 30 min, we would end up wasting a total of 80 pages for calibration only. Hence, any savings in wasted paper is considered useful.

For illustration, let us assume we decided to use $N = 10$ control patches for printer gray-balance calibration. Using the normal control-based technique with state-feedback approach (Section 8.6), we would print 10 patches per each iteration.

Let us assume that there are no replicas required to average out the noise. Using the predictive algorithm, we will print only half the number of patches per each iterations. We now describe the algorithm for predictive gray balance below.

(a) **Switching Strategy**: Assume that there are 10 target patches $t_1, t_2, \ldots, t_{10}$ and thus 10 different control loops are running in parallel. We divide these 10 loops into two sets: the odd set containing $t_1, t_3, \ldots, t_9$ and the even set containing $t_2, t_4, \ldots, t_{10}$. In the first iteration, we print and measure the patches corresponding to the odd set and use a predictive method to compute the CMY values (control actuation) corresponding to the patches in the even set. In the second iteration, we do the reverse that is print and measure for the even set and estimate the CMY values (control actuation) for the odd set. This kind of switching process is repeated till we converge to the true CMY value. That is, we alternate between a set of loops with actual state measurement by printing those control patches and assess which ones do not need printing, and can work well with predictive algorithm. This way, we can save printing of redundant patches. Other switching strategies include (a) dividing the N loops into M disjoint sets, and selecting one set at each iteration different from the others in a round-robin fashion (e.g., iteration $1 = $ set 1, iteration $2 = $ set $2, \ldots$, iteration $M = $ set M, iteration $M + 1 = $ set $1, \ldots$). The case described earlier is for $M = 2$ and each set containing $\frac{N}{M}$ loops. Yet, other strategy include first printing control patches for all N loops say for two iterations. After that print only those that show the largest errors.

(b) **Predictive Technique**: Ultimately, for each iterations and for each one of the N loops, we need to compute the CMY values. If for a given loop, we print and measure a patch, we use the normal iterative approach to update the CMY values as obtained by the control algorithm. If we do not print a patch for a control loop, then we use a prediction method to estimate their CMY values. From the new values of CMY already computed for other loops using the control algorithm, we estimate the new CMY values for the loops where we do not print a patch. Linear and nonlinear interpolation/extrapolation or polynomial least-square curve fitting algorithms can be used for estimating the required CMY values. As an example, let the cyan components of the five measured and controlled patches plus the two corner points be, as shown in Table 8.8, in one of the iteration cycles. Let the five unmeasured patches have the L^* target values of 20, 45, 55, 70, and 85. Since our gray-balance aim is the pure neutral axis, we assume $a^* = b^* = 0$.

A plot of the data in Table 8.8 with cubic interpolation is shown in Figure 8.25. From this interpolation, the predicted (or estimated) values of the cyan components of the five unmeasured patches corresponding to L^* values of 20, 45, 55, 70, and 85 are estimated respectively as 83.5, 52.1, 37.32, 25.85, and 15.28. In Table 8.9, we show an example of the ΔE_{ab}^* convergence response (ΔE_{ab}^* vs. iterations) for the case when we print all patches and for the case when we print half of them.

TABLE 8.8

Control Patches Used for Predictive Gray-Balance during One of the Iteration Cycles

L^*	100 C/255
0	100
40	60
50	44
60	32
80	20
90	10
100	0

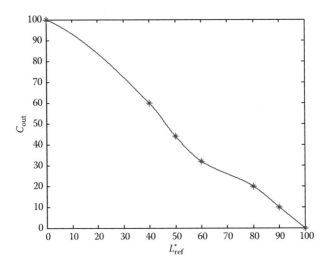

FIGURE 8.25 Cyan TRC shown with control and corner points (*Note:* to convert TRC to 0 to 255 scale use $x = (100 - L^*_{ref})2.55$ and $y = C_{out} \times 2.55$).

8.7 SPOT-COLOR CONTROL

There are many commercially available packages today that define specific colors called spot colors. They can be defined as a fixed set of colors, which may be Pantone colors (Figure 8.26), customer logo colors, colors in a customer's proprietary marked patterns, or customer-defined colors in the form of an index color table [8]. Spot colors are often used, or can be used, for large background areas, which may be the most color-critical portion of a particular page. Consistent color in these areas may make the difference between success and failure in meeting the color-consistency requirements.

TABLE 8.9

ΔE_{ab}^* vs. Iterations for the Case When We Print All Patches and for the Case When We Print Half of Them

	Loop 1	Loop 2	Loop 3	Loop 4	Loop 5	Loop 6	Loop 7	Loop 8	Loop 9	Loop 10
					ΔE_{ab}^* (Printing All Patches)					
Initial	9.5676	11.5172	7.9895	8.4248	8.8200	7.6253	5.3009	3.2287	4.3084	5.1856
Iteration 1	2.2099	4.0189	3.3067	1.2169	1.4411	1.8325	1.4351	0.6187	1.0427	1.6345
Iteration 2	0.1605	0.8001	0.7603	0.4459	0.2986	0.7680	0.4097	0.1893	0.4674	0.7890
Iteration 3	0.0498	0.1252	0.3053	0.0247	0.1293	0.2605	0.1106	0.0602	0.2205	0.4369
Iteration 4	0.0028	0.0197	0.1352	0.0059	0.0333	0.0789	0.0235	0.0193	0.0970	0.2398
Iteration 5	0.0010	0.0041	0.0386	0.0015	0.0054	0.0214	0.0067	0.0063	0.0374	0.1235
					ΔE_{ab}^* (Printing Half of the Patches)					
Initial	9.5676	11.5172	7.9895	8.4248	8.8200	7.6253	5.3009	3.2287	4.3084	5.1856
Iteration 1	2.2099	2.8726	3.3067	1.7312	1.4411	2.0814	1.4351	0.4473	1.0427	2.8275
Iteration 2	2.0284	0.4731	1.2237	0.5168	0.3374	0.4299	1.3054	0.4473	0.7570	1.2522
Iteration 3	0.0957	0.4731	0.4870	0.5168	0.3374	0.4299	0.2422	0.4473	0.7570	1.7131
Iteration 4	0.0957	0.4731	0.4870	0.5168	0.3374	0.4299	0.2422	0.4473	0.7570	0.7850
Iteration 5	0.0957	0.4731	0.4870	0.5168	0.3374	0.4299	0.2422	0.4473	0.7570	0.7850

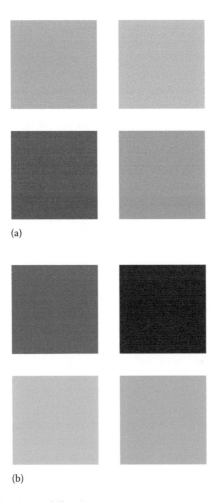

FIGURE 8.26 (See color insert following page 428.)   Examples of Pantone colors reproducible by a digital production printer on a coated stock: (a) top left: 108, top right: 1215,
bottom left: 1255, and bottom right: 135; and (b) top left: 7530, top right: 7546, bottom left:
cool gray 2, and bottom right: warm gray 2.

Many software packages have the ability to enter or create a cyan, magenta,
yellow, and black (*CMYK*) recipe using multidimensional profiles and store them for
future use or manually edit the recipe. The main problem with this kind of approach
is that the *CMYK* formula that was originally entered or obtained through manual
entry may not be very accurate at the time of use due to (a) variations in the print
engine state, (b) variability due to operator error associated with manual adjust-
ments/modifications to *CMYK* values, and (c) inaccurate recipes (to begin with). If
the recipes are created on a different print engine, then sensor errors, process drift,
and algorithm inaccuracies can lead to error in the recipe with respect to reproducing
spot color. Drift in the print-engine state, which generally occurs, is one of the major

difficulties faced by print shop owners regarding spot color reproduction. Manual adjustment of the *CMYK* values via trial-and-error process, is often used, which ends up consuming more time and can still lead to unacceptable results. Adjusting four variables (*CMYK*) manually, each interacting with the output color, can make it very difficult to find the best matching recipe. Consequently, the user has no assurance that print shop job can reliably produce colors unless every print job is reviewed by a well-trained expert who is very familiar with all the capabilities of the shop.

Because imaging can occur over a variety of different printing systems and practiced by a variety of different clients and customers, the colors may not always be consistent or accurate. We show the use of an automated closed-loop control method based on tracking to Pantone-defined or customer-defined target values that are meant to automate the determination of the appropriate *CMYK* recipe for a particular spot-color target at any point in time using sensed values of the spot color at that time. This kind of automated functionality is called automated spot-color editing [9]. A typical workflow is comprised of the following key steps:

(a) Selection and creation of appropriate target values for spot colors can be obtained from a Pantone color table or obtained from a dictionary or created using in-line color sensors. This step can be done off-line.
(b) Activation of the automated spot-color editing function at the user interface that determines the *CMYK* recipes for the target spot colors by executing spot-color closed-loop control algorithms based on sensor measurements.
(c) Acceptance/rejection of the resulting *CMYK* recipes.

Once the user accepts the *CMYK* recipe, the spot-color tag/name links the recipe to a particular spot color in the document. The acceptability criteria may be based on visual inspection of a proof copy with a printed color or simply the mean ΔE values between the target spot color and the measured spot color. The user then RIPs the image containing the new *CMYK* values for printing. In this workflow, the user still has to select the spot colors manually to create the recipe, initiate running of the closed-loop control algorithm automatically, and then insert the new recipe into a print job prior to printing.

Although, in the above workflow, the creation of the recipe is automatically generated, spot-color selection/definition and insertion of the recipe is still done manually. In a fully automated process that does not involve human intervention, the images may be checked for spot colors by monitoring the tags automatically in a preflight step after they are submitted to the printer but prior to being printed. The preflight step can detect the presence of spot colors in a submitted print job and activate the automated closed-loop control to correct for all spot colors found in the search. The spot-color detection routine looks for any standard document convention describing the use of spot colors and their names (as one example, names standardized by Pantone Inc.) in the document. For example, if the user's image is in PostScript® format, the comment %%DocumentCustomColors indicates the use of custom (spot) colors. An image processing application names these colors and their *CMYK* or *RGB* approximations through the %%CMYKCustomColor

TABLE 8.10

Example of a PostScript Document with Pantone Spot Colors

%!PS-Adobe-3.0 EPSF-3.0

%%Creator: Adobe Illustrator(R) 8.0

%%AI8_CreatorVersion: 8

%%For: (John Stanzione) (Spot Color Source)

%%Title: (solid to process.eps)

%%CreationDate: (5/24/01) (1:02 PM)

%%BoundingBox: 0 0 0 0

%%HiResBoundingBox: 0 0 0 0

%%DocumentProcessColors:

%%DocumentSuppliedResources: procset Adobe_level2_AI5 1.2 0

%%+ procset Adobe_ColorImage_AI6 1.3 0

%%+ procset Adobe_Illustrator_AI5 1.3 0

%%+ procset Adobe_cshow 2.0 8

%%+ procset Adobe_shading_AI8 1.0 0

%AI5_FileFormat 4.0

%AI3_ColorUsage: Black&White

%AI3_IncludePlacedImages

%AI7_ImageSettings: 1

%%CMYKCustomColor: 0 0 0.51 0 (Color name)

%%+ 0 0 0.79 0 (Color name)

%%+ 0 0 0.95 0 (Color name)

%%+ 0 0.03 1 0.38 (Color name)

%%+ 0 0.03 1 0.6 (Color name)

%%+ 0 0.07 1 0.5 (Color name)

%%+ 0 0.02 0.81 0 (Color name)

%%+ 0 0.04 0.62 0 (Color name)

%%+ 0 0.02 0.95 0 (Color name)

or %%RGBCustomColor comments in the body of the document. Table 8.10 provides an example of a PostScript document with Pantone spot colors and comments describing the Pantone colors. As shown in the table, %%CMYKCustomColor provides an approximation of the custom color (spot color) *CMYK* values specified by the color name in parentheses. The four components (cyan, magenta, yellow, and black) are specified as numbers from 0 to 1, representing the percentage of that process color. "%%+" identifies a continuation line. So, to find additional colors, the detection routine examines lines starting with "%%CMYKCustomColor" and lines immediately following that begin with "%%+." Alternatively, an approach described in Ref. [10], which describes some methods of spot-color sniffing from an encapsulated PostScript file, is also applicable. If no spot colors are identified in the image file, the image-processing computer RIPs the file and sends the image to production. If spot colors are present, the spot-color control algorithm is activated automatically. The spot-color control may run while another job is printing, or the print job may be held until the current print job completes. The advantages of running spot-color control automatically include reduction of long-term engine drift effects, spot-color stability, time savings, reduced operator error, and reduced operator intervention. In essence, this kind of automation would decrease the time needed to improve the accuracy and repeatability of spot colors in print jobs. Currently, such a system is still not commercially available, but is conceivable for digital printing systems of the future.

A spot-color control algorithm is comprised of the of following key steps:

(a) Determining whether or not the spot-color targets are inside the printer gamut
(b) Mapping out-of-gamut colors to printable colors using appropriate gamut mapping algorithms and determining the new target $L*a*b*$ values
(c) Selecting appropriate GCR (or gamut classes) for the target $L*a*b*$ values
(d) Applying closed-loop control algorithm for the GCR constraints set in Step c
(e) Selecting best $CMYK$ recipe out of multiple iteration steps

These steps are described in detail next.

8.7.1 Gamut Mapping for Spot-Color Control

One of the key components in finding a $CMYK$ recipe for a given spot color is determining whether the target $L*a*b*$ values are either inside/on-boundary or outside of a printer's gamut. Colors located very near or on the gamut's boundary could be mistakenly reported as outside the gamut by algorithms that are not sufficiently accurate. A color that is wrongly classified as outside the gamut will be handled by the gamut mapping algorithm. Thus, this mapping algorithm will, in turn, map the colors in reference to a point in the surface of the printer's gamut.

The consequence of this action may lead to the reproduction of a color that is slightly different from the original one; obviously, this could have been avoided since there was no need to map the original color. The ray-based control model described in Section 7.6.1 can be used to determine whether the spot-color targets are inside or outside the gamut surface. Any of the gamut clipping or gamut compression methods can be used to map out-of-gamut colors. ΔE_{2000} is one of the preferred [11] methods since it can map the out-of-gamut colors perceptually close to the gamut boundary.

8.7.2 Gamut Classes

The basic algorithm requires the use of GCR-constrained inversion for the target $L*a*b*$ values of the given spot color. A tricolor GCR is preferred for spot colors due to the advantages described in Section 7.5.4.3. The overall color printer gamut is represented as a composite of the gamut subclasses wherein each gamut subclass is comprised of a subset of $CMYK$. Selected spot-color targets are assigned to one of the gamut subclasses for efficiently calculating the CMYK recipe for a given spot-color target. An example of a tricolor gamut class is comprised of four gamut classes

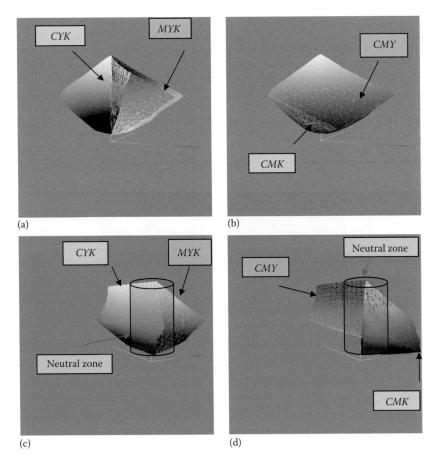

FIGURE 8.27 (a) $CYK \rightarrow L^*a^*b^*$ and $MYK \rightarrow L^*a^*b^*$ gamuts, (b) $CMK \rightarrow L^*a^*b^*$ and $CMY \rightarrow L^*a^*b^*$ gamuts, (c) $CYK \rightarrow L^*a^*b^*$ and $MYK \rightarrow L^*a^*b^*$ gamuts (alternate view shown with neutral zone), and (d) $CMK \rightarrow L^*a^*b^*$ and $CMY \rightarrow L^*a^*b^*$ gamuts (alternate view shown with neutral zone).

($CMY \rightarrow L^*a^*b^*$, $MYK \rightarrow L^*a^*b^*$, $CYK \rightarrow L^*a^*b^*$, and $CMK \rightarrow L^*a^*b^*$) and a neutral zone.

Figure 8.27a through d shows in $L^*a^*b^*$ space for the four gamut classes, $CMY \rightarrow L^*a^*b^*$, $MYK \rightarrow L^*a^*b^*$, $CYK \rightarrow L^*a^*b^*$, and $CMK \rightarrow L^*a^*b^*$, respectively. The overall gamut is the union of all of them. In plots c and d, Figure 8.27 shows an additional region classified as the neutral zone. This approach reduces the dimensionality of the four color process to three color groups. As a result of this reduction, a three-input three-output MIMO control algorithm can be implemented to achieve improved spot-color accuracy.

These gamut classes represent the color reproduction capability of a printer obtained by considering three separations at a time. This can be a forward printer model obtained via experimentation on a color printer or a mathematical

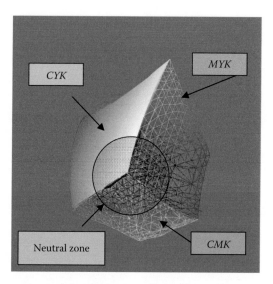

FIGURE 8.28 (See color insert following page 428.)   Top view of *CMK, MYK, CYK* gamuts and neutral zone (CMY gamut removed from the figure).

model of the type described in Section 7.4. There is a high degree of overlap between the *CMY* gamut with other classes that include the *K* separation. The overlap of the *CMY* gamut with the *CMK, CYK,* and *MYK* gamuts can cause problems while assigning spot colors to appropriate gamut classes. Whereas the overlap region of the *CMK, CYK,* and *MYK* gamuts is low. Figure 8.28 shows the *CMK, MYK,* and *CYK* gamuts along the chromatic axes when viewed from the top of the gamut.

To reduce the overlap with the *CMY* gamut class and to preserve the appearance of images, the neutral zone (region) is defined using the ΔE metrics. In some cases, this neutral zone may not be needed. This zone, as shown in Figure 8.27c and d, geometrically looks like a cylinder of circular cross section in 3-D space when the cylinder is specified with ΔE_{ab}^{*} as a radius from the neutral axis. It is much less cylindrical when the radius is specified in ΔE_{2000} or perceptual space. A classification algorithm is shown in Ref. [12].

Whenever there is overlap with the *CMY* gamut class, at a high level, this strategy has two components: (1) minimum black strategy and (2) maximum black strategy. For minimum black strategy, we use the *CMY* gamut class wherever possible. For maximum black strategy, we use black everywhere (i.e., no *CMY* gamut is used).

The use of tricolor gamut classification algorithm removes the degeneracy while finding the *CMYK* recipes. In specific terms, it reduces the dimensionality of the four-color process to three-color groups. In summary, this approach provides (a) improved spot-color accuracy with one of the separations always held to zero, (b) improved toner usage for high area-coverage spot-color printing by identifying the most toner efficient *CMYK* values for each spot color, (c) more room for the controller regarding the actuators when the spot colors under consideration are common to two or more gamut classes, and (d) a unique solution for each spot color once a gamut class has been chosen.

8.7.3 CONTROL ALGORITHM

Once a gamut class is selected, we know which of the three color separations to use to find the recipe for a given spot color. For out-of-gamut colors, choice of gamut class not only depends on the color, but also depends on the gamut mapping strategy used to map the colors to the printer's gamut. The ICI (Section 6.4.3) or moving-matrix (Section 6.3.2) algorithms can be used to find the estimated tricolor recipe associated with each mapped or in-gamut spot-color target using the printer model (analytical or empirical, obtained using experimental data) associated with each gamut class. Accuracy of this recipe depends on (a) accuracy of the printer model and (b) accuracy of the inversion algorithm. To further refine the accuracy, a three-input three-output MIMO state-feedback controller described in Section 7.5.2.2 is used. The approach is shown in Figure 8.29 for *CMY* gamut class, but the same method applies to the rest of the gamut classes. The recipe obtained from the ICI or moving-matrix algorithms are used as nominal *CMYK* values while working with the iterative feedback loop corresponding to the gamut class under use. Iterations can also be carried out directly on the printer with an in-line color sensor.

The control law can be designed using MIMO state-feedback methods or LQR or model predictive control [13] approaches. Thus, $u(k) = +Ke(k)$, where $e(k)$ is the error between the target and the measured $L^*a^*b^*$ at iteration k. The gain matrix, K, is derived based on the pole values specified such that the closed loop shown in Figure 8.29 is stable. This is achieved by assigning pole values in the range $[0, 1)$, which will place the eigenvalues of the closed-loop system inside the unit circle. During iterations, if the error, e (shown in Figure 8.29 for *CMY* gamut class), for the spot color at or near the gamut boundary is higher than the previous one, then a best *CMY* selection algorithm (see Figure 7.32) can be used because the *CMY* values near the boundary may reach saturation (i.e., the *CMY* values may go to outside the range 0 or 255).

8.7.4 CONTROL ALGORITHM WITH INK LIMITS

Often the tricolor gamut class does not give full gamut coverage in the dark part of the gamut since those colors cannot be produced with only three separations. To reproduce spot colors accurately with full gamut coverage, often a four-color gamut class is introduced. A GCR-constrained 4-to-3 control-based inversion, as described in Figure 7.41, can be used. Ink limiting constraints are often required in a four-color system due to limitation on the fuser. Constraints for ink limits can be applied during iteration while finding the recipe automatically. This would require more complex control algorithms.

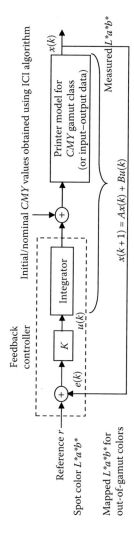

FIGURE 8.29 Closed-loop control algorithm with a gain matrix and the integrator as controller shown for spot-color control algorithm using *CMY* gamut class.

PROBLEMS

8.1 Ten yellow patches with digital input of $Y = 25.5i$, $i = 0, 1, 2, \ldots, 10$ and $C = M = 0$ are printed on a digital printer. The ΔE_{ab}^* difference between the output patches and paper are tabulated in Table 8.11. Find the yellow TRC for channel-independent linearization.

8.2 (a) Write a general MATLAB code to find the inverse of function $f(x)$ graphically from data given by a LUT.
  (b) Write a general MATLAB code to find the inverse of function $f(x)$ numerically from data given by a LUT.

8.3 Ten gray patches are printed on a digital printer and their $L*a*b*$ values are measured. The resulting data is shown in Table 8.12.
  (a) Find the cyan, magenta, and yellow gray-balanced TRCs for this printer.
  (b) What would the resulting measured $L*a*b*$ values be if the patches are printed before and after calibration?

8.4 Design a control loop for 1-D ΔE_{ab}^* from paper linearization using state feedback where the plant is modeled as combination of printer and ΔE_{ab}^* sensor. Linearize the overall system and check the stability of the system.

8.5 Use Neugebauer model, drive a closed-form expression for the sensitivity matrix B at a nominal CMY value.

TABLE 8.11

ΔE_{ab}^* Difference between the Output Patches and Paper

Y	0	25.5	51	76.5	102	127.5	153	178.5	204	229.5	255
ΔE_{ab}^*	0	0.8	3.3	7.3	12.8	20	29	39.5	51.5	65	80

TABLE 8.12

Gray Patches: $CMY \rightarrow L*a*b*$

C	M	Y	$L*$	$a*$	$b*$
0	0	0	92.5	0.19	0.09
25	25	25	88.7	0.06	0.12
51	51	51	76.9	0.87	0
76	76	76	67.22	0.13	0.19
102	102	102	55.61	0.09	0.05
127	127	127	47.33	0.08	0.09
153	153	153	36.89	0.12	0.08
178	178	178	27.78	0.32	0.08
204	204	204	18.72	0.13	0.11
229	229	229	7.98	0.07	0.08
255	255	255	3.45	0.09	0.08

8.6 Use the sensitivity matrix B obtained in Problem 8.5 to simulate the in-gamut spot-color convergence when the spot color is inside the *CMY* gamut class. Use the pole-placement algorithm and three different combinations of poles.

8.7 Perturb the sensitivity matrix with reasonable ΔB and show the robustness of spot-color recipe obtained in Problem 8.6 for *CMY* gamut class.

8.8 Repeat Problems 8.6 and 8.7 for *CYK*, *MYK*, and *CMK* gamut classes by choosing in-gamut colors in the respective gamut classes.

REFERENCES

1. A full GRACoL technical specification document, *Calibrating, Printing and Proofing to the G7 Method*, V4, Mar. 2006.
2. L.K. Mestha, P.A. Crean, M.S. Maltz, R.J. Rolleston, Y.R. Wang, E. Jackson, and T. Balasubramanian, On-line calibration system for a dynamically varying color marking device, US Patent 7,307,752, Dec. 11, 2007.
3. L.K. Mestha, R.E. Viturro, Y.R. Wang, and S.A. Dianat, Gray balance control loop for digital color printing systems, *NIP21: Proceedings of IS&T's International Congress on Digital Printing Technologies*, pp. 499–504, Baltimore, MD, Sep. 18–23, 2005.
4. G. Sharma, R. Bala, J.R.N. Van de Capelle, M. Malz, and L.K. Mestha, Two-dimensional calibration architectures for color devices, US Patent 7,355,752, Apr. 8, 2008.
5. P.K. Gurram, S.A. Dianat, L.K. Mestha, and R. Bala, Comparison of 1-D, 2-D and 3-D printer calibration algorithms with printer drift, *NIP21: Proceedings of IS&T's International Congress on Digital Printing Technologies*, pp. 505–510, Baltimore, MD, Sep. 18–23, 2005.
6. T.P. Sim and P.Y. Li, On coordination and stabilization of two xerographic printers, *Proceedings of 2008 American Control Conference*, Jun. 11–13, Seattle, WA, 2005.
7. L.K. Mestha and S. Dianat, Sensitivity matrix determination for adaptive color control, US Patent Application, Attorney Docket 20071048, Feb. 2008.
8. J. Bares, M.R. Furst, L.K. Mestha, S.J. Harrington, and E. Jackson, Accurate printing of proprietary mark patterns and colors, US Patent 7,110,143, Sept. 19, 2006.
9. J.D. Hancock, P.S. Fisher, L.K. Mestha, K.J. Mihalyov, T.L. Love, P.A. Crean, and M.F. Hoffmann, System and method for automated spot color editor, US Patent Application 20080043264, Feb. 21, 2008.
10. E.H. Ringness, Method for separating colors of encapsulated postscript images, US Patent 6,456,395, Sep. 2002.
11. A.E. Gil, L.K. Mestha, and M.F. Hoffmann, Spot color control system and method, US Patent Application, 20080043271, Feb. 2008.
12. L.K. Mestha, A.E. Gil, and M.F. Hoffmann, A method for classifying a printer gamut into subgamuts for improved spot color accuracy, US Patent Application, Attorney Docket 20061087, Dec. 2007.
13. A.E. Gil and L.K. Mestha, Spot color controls and method, US Patent Application, 20080043264, Feb. 21, 2008.

9 Internal Process Controls

9.1 INTRODUCTION

The output of digital printers drifts over time or deviates from the predetermined optimum standards due to a variety of factors. These factors include environmental conditions (temperature, relative humidity [RH], etc.), use patterns, the type of media, variations in media, aging of the components, variations from original models used in initialization, general wear, etc. To achieve predictable print quality time after time, important internal parameters (states of the machine) are controlled by applying feedback [1]. These loops maintain background, solid area development, and tone reproduction curves (TRCs) of the individual primaries by adjusting various internal process and image actuators that operate at varying frequency while making prints.

Many earlier color products only used solid area control for low- and high-density regions of the tone curve (e.g., Xerox product, 5775). This corresponds to one-point TRC control without any hierarchy. The control algorithms for set-point tracking were done through "on–off" rules; for example, when certain conditions are met, dispense the toner, otherwise not. Control loops were single-input single-output (SISO) although they were implemented on a coupled multiple-input multiple-output (MIMO) system. On some occasions, proportional, integral control was used in the SISO control loop. Today's digital production or entry production systems have become more complex and require a multivariable, modular design approach to implement stable feedback systems that can deal with a wide range of process parameters and ensure the accuracy, consistency, and stability of the internal states of the system. α time hierarchy is preferred for simplifying the implementation complexity of such a system.

In this chapter, theoretical methods are developed for analyzing the control performance of processes within the xerographic printing system. First, we describe how the system is represented in a system theoretic form, often called a state-space model. These models are used to design a time hierarchical control system for controlling all the processes that affect the images printed on paper. After that, design of the key process control loops and their effects on the final print quality are described in detail. Unique strategies and methods to overcome the adverse effects of loop interactions and avoid undesirable glitches in the controller performance are also presented.

9.2 PROCESS CONTROL MODELS—A GENERAL CONTROL VIEW

The compact characterization of a printer, described in Chapter 10, as an analytical MIMO system is very important for applications, and is something quite innovative at the time this book was written. However, this is not enough for control. For typical control applications, we need a representation of the system in the form of a vector differential equation in continuous time

$$
\begin{aligned}
\dot{x} &= g(x(t), u(t), t) \\
y &= h(x(t), u(t), t)
\end{aligned}
\tag{9.1a}
$$

or the discrete time version

$$
\begin{aligned}
x(k+1) &= g(x(k), u(k), k) \\
y(k) &= h(x(k), u(k), k)
\end{aligned}
\tag{9.1b}
$$

where
  x is the system state
  u is the input actuator
  y is the output quantity
  t and k represent continuous and discrete-time instants, respectively

In fact, we are especially interested in the linear version of the system given by

$$
\begin{aligned}
\dot{x} &= A(t)x + B(t)u \\
y &= C(t)x + D(t)u
\end{aligned}
\tag{9.2a}
$$

or

$$
\begin{aligned}
x(k+1) &= A(k)x(k) + B(k)u(k) \\
y(k) &= C(k)x(k) + D(k)u(k)
\end{aligned}
\tag{9.2b}
$$

If we can express the system in this form, there are well established results available in the literature for the control of such systems using state or output feedback control. If the linear system is also time invariant (i.e., when the matrices A, B, C, and D are constant), then simple design techniques are readily available for powerful controllers and there are simple criteria for controllability and observability. This version can be written in continuous time as

$$
\begin{aligned}
\dot{x} &= Ax + Bu \\
y &= Cx + Du
\end{aligned}
\tag{9.3a}
$$

or in discrete domain as

$$x(k + 1) = Ax(k) + Bu(k)$$
$$y(k) = Cx(k) + Du(k)$$

(9.3b)

The question is how to convert the nonlinear static MIMO representation modeled in Chapter 10 to a suitable linear time-invariant (LTI) form discussed above. The answer is simple, through the *Jacobian matrix* of the system which is by definition the derivative of the outputs with respect to the inputs. Let x be the state vector which is also the output, so $y = x$, and let u the input vector to the nonlinear MIMO system. For the case $x = f(u)$, the Jacobian is given by

$$J(t) = \frac{df(u(t))}{du}$$

(9.4)

where
  $f(u)$ is an $m \times 1$ vector
  $u(t)$ is an $n \times 1$ vector

that is

$$f(u) = \begin{pmatrix} f_1(u) \\ f_2(u) \\ \vdots \\ f_m(u) \end{pmatrix}$$

(9.5)

$$u(t) = \begin{pmatrix} u_1 \\ u_2 \\ \vdots \\ u_n \end{pmatrix}$$

(9.6)

The Jacobian matrix can then be expressed as the following $m \times n$ matrix:

$$J(u) := \begin{pmatrix} \frac{\partial f_1(u)}{\partial u_1} & \frac{\partial f_1(u)}{\partial u_2} & \cdots & \frac{\partial f_1(u)}{\partial u_n} \\ \frac{\partial f_2(u)}{\partial u_1} & \frac{\partial f_2(u)}{\partial u_2} & \cdots & \frac{\partial f_2(u)}{\partial u_n} \\ \vdots & \vdots & \ddots & \vdots \\ \frac{\partial f_m(u)}{\partial u_1} & \frac{\partial f_m(u)}{\partial u_2} & \cdots & \frac{\partial f_m(u)}{\partial u_n} \end{pmatrix}$$

(9.7)

Applying the chain rule of calculus we obtain

$$\dot{x} = \frac{\partial f(u(t))}{\partial u} \frac{du}{dt} = J(u(t))\dot{u}$$

(9.8a)

The discrete-time case is similar with t replaced by k

$$x(k + 1) = x(k) + J(u(k))\Delta u(k)$$

(9.8b)

where

$$\Delta u(k) = u(k+1) - u(k) \tag{9.8c}$$

In the rest of this chapter, we will focus on the discrete-time domain, where k refers to index of the print. The index k can also refer to the sample measurements within a single page, if the entire sensing-processing-actuation cycle is implemented within the printed page.

This model is trivial to put into the state-space form of Equation 9.3b. First define

$$v(k) := \Delta u(k) = u(k+1) - u(k) \tag{9.9}$$

which is mathematically equivalent to the use of a discrete-time integrator for the calculation of $u(k)$ from $v(k)$ (since we have access to the input $u(k)$ for control), that is

$$u(k+1) = u(k) + v(k) \tag{9.10}$$

Substituting into Equation 9.8b, we get

$$x(k+1) = x(k) + J(u(k))v(k) \tag{9.11}$$

which is a linear time-varying state-space model of the form

$$\begin{aligned} x(k+1) &= A(k)x(k) + B(k)v(k) \\ y(k) &= C(k)x(k) + D(k)v(k) \\ A(k) &= I, \quad B(k) = J(u(k)), \quad C(k) = I, \quad D(k) = 0 \end{aligned} \tag{9.12}$$

where, the only time-varying component is the system's Jacobian matrix, which has to be reevaluated at each step. Thus, Equation 9.12 is a generalized vector linear time-varying representation of a printing system with actuators (inputs) and outputs.

It is vital to recognize that for this approximation to hold true, with respect to the reference point of linearization, $x(k)$ and $v(k)$ must have relatively small deviations between consecutive time instants. The size of the deviation depends on the local nonlinearity of the model.

The Equation 9.8b represents a typical linearization scheme for a nonlinear discrete-time dynamic system around a time-varying point. For the special case where the system is linearized about a fixed, time-invariant nominal value of the actuators, namely u_0 (which gives in turn a nominal value of the output, namely x_0) Equation 9.8b becomes

$$x(k+1) = x_0 + J\Delta u(k) \tag{9.13}$$

where matrix J is the Jacobian computed at $u = u_0$

$$J = J(u_0) \tag{9.14}$$

and

$$\Delta u(k) = u(k + 1) - u_0 \tag{9.15}$$

However, this equation is still not in the typical state-space form of Equation 9.3b, which is the most desired form. To get an equation of the form of Equation 9.3b, we use a discrete integrator for $\Delta u(k)$ and define

$$\Delta u(k) = \Delta u(k - 1) + v(k) \tag{9.16}$$

or equivalently (by the definition of $\Delta u(k)$ in Equation 9.15)

$$u(k + 1) = u(k) + v(k) \tag{9.17}$$

Substituting Equation 9.16 into Equation 9.13, we get

$$x(k + 1) = x_0 + J(\Delta u(k - 1) + v(k)) \tag{9.18}$$

Replacing $k + 1$ with k in Equation 9.13, we obtain

$$x(k) = x_0 + J\Delta u(k - 1) \tag{9.19}$$

Solving Equation 9.19 for x_0 and then substituting it into Equation 9.18 we get

$$x(k + 1) = x(k) + Jv(k) \tag{9.20}$$

which is of the desired form. Putting everything together, we have

$$\begin{aligned} x(k + 1) &= Ax(k) + Bv(k) \\ y(k) &= Cx(k) + Dv(k) \end{aligned} \tag{9.21}$$

where $A = I$, $B = J$, $C = I$, and $D = 0$. Matrix I is a 3×3 identity matrix.

From the analysis above, we observe that the use of a discrete integrator results in a very neat and elegant linear state-space representation. This is schematically presented in Figure 9.1 below for an example system from Chapter 8. Note that as shown $v(k)$ is defined as $v(k) = ke(k)$.

Although Equation 9.21 is fairly simple from a theoretical perspective, it is difficult to apply in practice. This is so because the models describing the printing system are very complicated and have some very challenging nonlinearities. Most of the difficulty stems from the fact that the process models for each separation are not independent and additionally because the color model employed presents a very complex dependency on the toner masses. These are the same reasons why it would

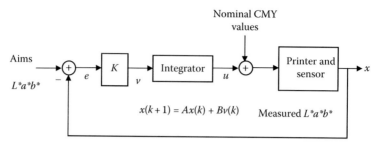

FIGURE 9.1   MIMO state-space model with the integrator used in spot color control model for CMY separations.

be very difficult to derive the Jacobian theoretically. However, it is possible to find the Jacobian using a computer application like MATLAB symbolic toolbox. This process is not trivial, but an algorithm can be implemented that derives the Jacobian pretty fast (~1 s). This difficulty, however, is easily avoided by making the assumption that $F = 0$ or $F = 1$ with the proper control of V_a in the transfer model (Table 10.8). This is quite typical in practice and is equivalent to ignoring the transfer model from the system and substituting it by, for example, a lookup table (LUT).

The sensitivity analysis is very important, because it opens the way toward the design of sophisticated real-time state-based feedback controllers and the use of model predictive control (MPC), all designed for increased performance.

Apart from that, the Jacobian provides invaluable information about two important properties of the system as explained below:

1. *Controllability*: Because of the special form of the state-space model, it is shown in the following section that the LTI system approximation is controllable if and only if the Jacobian matrix is full rank. A heuristic approach has been followed, which uses an algorithm to minimize the minimum absolute singular value or, equivalently, to see for what values this quantity approaches 0, which in turn means that the system is no longer full rank.
2. *Robustness*: Robustness with respect to measurement noise is a very important property since it provides insight regarding the behavior of the system in the presence of noisy measurements. The magnitude of each of the entries in the Jacobian is an index of robustness of the system. The Jacobian also helps answer questions like the two that follow. What are the limits of uncertainty over its operational boundary? Can we design the feedback system robust enough by considering the bounds on the Jacobian so that the actuator latitudes are preserved during a variety of operational scenarios with different media and images acting as disturbances to the control system?

Section 9.3 includes a discussion of the Jacobian for different process control hierarchies.

9.3   TIME HIERARCHICAL PROCESS CONTROL LOOPS

The controls required for color printers are significantly more complex as compared with those required for black and white printers, not only due to the use of multiple primaries, but also due to high print quality requirements for color due to the eye's sensitivity to small color variations. Given the traditional difficulties associated with the control of color, a multilevel modular control architecture was proposed [2,3] that is conceptually very similar to other hierarchical control systems [4]. Time hierarchy comes from the "reduction of complexity" rule used to design complex control systems, which transforms the printing system to many simpler subsystems while preserving the overall performance goals. Each controller sees the controllers below it as a virtual body from which it gets information and sends commands. In a control hierarchy, the lower level controllers run faster than the higher level loops, controlling a group of subsystem variables at a higher rate. The lower level controllers deliver simpler view to higher level controls. The higher level controls coordinate commands to subsystems at a much lower rate. In Xerox, levels 1, 2, 3, and 4 controls are used to describe the time hierarchy of process controls. Level 1 includes the lower level subsystem controls such as the "charge control," "toner concentration control," etc., level 2 the controls between subsystems (e.g., "charge and development" systems), level 3 the image control for each separation tone adjustments (e.g., 1-D tone reproduction control), and level 4 the image control between multiple separation tone adjustments (e.g., 2-D LUT and 3-D profiles) to minimize the interactions between colorants that cause color shift in the output. The control functions managing the job scheduling and managing set points based on media attributes (on a sheet-by-sheet basis) are done by constraint-based schedulers that are executed at a different level. Control functions that require "human/operator-in-the loop," for example, redirecting jobs to an available printer when another fails, are several layers higher than the subsystem level controls.

Each layer in the architecture is characterized by the nature of its sensor input, actuator output, and algorithm properties. An abstraction of the architecture for levels 1, 2, and 3 is presented for process control loops schematically in Figure 9.2a for an image-on-image (IOI) printing system with a belt photoconductor. Similar abstraction is applicable to a typical printing system that is non-IOI, as in the drum photoconductor system.

9.4   LEVEL 1 ELECTROSTATIC CONTROL SYSTEM

The level 1 controller is closely coupled to its related subsystem. It operates at the subsystem level to control subsystem parameters directly. Both its sensing and actuations occur locally. At this level, the actuations and the sensed parameters are coupled by a single process step. An important feature of the level 1 controller is that there is a simple and direct relationship between the sensed and controlled parameters. The algorithms of these lower level controls must be both noise immune and able to respond rapidly to changes in set points since their set points will be altered by the higher level algorithms.

It is important to understand the principal subsystems involved in the design of the level 1 controller. The xerographic process is centered on the photoconductor, which is a multilayer belt or drum that retains charge in the dark but discharges when

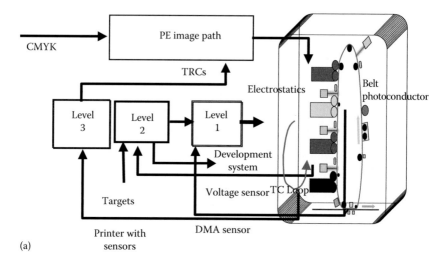

(a)

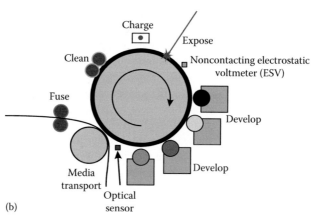

(b)

FIGURE 9.2 (a) Abstract representation of multilevel control architecture. (b) A typical xerographic print engine with a drum photoconductor showing principal subsystems.

exposed to light. The first subsystem begins with an uncharged photoconductor and finishes with the exposed customer page (the mirror image, also called the latent image) expressed as a pattern of discharged dots. Therefore, the level 1 controller, in its simple form, maintains the charge on a moving photoconductor surface at desired set points, for both unexposed (charged) and exposed (discharged) regions. The actuators that are normally used for controlling the photoconductor surface potential in this process are the grid voltage, V_g, and the exposure intensity, X (Table 10.3) [5–7]. The sensor used to measure the voltage levels on the photosensitive surface is a nonconducting electrostatic voltmeter (ESV) or electrometer; sensor examples are available in Ref. [8]. The electrometer is generally rigidly secured to the printer, adjacent to the moving photoconductor surface, and measures the voltage level of the photoconductor surface as it traverses under an ESV probe before and after the exposure. A typical ESV is controlled by a switching arrangement that provides the measuring condition in which charge is induced on a probe electrode corresponding to the sensed voltage level on the photoconductor. The induced charge is proportional to the sum of the internal capacitance of the probe and its associated circuitry relative to the probe-to-measured surface capacitance. A DC measurement circuit is combined with the ESV circuit for providing an output that can be read by a conventional test meter or used as input to the feedback control computer. The surface voltage is a measure of the density of the charge on the photoconductor, which is related to the quality of the printed output. There is typically a routine within the operating system of the printer to periodically create test areas (called patches) [9,10] at predetermined locations on the photoconductor by deliberately causing the exposure system to charge or discharge as necessary for measurement with the ESV. To avoid negative impact on productivity, color patches are created in small unobtrusive areas, such as the inter document zone (IDZ), between the latent images as illustrated in Figure 9.3.

A block diagram representation of the level 1 control system is shown in Figure 9.4. The exposed patch area voltage level, V_1, and an unexposed patch area voltage level, V_h, are shown in the diagram with the actuators, the grid voltage, U_g (also denoted by V_g for later use) , and the exposure intensity, U_1 (also denoted by X for later use). The desired goals for the feedback system is to maintain the charge (and hence voltages V_h and V_1) on a moving photoconductor surface to desired set points (V_h^T and V_1^T respectively).

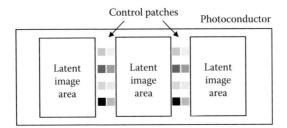

FIGURE 9.3 IDZ patches (shows typical patches used for measuring exposed and unexposed areas of the *CMYK* separations using four ESVs).

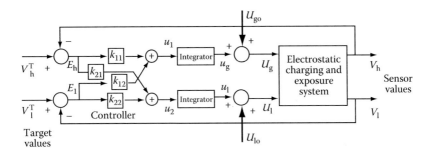

FIGURE 9.4 Block diagram representation of the level 1 electrostatic control loop.

The model-based curves for a typical electrostatic system are shown at a nominal operating point in Figure 9.5a through c with notations $V_{go} = U_{go}$, $X_{lo} = U_{lo}$, $V_g = U_g$, $X = U_1$. Figure 9.5a and c plot photoconductor potentials as a function of the grid voltage for the unexposed and exposed regions, respectively. Figure 9.5b is basically a laser power curve, a photo induced discharge curve (PIDC) at the fully charged photoconductor potential. Points on these figures marked by "x" indicate a nominal operating point (chosen at random to illustrate the approach). V_{ho} and V_{lo} are voltages at the nominal operating point "x." When there is no feedback, V_{ho} is the voltage on the unexposed photoreceptor, for the grid voltage set to V_{go}. With the grid voltage remaining at V_{go}, if the laser power is set equal to X_{lo}, then the photoconductor will be exposed to V_{lo} volts (shown in Figure 9.5b).

Let b_{11} be the slope of the curve in Figure 9.5a at the point marked "x" at $\{V_{go}, V_{ho}\}$. Let ΔU_g (used synonymously with u_g in Figure 9.4) be the deviation around V_{go} which would be generated by the controller when the charging control loop is closed. Let b_{22} be the slope of the curve at point "x" in Figure 9.5b that has the coordinates $\{X_{lo}, V_{lo}\}$. Similarly, let b_{21} be the slope of the curve at point "x" in Figure 9.5c that has the coordinates $\{V_{go}, V_{lo}\}$. Let ΔU_1 (used synonymously with u_1 in Figure 9.4) be the deviation about X_{lo}, caused by the electrostatic controller. The expressions for the deviations in photoreceptor voltages, ΔV_h and ΔV_l, can be written in terms of the small signal deviations $\{\Delta U_g, \Delta U_1\}$ $\{u_g, u_1\}$, as follows:

$$\Delta V_h = b_{11}\Delta U_g \quad \Delta V_l = b_{21}\Delta U_g + b_{22}\Delta U_1 \qquad (9.22)$$

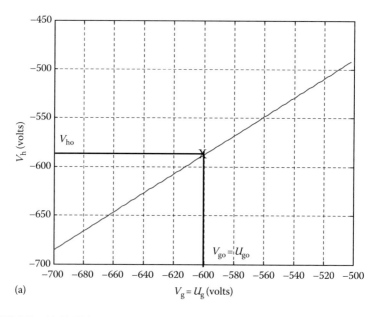

(a)

FIGURE 9.5 (a) $V_g(U_g)-V_h$ curve gives slope b_{11} at point "x."

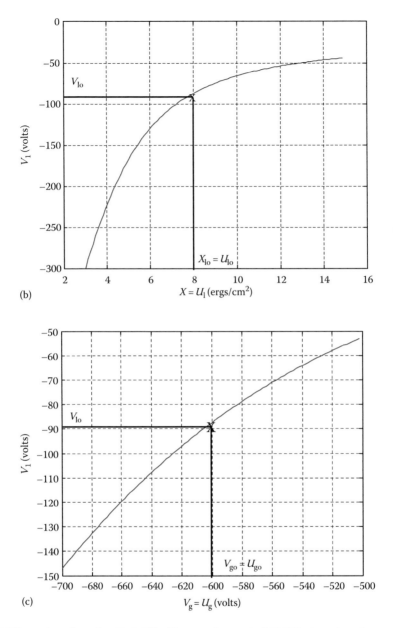

FIGURE 9.5 (continued) (b) $X\,(U_1)-V_1$ curve for nominal $V_g(U_g)$ gives slope b_{21} at point "x." (c) $V_g(U_g)-V_1$ curve for nominal $X(U_1)$ gives slope b_{22} at point "x."

Note that we assume small deviations so we can ignore all the second and higher order terms in Equation 9.22 so that the system equations are simple and linear [11]. The linear electrostatic system described by Equation 9.22 for the nominal operating point is shown in block diagram form in Figure 9.6. The actuator signals before and

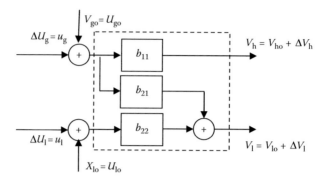

FIGURE 9.6 Block diagram representation of linear electrostatic system (Equation 9.22) shown with nominal actuator values.

after the integrator are $\{u_1, u_2\}$ and $\{u_g, u_1\}$, respectively. The control system is modeled in state-space form (Equation 9.21) by introducing the integrator, where $A = I$ is the 2×2 identity matrix, B is the 2×2 Jacobian matrix given by

$$B = \begin{bmatrix} b_{11} & 0 \\ b_{21} & b_{22} \end{bmatrix} \tag{9.23}$$

The state vector $x(k)$, control vector $v(k)$, and the output vector $y(k)$ are given by

$$x(k) = \begin{bmatrix} V_h \\ V_l \end{bmatrix}; \quad v(k) = \begin{bmatrix} u_g = \Delta U_g \\ u_1 = \Delta U_1 \end{bmatrix}; \quad y(k) = x(k) \tag{9.24}$$

It is clear from Figure 9.6 that if the slope, b_{22}, is equal to zero, the exposed voltage, V_1, is not affected by the laser intensity. In other words, when $b_{22} = 0$ the states $x(k)$ are not fully controllable using only the two electrostatic actuators. As the photo-conductor charge is reduced in magnitude or increased from more negative to positive value (see Figure 9.7 at nominal $X(U_1) = 8$ ergs/cm^2), the slope, b_{22}, decreases rapidly. If the slope b_{22} is too small, it can result in loss of controllability as described above.

Example 9.1

Test the following electrostatic control system for controllability. The system is at the nominal operating point $\{V_{go} = -600V, X_{lo} = 8 \text{ ergs/cm}^2\}$.

$$x(k+1) = \begin{bmatrix} 1 & 0 \\ 0 & 1 \end{bmatrix} x(k) + \begin{bmatrix} 0.9798 & 0 \\ 0.3315 & 11.1026 \end{bmatrix} v(k)$$

$$y(k) = \begin{bmatrix} 1 & 0 \\ 0 & 1 \end{bmatrix} x(k)$$

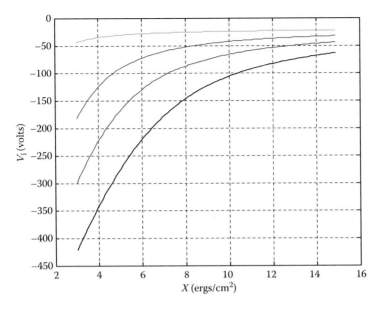

FIGURE 9.7 $X(U_1)$–V_1 curves for $V_g(U_g) = [-300\,\text{V}$ (upper curve), $-500\,\text{V}$, $-600\,\text{V}$, $-700\,\text{V}$ (lower curve)].

SOLUTION

The controllability matrix (Equation 4.121) is given by

$$Q = [B \quad AB] = \begin{bmatrix} 0.9798 & 0 & 0.9798 & 0 \\ 0.3315 & 11.1026 & 0.3315 & 11.1026 \end{bmatrix}$$

Since the rank of the controllability matrix is two, the electrostatic system is fully controllable at the nominal operating point. Since A is identity matrix, the same conclusion can be reached by finding the rank of the Jacobian matrix (B).

9.4.1   ELECTROSTATIC CONTROLLER DESIGN

From the theory of linear systems, it is known that two of the most important properties to examine before designing a controller are controllability and observability (see Chapters 4 and 5). We now check these conditions for the LTI system of Equation 9.21. While observability is, by definition, guaranteed, since the output is equal to the state vectors ($C = I$), we need to find a condition for controllability. Since $A = I$, this condition is trivial and simply requires that matrix B is full rank. This is an important condition for control to be successful and is usually satisfied everywhere except at the boundary limits of the actuators. If this is satisfied, then state-feedback control (equivalent to output-feedback control in this case) is feasible and arbitrary pole placement of the closed-loop system can be carried out via many techniques through the appropriate selection of the feedback gain matrix K (see Figure 9.1).

In the case of tracking of a reference value, say x_d, we can define the error signal as

$$e(k) = x_d - x(k) \qquad (9.25)$$

Then, via Equation 9.21 we have the following equations for the closed-loop system

$$\begin{aligned} e(k+1) &= e(k) - Bv(k) \\ v(k) &= +Ke(k) \end{aligned} \qquad (9.26)$$

or equivalently

$$e(k+1) = (I - BK)e(k) \qquad (9.27)$$

Therefore, we end up having an autonomous linear, time-invariant state-space model for the error and we can control the decay rate of the components of vector $e(k)$ by properly selecting the gain matrix K to place the eigenvalues (poles) of the matrix $(I - BK)$ at positions inside the unit circle in the complex z-plane to meet the required stability specifications of the closed-loop system (Section 5.2.3). A robust pole-placement algorithm for a MIMO electrostatic system is called "place," which can be found in the MATLAB Control System Toolbox [12], which uses an extra degrees of freedom to find a robust solution for gain matrix K. It minimizes the sensitivity of closed-loop poles to uncertainties in the A and B matrices (see Example 5.5), which in our case are the uncertainties in the elements of the Jacobian matrix. Reference [13] gives an additional procedure for assigning poles to a closed-loop MIMO system. A simple approach described in Section 9.8 for designing level 2 controller gain matrix is also applicable to the level 1 controller gain matrix. Other aspects of closed-loop performance can be also used in designing K (see Section 5.3), which will be discussed later in this chapter.

Example 9.2

For the open-loop electrostatic system shown in Example 9.1, find the gain matrix K to place the closed-loop poles within the unit circle on the real axis at 0.2 and 0.3.

SOLUTION

Using $A = I, B = \begin{bmatrix} 0.9798 & 0 \\ 0.3315 & 11.1026 \end{bmatrix}$ and $P = [0.2 \ 0.3]$ in MATLAB, command $K = \text{place } (A, B, P)$. The gain matrix K is computed to be

$$K = \begin{bmatrix} 0.8165 & 0 \\ -0.0244 & 0.0630 \end{bmatrix}$$

This gain matrix will give rise to stable performance.

Example 9.3

Using the gain matrix of Example 9.2 and the electrostatic model described in Chapter 10, simulate the transient performance of the closed-loop linear

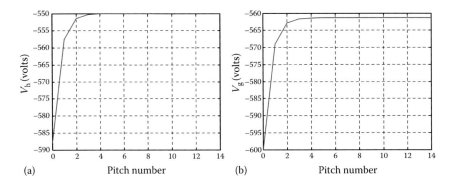

FIGURE 9.8 (a) Convergence plot of V_h with respect to pitch number. (b) Plot of actuator $V_g(U_g)$ with respect to pitch number.

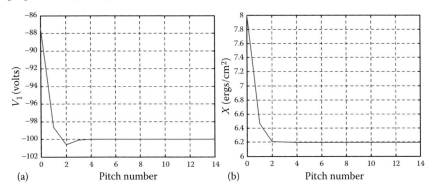

FIGURE 9.9 (a) Convergence plot of V_l with respect to pitch number. (b) Plot of actuator $X(U_l)$ with respect to pitch number.

electrostatic control system when the desired unexposed and exposed voltages on the photoconductor are given by $x_d = [-550 \; -100]^T$ volts.

SOLUTION

Figures 9.8 and 9.9 show the convergence of the unexposed and exposed voltages as the actuators (grid voltage and laser intensity) are changed every photoconductor pitch. Clearly, except for a small overshoot in V_l, the transient performance is similar to the performance expected by a system with closed-loop poles at 0.2 and 0.3.

The length of the time that the charge is retained on a photoconductor is determined by the decay rate in the dark. Charge and exposure do not take place instantaneously. Sometimes the period between charge and exposure can lead to significant loss of charge. Automatic control systems, if designed to be stable, that is, with closed-loop eigenvalues within the unit circle, can maintain the exposure level to the desired value inspite of the dark decay as seen in the next example.

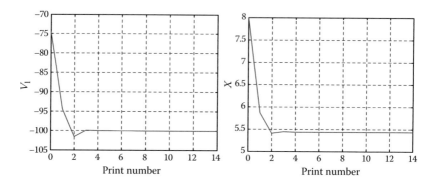

FIGURE 9.10 Convergence plot of V_1 and actuator $X(U_1)$ as a funtction of print number for a one pitch photoconductor with dark decay of 30 V at the exposure station.

Example 9.4

Let the charge on the photoconductor at the exposure station be decayed by 30 V (i.e., V_h at the exposure station is equal to $-550 + 30$ V). Simulate the transient performance of the controller shown in Example 9.3.

SOLUTION

V_h and V_g plots do not change. V_l has the desired final value, whereas the starting value is different because the photoconductor was charged to a lower potential due to dark decay. X has the same starting value but a different final value when compared to Example 9.9. Interestingly, the electrostatic control system is automatically able to maintain the photoconductor voltages to the desired target (Figure 9.10).

It is interesting to note, at this point, that all the pole-placement algorithms are designed for continuous-time linear systems of the form shown in Equation 9.2a. In the next section, we show that they are equivalent to pole-placement techniques for the discrete model Equation 9.2b which is being used extensively in this study. Note that it is known from linear system theory that for arbitrary pole placement to be possible, a necessary and sufficient condition is controllability, which in our case reduces to the simple condition that the Jacobian matrix (at a nominal operating point) has full rank.

9.5 STATE SPACE TO TRANSFER FUNCTION CONVERSIONS

In this section, we demonstrate how to transform a system from state space to transfer function form. The system is assured to be LTI. This is done for both continuous and discrete-time systems using the Laplace transform (s-domain) and z-transform, respectively. The result is well known, particularly for the continuous-time case, but the motivation of this exercise is to prove that pole-placement algorithms for both cases have an identical form. Of course, we know from linear system theory that discussion on arbitrary pole placement is only possible if the controllability condition is satisfied.

In recapitutation, an LTI continuous-time system can be written as

$$\dot{x} = Ax + Bu$$
$$y = Cx + Du$$

(9.28)

or in the discrete-time

$$x(k + 1) = Ax(k) + Bu(k)$$
$$y(k) = Cx(k) + Du(k)$$

(9.29)

Taking the Laplace transform of both sides of Equation 9.28, we get

$$sX(s) - x(0) = AX(s) + BU(s)$$
$$Y(s) = CX(s) + DU(s)$$

$\Rightarrow$

$$X(s) = (sI - A)^{-1}X(0) + (sI - A)^{-1}BU(s)$$
$$Y(s) = CX(s) + DU(s)$$
$$= C(sI - A)^{-1}x(0) + (C(sI - A)^{-1}B + D)U(s)$$

(9.30)

where $X(s)$, $Y(s)$, and $U(s)$, denote the Laplace transforms of the signals $x(t)$, $y(t)$, $u(t)$, respectively, and where the inverse $(sI - A)^{-1}$ exists for all real values of s except at the eigenvalues of matrix A.

Similarly, applying one-sided z-transform on Equation 9.29, we get

$$zX(z) - x(0) = AX(z) + BU(z)$$
$$Y(z) = CX(z) + DU(z)$$

$\Rightarrow$

$$X(z) = (zI - A)^{-1}x(0) + (zI - A)^{-1}BU(z)$$
$$Y(z) = CX(z) + DU(z)$$
$$= C(zI - A)^{-1}x(0) + (C(zI - A)^{-1}B + D)U(z)$$

(9.31)

where $X(z)$, $Y(z)$, and $U(z)$ denote the one-sided z-transforms of the signals $x(k)$, $y(k)$, $u(k)$, respectively, and where the inverse $(zI - A)^{-1}$ exists for all values of z except at the eigenvalues of matrix A. This shows that the two (matrix) transfer functions have exactly the same form, only s is replaced by z. Therefore, a pole-placement algorithm to design state feedback for a continuous-time system can also be used for the discrete case, since the transfer function has the exact same form in both cases.

Hence, taking $u(t) = +Kx(t)$, where K is a constant gain matrix chosen to satisfy the specifications (i.e., so that the closed-loop system has the desired eigenvalues) we get

$$X(s) = (sI - A - BK)^{-1}x(0)$$
$$Y(s) = (C + DK)(sI - A - BK)^{-1}x(0) \tag{9.32}$$

or in the discrete case:

$$X(z) = (zI - A - BK)^{-1}x(0)$$
$$Y(z) = (C + DK)(zI - A - BK)^{-1}x(0) \tag{9.33}$$

that is, again s is replaced with z. Therefore, we can use the same pole-placement algorithms developed for continuous-time LTI systems. Of course, to achieve stability of the closed-loop system, the poles have to be placed in the left-hand side of complex s-plane for continuous-time systems and inside the unit circle of the complex z-plane for discrete-time systems.

9.6   LEVEL 2 DEVELOPABILITY CONTROLLER

Control of the field strength and the uniformity of the charge on the photoconductor is very important because high-quality prints are produced when a uniform charge with the desired magnitude is maintained on the photoconductor. If the photoconductor is not well controlled for both fully exposed (V_1—the voltage on the discharged region) and unexposed levels (V_h—the voltage on the charged area), then the electrostatic latent image obtained upon exposure of the image (e.g., the development voltage, for a discharged area development, V_{dev}, shown in Equations 10.29 and 10.30) will be relatively weak and the resulting deposition of toner material will be weakened. As a result, the copy produced by an underexposed photoconductor would look faded since V_{dev} will not be sufficient for good development. If, however, the photoconductor is overcharged, there will be too much developer material that is deposited on the photoconductor. This results in a higher cleaning voltage, V_{clean} which affects the background and dot growth of the image on the photoreceptor. A copy produced by an overcharged photoconductor will have a gray or dark background instead of the white background of the paper. In addition to the background effect, areas intended to be gray will be black and the tone reproduction will be poor. Moreover, if the photoconductor is excessively overcharged, it can be permanently damaged.

In a digital printing machine such as an electrophotographic engine, each tone of a contone image is produced by a certain spatial combination of the available tone levels produced by the halftone screening process. A binary printer uses two levels with a dot "on" or "off"; that is, toner is developed on an exposed dot to create a solid cyan, magenta, yellow, or black dot for a discharge area development. These levels are used to reproduce the contone image. In some machines multiple exposure levels are obtained by performing intensity modulation during exposure.

Although halftoning process is well defined and repeatable, the quality of the developed image on the photoconductor depends on the system's ability to reproduce the tones in the presence of exposure and development uncertainties. The developed mass of toner per unit area (DMA) is normally used to quantify the macro level toner reproducibility on the photoconductor. This toner mass is a function of numerous parameters of which many are fixed by design. Parameters that contribute to the variability of DMA within print runs can lead to print quality problems and increase the number of misprints. The most important parameters in a two component developer system are (1) toner concentration (TC)—the ratio of the amount of toner to the amount of carrier available in the development system and (2) the toner charge per unit mass (Q/M ratio). The TC changes during engine operation due to a varying amount of toner depletion (or addition) caused by image development. The charge-to-mass ratio of the toner is dependent on the triboelectric properties of the toner and carrier, TC, relative humidity (RH) of the air in the development substation, and toner agitation mechanism in the developer. Generally, in xerographic engines, TC is maintained to some constant level using a sensor in a separate multilevel asynchronous control loop described in Section 9.11. Although development can be affected directly by varying the Q/M ratio, a direct control of Q/M ratio cannot be implemented easily without increased hardware cost. Another way of reducing the variability in development is to measure the DMA with sensors and adjust the tone levels using area modulation as in the level 3 control loop described later in Section 9.9 and/or by adjusting photoconductor charge, exposure, and development bias at some periodic intervals.

A common technique for measuring DMA is to artificially create test patches of reasonable size (e.g., 1 in.^{2} on a photoconductor belt) in the IDZ at predetermined area coverages that act as a surrogate measurement for capturing effects on the customer image area [9,10]. These test patch areas are charged, exposed, and developed by the laser system as necessary. The developed test patches are read by the reflected signal from the patch area [14]. Spatial effects are captured as an integrated signal in the sensor output and they contribute toward noise in the measurement. The number of test patches used depends on how many tone levels in the tone curve have to be sampled to generate the control actuation and the approach used to implement the control function. For example, in Ref. [15], for time-sequential sampling of tone curves, one tone level (area coverage) is proposed during a periodic actuation instance. The patch area coverage is varied over time using a time-sequential sampling strategy. Other more feasible implementations (e.g., iGen3) use fixed area coverages; one at high area coverage (90%–100%), one at low area coverage (0%–20%) and one at mid-tone (around 50%), which is considered adequate to sample the critical regions of the tone curve and affect the entire curve through feedback to process actuators such as charge, exposure intensity, and development bias [1,16]. A three patch sampling is required at reasonably high print frequency to maintain the stability of mid-tones in image-on-image engines. Some printers, such as the DC2060 and DC6060, which develop the image on four individual drums, use a two patch sampling strategy.

During control, the DMA measurement vector from the sensor is filtered using an appropriate algorithm to remove noise components and obtain a true

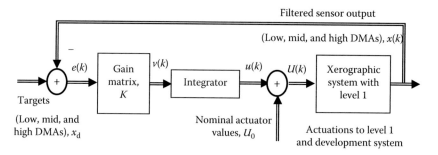

FIGURE 9.11   MIMO level 2 developability controller.

signal. The filtered DMA vector is compared to the desired DMA target vector and the error vector is processed by the feedback controller to generate the actuator values. The choice of actuators depends on the control hierarchy and the ease of implementation. In a time hierarchical control architecture, since the electrostatic control loop is fully settled before closing the outer level 2 control loop, the set points for the level 1 loop $\{V_h$ and $V_l\}$ act as two actuators. The developer bias, V_{bias}, which affects how much toner is transferred to the photoconductor, is used as the third actuator. Numerous additional architectures use combinations of process parameters as actuators [16–21] in order to optimize for cost and performance. For a three patch sampling system, in which the sensed signals are coupled to all of the actuators, the simplest controller would be a three-input three-output state feedback controller whose gains are determined from the Jacobian matrix. Elements in the matrix represent the first-order derivatives of the DMA of the three patches with respect to the three actuators. The design of a MIMO controller is described in Figure 9.11.

Let $D_l(k)$, $D_m(k)$, and $D_h(k)$ represent the three different DMA measurements (output of the filter) at the sample time defined by the parameter k. These DMA measurements are basically identified as three different points on the TRC. k is a parameter used to describe the iteration of the level 2 control loop. This parameter is different from the iteration number of the level 1 loop. The hierarchical structure contains nested loops: an inner loop (level 1) operating within the body of an outer loop (level 2). Several iterations of the inner loop are required to reach steady performance before the outer loop is closed. One iteration pass for a level 2 loop comprises of a sensing-processing-actuation update cycle. Normally, 10–30 prints are made for every level 2 iteration.

Let the corresponding target values of the toner mass be represented by the vector, $x_d = [D_l^T \quad D_m^T \quad D_h^T]^T$. The target vector does not depend on k since there is no higher level loop used to update the target vector. The actuator signals $U_h(k) = V_h^T(k)$, $U_l(k) = V_l^T(k)$, and $U_b(k) = V_{bias}(k)$ are captured by the actuator vector $U(k) = [U_h(k) \quad U_l(k) \quad U_b(k)]^T$. Let $U_0 = [U_{ho} \quad U_{lo} \quad U_{bo}]^T$ be the corresponding static nominal actuator vector that does not depend on the sample time. The linear state feedback controller will have the architecture shown in Figure 9.11.

The actuator updates are obtained from

$$U(k) = U_0 + u(k)$$
$$u(k + 1) = u(k) + v(k)$$

(9.34)

This closed-loop system equation is very similar to Equation 9.26.

9.6.1 JACOBIAN MATRIX FOR DEVELOPABILITY CONTROL

Development of the Jacobian matrix for level 2 control loop is very challenging because there are three actuators being coupled to the sensor signals as apposed to two actuators in the case of electrostatic control. We can think of three methods for extracting the Jacobian matrix. They are (1) an analytical method, (2) design of experiments (DoE) methods [22,23], and (3) direct numerical methods. In the analytical method, at first the Jacobian matrix of a development system is expressed in terms of analytical expressions for the separation of interest using a charging, exposure, and development model. The Jacobian with respect to the actuators of the process models can be calculated using the symbolic toolbox in MATLAB. Apart from the Jacobian matrix, this approach gives invaluable information about the robustness of the model with respect to key process parameters of the print engine. This information can be used to satisfy several other specifications and improve the system performance. The direct experimental method involves running carefully designed experiments to extract the nonlinear surfaces in the DMA outputs with respect to actuator values. The DoE method allows systematic ways to select actuator values in the vector, $U_0 = [U_{ho} \quad U_{lo} \quad U_{bo}]'$, between their operating limits, so that they are orthogonal and yet minimize the number of experiments required to determine the Jacobian matrix. We will describe the numerical method in detail below.

The numerical method will be useful to understand the available operating space for the DMA vector in a static machine when actuators are varied within their limits. Figure 9.12 shows a three-dimensional (3-D) space covered by the 16^3 actuator values for combinations sampled uniformly for a tandem printer example. Each point on this plot represents the actuator vector, U. In DoE language, this type of design matrix is called a "full factorial design." In Operations Research, this is known as an exhaustive search. Three patches with low, mid, and high area coverages are developed under different actuator values for U spread over their limits. They are measured with the optical DMA sensor, if actual printer is used (or calculated when a virtual printer model is used). Figure 9.13 shows the volumetric space covered by the DMA vector for the actuator values of Figure 9.12 shown for four different orientations of the DMA axes. Figure 9.14 shows the full development TRC for the same actuator values. Clearly, the available space for controlling the DMA using the hierarchical level 2 controls is quite limited. The high DMA values show a larger change, meaning higher sensitivity to actuator changes. However, careful selection of actuators can further increase the available space, which will be covered briefly in Example 9.7 and Section 9.14.

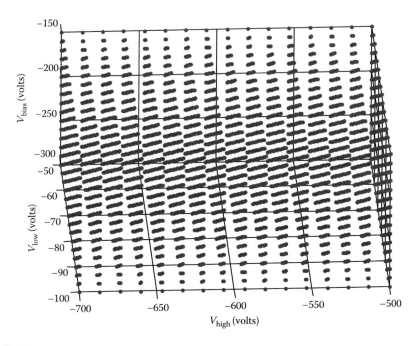

FIGURE 9.12 Actuator space used for identifying the numerical DMA model for a static printer.

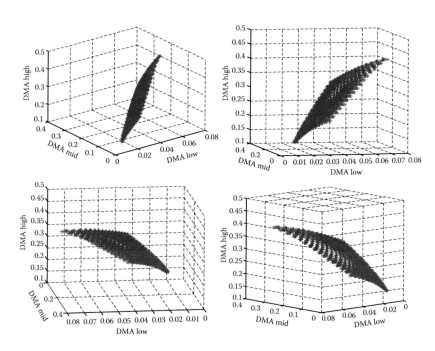

FIGURE 9.13 DMA vector shown as points along the DMA low, mid, and high axes (DMA values are shown in mg/cm^2 units).

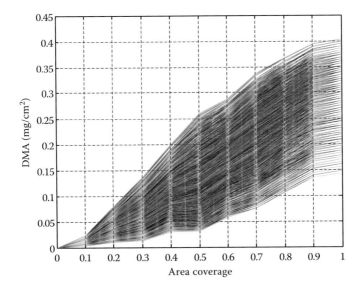

FIGURE 9.14 Development TRCs for the actuator space shown in Figure 9.12.

The input–output data is used for determining the Jacobian matrix. A numerical 3-D model is constructed through interpolation. Nearest neighbor, linear, spline, or cubic interpolation methods can be used, depending on the input–output data. Key steps in determining the Jacobian matrix at the nominal operating points are

1. Set the actuator vector, $U_0 = [U_{ho} \quad U_{lo} \quad U_{bo}]'$ to the desired nominal value.
2. Vary U_h to some value about the nominal value of U_{ho} (increase by that value and decrease by that value) while keeping other actuator values at their nominal operating points. Record the difference DMA vector for these actuator settings obtained using the numerical model. This step gives the first column vector of the Jacobian matrix.
3. Vary U_l to some value about the nominal value of U_{lo} while keeping other actuator values at their nominal operating points. Record the difference DMA vector for these actuator settings obtained using the numerical model. This step gives the second column vector of the Jacobian matrix.
4. Vary U_b to some value about the nominal value of U_{bo} while keeping other actuator values at their nominal operating point. Record the difference DMA vector for these actuator settings obtained using the numerical model. This step gives the last column vector of the Jacobian matrix.

Example 9.5

Test the following developability control system for complete controllability. The system is given at the nominal operating point $\{U_{ho} = -600 \text{ V}, U_{lo} = -100 \text{ V}, U_{bo} = -300 \text{ V}\}$.

$$x(k+1) = \begin{bmatrix} 1 & 0 & 0 \\ 0 & 1 & 0 \\ 0 & 0 & 1 \end{bmatrix} x(k) + 10^{-3} \begin{bmatrix} 0.0282 & 0.5024 & -0.2633 \\ 0.1413 & 1.6703 & -0.9165 \\ 0 & 1.2781 & -1.2854 \end{bmatrix} v(k)$$

$$y(k) = \begin{bmatrix} 1 & 0 & 0 \\ 0 & 1 & 0 \\ 0 & 0 & 1 \end{bmatrix} x(k)$$

SOLUTION

The controllability matrix is given by

$$Q_o = \begin{bmatrix} B & AB & A^2B \end{bmatrix}$$

$$= 10^{-3} \begin{bmatrix} 0.0282 & 0.5024 & -0.2633 & 0.0282 & 0.5024 & -0.2633 & 0.0282 & 0.5024 & -0.2633 \\ 0.1413 & 1.6703 & -0.9165 & 0.1413 & 1.6703 & -0.9165 & 0.1413 & 1.6703 & -0.9165 \\ 0 & 1.2781 & -1.2854 & 0 & 1.2781 & -1.2854 & 0 & 1.2781 & -1.2854 \end{bmatrix}$$

The rank of the controllability matrix is equal to 3. That is, the developability system with level 1 loop is fully controllable at the given nominal operating point. In other words, there are three independent factors involved in this process (due to rank of the Jacobian matrix), and all of them can be controlled using three actuators in vector $v(k)$. That means, a gain matrix K of size 3×3 can be designed to affect all three DMAs using an arbitrary pole-placement strategy.

9.7 STEADY-STATE ERROR

An important part of the level 1 and 2 controllers is that the closed system has the ability to give zero (or minimum) steady-state error, theoretically, in a single measurement-actuation update cycle. The steady-state error is defined as the error between the measured (filtered) values and the desired target values. Due to the use of an integrator in the loop we show mathematically below how the steady-state error is driven to zero with multiple iterations.

Consider the open-loop state Equation 9.21, error Equation 9.25, and the feedback Equation 9.26. The state-space equations are very similar in structure for both level 1 and 2 controllers, but they have different states, Jacobian, dimensionality, output, etc. Substituting Equation 9.26 in Equation 9.21 we get the closed-loop state equation as follows:

$$x(k+1) = x(k) + B[Ke(k)] \tag{9.35}$$

Substituting the error vector, Equation 9.25, in Equation 9.35, we get

$$x(k+1) = (I - BK)x(k) + BKx_d \tag{9.36}$$

Taking the z-transform of Equation 9.36, we have

$$zX(z) - zx_0 = (I - BK)X(z) + BKX_d(z) \tag{9.37}$$

where x_0 is shown in Equation 9.13, and is the value of the state vector when the actuators are set equal to the nominal values, u_0. $X_d(z)$ is the z-transform of the desired target vector. Now, by rearranging Equation 9.37, we can write the expression for $X(z)$ as follows:

$$X(z) = [zI - (I - BK)]^{-1}BKX_d(z) + z[zI - (I - BK)]^{-1}x_0 \qquad (9.38)$$

where matrix I is the identity matrix. Now the error vector can be written in z-domain as follows:

$$\begin{aligned} E(z) &= X_d(z) - X(z) \\ &= [I - \{zI - (I - BK)\}^{-1}BK]X_d(z) - z[zI - (I - BK)]^{-1}x_0 \qquad (9.39) \end{aligned}$$

The reference vector containing desired target values can be written in z-domain as follows:

$$X_d(z) = \frac{z}{z-1}X_d \qquad (9.40)$$

Using final value property, Equation 3.70, we can show that the steady-state error is driven to zero after multiple actuations.

$$e_{ss}(k)|_{closed} = \lim_{z \to 1}(z - 1)E(z) \qquad (9.41)$$

Substituting Equations 9.39 and 9.40 into Equation 9.41 and applying the limit as $z \to 1$, the steady-state error will make its way to zero. This clearly shows that when the system is stable (very important condition!) the output will settle to the desired target with minimum error equal to or close to zero. This is true regardless of the Jacobian/gain matrix and even the initial state, x_0 as long as the feedback loop is stable.

In contrast, for the open-loop case, the following analysis shows that the steady-state error is not independent of the initial state, x_0, because the actual states will drift with time (i.e., actual $x(k)$ will be different from x_0).

Applying z-transform on the open-loop state equation, we can write the state vector in z-domain as follows:

$$X(z) = \frac{B}{z}V(z) + x_0\frac{z}{z-1} \qquad (9.42)$$

For the open-loop case the vector, $V(z) = 0$.

Now the expression for the error vector between the desired inputs to the outputs without feedback can be written as follows:

$$\begin{aligned} E(z) &= X_d(z) - X(z) \\ &= (X_d - x_0)\frac{z}{z-1} - \frac{B}{z}V(z) \qquad (9.43) \end{aligned}$$

Using final value property, we have

$$e_{ss}(k)|_{\text{open}} = X_d - x_0 \tag{9.44}$$

We find that the steady-state error is a function of the state vector, x_0 and is not equal to zero unless u_0 creates the same value as the desired vector, X_d, which is hardly the case since there is always drift/media change happening during system operation.

9.8 DESIGN OF THE GAIN MATRIX

The design of the gain matrix is important for two reasons: (1) for controlling the convergence rate (or number of iterations) for the states to a steady-state value and (2) for staying within the stability bounds so that even when the toner mass deviates due to various disturbances (e.g., toner usage change, media change, developer aging, dark decay, etc.) in the print engine, the closed-loop system should move the states to the desired steady-state value. In this section, the design of gain matrix using a limited pole-placement design technique is discussed.

There is no simple generalized procedure for pole-placement design of MIMO systems. Since the degree of freedom available to choose the right kind of gains is sufficiently large, one can assign performance measures to pick the gains (e.g., to improve the output states in the presence of noise), to assign eigenvectors that would respond according to the natural modes of the system. Such techniques are more elaborate and difficult. Optimal controls and various robust pole assignment techniques are readily available in the literature [13,24] and discussed in Chapter 5 to perform those calculations using the state variable formulation described above. For a simple pole-placement design for level 2 system, let us choose the gain matrix in such a way that the eigenvalues of the closed-loop system matrix, $A-BK$, shown in Equation 9.36 has the desired closed-loop poles. To select desired poles for a discrete system, see Refs. [13,24,25]. The location of the closed-loop poles dictates the required number of iterations for driving the error states toward zero. Since $A = I$ for the level 1 and 2 system, the gain matrix can be easily written in terms of closed-loop poles as

$$K = B^{-1}\sigma \tag{9.45}$$

where σ is a 3×3 matrix determined by the closed-loop poles. For a stable system, all the poles should be assigned values between 0 and 1. For a three-input three-output control system, σ is written in terms of three poles (p_1, p_2, p_3) as follows:

$$\sigma = \begin{bmatrix} 1 - p_1 & 0 & 0 \\ 0 & 1 - p_2 & 0 \\ 0 & 0 & 1 - p_3 \end{bmatrix} \tag{9.46}$$

Theoretically, for the system with $A = I$, we can show that when all three closed-loop poles are chosen real and equal, the gain matrix will result in a more robust closed-loop performance.

Since the Jacobian matrix is a function of the actuator for the level 2 system, the gain matrix can be computed for different regions of the actuator operating space. That is, the gain matrix will be actuator dependent, which can be written as

$$K_j(U_j) = B_j^{-1}(U_j)\sigma \qquad (9.47)$$

where suffix j represents the discrete nature of the actuator vector. It also means that the operating actuator space is quantized to $j = 1, 2, 3, \ldots, J$ regions.

This type of pole assignment will work when the Jacobian matrix is invertible. If faster convergence is required, then it is normal practice to assign the poles (p_1, p_2, p_3) close to 0. If the poles are assigned equal to zero, then $\sigma = I$, and the performance will be theoretically "dead beat." That means, due to the multivariable nature of the system, the theoretical minimum for convergence to zero steady-state error will be "one iteration" (i.e., one measurement-actuation update). If the poles are assigned closer to unity (but not exactly unity, because unit pole values will lead to marginal stability, meaning no convergence), then number of iterations will be greater than three (about 7 to 8). As a result, the performance of the closed-loop system in the presence of system or sensor noise can be improved.

Example 9.6

Let the desired closed-loop poles for level 2 controller be located at p_1, p_2, p_3.

(i) Using Equation 9.45 for the Jacobian matrix of Example 9.5 find the gain matrix that will provide equal poles $p_1 = p_2 = p_3 = 0.3$. Use the MATLAB pole-placement algorithm from Ref. [12]. Find the gain matrix using both methods for poles at $p_1 = 0.3$; $p_2 = 0.3$; $p_3 = 0.7$. What do you see?

(ii) Calculate the step responses for case (i). Compare the results.

SOLUTION

Case (i): Gain matrix from Equation 9.45 for pole values $p_1 = p_2 = p_3 = 0.3$ is given by

$$K = 10^4 \begin{bmatrix} -4.223 & 1.3384 & -0.0891 \\ 0.7863 & -0.1570 & -0.0492 \\ 0.7818 & -0.1561 & -0.1033 \end{bmatrix}$$

The MATLAB pole-placement algorithm gives the same gain values for equal poles. Equation 9.45 and the MATLAB pole-placement algorithm give the same gain matrix for pole values $p_1 = 0.3$; $p_2 = 0.3$; $p_3 = 0.7$.

$$K = 10^4 \begin{bmatrix} -4.223 & 1.3384 & -0.0382 \\ 0.7863 & -0.1570 & -0.0211 \\ 0.7818 & -0.1561 & -0.0433 \end{bmatrix}$$

Since the system matrix, A, is 3×3 and is equal to identity, Equation 9.45 gives a robust gain matrix with all the benefits of the pole-placement algorithm, place().

This is true even for the level 1 controller gain matrix. Also, notice the changes between two gain matrices. They are in the last column of the gain matrix which is due to change in p_3. The third pole p_3 controls how the DMA transients are affected due to the actuator, V_{bias}.

Case (ii): Step responses for the DMA states were created by changing the DMA vector from [0.0384 0.1301 0.3398] to [0.0384 0.1301 0.3738] for two different gain matrices shown in case (i). Level 2 controller has sensing-processing-actuation updates at every 30 pitch, whereas the inner level 1 loop is updated every pitch. DMA vector shown in Figure 9.15a converges to the desired steady-state value after few updates. The actuator response is shown in Figure 9.15b. Figure 9.15c shows the response of the inner loop when the level 2 control loop is in its last iteration pass.

Example 9.7

The cost of the electrostatic sensing system can be reduced by removing the charge sensor (ESV) and associated level 1 control loop. It is now required to design a three-input three-output developability control loop using three process actuators, the grid voltage, $U_g(k) = V_g(k)$, the exposure intensity, $U_l(k) = X(k)$, the development bias, $U_b(k) = V_{bias}(k)$, and measurements from the DMA sensor. Let $D_l(k)$, $D_m(k)$, and $D_h(k)$ represent the three different DMA measurements shown in the state vector, $x(k)$, measured every pitch, indicated by the parameter, k. The linear state variable description of the control system is characterized by the Jacobian matrix at the nominal operating point $\{U_{go} = -600$ V, $U_{lo} = 8$ erg/cm^2, $U_{bo} = -300$ V$\}$ as shown below:

$$x(k+1) = \begin{bmatrix} 1 & 0 & 0 \\ 0 & 1 & 0 \\ 0 & 0 & 1 \end{bmatrix} x(k) + 10^{-3} \begin{bmatrix} 0.2789 & -0.2777 & 8.7889 \\ 0.9761 & -1.0703 & 29.7035 \\ 0.4225 & -1.2727 & 16.8863 \end{bmatrix} v(k)$$

$$y(k) = \begin{bmatrix} 1 & 0 & 0 \\ 0 & 1 & 0 \\ 0 & 0 & 1 \end{bmatrix} x(k)$$

where $x(k) = \begin{bmatrix} D_l(k) \\ D_m(k) \\ D_h(k) \end{bmatrix}$; $v(k) = \begin{bmatrix} \Delta U_g(k) \\ \Delta U_l(k) \\ \Delta U_b(k) \end{bmatrix}$; $y(k) = x(k)$. The controller uses a gain matrix and the integrator modeled by following equation:

$$e(k+1) = e(k) - Bv(k)$$
$$v(k) = +Ke(k)$$

where $e(k) = x_d - x(k)$, and x_d is the desired state vector.

i. Find the gain matrix for placing the poles at location [0.3, 0.3, 0.3] using Equation 9.45.
ii. Show the time evolution of the states for a step response as a function of pitch number using recursive solution of equation (Section 4.8.3), the measurement-actuation updates are executed at every pitch.

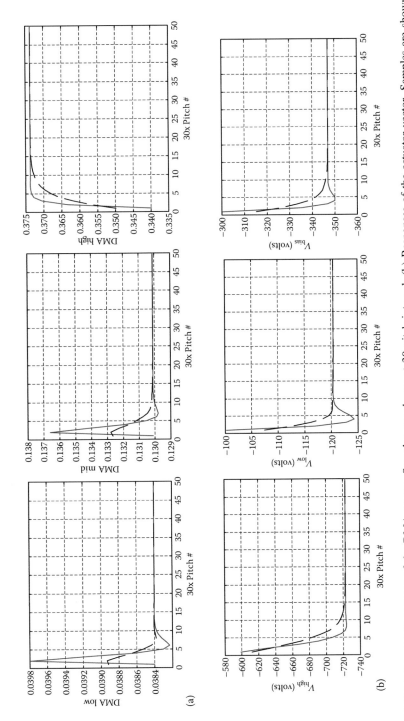

FIGURE 9.15 (a) Responses of the DMA vector. Samples are shown at 30 pitch interval. (b) Responses of the actuator vector. Samples are shown at 30 pitch interval. Solid curve represents $P_1 = P_2 = P_3 = 0.3$ and dashed curve represents $P_1 = P_2 = 0.3$ and $P_3 = 0.7$. All DMA values are shown in mg/cm^2 units.

(continued)

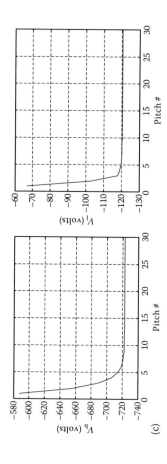

FIGURE 9.15 (continued)   (c) Responses of the level 1 state vector (V_h, V_l) with respect to every pitch when level 2 is in its last iteration pass.

iii. Compare the results of step ii by running the control simulation with the charge and development models from Chapter 10.

iv. Are there other reasons why this type of control approach is more suitable than level 1 and 2 architecture? Justify your answers.

SOLUTION

Case (i): Gain matrix from Equation 9.43 for pole values $p_1 = p_2 = p_3 = 0.3$ is given by

$$K = 10^4 \begin{bmatrix} -3.9405 & 1.2977 & -0.2317 \\ 0.7855 & -0.1988 & -0.0592 \\ 0.1578 & -0.0475 & 0.0055 \end{bmatrix}$$

Gain matrix from Equation 9.43 for pole values $p_1 = 0.3$; $p_2 = 0.3$; $p_3 = 0.7$ is given by

$$K = 10^4 \begin{bmatrix} -3.9405 & 1.2977 & -0.0993 \\ 0.7855 & -0.1988 & -0.0254 \\ 0.1578 & -0.0475 & 0.0023 \end{bmatrix}$$

Case (ii): Step responses for the DMA states were created by changing the DMA vector from [0.0384 0.1301 0.3398] to [0.0384 0.1301 0.3738]. The DMA vector shown in Figure 9.16a converges to the desired steady-state value for the two different gain matrices. The actuator response is shown in Figure 9.16b.

Solutions for cases (iii) and (iv) are left to the reader.

9.9 LEVEL 3 CONTROL LOOPS

At level 3, the TRCs of the individual separation are maintained in such a way that they correspond to a reference tone curve on the photoconductor or on the paper with the fused toner. The tone curve on the photoconductor (called "developed reproduction curve") is a function that maps digital contone values of a color separated image to DMA values on the photoconductor. Similarly, the tone curve on the paper is a function that maps a color separated image to an optical density on the paper after fusing. For black and white printers, what matters is the tone curve on the paper. To achieve good reproduction, it is important that the desired curve be linear. Often, users request different tone curves depending on the overall appearance of the image. For four-color digital reproduction, the complex interaction of each of the tone curves on the paper makes it difficult to specify the exact shape of the desired tone curves.

A linear reference curve is more widely used in the industry. By design, the DMA at 100% area coverage is generally limited to a set point and density control loops try to maintain solids at those points. As a result, the available control space is externally limited for level 2 process controls, which involves achieving the desired linear tone response on the photoconductor or paper. The controllable space is illustrated conceptually in Figure 9.17. In this figure, level 2 targets are in the controllable space that is outside the space where the desired reference TRC curve, which is linear in this illustration. Thus, the desired tone curve is uncontrollable by levels 1 and 2 actuators. For such situations, a preferred way to achieve the desired tone response is through

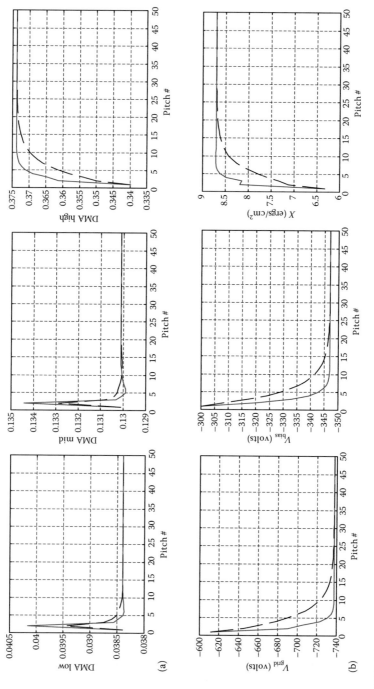

FIGURE 9.16 (a) Responses of the DMA vector (all DMA values are shown in mg/cm² units). (b) Responses of the actuator vector.

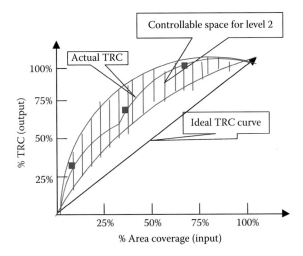

FIGURE 9.17 Schematic diagram representing the TRC with controllable space using process actuators (charge and development bias).

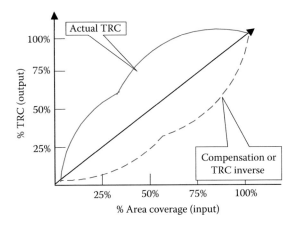

FIGURE 9.18 Schematic diagram representing the inverse TRC obtained to achieve the desired reference TRC.

level 3 controls, since these higher level controls can provide extra degrees of freedom that lower level (charge and developability) controls cannot. The level 3 control system generates the inverse curve (Figure 9.18) for each separation [26]. This curve can be obtained by inverting the measured TRC over a reference TRC.

In another scenario presented in Figure 9.19, it appears as though there is sufficient actuator space to obtain the desired tone curve. However, we are still not able to reach full inversion with three point level 2 controls because of insufficient sampling; sampling of only three (low, mid, and high area coverages) is sufficient for inversion. A time-sequential sampling proposed in Ref. [15] may provide reliable control for such systems for a given area coverage (or gray level); during each

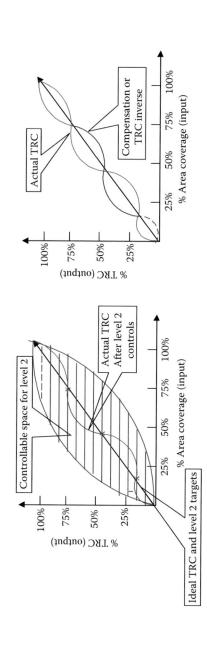

FIGURE 9.19 Left: Schematic diagram representing the TRC with controllable space covering the reference TRC using process actuators (charge and development bias). Right: Schematic diagram representing the inverse TRC achievable using contone actuators.

measurement-process-actuation cycle, measurements can be carried out at different area coverages leading to a higher bandwidth tone correction with control of the entire tone curve after sampling all area coverage levels. Correcting one gray level at a time using process actuators has the risk of affecting other gray levels if the actuators are not simultaneously optimized for all levels. Also, complexity and the cost of implementation may prevent the use of such control loops. Whereas, the use of level 3 control can easily overcome such barriers and, hence, is the most widely used control system for linearizing tone levels in many digital printers.

9.9.1 STATIC TRC INVERSION PROCESS

To completely measure the TRC, every possible area coverage that can be printed must be measured, but this results in too many patches and may be unnecessary when the TRC has less structure. The printer tone curve is generally sensed at 7 to 10 different area coverages, depending on the measurement resources supplied by the print engine. A curve fitting algorithm [27] can be used to reconstruct the TRC for all area coverages between 0 and 255. In addition, noise in the measurements must be filtered out to obtain the final best estimate of the actual TRC. Also, the curve is forced to pass through 0% and 100% area coverage, which is consistent with the fact that these points always remain fixed, thanks to levels 1 and 2 controls.

If the reconstructed TRC has values available at all area coverages and is monotonic, then the inverse TRC is obtained using the "static inverse" techniques illustrated graphically in Figures 9.20 and 9.21. In these figures, the best estimate of the reconstructed TRC that passes through all the measurements is shown schematically by the continuous solid curve. The normalized DMA or ΔE from paper (referred to as the output tone value) can have values lower or higher than the reference curve as illustrated in Figure 9.19 (right). To demonstrate how the "static inverse" works, let us consider these cases separately as illustrated in Figures 9.20 and 9.21. Let the

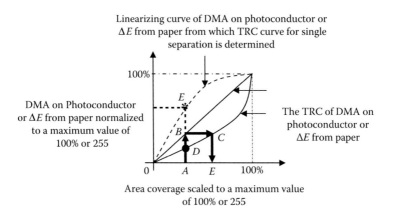

FIGURE 9.20 Schematic diagram illustrating graphically the static TRC inversion process shown for one measurement point at area coverage A when the measured tone value is smaller than the reference tone value.

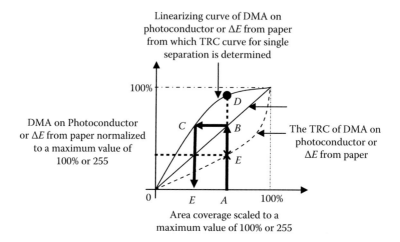

Linearizing curve of DMA on
photoconductor or ΔE from paper
from which TRC curve for single
separation is determined

DMA on Photoconductor
or ΔE from paper normalized
to a maximum value of
100% or 255

The TRC of DMA on
photoconductor or
ΔE from paper

Area coverage scaled to a
maximum value of 100% or 255

FIGURE 9.21 Schematic diagram illustrating graphically the static TRC inversion process shown for one measurement point at area coverage A when the measured tone value is greater than the reference tone value.

input area coverage A at which the measured tone value on the TRC is equal to D and the output tone value on the reference TRC is calculated as B. The tone value on the reference TRC is calculated by simply drawing a vertical line until it intercepts the reference curve. Now, to find the tone value on the inverse TRC for area coverage A, draw a horizontal line (toward right in Figure 9.20 and toward left in Figure 9.21) until it intercepts the actual TRC. The output tone value at this intercept is indicated by C (this is also equal to B), whose input area coverage (or tone value) will be equal to E. Thus, E becomes the tone value on the inverse TRC at area coverage A. This process is repeated until all the input area coverages have tone values on the inverse TRC.

This type of inverse TRC has benefits for full range of tone control from 0% to 100%. In a digital printing system, since every pixel in the image that contains input contone values is processed through the inverse TRC of the level 3 controller, the final curve should be smooth and free of undesirable curvatures or artifacts introduced during the inversion process. If not, image quality defects or image noise in the output images may occur, and these are not acceptable for high-quality color printing. While the process of constructing "static inverse TRC" looks simple, many practical considerations require attention. For example, the measurement noise could be different for low area coverage (highlight) patches as compared to high area coverage (shadow) patches, which can induce undesirably high-frequency structure in the inverted TRC. The rounding of values between 0 and 255 can also introduce errors which can further sacrifice inversion accuracy. Since the real goal of level 3 controller is to linearize the tone response on paper to control the appearance of images on paper, paper-based measurements may be more desirable than those on the photoconductor. The area coverage selection for paper-based measurements may be different from those used for measurements on the photoconductor. We will show later how an optimal approach can be used to find the right area coverages for measurements. In all these cases, we need to ensure monotonicity in the measured (or reconstructed) tone response. Otherwise, the static inverse methods described above will not work. In Section 9.9.2 we show a control-based inversion technique, which can be used to accurately construct the inverse tone response curve in the absence of the monotonicity condition.

Example 9.8

Let the vector U_0 contain the input patch gray levels (or area coverages) for controlling DMA on the photoconductor for the magenta separation shown in Table 9.1 below. The vector x contains the normalized values of the measurements corresponding to each gray level. This data was produced for a maximum DMA value of 0.55 units with levels 1 and 2 controls enabled. Three patch DMA controls were used in the level 2 loop. The targets were set to [0.0384 0.1301 0.3738] DMA units. Three reference TRCs are sampled at the gray levels contained in the U_0 vector. They are shown in columns $(x_d)_1$, $(x_d)_2$, and $(x_d)_3$. The shape of the reference TRCs for all gray levels are shown in Figure 9.22.

a. Construct the complete inverse TRC using static inversion process described in Section 9.9.1 with end points fixed at gray levels 0 and 255 for all three reference TRCs.
b. Show how the gray-level image appears when the image pixels are processed with and without the inverted TRCs obtained with each reference.

SOLUTION

Case (a): The data shown in vector x are first interpolated are (see Chapter 6) to form a smooth curve for each gray levels. The smooth function is then inverted around the reference TRC using the static inversion technique described above. Figure 9.23 shows three plots with inverted TRCs for each reference TRCs.

TABLE 9.1
TRC and Reference Samples at Discrete Input Gray Levels

Number of Patches	U_0	x	$(x_d)_1$	$(x_d)_2$	$(x_d)_3$
0	0.00	0.00	0.00	0.00	0.00
1	26.00	8.37	26.00	45.17	47.80
2	51.00	26.20	51.00	62.76	91.70
3	77.00	43.73	77.00	64.85	133.28
4	102.00	75.77	102.00	82.98	167.90
5	128.00	88.76	128.00	128.49	197.34
6	153.00	128.61	153.00	172.02	218.99
7	179.00	161.31	179.00	190.35	234.94
8	204.00	203.92	204.00	192.24	244.87
9	230.00	242.89	230.00	211.14	251.11
10	255.00	255.00	255.00	255.00	255.00

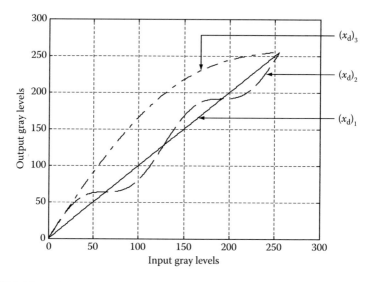

FIGURE 9.22 Reference TRCs for 256 input gray levels.

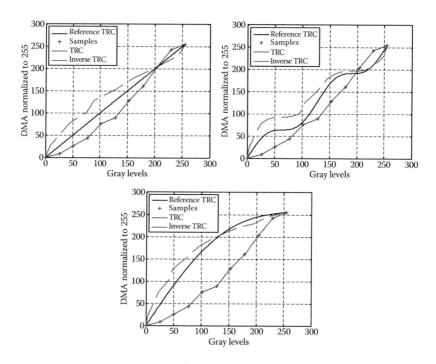

FIGURE 9.23 Inverse TRCs for 256 input gray levels shown for the three reference TRCs.

Case (b): In Figure 9.24a through c, the cyan sweep is displayed by processing the cyan separation through three different reference TRCs. The cyan sweep when processed through the printer model (Chapter 10) without any level 3 correction is

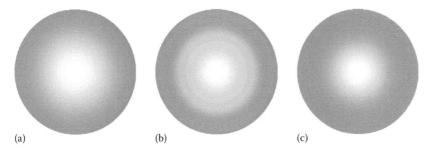

(a)									(b)									(c)

FIGURE 9.24   Circular cyan sweeps through (a) identity reference TRC, (b) reference TRC 2, and (c) reference TRC 3. No printer model is involved in this simulation.

FIGURE 9.25   Circular cyan sweeps through the printer model without any inversion.

shown in Figure 9.25. Figure 9.26 contains images with and without level 3 compensation. Images shown on the right are processed through level 3 TRCs and the printer model. They show matching with corresponding figures in Figure 9.24, as expected by the TRC linearization process.

9.9.2 Control-Based TRC Inversion Process

A control-based TRC inversion process is made possible via a state-space representation of the tone reproduction at each of the measurement patch area coverages. If the gradient of the tone curve at each of these patch area coverages is known, then a state-space model with an integrator in the control loop can be written as follows:

$$x(k+1) = Ax(k) + Bv(k)$$
$$y(k) = Cx(k) + Dv(k)$$

(9.48)

where $x(k) = \begin{bmatrix} x_1(k) \\ x_2(k) \\ \cdot \\ \cdot \\ \cdot \\ x_N(k) \end{bmatrix}$, $v(k) = \begin{bmatrix} \Delta U_1(k) \\ \Delta U_2(k) \\ \cdot \\ \cdot \\ \cdot \\ \Delta U_N(k) \end{bmatrix}$, $y(k) = x(k)$, $A = I$, $B = J$, $C = I$, and

$D = 0$. The controller uses a gain matrix and an integrator modeled by the following equation (Section 9.2):

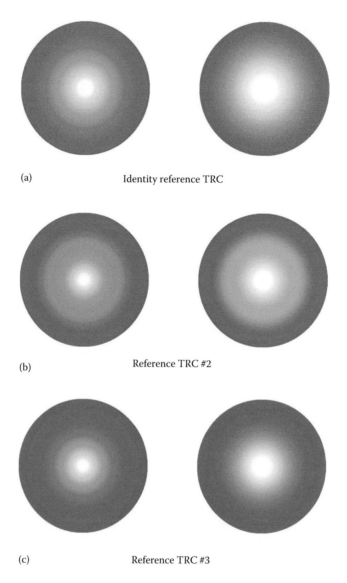

(a)   Identity reference TRC

(b)   Reference TRC #2

(c)   Reference TRC #3

FIGURE 9.26   Circular cyan sweeps through the inverted TRC obtained for reference TRCs and then through the printer model with output TRC shown in Figure 9.23. The images on the right should match the images shown in Figure 9.24.

$$U(k) = U_0 + u(k)$$
$$u(k+1) = u(k) + v(k) \qquad (9.49)$$
$$v(k) = +Ke(k)$$

where the vector, U_0, contains the patch area coverages, $U_1^0, U_2^0, \ldots, U_N^0$, the error vector, $e(k) = x_d - x(k)$, and x_d is a column vector containing the desired tone values $r_1, r_2, \ldots, r_N$, for the reference TRC at the patch area coverages. That is,

$$U_0 = \begin{bmatrix} U_1^0 \\ U_2^0 \\ . \\ . \\ . \\ U_N^0 \end{bmatrix} \quad \text{and} \quad x_d = \begin{bmatrix} r_1 \\ r_2 \\ . \\ . \\ . \\ r_N \end{bmatrix} \tag{9.50}$$

Also, in this case, the Jacobian matrix, J, contains the first derivatives of the tone values at each of the patch area coverages and the structure of the matrix is diagonal because during level 3 control each patch tone value can be varied without interacting with the other patches. That is, development of each patch can be regarded independent of the other patch. This assumption holds good, provided the level 2 actuators do not change (i.e., have fully settled) while level 3 control loop is running. The measurement-process-actuation interval for level 3 control is assumed quite high as compared with level 1 and 2 controls, preserving the time hierarchy described earlier in the chapter.

The derivatives are obtained at the patch area coverages for nominal operating conditions. For N number of level 3 control patches with area coverages $U_1^0, U_2^0, \ldots, U_N^0$, the Jacobian matrix is written as

$$J = \begin{bmatrix} \frac{\partial x_1}{\partial U_1^0} & 0 & \cdots & 0 \\ 0 & \frac{\partial x_2}{\partial U_2^0} & \cdots & 0 \\ \vdots & \vdots & \ddots & \vdots \\ 0 & 0 & \vdots & \frac{\partial x_N}{\partial U_N^0} \end{bmatrix} \tag{9.51}$$

Since the Jacobian is diagonal for level 3 control, the gain matrix (Equation 9.49) will be diagonal, meaning the dynamic control loop will have a SISO form, that is, each reference tone value is controlled independent of the others. This independence is exactly the reason why a nonmonotonic TRC can be inverted using control-based approach. Once the level 3 loop converges to a stable state with a near zero error vector, $e(k) = x_d - x(k)$, the measured tone values converge to the reference tone values. Since the measured tone values are present in the control vector, $U(k)$, the inverse tone values can also be obtained from the same vector. A smooth curve is constructed to pass through the tone values in the control vector $U(k)$ and the end points (0 and 255). This smooth curve represents the final inverse TRC used to process the pixels in the images. The approach is described below with a numerical example.

Example 9.9

Let the vector, U_0, contain the input patch gray levels (or area coverages) for controlling DMA on the photoconductor for cyan separation shown in Table 9.1. The vector x contains the normalized values of the measurements, the vector x_d contains the tone values of the desired reference TRC. Both vectors are shown as a function of the patch gray level.

a. Find the derivatives at the patch values and the gain matrix for pole
   values equal to 0.4. Use Equation 9.45 to find the gain matrix.
b. Obtain the inverted tone values for (i.e., $U(k)$) iterations, $k = 1, 2, ..., 9$.
   Plot them with respect to iteration number, k.
c. Using end points 0 and 255 and a suitable curve fitting algorithm complete the full inverse TRC. Plot the inverted TRC. Mark the values of the
   reference vector, x_d, and the normalized values of the measurements, x,
   corresponding to the patch gray levels.
d. Complete steps a through c for the new reference tone values.

SOLUTION

Inverted tone values for (i.e., $U(k)$) iterations, $k = 1, 2, ..., 10$ are plotted in Figures
9.27a, 9.28a, and 9.29a for three different reference TRCs as a function of the
iteration number, k. Corresponding full TRCs are shown in Figures 9.27b, 9.28b,
and 9.29b, respectively. Reference vector, x_d, and the normalized values of the
measurements, x, are shown in these figures corresponding to each patch gray level.

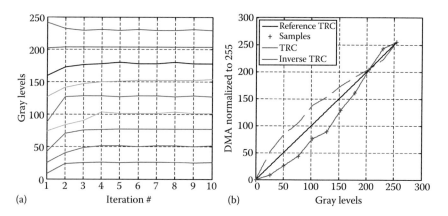

FIGURE 9.27   (a) Convergence plot (poles $= 0.4$) for the inverse tone values. (b) Inverse
TRCs for 256 input gray levels shown for reference TRC 1.

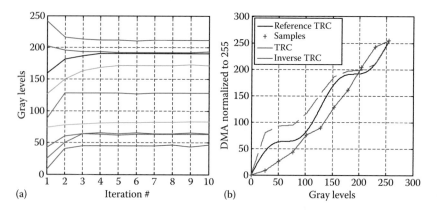

FIGURE 9.28   (a) Convergence plot (poles $= 0.4$) for the inverse tone values. (b) Inverse
TRCs for 256 input gray levels shown for reference TRC 2.

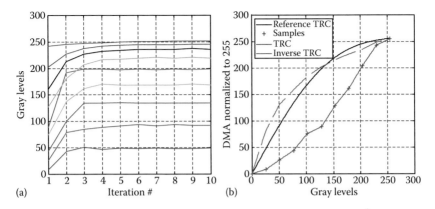

(a)   Iteration #   (b)   Gray levels

FIGURE 9.29 (a) Convergence plot (poles $= 0.4$) for the inverse tone values. (b) Inverse TRCs for 256 input gray levels shown for reference TRC 3.

9.10 DEAD BEAT RESPONSE

Dead beat response is the response of a control loop in which the error is dead in one beat. This implies that the desired response of the control system is obtained in a minimum number of measurement-process-actuation update steps [13]. In other words, the target is reached in the fewest number of time steps. In the linear discrete case such as for level 1, 2, or 3 controls, the dead beat response can be achieved by making the closed-loop transfer function have poles at the origin (i.e., zero). In this section we will determine the minimum number of steps required to achieve dead beat response for the both SISO and MIMO systems. There is no such thing as a dead beat response for continuous control systems. Hence, the theory is developed only for a discrete system.

Consider the state variable description of Equation 9.48 and the feedback equation with gain matrix of Equation 9.49. The closed-loop state Equation is equal to

$$x(k+1) = (I - BK)x(k) + BKx_d$$
$$= \hat{B}x(k) + BKx_d \qquad (9.52)$$

From Chapter 4, we can write the solution to the above equation as

$$x(k) = \hat{B}^k x_0 + BKx_d \qquad (9.53)$$

We know that the feedback control law that assigns all the closed-loop poles to the origin is a deadbeat control law. Therefore, all eigenvalues of the closed-loop system (Equation 9.52) will be at zero for deadbeat control. That is, the closed-loop system matrix $\hat{B}$ will have all eigenvalues equal to zero. Let $\lambda_1 = \lambda_2 = \cdots = 0$. For the SISO case (e.g., Section 9.9.2), the $\hat{B}$ matrix will be diagonal with all elements zero for deadbeat control and $K = B^{-1}$. Thus, for a SISO system,

$$x(1) = x_d \tag{9.54}$$

It can be seen from Equation 9.54 that the deadbeat control law can provide control in one measurement-process-actuation cycle.

For a MIMO system, as in level 1 and 2 control, all elements of the closed-loop system matrix $\hat{B}$ will not be equal to zero and is not diagonal. To determine the number of measurement-process-actuation cycles required to achieve deadbeat control, the matrix $\hat{B}$ has to be expressed in diagonal or pseudo-diagonal (i.e., Jordan) form (Chapter 3). Since there will be repeated eigenvalues during deadbeat control, the $\hat{B}$ matrix can be written as follows:

$$M^{-1}\hat{B}M = J \tag{9.55}$$

Substituting Equation 9.55 in Equation 9.53

$$x(k) = MJ^k M^{-1}x_0 + BKx_d \tag{9.56}$$

To make this problem simple, let us consider that the $\hat{B}$ matrix is of size 4×4. Now, if $\lambda_1 = \lambda_2 = \lambda_3 = \lambda_4$ are eigenvalues of matrix $\hat{B}$, then from linear algebra [28], J^k can be expressed as follows:

$$J^k = \begin{bmatrix} \lambda_1^k & k\lambda_1^{k-1} & (k-1)\frac{\lambda_1^{k-2}}{2!} & (k-2)\frac{\lambda_1^{k-3}}{3!} \\ 0 & \lambda_1^k & k\lambda_1^{k-1} & (k-1)\frac{\lambda_1^{k-2}}{2!} \\ 0 & 0 & \lambda_1^k & k\lambda_1^{k-1} \\ 0 & 0 & 0 & \lambda_1^k \end{bmatrix} \tag{9.57}$$

when $k = 1$, from Equation 9.57

$$J^1 = \begin{bmatrix} 0 & 1 & 0 & 0 \\ 0 & 0 & 1 & 0 \\ 0 & 0 & 0 & 1 \\ 0 & 0 & 0 & 0 \end{bmatrix}, \quad J^2 = \begin{bmatrix} 0 & 0 & 1 & 0 \\ 0 & 0 & 0 & 1 \\ 0 & 0 & 0 & 0 \\ 0 & 0 & 0 & 0 \end{bmatrix}, \quad J^3 = \begin{bmatrix} 0 & 0 & 0 & 1 \\ 0 & 0 & 0 & 0 \\ 0 & 0 & 0 & 0 \\ 0 & 0 & 0 & 0 \end{bmatrix}, \quad \text{and} \quad J^4 = \begin{bmatrix} 0 & 0 & 0 & 0 \\ 0 & 0 & 0 & 0 \\ 0 & 0 & 0 & 0 \\ 0 & 0 & 0 & 0 \end{bmatrix} \tag{9.58}$$

For $k = 4$, we see that the states, $x(4) = x_d$. This implies that for a four-dimensional (4-D) system, if $\hat{B}$ is diagonalizable to pseudo-diagonal form such as the Jordan canonical form, a minimum of four measurement-process-actuation cycles are required to reach the final desired values. In most cases, in printers, the matrix $\hat{B}$ is diagonalizable. Hence, theoretically speaking, it is possible to reach deadbeat in one measurement-process-actuation cycle. However, in practice, noise and uncertainties will make it difficult to achieve deadbeat. Loops must be designed away from the deadbeat response.

9.11   TC CONTROL LOOP

In a typical two-component development housing, toners and carriers are mixed in the sump at a specified ratio known as the TC (Section 10.2.3). The toner particles consist of dyed or colored thermoplastic powder that is mixed with coarser carrier granules, such as ferromagnetic granules. The toner particles and carrier granules are selected such that the toner particles acquire the appropriate charge during mixing. This mixture is transported to the development zone where the toner particles are presented to the latent image on the photoconductor. When they are brought into contact with the charged photoconductive surface, the greater attractive force between the electrostatic latent image and the toner particles causes the toner particles to transfer from the carrier granules and adhere to the electrostatic latent image. In some of the multicolor electrophotographic print engines, the mixture is picked up by a developer roll, metered using a trim bar to achieve a uniform thickness with toner particles alone and transported to the development zone.

The concentration of toner in the development housing has an effect on the amount of toner attracted to charged portions of the photoconductor. For instance, the higher the concentration of toner in the housing, the more toner is attracted to the photoconductor (Equations 10.30 and 10.31). TC can be controlled by controlling the rate at which toner from the toner supply is delivered to the developer housing. TC sensors [29–36] are used to sense, for example, the magnetic reluctance associated with magnetic carrier particles in the developer housing. As the TC grows higher, the average spacing between the carrier beads gets larger and the reluctance becomes lower. The magnetic reluctance signal is used to measure the TC in the developer. Developer granule size, developer age, tribo, and, possibly, the temperature and RH affect the TC measurements. These factors are referred to as sensor noise, and some of them can be carefully measured, estimated (e.g., toner age [31,32]), and isolated from the actual signal depending on the complexity and cost associated with the system.

Changing the TC in the developer housing can affect the lightness or darkness of a rendered image. Hence, the TC can be used as one of the xerographic process actuators [16] in addition to those used in level 2 controls as described in Section 9.6. The design of such a feedback system could be inherently simple if we didn't have the complexity of the transport delay in dispensing the toner to the developer housing, that is, the rate limitation in dispense and consumption of toner as images are rendered at production speed. In Section 9.13, we show a systematic way to realize a feedback system with a transport delay using state feedback control approaches. In this section, by maintaining TC constant at a set point, we minimize the number of factors affecting the developability, and the toner dispense rate is used as the actuator [35].

9.11.1   OPEN-LOOP TC MODEL

TC is the ratio of toner mass to the carrier mass. In simple terms, the toner mass (t_{m}) at cycle k can be modeled by the following difference equation

$$t_{\mathrm{m}}(k + 1) = t_{\mathrm{m}}(k) + m_{\mathrm{d}}(k - \mu) - m_i(k) \tag{9.59}$$

where

$t_m(k)$ stands for the mass of the toner at cycle k in grams

$m_d(k)$ is the mass of toner dispensed from the dispenser at cycle k in grams

$m_i(k)$ is the mass of toner which is used for the image (based on the consumption profile at cycle k in grams)

μ is the transport delay in toner dispense (in number of cycles) that includes the delay in the mixing system

The index k denotes the iteration number (or measurement-process-actuation cycle). For simplicity, we assume that the TC cycle is synchronous with the measurement-process-actuation cycle of the level 2 control system. This assumption is unrealistic for most of the practical development systems.

The mass dispensed is calculated by using the duty cycle which will be derived by processing the measured TC signal inside the feedback controller.

$$m_d(k) = d(k) \times R_{max} \times \tau \qquad (9.60)$$

where

R_{max} is the maximum dispense rate at which the toner can be dispensed from the toner reservoir

τ is the measurement-process-actuation period in seconds

$d(k)$ is the duty cycle at cycle k

Generally, a pixel counter is included in many modern printing systems that can be used to calculate the total area coverage of the image in which the toner will be developed. The $m_i(k)$ is calculated by using the area coverage $a(k)$ of the image. It is given by

$$m_i(k) = a(k) \times R_{max} \times \tau \qquad (9.61)$$

From the definition of TC, if m_c is the carrier mass, then we have

$$t_c(k) = \frac{t_m(k)}{m_c} \qquad (9.62)$$

Rewriting Equation 9.59 in terms of TC

$$t_c(k+1) = t_c(k) + g[u(k-\mu) - v(k)] \qquad (9.63)$$

where $u(k)$ is the dispensed mass, $m_d(k)$, and $v(k)$ is the developed mass, $m_i(k)$, and g is a scale factor given by

$$g = \frac{1}{m_c} \qquad (9.64)$$

The Equation 9.63 can be transformed to the z-domain by applying the z-transform. After the transformation we get

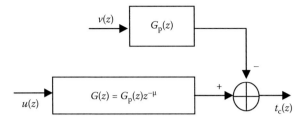

FIGURE 9.30 Open-loop transfer function of TC control system.

$$t_c(z) = \frac{g}{z-1}(z^{-\mu}u(z) - v(z)) \tag{9.65}$$

or

$$t_c(z) = G_p(z)(z^{-\mu}u(z) - v(z)) \tag{9.66}$$

where $G_p(z) = g/(z-1)$. The block diagram of the open-loop TC system of Equation 9.66 is shown in Figure 9.30.

Clearly, the open-loop TC system becomes unstable if the developed mass $v(k)$ is zero, this condition occurs when the developer housing is being cycled to fill the toner without printing images. The system of Equation 9.66 contains an integrator and a time delay. For now, if we ignore the time delay, then a simple proportional integral (PI) controller can be used to stabilize the system.

9.11.2 Design of a TC Control Loop Using a PI Controller

Pole-placement design is applied to bring the TC from the initial state to the desired state in few cycles. Note that the TC sensor model is ignored in this design.

Figure 9.31 shows the PI controller in the z-domain. It is important to note that a simple integral control (i.e., with $K_p = 0$) will give stable performance provided the gains are designed correctly. The parameter $\tilde{u}(z) = u(z)/\tau$ represents the rate of toner dispense and τ is the TC measurement-process-actuation interval in seconds. The intermediate quantity, $w(z)$, and the error signal, $e_s(z)$, are related by the following transfer function:

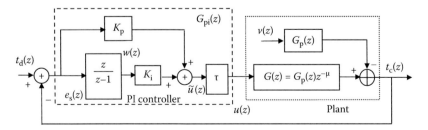

FIGURE 9.31 TC feedback system with PI controller.

$$w(z) = \frac{z}{z - 1} e_s(z) \tag{9.67}$$

The gains of the PI controller are designed by first expressing Equation 9.66 in state-space form as follows:

Substituting the error signal $e_s(z) = t_d(z) - t_c(z)$ in Equation 9.67 and taking the inverse z-transform, we get the following difference equation:

$$w(k + 1) = w(k) + t_d(k + 1) - t_c(k + 1) \tag{9.68}$$

For a TC system with no dispenser lag (i.e., $\mu = 0$) and no consumption (i.e., $v(k) = 0$), Equation 9.63 can be written as

$$t_c(k + 1) = t_c(k) + gu(k) \tag{9.69}$$

Now, Equations 9.68 and 9.69 are arranged in matrix form as follows:

$$\begin{bmatrix} t_c(k + 1) \\ w(k + 1) \end{bmatrix} = \begin{bmatrix} 1 & 0 \\ -1 & 1 \end{bmatrix} \begin{bmatrix} t_c(k) \\ w(k) \end{bmatrix} + \begin{bmatrix} g \\ -g \end{bmatrix} u(k) + \begin{bmatrix} 0 \\ 1 \end{bmatrix} t_d(k + 1) \tag{9.70}$$

Equation 9.70 can be written in more general matrix difference (state space) form for a single-input system as

$$x(k + 1) = Ax(k) + Bu(k) + Er(k + 1) \tag{9.71}$$

with $x(k) = \begin{bmatrix} t_c(k) \\ w(k) \end{bmatrix}$, $A = \begin{bmatrix} 1 & 0 \\ -1 & 1 \end{bmatrix}$, $B = \begin{bmatrix} g \\ -g \end{bmatrix}$, $E = \begin{bmatrix} 0 \\ 1 \end{bmatrix}$, and $r(k) = t_d(k)$.

The desired TC value is considered constant (i.e., $t_d(k + 1) = t_d(k) = \text{constant}$). In some systems, the TC target is adjusted at a lower rate as a function of toner age in order to achieve better control for low area coverage images. Due to the repeated mechanical stresses during the mixing process, the toner will start to lose its charging property. This is called toner aging [31], which has a negative impact on development and transfer performance. Fresh toner has higher additive coverage whereas the aged toner has almost no additives on the toner surface. As a result of this, there is a significant difference in toner transfer to the paper.

For simplicity, let us assume a constant TC target.

For a constant TC target, the amount of dispensed toner can be expressed as a function of the controller gains:

$$u(k) = \tau \big(K_p e_s(k) + K_i w(k) \big) \tag{9.72}$$

Equation 9.72 can then be written in matrix form as

$$u(k) = -\tau \begin{bmatrix} K_p & -K_i \end{bmatrix} \begin{bmatrix} -e_s(k) \\ w(k) \end{bmatrix} \tag{9.73}$$

or in a more compact form as

$$u(k) = -Ky(k) \tag{9.74}$$

where
$K = [\tau K_p \quad -\tau K_i]$ is the gain matrix
$y(k)$ is the output vector for the system of Figure 9.31

The vector $y(k)$ can be written in following form:

$$y(k) = \begin{bmatrix} -e_s(k) \\ w(k) \end{bmatrix} = \begin{bmatrix} 1 & 0 \\ 0 & 1 \end{bmatrix} \begin{bmatrix} t_c(k) \\ w(k) \end{bmatrix} + \begin{bmatrix} -1 \\ 0 \end{bmatrix} t_d(k)$$

$$= Cx(k) + Fr(k) \tag{9.75}$$

where $C = \begin{bmatrix} 1 & 0 \\ 0 & 1 \end{bmatrix}$, $x(k) = \begin{bmatrix} t_c(k) \\ w(k) \end{bmatrix}$, $F = \begin{bmatrix} -1 \\ 0 \end{bmatrix}$, and $r(k) = t_d(k)$

Equations 9.71 and 9.75 represent the TC control system in state variable form and Equation 9.74 the output feedback equation. Figure 9.32 shows these equations in block diagram form after the equations are transformed to z-domain. Note that although the original system was SISO, the control system is SIMO (single input multiple output). A systematic approach is required to design the controller gains.

The controller gains can be designed using pole placement or linear quadratic regulator (LQR) techniques (Section 5.2 or 5.3) [13]. Below we show the pole-placement design.

Substituting Equation 9.74 in Equation 9.71 and assuming $r(k+1) = r(k)$, following closed-loop state equation is obtained:

$$x(k+1) = (A - BKC)x(k) + (E - BKF)r(k) \tag{9.76}$$

The matrix, $A_c = A - BKC$ in Equation 9.76 represents the closed-loop system matrix of Figure 9.32. Hence, the closed-loop characteristic matrix is $zI - A_c = 0$. Recall that

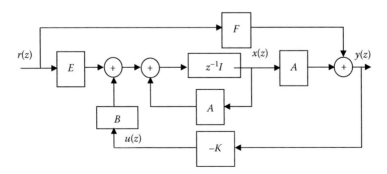

FIGURE 9.32 Block diagram of the feedback system with PI controller.

the gain matrix K places the poles (or roots) of the characteristic equation. In other words, K is chosen by assigning the closed-loop poles. See Example 9.10 for more details.

Example 9.10

State equations for a closed-loop TC control system are given below. Design the gain matrix for placing the poles at (a) $\lambda_1 = 0.2$, $\lambda_2 = 0.25$ and (b) $\lambda_1 = 0.6$, $\lambda_2 = 0.65$. Assume TC target in percent. Carrier mass $= 800$ g. The state equation is

$$x(k + 1) = Ax(k) + Bu(k) + Er(k + 1) \tag{9.77a}$$

And the output equation is

$$y(k) = Cx(k) + Fr(k) \tag{9.77b}$$

The actuator with output linear feedback is given by

$$u(k) = -Ky(k) \tag{9.78}$$

where

$$x(k) = \begin{bmatrix} t_c(k) \\ w(k) \end{bmatrix}, \ A = \begin{bmatrix} 1 & 0 \\ -1 & 1 \end{bmatrix}, \ B = \begin{bmatrix} g \\ -g \end{bmatrix}, \ E = \begin{bmatrix} 0 \\ 1 \end{bmatrix}, \ r(k) = t_d(k),$$

$$C = \begin{bmatrix} 1 & 0 \\ 0 & 1 \end{bmatrix}, \ F = \begin{bmatrix} -1 \\ 0 \end{bmatrix}, \ \text{and } K = \begin{bmatrix} \tau K_p & -\tau K_i \end{bmatrix}$$

Solution

We use pole-placement design for SISO system discussed in Section 5.2.2.

Case (a): Since TC target is to be calculated as a percentage and the carrier mass $= 800$ g, we can write $g = 100/800 = 1/8$. The open-loop TC equation is given by

$$x(k + 1) = \begin{bmatrix} 1 & 0 \\ -1 & 1 \end{bmatrix} x(k) + \begin{bmatrix} \frac{1}{8} \\ -\frac{1}{8} \end{bmatrix} u(k) + \begin{bmatrix} 0 \\ 1 \end{bmatrix} r \tag{9.79}$$

where r is the constant value of TC expressed in percent. The actuator equation can be written as follows:

$$u(k) = -Ky(k) = -K[Cx(k) + Fr(k)] = -KCx(k) - KFr = -Kx(k) - KFr \tag{9.80}$$

Substituting Equation 9.80 in Equation 9.77, we get

$$x(k + 1) = (A - BK)x(k) - BKFr + Er \tag{9.81}$$

Now, designing K to place the eigenvalues of the closed-loop matrix $(A - BK)$ at locations $\lambda_1 = 0.2$, $\lambda_2 = 0.25$ will lead to a stable solution provided the controllability matrix (Equation 4.121)

$$Q = [B \quad AB] = \frac{1}{8}\begin{bmatrix} 1 & 1 \\ -1 & -2 \end{bmatrix} \tag{9.82}$$

is of full rank, which we can see by inspection (the two columns of Q are independent) is the case which is full. Hence both states can be controlled using a single actuator, $u(k)$. To find the gains let us use the procedure outlined in Example 5.4. The characteristic polynomial of the open-loop system is

$$P(\lambda) = |\lambda I - A| = \begin{vmatrix} \lambda & 0 \\ 1 & \lambda - 1 \end{vmatrix} = (\lambda - 1)(\lambda - 1) = \lambda^2 - 2\lambda + 1 = 0 \tag{9.83}$$

Comparing the open-loop characteristic polynomial, Equation 9.83, with Equation 5.21, $\beta_1 = -2$, $\beta_2 = 1$. The characteristic polynomial of the desired closed-loop system is

$$P_c(\lambda) = (\lambda - 0.2)(\lambda - 0.25) = \lambda^2 - 0.45\lambda + 0.05 \tag{9.84}$$

Comparing the closed-loop characteristic polynomial, Equation 9.84, with Equation 5.19, $\alpha_1 = -0.45$, and $\alpha_2 = 0.05$. The transformation T (Equation 5.30) is given by

$$T = QW = [B \quad AB]\begin{bmatrix} \beta_1 & 1 \\ 1 & 0 \end{bmatrix} = \frac{1}{8}\begin{bmatrix} 1 & 1 \\ -1 & -2 \end{bmatrix}\begin{bmatrix} -2 & 1 \\ 1 & 0 \end{bmatrix} = \frac{1}{8}\begin{bmatrix} -1 & 1 \\ 0 & -1 \end{bmatrix} \tag{9.85}$$

The gain vector K (Equation 5.32) becomes

$$K = [\alpha_2 - \beta_2 \quad \alpha_1 - \beta_1]T^{-1} = [-0.95 \quad 1.55]\begin{bmatrix} -\frac{1}{8} & \frac{1}{8} \\ 0 & -\frac{1}{8} \end{bmatrix}^{-1} = [7.6 \quad -4.8] \tag{9.86}$$

Check: To know whether our computation is correct, find the eigenvalues of $(A - BK)$ using above gain vector. It should give $\lambda_1 = 0.2$, $\lambda_2 = 0.25$.

Case (b): Applying similar pole-placement techniques for SISO design, gain matrix is calculated to place the closed loop eigenvalues at $\lambda_1 = 0.6$, $\lambda_2 = 0.65$. The characteristic polynomial of the desired closed-loop system is

$$P_c(\lambda) = (\lambda - 0.6)(\lambda - 0.65) = \lambda^2 - 1.25\lambda + 0.39 \tag{9.87}$$

Hence $\alpha_1 = -1.25$, and $\alpha_2 = 0.39$. The transformation T is the same as in Equation 9.85. Hence the gain matrix K is given as

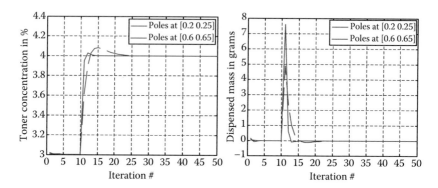

FIGURE 9.33 TC convergence plot for a step input (left), corresponding actuator signal (right).

$$K = [\alpha_2 - \beta_2 \quad \alpha_1 - \beta_1]T^{-1} = [-0.61 \quad 0.75] \begin{bmatrix} -\frac{1}{8} & \frac{1}{8} \\ 0 & -\frac{1}{8} \end{bmatrix}^{-1} = [4.88 \quad -1.12]$$

$$(9.88)$$

For this gain vector, clearly the eigenvalues of $(A - BK)$ are $\lambda_1 = 0.6$, $\lambda_2 = 0.65$.

Figure 9.33 shows the TC convergence plot and the corresponding actuator curve for a step change in TC from 3% to 4% for case (a) and (b). Actuator limits have not been considered in this calculation. Hence, the actuator signal shows a very high dispense magnitude to meet the demand. The desired eigenvalues can be adjusted to reduce the magnitude if it violates the actuation limits.

9.11.3 DESIGN OF A TC CONTROL LOOP WITH A TIME DELAY USING A PI CONTROLLER

For a TC system with a dispenser lag, the time delay component $\mu \neq 0$. This delay can affect the stability and control of the TC system. It can induce oscillations, and eventually destabilize the system even though the closed-loop poles are assigned to ensure stability. It can also lead to chaotic behavior when the maximum dispenser rate is limited by the admix of the material package (i.e., when the actuator, dispenser rate rails at its maximum limit). Actuator saturation can make the TC system nonlinear. All of these aspects motivate the need to compensate for the delay effects.

There are many methods and techniques for the analysis and control of dynamic systems in the presence of time delay [37]. The Smith predictor-based control scheme is the most commonly used technique in industrial systems [38–40]. Here we present the design of a Smith predictor for a fixed time delay. For systems with a varying time delay, underestimating or overestimating the time delay significantly degrades the control quality. Reference [39] shows a design method of the Smith predictor with varying time delay.

Assuming zero consumption, that is, $v(k) = 0$, if $G_{pi}(z)$ is the transfer function of the PI controller shown in Figure 9.31, and $G(z) = G_p(z)z^{-\mu}$ is the transfer function

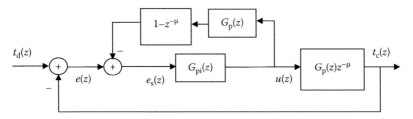

FIGURE 9.34 Block diagram of the feedback system with Smith predictor.

of the open-loop TC system with a dispenser lag, μ cycles, then the closed-loop transfer function is given by

$$G_c(z) = \frac{G_{pi}(z)G_p(z)z^{-\mu}}{1 + G_{pi}(z)G_p(z)} \tag{9.89}$$

Equation 9.89 implies that the closed-loop system response of a PI controller designed by ignoring time delay, as in Section 9.11.2, is governed by the transfer function of the controller-plus-undelayed plant while the actual response of the controller is subjected to a time delay of μ cycles. With the Smith predictor, input to the PI controller is modified by a feedback term $(1 - z^{-\mu})\,G_p(z)$ as shown in Figure 9.34.

To implement the Smith predictor, the block diagram of Figure 9.34 is converted to difference equations as follows.

First, let us write the important equations of the system with Smith predictor and PI controller in the z-domain, which is easy to interpret. From Figures 9.31 and 9.34, we can write

$$e_s(z) = e(z) - (1 - z^{-\mu})G_p(z)u(z) \tag{9.90}$$

$$\tilde{u}(z) = K_p e_s(z) + \frac{z}{z-1}K_i e_s(z) \tag{9.91}$$

$$u(z) = \tau\tilde{u}(z) \tag{9.92}$$

A useful form to implement in the computer is the difference equation form which contains a sequence of numbers iterated from each measurement-process-update cycle. To do this, the transfer function, $G_p(z)$, is replaced by $g/(z-1)$ in Equation 9.90 and inverse z-transform is applied on the resulting equation to yield the difference equation form:

$$e_s(k) = e_s(k-1) + e(k) - e(k-1) - gu(k-1) + gu(k-\mu-1) \tag{9.93}$$

$$\tilde{u}(k) = \tilde{u}(k-1) + (K_p + K_i)e_s(k) - K_p e_s(k-1) \tag{9.94}$$

$$u(k) = \tau\tilde{u}(k) \tag{9.95}$$

$$e(k) = t_d(k) - t_c(k) \tag{9.96}$$

Equations 9.93 through 9.96 are recursive. For computation, we first start with Equations 9.96 and 9.95 with known desired TC and measured TC values for a known dispense process. After this, we compute $e_s(k)$ and $\tilde{u}(k)$ from Equations 9.93

and 9.94, respectively, with appropriate initial conditions. The dispense rate $\tilde{u}(k)$ is used to actuate the dispenser.

Example 9.11

Use Equation 9.63 as the TC system and the PI controller with closed-loop poles $\lambda_1 = 0.6$, $\lambda_2 = 0.65$ and Smith predictor Equations 9.93 through 9.96 to simulate the performance of TC system with a dispenser lag of (a) 3 cycles and (b) 10 cycles. Assume no images are printed during the entire simulation.

SOLUTION

We show the results in Figures 9.35 through 9.38. Clearly the feedback system is unstable for delay of 3 and 10 cycles as shown in Figures 9.35 and 9.37. The Smith

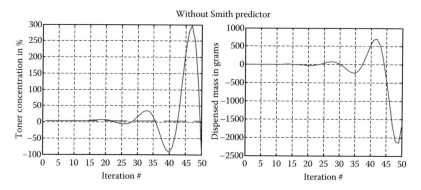

FIGURE 9.35 TC and actuator response for a step change to TC target with time delay of three cycles and without Smith predictor (dashed curve: TC target input; solid curve: TC output [left], dispensed mass [right]).

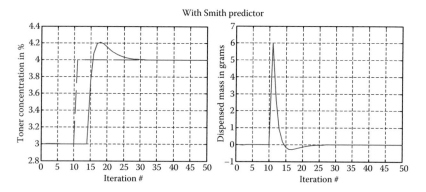

FIGURE 9.36 TC and actuator response for a step change to TC target with time delay of 3 cycles and with Smith predictor (dashed curve: TC target input; solid curve: TC output [left], dispensed mass [right]).

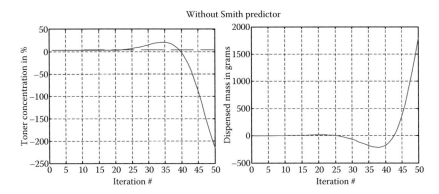

FIGURE 9.37   TC and actuator response for a step change to TC target with time delay of 10 cycles and without Smith predictor (dashed curve: TC target input; solid curve: TC output [left], dispensed mass [right]).

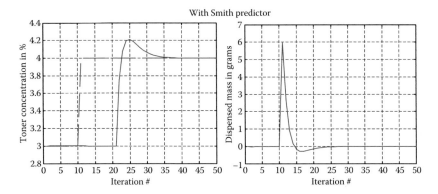

FIGURE 9.38   TC and actuator response for a step change to TC target with time delay of 10 cycles and with Smith predictor (dashed curve: TC target input; solid curve: TC output [left], dispensed mass [right]).

predictor improves the stability of the loop (Figures 9.36 and 9.38), while at the same time injects some overshoot when compared to the response without any delay as in Figure 9.33.

9.11.4   FEEDFORWARD COMPENSATION FOR IMAGE DISTURBANCE

Generally, pixel count is used to estimate the amount of toner required for developing the images. Since images can have varying degrees of area coverage, it can represent a large disturbance to the TC control system, depending on how the area coverages are distributed while printing. Particularly, if the jobs contain a mixture of low and high area coverages and the change is rapid, the system can be overloaded with disturbance before it has time to recover. This can induce undesirable fluctuations in TC and even a loss of stability. Proper feedforward compensation of the control input, $\tilde{u}$, from the pixel counter can potentially minimize the impact.

We first compute the area coverage from the pixel counter. The estimated amount of developed mass, $v(k)$, in grams, can be found using following equation:

$$v(k) = \frac{\text{image size} \times \text{ toner mass}}{1000} \times \text{area coverage } (k) \qquad (9.97)$$

where toner mass is the amount of toner developed in milligrams/square centimeters. Image size is the size of the image (e.g., $21.59 \times 27.94 \times 10$ square centimeters for a $8.5'' \times 11''$ page with 10 pages being printed per one TC cycle). Area coverage $= P_c$, is the ratio of the total area where there is toner to the total image area, which is calculated for each image using the pixel counter. In the actual implementation of the feedforward compensation function, the pixel counter data is multiplied by a constant, K_{ff}, and the resulting signal is advanced by μ cycles to account for the dispenser lag. If the area coverage of each image is known μ cycles ahead of actual print, and the estimated toner mass is known accurately (which is included in the feedforward gain, K_{ff}), perfect cancellation for image disturbance can be achieved. The control input (Equation 9.91) is modified with feedforward compensation as shown below:

$$\tilde{u}(z) = K_p e_s(z) + \frac{z}{z-1} K_i e_s(z) + K_{ff} z^\mu p_c(z) \qquad (9.98)$$

9.11.5 DESIGN OF TC CONTROL LOOP WITH STATE FEEDBACK CONTROLLER AND STATE ESTIMATOR

To improve the overall TC performance, the following three major system components need to be optimized: (1) the open-loop toner replenishment (dispense) and mixing system (i.e., plant itself), (2) the TC measurement system (i.e., sensor), and (3) the closed-loop control algorithm. Optimization of the toner replenishment system and measurement system is related to the system hardware. Optimization of the feedback controller is a software function which is easy to be incorporated even after the hardware system is fully commissioned.

In this section, a new state-based feedback control algorithm (i.e., a set of instructions implemented via a computer code) is discussed. In this algorithm, the time delay in the toner dispense is replaced by states, and each state is estimated using measurements from the TC sensor. States are defined by creating an open-loop state-based model of the system. States are estimated from the sensor output using a state estimator. As the system and each of the states is controlled, an improved dynamic performance is expected as compared with the Smith predictor implementation. In addition to this benefit, since the controller does not use an integrator, an antiwindup compensator will not be required if the TC loop were to operate under limited actuator (Section 9.12).

The system described by Equation 9.63 has a delay of μ cycles. This is split into μ states as in Figure 9.39. Formations of these states can be understood when the system equation is expressed in terms of z-transforms.

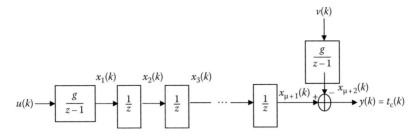

FIGURE 9.39 Block diagram of open-loop TC model with time delay.

In physical terms, for the TC system, states can be defined as quantities related to the toner mass at discrete delay cycle. From the block diagram shown in Figure 9.39, the state equations can be written as

$$x_1(k+1) = x_1(k) + gu(k)$$
$$x_2(k+1) = x_1(k)$$

$$\vdots \tag{9.99a}$$

$$x_{\mu+1}(k+1) = x_\mu(k)$$
$$x_{\mu+2}(k+1) = x_{\mu+2}(k) + gv(k)$$

and the output equation is given by

$$y(k) = t_c(k) = x_{\mu+1}(k) - x_{\mu+2}(k) \tag{9.99b}$$

It is important to note that, in Equation 9.99, the states represent TC at a discrete delay cycle, which in turn is related to the toner mass.

Total number of states depends on the delay cycles. If the time delay is zero, that is, when the toner dispense is instantaneous, number of states will be reduced to two (one due to the mass dispensed and another due to the mass developed) as seen from the state equation below. The above equations can be grouped into matrix form as

$$x(k+1) = Ax(k) + Bu(k)$$
$$x_{\mu+2}(k+1) = x_{\mu+2}(k) + gv(k) \tag{9.100}$$
$$y(k) = Cx(k) - x_{\mu+2}(k)$$

where $x(k) \in R^{\mu+1}$, $x_{\mu+2}(k) \in R^1$, $u(k) \in R^1$, $A \in R^{(\mu+1) \times (\mu+1)}$, $B \in R^{\mu+1}$, and $C \in R^{\mu+1}$. These vectors and matrices are

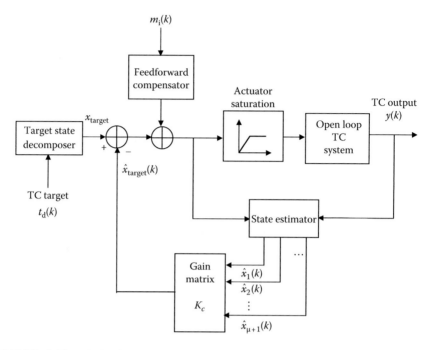

FIGURE 9.40 Block diagram of open-loop TC model with state estimator and state controller.

$$x(k) = \begin{bmatrix} x_1(k) \\ x_2(k) \\ \vdots \\ x_{\mu+1}(k) \end{bmatrix}, \quad A = \begin{bmatrix} 1 & 0 & 0 & \cdots & 0 \\ 1 & 0 & 0 & \cdots & 0 \\ 0 & 1 & 0 & \cdots & 0 \\ \vdots & \vdots & \vdots & \vdots & \vdots \\ 0 & 0 & \cdots & 1 & 0 \end{bmatrix}, \quad B = \begin{bmatrix} g \\ 0 \\ \vdots \\ 0 \end{bmatrix} \quad \text{and}$$

$$C = \begin{bmatrix} 0 & 0 & \cdots & 0 & 1 \end{bmatrix} \qquad (9.101)$$

This state-space equation is used to develop the state feedback estimator and state feedback controller as follows.

The block diagram of the closed-loop TC control system with state estimator and state controller is shown in Figure 9.40. The states of the open-loop TC system can be estimated using a Kalman filter (an optimal observer) or a linear full-order observer (a suboptimal observer). A full-order observer (Section 5.4.2) [41], Equation 9.102, with K_0 as the observer gain is considered as the state estimator.

$$\hat{x}(k+1) = A\hat{x}(k) + Bu(k) + K_0\big[y(k) - C\hat{x}(k) + x_{\mu+2}(k)\big] \qquad (9.102)$$

The gain K_0 controls the rate at which the error signal between the true states and the estimated states, $e(k) = x(k) - \hat{x}(k)$ reaches zero steady state. A difference equation can be obtained for the error signal by using Equations 9.100 and 9.102 as follows:

$$e(k + 1) = x(k + 1) - \hat{x}(k + 1) = (A - K_0C)(x(k) - \hat{x}(k)) \tag{9.103}$$

Since $e(k) = x(k) - \hat{x}(k)$ Equation 9.103 can be written as

$$e(k + 1) = (A - K_0C)e(k) \tag{9.104}$$

The observer is in the form of a feedback system with the error dynamics satisfied by Equation 9.104. The matrix $A - K_0C$ is called the observer matrix. For the error to go to zero asymptotically, all the eigenvalues of the observer matrix must lie inside the unit circle in the z-domain. Determination of the gain matrix that accomplishes this is a pole-placement task. Therefore, we chose the gain matrix K_0 to place the eigen-values of the observer matrix within the unit circle so that the observer states asymptotically converge to the true state. Note that, in order to place the poles of the observer matrix, it is necessary that the observability matrix (Equation 4.126)

$$P = \begin{bmatrix} C' & A'C' & .. & (A')^{\mu+1-1}C' \end{bmatrix} \tag{9.105}$$

be full rank. We can use the Bass–Gura formula or Ackermann formula, Equation 5.84 to design the observer gain matrix.

By the separation theorem, the observer and the state feedback designs can be decoupled and designed independently. The dynamic TC system, Equation 9.100, is in the linear state-space form, and we designed the state feedback control law, $u(k) = -Kx(k)$ with $K = [K_1 \quad K_2 \quad \cdots \quad K_{\mu+1}]$ under the assumption that the state vector, $x(k)$, is accessible for measurement. But, because of our inability to measure the system states, they are estimated using a linear observer (Equation 9.102) by measuring the TC, $y(k)$, with a TC sensor and estimating the feedforward signal, $x_{\mu+2}(k+1) = x_{\mu+2}(k) + gv(k)$, based on the area coverage information. After the states are estimated, the control input is calculated using the controller gains (row vector) and estimated states as follows:

$$u(k) = -K\hat{x}(k) \tag{9.106}$$

The system described by Equation 9.100 is a SISO system unlike our augmented system (Equation 9.75) with the PI controller. The number of gains is equal to the number of states that the control input $u(k)$ affects during feedback. The gain vector $K = [K_1 \quad K_2 \quad \cdots \quad K_{\mu+1}]$ can be obtained by using the pole-placement or optimal control techniques.

Having determined the observer gain matrix and the controller gain matrix, these processing units have to be interconnected with the TC target to derive proper control input. The target state decomposer (shown in Figure 9.40) is designed for creating a vectorized version of the TC target so that the output from the state feedback controller can be subtracted correctly. It is a simple linear transformation formed by multiplying the controller gain matrix with elements arranged along the diagonal with the desired TC value. That is,

$$x_{\text{target}} = K_c x_{\text{desired}} \qquad (9.107)$$

where $x_{\text{desired}} \in R^{\mu+1}$, $x_{\text{target}} \in R^{\mu+1}$, and $K_c \in R^{(\mu+1)\times(\mu+1)}$ is the controller gain matrix with elements along the diagonal.

$$x_{\text{desired}} = \begin{bmatrix} t_d \\ t_d \\ \cdot \\ \cdot \\ t_d \end{bmatrix}, \quad K_c = \text{diag}\begin{bmatrix} K_1 & K_2 & \cdots & K_{\mu+1} \end{bmatrix} \qquad (9.108)$$

9.12 PROCESS CONTROLS UNDER LIMITED ACTUATION

In modern electrophotographic printers, exposure, development, and TC control systems use some form of integral control to reduce steady-state errors. This is because, over time, an integral controller integrates even small variations of the error signal between the sensed value and the target value, to provide an increasing control signal that brings the process output back to the desired target.

There are, however, several aspects of an integral controller that can potentially have ill effects on the performance of the system when the dynamic range of the actuators is bounded by some practical limits [42–48]. For example, in the TC control system, dispense rate is limited to low and high values. At the low end, the rate is set to a minimum value to avoid the small rates that cannot be provided reliably by the dispenser motor. At the high end, the maximum rate is limited by the admix of the material package. Similarly, practical limits are set for charge voltage, exposure raster output scanner (ROS) intensity and the development bias voltage due to cost and other considerations. When an integrator is used in the control loop, a phenomenon known as integrator windup occurs whenever the actuator signal is driven to its maximum limit by the controller in an effort to bring the output to its desired target. If the output does not return to the desired target, there will be a nonzero error that will continue to be integrated by the integrator and this phenomenon is known as "windup." As a result, the actuator will remain at its limit until the system output crosses the target value and sufficient error of opposite sign is integrated to remove the windup. Windup can decrease the overall performance of the control system and cause large overshoot, longer settling time, and large steady-state errors. It can also occur in other situations like when the loop dynamics are changed abruptly, are switched off for some valid reason, and when a sensor fails or when there is inaccurate feedforward compensation due to measurement errors, etc. These short comings of the integral controllers can be minimized by including an antiwindup compensation. This is necessary to sustain reliable operation of production printing systems.

Although an optimization-based antiwindup design for MIMO control system is still an active area of research and many questions remain unanswered, many

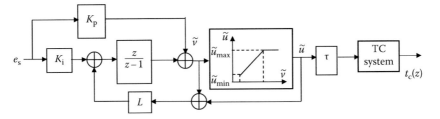

FIGURE 9.41 PI controller with antiwindup compensation for a TC system.

systems use a trial and error-based approach for compensation. Figure 9.41 shows a typical antiwindup scheme used in printers for a SISO TC system.

In the absence of actuator limits, the actuator signal $\tilde{v}$ will be same as $\tilde{u}$. When $\tilde{v}$ saturates to $\tilde{u}_{max}$ during the course of the loop operation, which may happen when the development voltage is changing rapidly, for example, the error signal e_s will be nonzero. The integration of the error signal will lead to a larger $\tilde{v}$ which will have no effect on the TC output since the actuator is limited to $\tilde{u}_{max}$. If the integrating effect continues further in the same direction, the integrator output may become excessively large. To compensate for these effects, we need to create a situation where the sign of the error is inverted (i.e., opposite to what was happening when $\tilde{v}$ saturates to $\tilde{u}_{max}$). This is achieved by inserting a high gain dead zone feedback using another error signal obtained by subtracting $\tilde{u}_{max}$ from $\tilde{v}$. In this scheme, when the integrator saturates, the antiwindup feedback obtained through the high gain, L, at the integrator becomes active to force the integrator windup error to zero. During this time, the integrator acts like a fast first-order lag with a transfer function that has a single pole created by the antiwindup compensator gain L. The value of the compensator gain determines how fast the integral action is pulled back from actuator saturation. When $\tilde{v}$ is within the actuator limits (i.e., between $\tilde{u}_{min}$ and $\tilde{u}_{max}$), a dead zone is created by disabling the high gain feedback automatically. The output of the integrator $\tilde{v}$ is the input to the high gain feedback. The implementation of the actuator limits would involve the use of a saturation function to generate the control signal $\tilde{u}$, which is the input to the high gain antiwindup compensator loop. The high gain antiwindup compensator can also be applied in Figure 9.41 before the integrator constant, K_i, as an alternate feedback structure.

Example 9.12

Consider the design of a TC feedback system with a transport delay using a state feedback estimator and controller. Design the observer gain matrix and the controller gain matrix using pole-placement techniques. Apply the area coverage disturbance shown in Figure 9.42. Use actuator saturation. Compare the TC response using the conventional approach with a PI controller, a Smith predictor, and an antiwindup compensator. Construct a figure of merit to compare the results.

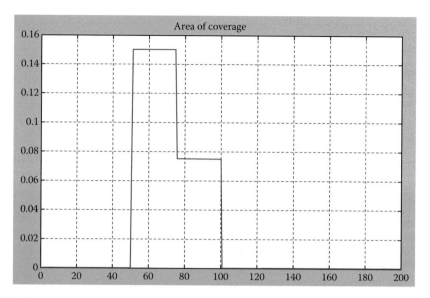

FIGURE 9.42   Area coverage disturbance shown with respect to TC update cycle.

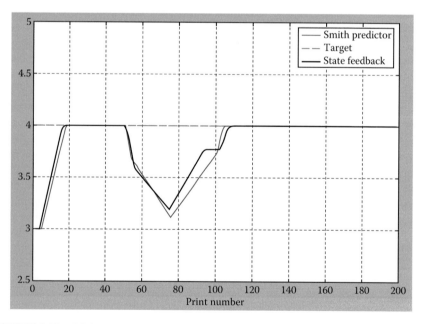

FIGURE 9.43   TC Response for a step input (3%–4%) in the presence of time delay, actuator limitation, and area coverage disturbance.

SOLUTION

The response to a step change in TC between 3% and 4% with area coverage disturbance is shown in Figures 9.43 through 9.47 in the presence of time delay, actuator saturation, area coverage disturbance, and noise. The PI controller

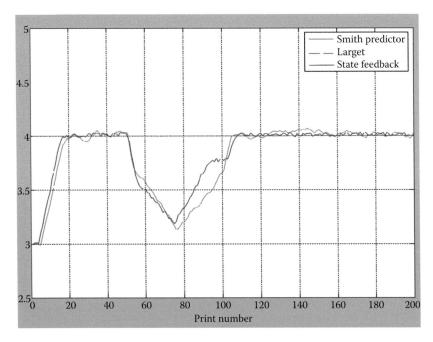

FIGURE 9.44   TC Response for a step input (3%–4%) in the presence of time delay, actuator limitation, and area coverage disturbance and 1% sensor noise.

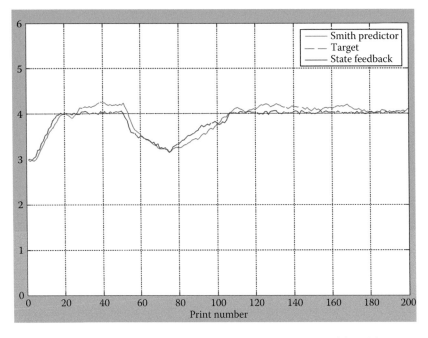

FIGURE 9.45   TC Response for a step input (3%–4%) in the presence of time delay, actuator limitation, and area coverage disturbance and 2% sensor noise.

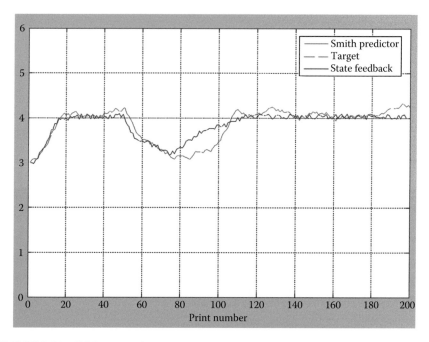

FIGURE 9.46   TC Response for a step input (3%–4%) in the presence of time delay, actuator limitation, and area coverage disturbance and 3% sensor noise.

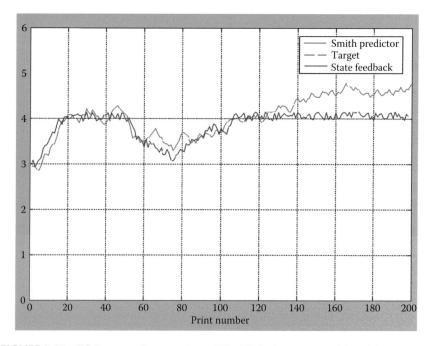

FIGURE 9.47   TC Response for a step input (3%–4%) in the presence of time delay, actuator limitation, and area coverage disturbance and 5% sensor noise.

without a Smith predictor gives an overshoot (not shown in the figures). This is because of the integral part of the controller. By including the Smith predictor, the overshoot due to time delay has been reduced, but the sensor noise can give rise to increased steady-state error, meaning the final TC will not be equal to what was requested (in this case 4%). But the state feedback gives reduced steady-state error under those noise uncertainties. The state feedback technique works well because the output is fed back to a number of states instead of just one as in PI controller. All states are estimated and controlled, thus, the error correction is more favorable with state feedback.

In Figure 9.48, we show the area coverage disturbances expected from a sample print job. Clearly it appears as noise without any structure. Figure 9.49 shows an

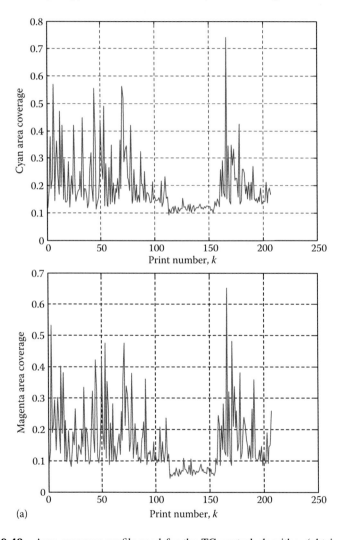

(a)

FIGURE 9.48   Area coverage profile used for the TC control algorithm (obtained using a scanner from magazine samples).

(*continued*)

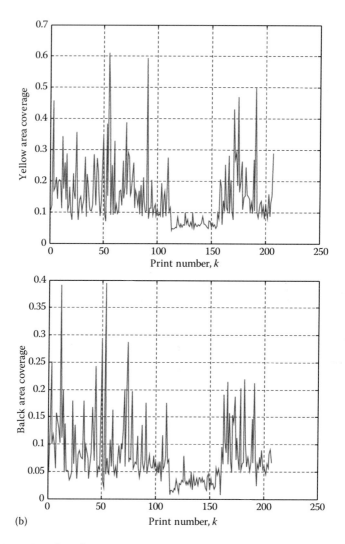

FIGURE 9.48 (continued) Area coverage profile used for the TC control algorithm (obtained using a scanner from magazine samples).

improved steady-state error as compared with the PI controller with Smith predictor and a nonoptimized model predictive controller (not described in this book).

The figure of merit, which is defined as the ratio of the area under the curve with disturbances and uncertainties to that without any such disturbance, is calculated by measuring the area under the response curve with respect to the target. In our comparison, the ideal figure of merit is a step response from 3% to 4%. When the TC response is above the target value, then it is subtracted from the required area. The figure of merit in Table 9.2 shows overall improvement can be achieved with state feedback methods. More advanced methods such as model predictive controller can be further optimized to work similar to or better than the state feedback method for the same problem.

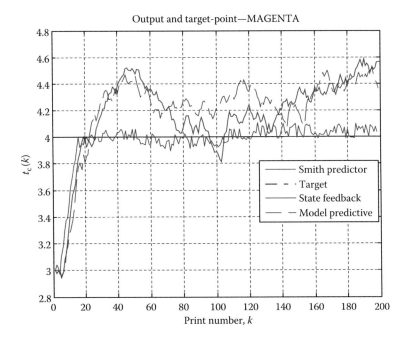

FIGURE 9.49   TC Response for a step input (3%–4%) in the presence of time delay, actuator limitation, and area coverage disturbance, 5% sensor noise and area coverage disturbance of Figure 9.48.

Having covered how the antiwindup compensator can be designed and implemented for a SISO loop, let us now focus our attention to the design of a similar compensation strategy for a three-input three-output MIMO level 2 system, which is obviously more complex than the MIMO level 1 system. We describe a general technique by vectorizing Figure 9.41 so that it is applicable to many of the control loops described in this book.

The uncompensated MIMO PI controller is shown in Figure 9.50. The controller state-space equation with the parameter matrices, A, B, C, and D, the input error vector $e(k)$, the output of the controller $v(k)$, and the controller state vector $x(k)$ is as follows:

$$x(k + 1) = Ax(k) + Be(k)$$
$$v(k) = Cx(k) + De(k)$$

(9.109)

The saturation function can be written in mathematical form as $v = u$, the actuator input when $u_{min} \leq v \leq u_{max}$, $v = u_{min}$ when $v < u_{min}$, $v = u_{max}$ when $v > u_{max}$, where u_{min} and u_{max} are the minimum and maximum allowable actuator vectors, respectively. Clearly, in a SISO system, the PI controller of Figure 9.41, has parameters, $A = 0$, $B = K_i$, $D = K_p$, and $C = 1$, $e(k) = e_s(k)$, $v(k) = \tilde{v}(k)$. When the integrator

TABLE 9.2

Figures of Merit Comparison between Three Different Methods (a) Smith Predictor with PI Controller, (b) State Feedback, and (c) Advanced Nonlinear Controller Obtained for a TC Response with a Step Input of 3%–4%

Percentage of Noise (%)	Color Band	Figure of Merit (%)		
		State Feedback	Smith Predictor	Model Predictive Controller (Not Described)
0	C	97.1473	97.3953	96.3173
	M	97.4564	97.6286	96.8857
	Y	97.5516	97.7089	96.9879
	K	98.1352	98.1406	97.7218
1	C	97.0907	97.2207	96.4488
	M	97.4071	97.6397	97.1076
	Y	97.4999	97.7065	97.3143
	K	98.0873	97.8153	97.1949
2	C	97.0717	97.1049	95.9487
	M	97.3869	97.1762	96.3570
	Y	97.4825	97.5682	96.4380
	K	98.0096	97.1548	95.1828
3	C	97.0907	96.8273	95.1031
	M	97.3146	96.1433	95.3468
	Y	97.4069	96.3993	94.3575
	K	97.7942	96.9666	92.7476
5	C	96.9215	94.7087	93.6156
	M	97.0455	92.7874	92.8630
	Y	97.1700	95.5060	89.9066
	K	97.4503	90.5272	86.4776

Note: Uses area coverage, time delay actuator saturation, and sensor noise as various system elements affecting the TC response.

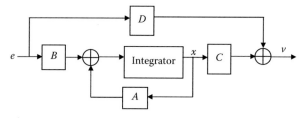

FIGURE 9.50 Block diagram of the state-space form for a MIMO PI controller.

saturates to the maximum allowable actuator limit, $v = u_{max}$. The actuator windup occurs when the controller matrix A is not stable and error vector $e(k)$ is nonzero. The antiwindup compensator, with a high gain feedback should be able to make the controller matrix stable again.

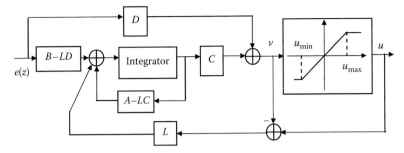

FIGURE 9.51 Block diagram representation of a MIMO antiwindup compensator with PI controller.

When the integrator is not saturated, $v(k) = u(k)$, that is, from Equation 9.109

$$u(k) - Cx(k) - De(k) = 0 \qquad (9.110)$$

Multiplying Equation 9.110 by the gain matrix, L, and adding the resulting equation with Equation 9.109 we get a new controller state equation

$$x(k+1) = (A - LC)x(k) + (B - LD)e(k) + Lu \qquad (9.111)$$

Proper choice of matrix L can make the controller matrix $(A - LC)$ stable. This expression represents a combined system of an antiwindup compensator and a PI controller, which is very similar in structure to a state observer (Section 5.4) for which the closed-loop dynamics are dependent on the eigenvalues of the matrix $(A - LC)$. The eigenvalues can be placed at arbitrary locations by selecting the anti-windup gain matrix, L, if the observability matrix $P = [C' \quad A'C' \quad \cdots \quad (A')^{n-1} \quad C']$ (with n number of states) is of full rank. Figure 9.51 shows the block diagram representation of the MIMO PI controller with antiwindup compensator. The compensated controller has new parameter matrices, A-LC and B-LD.

9.13 OPTIMAL CONTROLS FOR SELECTIVE STATES

As the photoconductor and developer materials age, or the machine environment changes, the TC target needs to be adjusted. These adjustments can be provided at a reasonably lower update cycle compared with the basic TC loop described in Section 9.11. In some xerographic printers, toner age correction is done on both the TC target and the TC sensor. This is because the desired TC value is adjusted as a function of toner age in order to stabilize the development system, particularly when images with low area coverage are printed. Another common approach is to augment the level 2 process actuators with a TC target and then develop a four-input three-output control system with low-, mid-, and high-density DMA measurements. A practical developability controller operating in Xerox black and white printers through the use of LQR and adaptive parameter estimator is described in Ref. [16].

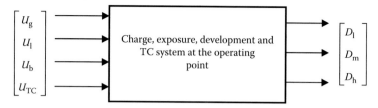

FIGURE 9.52   Four-input three-output MIMO process control system.

Referring to the block diagram of Figure 9.52, if, $x_\mathrm{d} = \begin{bmatrix} D_\mathrm{l}^\mathrm{T} & D_\mathrm{m}^\mathrm{T} & D_\mathrm{h}^E \end{bmatrix}'$ is the desired target DMA vector, if $U(k) = [U_\mathrm{g}(k) \quad U_\mathrm{l}(k) \quad U_\mathrm{b}(k) \quad U_\mathrm{TC}(k)]'$ is the hard process actuator vector with charge, ROS intensity, developer bias, and the TC target [49], respectively, and $x(k) = [D_\mathrm{l}(k) \quad D_\mathrm{m}(k) \quad D_\mathrm{h}(k)]'$ is the DMA measurement vector, then for the purpose of developing a four-input three-output stable process control system, a linear state-space model can be written as

$$x(k+1) = Ax(k) + Bu(k)$$
$$y(k) = Cx(k) \tag{9.112}$$

where
   C is the identify matrix
   $y(k)$ is same as $x(k)$

The system matrix A, and sensitivity matrix B are expressed as

$$
A = \begin{bmatrix} 1 & 0 & 0 \\ 0 & 1 & 0 \\ 0 & 0 & 1 \end{bmatrix}, \quad
B = \begin{bmatrix} \frac{\partial D_\mathrm{l}}{\partial U_\mathrm{g}} & \frac{\partial D_\mathrm{l}}{\partial U_\mathrm{l}} & \frac{\partial D_\mathrm{l}}{\partial U_\mathrm{b}} & \frac{\partial D_\mathrm{l}}{\partial U_\mathrm{TC}} \\ \frac{\partial D_\mathrm{m}}{\partial U_\mathrm{g}} & \frac{\partial D_\mathrm{m}}{\partial U_\mathrm{l}} & \frac{\partial D_\mathrm{m}}{\partial U_\mathrm{b}} & \frac{\partial D_\mathrm{m}}{\partial U_\mathrm{TC}} \\ \frac{\partial D_\mathrm{h}}{\partial U_\mathrm{g}} & \frac{\partial D_\mathrm{h}}{\partial U_\mathrm{l}} & \frac{\partial D_\mathrm{h}}{\partial U_\mathrm{b}} & \frac{\partial D_\mathrm{h}}{\partial U_\mathrm{TC}} \end{bmatrix} \tag{9.113}
$$

Note that the above linear representation can also be used for controlling a paper-based density or L^* or ΔE_{ab}^* measurements (extracted with respect to paper) as pointed out in Section 9.9. We use the LQR for designing the feedback controller. The linear quadratic controller minimizes a selected quadratic objective function. For example, if our goal is to minimize the error vector between the target vector and the measured vector, then the objective function should be formed with the sums of the squares of the weighted error (or the state) vector. Another term can be included with the sums of the squares of the actuator values in the objective function to appropriately weigh the desired actuator.

Let the performance objective be defined in terms of the following function:

$$\tilde{J} = \tfrac{1}{2} \sum_{k=0}^{N-1} \alpha^{2k} \left[x^\mathrm{T}(k)Qx(k) + u^\mathrm{T}(k)Ru(k) \right] \tag{9.114}$$

The scalar parameter α is a positive weight that is chosen by the customer/designer to constrain all the eigenvalues of the overall system to lie inside a circle of radius $\beta \leq 1$ in the z-domain, where $\beta = 1/\alpha$. The index N indicates the length of a finite process. This will guarantee that the largest eigenvalue will be less than β in magnitude or the transient response will decay faster than β^k. N can be chosen to be equal to number of cycles over which the optimization function is to be minimized. The importance of states and actuators is adjusted by changing the matrices, Q and R, respectively. When we increase the weights in Q corresponding to the state x_3—the high density state will be emphasized more and affected more (Problem 9.14).

Let $\tilde{x}(k) = \alpha^k x(k)$ and $\tilde{u}(k) = \alpha^k u(k)$. Then the new performance index becomes

$$\tilde{J} = \frac{1}{2} \sum_{k=0}^{N-1} [\tilde{x}^{\mathrm{T}}(k)Q\tilde{x}(k) + \tilde{u}^{\mathrm{T}}(k)R\tilde{u}(k)] \tag{9.115}$$

$$\tilde{x}(k+1) = \alpha^{k+1}x(k+1) \tag{9.116}$$

Substituting Equation 9.112 in Equation 9.116 and using the definition of $\tilde{x}(k)$, we get

$$\tilde{x}(k+1) = \alpha A\tilde{x}(k) + \alpha B\tilde{u}(k)$$
$$= \tilde{A}\tilde{x}(k) + \tilde{B}\tilde{u}(k) \tag{9.117}$$

where $\tilde{A} = \alpha A$ and $\tilde{B} = \alpha B$. Optimal gain can be obtained by minimizing $\tilde{J}$ subject to the constraint shown by system model (Equation 9.117). This type of minimization problem will yield the feedback law

$$\tilde{u}(k) = -\tilde{K}(k)\tilde{x}(k) \tag{9.118}$$

This can be solved using the standard optimization method involving a Lagrange multiplier. We do not show a proof of this method, since the basic techniques are already described in Section 5.3.2. Instead, the final set of recursive algorithms are shown below.

Control law:

$$\tilde{u}(k) = -\tilde{K}(k)\tilde{x}(k) \tag{9.119}$$

Gain matrix:

$$\tilde{K}(k) = \left[R + \tilde{B}^{\mathrm{T}}P(k+1)\tilde{B}\right]^{-1}\tilde{B}^{\mathrm{T}}P(k+1)\tilde{A} \tag{9.120}$$

Recursive equation:

$$P(k) = \tilde{A}^{\mathrm{T}}P(k+1)\tilde{A} - \tilde{A}^{\mathrm{T}}P(k+1)\tilde{B}\left[R + \tilde{B}^{\mathrm{T}}P(k+1)\tilde{B}\right]^{-1}\tilde{B}^{\mathrm{T}}P(k+1)\tilde{A} + Q \tag{9.121}$$

Boundary condition:

$$P(N) = 0 \tag{9.122}$$

Steady-state solution:
If the control process is finite $(N < \infty)$, the feedback gain $K(k)$ is time-varying. If the process is an infinite stage process $(N \to \infty)$, the feedback gain K attains a constant value. If the steady-state form of matrix $P(k)$ is defined as P and the steady-state form of gain matrix $K(k)$ is defined as K, then P and K can be determined by solving the following algebraic equations:

Gain matrix:

$$\tilde{K} = \left[R + \tilde{B}^T P \tilde{B}\right]^{-1} \tilde{B}^T P \tilde{A} \tag{9.123}$$

Recursive Equation:

$$P = \tilde{A}^T P \tilde{A} - \tilde{A}^T P \tilde{B} \left[R + \tilde{B}^T P \tilde{B}\right]^{-1} \tilde{B}^T P \tilde{A} + Q \tag{9.124}$$

Control law:

$$\tilde{u}(k) = -\tilde{K}\tilde{x}(k) \tag{9.125}$$

From the above equations, the gains are clearly dependent on the weights. The higher the weights, the tighter the tolerance on the control variables. Equation 9.124 is nonlinear. Therefore, a close form solution is not feasible. Usually, instead of solving an algebraic equation like Equation 9.124, we solve the recursive equation, in this case, Equation 9.121, by marching backward in cycles from $k = N$ to 0. Then the steady-state solution is $P(0)$. Once P is found, Equation 9.123 can be solved to obtain a steady-state gain matrix K.

Selection of weight matrices:
One way of selecting the weights in the quadratic performance equation is shown below. Let α_1, α_2, and α_3 be the maximum percentage change deviations of the states x_1, x_2, and x_3 from their nominal values. Assume matrix Q to be diagonal (which is more realistic), then the total cost is

$$x^T Q x = Q_{11} x_1^2 + Q_{22} x_2^2 + Q_{33} x_3^2 \tag{9.126}$$

Now, choose Q such that the overall cost is uniformly distributed between the three states. This means that

$$Q_{11} x_1^2 = Q_{22} x_2^2 = Q_{33} x_3^2 \tag{9.127}$$

or

$$Q_{11} \alpha_1^2 \beta^2 = Q_{22} \alpha_2^2 \beta^2 = Q_{33} \alpha_3^2 \beta^2 = \beta^2 \tag{9.128}$$

Thus

$$Q_{11} = \frac{1}{\alpha_1^2}, \quad Q_{22} = \frac{1}{\alpha_2^2}, \quad Q_{33} = \frac{1}{\alpha_3^2} \tag{9.129}$$

Similarly, we choose matrix R to be diagonal, with diagonal elements chosen as

$$R_{11} = \frac{1}{u_{1\,max}^2}, \quad R_{22} = \frac{1}{u_{2\,max}^2}, \quad R_{33} = \frac{1}{u_{3\,max}^2}, \quad R_{44} = \frac{1}{u_{4\,max}^2}$$

Recursive algorithm (steady-state):
If N is chosen to be sufficiently large, the feedback gain matrix becomes a constant matrix. This constant gain can be obtained by either solving the algebraic Equations 9.123 and 9.124, or the recursive equations given by Equations 9.120 and 9.121. Since the algebraic equations are nonlinear, it would be computationally more efficient and easier to solve the recursive Equations 9.120 and 9.121. We start with the boundary condition $P(N) = 0$ and solve Equation 9.121 backward from $P(N)$ to $P(0)$. The steady-state solution for P is the initial value $P(0)$. Once P is found, Equation 9.123 can be used to compute the gain matrix K.

9.14 OPTIMAL MEASUREMENTS

TRCs are updated over time by periodically printing test patches at various gray levels. Printed test patches are sensed by the sensor to determine the appropriate compensation so that the new TRCs provide compensation for the current state of the print engine. TRCs in level 3 control are used for linearizing each separation. Whereas, the spatial TRCs, although serving a similar purpose, are generally used to adjust the pixel values (0–255) at a relatively fine resolution to compensate for the characteristics of the print engine so that images are rendered uniformly [50–52]. In such systems, one may select an appropriate compensating TRC for a pixel location in a rendered image space based on the contone value of the input pixel.

A problem usually encountered in such a control system is the ability to define a minimal sampling procedure for creating and updating the compensation TRCs. In particular, there is a need to determine the optimum minimal set of gray levels or colors for use during the calibration process. Defining a minimal set of samples becomes increasingly important when the images have to be controlled under multitude of variables, particularly for production printers, under TRC-based control methodologies (e.g., colors, halftones, media, printers, etc.). The amount of unnecessary measurement time spent by the printer and the sensing system can be otherwise utilized for making prints useful to the customer. Minimizing the number of samples reduces the computation, time, toner/ink usage, and memory requirements of the system, as well as the spatial extent of the test patches when the test patches are rendered in the IDZ areas. This results in a reduction in cost and enables more frequent/efficient sampling, particularly when the method corrects time-dependent print quality errors in production color printers.

In this section, a theoretically optimal (or minimal) test patch selection algorithm is developed. The method uses techniques outlined in Ref. [53], in which a measured TRC is approximated in terms of a linear combination of basis functions obtained from experimental data. It can be used to model the imaging system with respect to the actuators required for producing compensating TRCs. The basis vectors which describe the characteristics of the printer are used to determine the optimum minimal set of gray levels or colors for use in calibration or TRC control process [54].

For simplicity, let us assume that the test patches are gray levels between 0 and 255 digital count. It is also assumed that the output drift in the system can be modeled as the weighted sum of K eigenvectors,

$$o = \sum_k a(k)v(k) + n \tag{9.130}$$

where
   o is the above-mentioned output drift vector with G entries
   G being the number of gray levels
   $v(k)$ is the kth eigenvector $(G \times 1)$
   $a(k)$ is the kth model parameter or weight
   n is the error vector with independent identical distribution

Equation 9.130 can also be written in matrix form as

$$o = Va + n \tag{9.131}$$

where V is an eigenvector matrix of size $G \times K$ whose kth column is $v(k)$, and a is $K \times 1$, whose kth entry is $a(k)$. The eigenvector matrix V is constructed via singular value decomposition (SVD) (Section 3.10) on numerous TRC data samples obtained from a series of controlled experiments in the laboratory, or obtained from machine population in the field. The parameter vector a can be estimated by the least square method as

$$a^* = \left(H^T H\right)^{-1} H^T y \tag{9.132}$$

where
   superscript T denotes matrix transpose
   H is an $S \times K$ matrix
   S is the number of sample patches used in estimation
   y is the data vector $(S \times 1)$ that is obtained from experimental data during the
      runtime control/calibration process

The rows of H are selected from the V matrix rows. Specifically, if the ith patch has a gray level of $g(i)$, the ith row of H is then a replica of the $g(i)$ th row of matrix V.

If the mean-square error is defined as

$$E = E\left[(o^* - o)^{\mathrm{T}}(o^* - o)\right] \tag{9.133}$$

where o^* is the estimation of o and $E[\cdot]$ stands for the expectation operator, then the problem of gray-level patch selection for measurement purpose can be formulated as a method to find the matrix H (or equivalently to find $g(i)$ for $i = 1, 2, \ldots, S$) that yields the least error E. There are three factors that contribute to the error E: (1) the noise n, (2) the low rank approximation if not all of the eigenvectors are used, and (3) the estimation error of a. The first two of these factors are not related to the gray-level sample selection and, therefore, only the estimation error of a is important.

The set of gray levels used for measurement that possess the maximum dispersion provides the minimum error E which is equivalent to minimizing the trace of a matrix expressed as

$$\text{Optimal gray levels} = S = \min_S \left\{ \text{tr}\left[H^{\mathrm{T}}H\right]^{-1} \right\} \tag{9.134}$$

For example, if the $G \times 1$ vector e is defined as the error introduced by estimation inaccuracy, it can be evaluated as

$$e = V \Delta a \tag{9.135}$$

where $\Delta a = a^* - a$ is a $K \times 1$ estimation error vector. Note that V is orthonormal, estimation a^* is unbiased, and the expected error energy can be evaluated as

$$E\left[e^{\mathrm{T}}e\right] = E\left[\Delta a^{\mathrm{T}}V^{\mathrm{T}}V\Delta a\right] = E\left[\Delta a^{\mathrm{T}}\Delta a\right] = \sum_k \text{Var}[a^*(k)] \tag{9.136}$$

The covariance matrix of a^*, as described in Ref. [54] can be evaluated as

$$\text{Cov}(a^*) = c\left(H^{\mathrm{T}}H\right)^{-1} \tag{9.137}$$

where c is a constant, yielding

$$\sum_k \text{Var}[a^*(k)] = \text{tr}[\text{Cov}(a^*)] = \text{tr}\, c\left[\left(H^{\mathrm{T}}H\right)^{-1}\right] \tag{9.138}$$

To reduce the computation time, the potential candidates are first pruned. In matrix V, if two columns are proportional, specifically, if $v_{kj} = \alpha v_{ki}$ for $k = 1, 2, \ldots, K$, and the proportional factor α is less than 1, column j will be eliminated from the selection process. This is due to the fact that if column j were selected, replacing it with column i will always reduce $\text{tr}[(H^{T}H)^{-1}]$. Therefore, column j can be removed from the matrix. When S is small, a brute force search may find the sample gray levels that minimize $\text{tr}[(H^{T}H)^{-1}]$. The calculation is to compute $\text{tr}[(H^{T}H)^{-1}]$ R^{S} number of times (assuming two or more samples may share the same gray level), where R is the number of the rows in V that have survived pruning. When S is larger, an iterative method such as Newton–Raphson or a genetic method may be applied. One simple method is as follows:

a. Select an initial set of gray levels $g(1)$, $g(2)$, ..., $g(S)$.
b. For each iteration, perform steps c and d below.
c. For $i = 1$ to S:
   Calculate the change of $\text{tr}[(H^{T}H)^{-1}]$ with the replacement of $g(i)$ with $g(i-1)$.
   Calculate the change of $\text{tr}[(H^{T}H)^{-1}]$ with the replacement of $g(i)$ with $g(i+1)$.
d. From the calculation in step c, find the one that minimizes $\text{tr}[(H^{T}H)^{-1}]$. The iteration ends when no replacement can make any improvement. Otherwise, update the H matrix based on another set of S gray levels and repeat step c until there are no more remaining sets of S gray levels to select.

The calculation in step c can be substantial, especially when K (model rank) is large, because inversion is computationally expensive for large matrices. However, we can take advantage of the fact that, on each pass, only one vector in H is replaced, and the calculation can therefore be simplified. If U denotes the updated matrix of $H^{T}H$ after $g(i)$ is replaced by $g(j)$, then

$$U^{-1} = \left(H^{T}H - h(i)h(i)^{T} + h(j)h(j)^{T}\right)^{-1} \tag{9.139}$$

where $h(i)$ is the ith column of V. Using the Woodbury identity twice, Equation 9.139 can be simplified as

$$U^{-1} = W^{-1} - \left[W^{-1}h(j)h(j)^{T}W^{-1}\right] / \left[1 + h(j)^{T}W^{-1}h(j)\right] \tag{9.140}$$

where

$$\begin{aligned}
W^{1} &= \left(H^{T}H - h(i)h(i)^{T}\right)^{-1} \\
&= \left[\left(H^{T}H\right)^{-1}h(i)h(i)^{T}\left(H^{T}H\right)^{-1}\right] / \left[1 - h(i)^{T}\left(H^{T}H\right)^{-1}h(i)\right]
\end{aligned} \tag{9.141}$$

Since $(H^T H)^{-1}$ is known from the results of the final step in the iteration, the above calculation essentially contains only two matrix–vector multiplications. No matrix inversion is performed.

In many typical control/calibration applications, the eigenvectors are smooth functions, and the selection process may be conducted in a hierarchical manner to further reduce computation. Also, intuition can be used in selecting combinations of gray levels so that the number of combinatorial searches required can be minimized. Specifically, the optimization procedure first executes the above-described procedural steps a through d with a reduced resolution. The solution for the reduced resolution is used as the initial condition to run the same procedural steps a through d again for the full resolution calculation.

Example 9.13

As an example, we find the optimal gray levels by solving Equation 9.134 for spatial TRC data. The quality of TRCs derived from equispaced levels is compared with the quality derived from levels determined by using the optimal gray level method. A system is assumed with a small number of gray levels (32, rather than 256) to make the demonstration tractable with a reasonable amount of effort and a small number of test prints and measurements.

A first printing is performed using all of the 32 gray levels in a calibration target where a gray strip for each level spans across the page (Figure 9.53a). Measurements are taken of the spatial nonuniformity using a scanner for each gray level to obtain spatial TRC data. These first printing and measurement results are used to

1. Derive TRC eigenvector matrix that characterize the system. In our example, a reduced set of two basis vectors is selected as the system characterization. Prior experience shows that this number is adequate when significant drift in the print engine response is not present. More basis vectors can be included depending on the level of accuracy needed.
2. Derive the optimal gray levels for future calibration updates. Optimal gray levels are derived for three separate cases: 2 levels, 4 levels, and 8 levels.

A second printing of the 32 levels is performed and used for the following steps:

1. Complete print engine state (spatial TRCs) is measured for $t > 0$. These measurements serve as a reference for comparison of TRCs derived from subsets of gray levels.
2. TRCs are generated using several equispaced subsets of gray levels, namely 2, 4, and 8 levels.
3. TRCs are generated using the optimal levels derived above.
4. Derived TRCs are compared to the fully measured TRCs.

The gray levels used in the procedure are provided in Table 9.3 below.

The error is calculated by first deriving all gray levels (32) from a given subset through the use of basis vectors. The derived 32-level TRCs are then

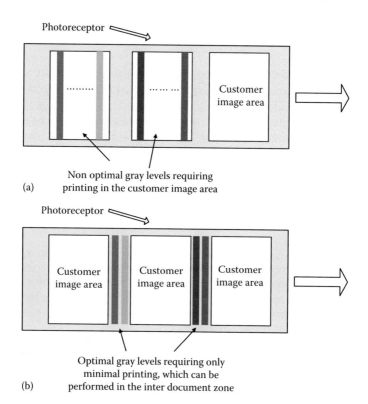

FIGURE 9.53 Photoreceptor showing calibration strips for on-belt sensing for calibrating streak compensating TRCs.

TABLE 9.3
Optimal and Equispaced Gray Levels (0–255)

Number of Patches	Optimal Values	Equispaced Values
2	137, 233	65, 193
4	137, 169, 225, 233	33, 97, 161, 225
8	113, 137, 153, 169, 201, 225, 233, 241	17, 49, 81, 113, 145, 177, 209, 241

compared to the measured 32 levels. For each subset, the mean-square error (MSE) over the 32 levels is plotted.

In Figure 9.54, the MSE for each of the 2-, 4-, and 8-level subsets is calculated across the spatial dimension for all gray levels. The resulting data points are plotted and it can be observed that the optimal-set error incurred by using the optimal gray levels is roughly half of the equispaced-set error incurred by using the equispaced levels.

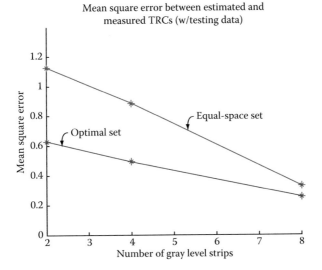

FIGURE 9.54   Mean-square error from the two gray-level selection methods for 2, 4, and 8 patches.

PROBLEMS

9.1 Let the state-space equations for the level 1 electrostatic system be

$$x(k+1) = \begin{bmatrix} 1 & 0 \\ 0 & 1 \end{bmatrix} x(k) + \begin{bmatrix} 0.9798 & 0 \\ 0.4574 & 14.9315 \end{bmatrix} v(k)$$

$$y(k) = \begin{bmatrix} 1 & 0 \\ 0 & 1 \end{bmatrix} x(k)$$

Use the control law shown in Equation 9.26 with the feedback gain matrix given as $K = \begin{bmatrix} 0.8165 & 0 \\ -0.025 & 0.0469 \end{bmatrix}$.

a. Show the evolution of the states for $x_d = [-550 \ -100]^T$ and $x(0) = [-588 \ -87]^T$ with respect to 10 pitch numbers.

b. Show that the steady-state error is held to zero, even when one of the elements of the Jacobian matrix, say b_{22}, is lowered by 20%.

c. Calculate the steady-state values of the actuator vector, $v(k)$, for (a) and (b).

9.2 Determine the controllability and observability matrices and then ranks for Problem 9.1. What are the eigenvalues of the open-loop system with and without the integrator in the loop? Is the system open loop stable in both cases? Close the loop with the gain matrix of Problem 9.1. Determine the eigenvalues. Comment on the stability.

9.3 Consider the electrostatic control system of Figure 9.4. The charging system is characterized by the Jacobian matrix B, at the nominal operating point.

a. Write the open- and closed-loop transfer functions of the system. Use any convenient signals as inputs and outputs.

b. Determine the steady-state error between the desired exposed and unexposed voltages to the measured voltages.

c. Replace the integrator with the unity gain. Determine the open and closed-loop transfer functions for the new structure.

d. Assume the gain matrix as $K = B^{-1}$ when the integrator is replaced with the unit gain. Determine the steady-state error.

e. Compare the results of b with d. Which controller gives the lowest steady-state error?

f. If all the elements of matrix B are increased by 20% and the gain matrix, K, is equal to that used in d, determine the percentage change in steady-state errors with and without integrators in the loop.

g. If B matrix has not changed, but the voltages at the nominal operating point, V_{ho} and V_{lo}, have changed by 20%, determine the percentage change in steady-state errors with and without the integrators.

9.4 Consider the development control system of Example 9.5. The development system is characterized by the Jacobian matrix B, at the nominal operating point. Find the expression for the steady-state actuator U. Is the steady-state actuator vector dependent on the gain matrix? Explain the reasons for your answer.

9.5 The cost of an electrostatic sensing system can be reduced by removing the charge sensor (ESV) and associated level 1 control loop. It is now required to design a three-input three-output developability control loop using three process actuators, the grid voltage, $U_g(k) = V_g(k)$, the exposure intensity, $U_1(k) = X(k)$, and the development bias, $U_b(k) = V_{bias}(k)$, and measurements from the DMA sensor. Let $D_1(k)$, $D_m(k)$, and $D_h(k)$ represent the three different DMA measurements shown in the state vector, $x(k)$, measured every pitch, indicated by the parameter, k. The linear state variable description of the control system is characterized by the Jacobian matrix at the nominal operating point $\{U_{go} = -600 \text{ V}, U_{lo} = 4 \text{ ergs/cm}^2, U_{bo} = -200 \text{ V}\}$ as shown below:

$$x(k+1) = \begin{bmatrix} 1 & 0 & 0 \\ 0 & 1 & 0 \\ 0 & 0 & 1 \end{bmatrix} x(k) + 10^{-3} \begin{bmatrix} 0.06813 & -0.06047 & 5.69587 \\ 0.27346 & -0.23242 & 21.52804 \\ 1.55592 & -1.43732 & 94.84767 \end{bmatrix} v(k)$$

$$y(k) = \begin{bmatrix} 1 & 0 & 0 \\ 0 & 1 & 0 \\ 0 & 0 & 1 \end{bmatrix} x(k)$$

where, $x(k) = \begin{bmatrix} D_l(k) \\ D_m(k) \\ D_h(k) \end{bmatrix}$; $v(k) = \begin{bmatrix} \Delta U_g(k) \\ \Delta U_i(k) \\ \Delta U_b(k) \end{bmatrix}$; $y(k) = x(k)$. The controller uses a

gain matrix and the integrator modeled by the following Equation:

$$e(k+1) = e(k) - Bv(k)$$
$$v(k) = +Ke(k)$$

where $e(k) = x_d - x(k)$ and x_d is the desired state vector.

a. Find the gain matrix for placing the poles at location [0.2, 0.2, 0.2] using Equation 9.45.

b. Show the time evolution of the states for a step response with respect to pitch number using recursive solution of Equation 4.118, the measurement-actuation updates are executed at every pitch.

c. Compare the results of step b by running the control simulation with the charge, development models from Chapter 10.

d. Are there other reasons why this type of control approaches are more suitable than level 1 and 2 architecture? Comment.

9.6 For a level 2 controller (Figure 9.8), design the antiwindup compensator gain matrix and show the performance with and without the compensator.

9.7 Let the vector U_0 contain the input patch gray levels (or area coverages) for controlling ΔE_{ab}^* from paper for magenta separation shown in Table 9.4. They are shown in the second column. In the third column, the vector, x, contain the normalized values transformed to gray levels of the ΔE_{ab}^* from paper measurements are shown. Use identity reference TRC and direct inversion process. Process magenta separated image through the inverted TRC and printer model.

TABLE 9.4
ΔE_{ab}^* from Paper (Normalized) with Respect to Gray Levels

Number of Patches	U_0	ΔE_{ab}^* with Respect to Paper
0	0.00	0.00
1	26.00	8.37
2	51.00	26.20
3	77.00	43.73
4	102.00	75.77
5	128.00	88.76
6	153.00	128.61
7	179.00	161.31
8	204.00	203.92
9	230.00	242.89
10	255.00	255.00

9.8 Obtain an inversion using control-based inversion for Problem 9.7. Use identity as the reference TRC. Process magenta separated image through the inverted TRC and printer model.

9.9 Consider a scenario where the level 1 loop of Example 9.1 is closed for every print and the level 2 loop of Example 9.6 is closed for every 5 prints. Let level 1 loop be designed for a dead beat response (i.e., closed-loop eigenvalues $= 0$) and level 2 loop has the following closed-loop poles:
a. $p_1 = 0$, $p_2 = 0$, $p_3 = 0$
b. $p_1 = -0.5$, $p_2 = -0.5$, $p_3 = -0.5$
Give a step change in DMA targets. Sketch the expected convergence plots for case a and b for state variables $\{V_h, V_l, D_h, D_m, D_l\}$ at every print for a total of 30 prints.

9.10 Design the gain matrix for the PI controller of Example 9.10 using the LQR method. Compare your TC performance with the pole-placement technique. Comment on the results.

9.11 For TC system with PI controller $\tilde{u}_{min} = 0$ and $\tilde{u}_{max} = $ a reasonable limit, simulate the situation with integrator windup. Design a antiwindup compensator with high gain L and simulate the integrator recovery and the TC performance with antiwindup compensator.

9.12 Simulate the TC performance for a system with the state feedback controller and estimator in Example 9.10 when the TC system has a fixed transport lag of 3 and 10 cycles. Apply the Smith predictor and a PI controller and simulate the performance again. Comment on the results.

9.13 Describe the relationship between TC control and a typical inventory management system. Compare the controllers.

9.14 For a four-input three-output controller with TC target as one of the actuator, design the LQR to emphasize the TC target. Show the equation for the steady-state value and plot it as a function of weights used in TC target, with all other weights unchanged.

REFERENCES

1. L.K. Mestha, Control advances in production printing and publishing systems, *NIP20: Proceedings of IS&T's International Congress on Digital Printing Technologies*, pp. 578–585, Salt Lake City, UT, Oct. 31–Nov. 5, 2004.
2. T.E. Thieret, T.A. Henderson, and M.A. Butler, Method and control system architecture for controlling tone reproduction in printing device, US Patent 5,471,313, Nov. 28, 1995; Other US Patents 5,087,940; 5,185,673; 5,189,441; 5,231,452; 5,255,085; 5,345,315; 5,461,462; 7,162,172.
3. L.K. Mestha, Y.R. Wang, M.A. Scheuer, and T.E. Thieret, A multilevel modular control architecture for image reproduction, in *Proceedings of the IEEE International Conference on Control Applications*, Trieste, Italy, Sep. 1–4, 1998.
4. P. Varaiya, Towards a layered view of control, UC Berkeley, CA 94720.
5. L.K. Mestha and P. Padmanabhan, Electrostatic control with compensation for coupling effects, US Patent 5,754,918, May 19, 1998.

6. M.A. Parisi, Image forming apparatus with predictive electrostatic process control system, US Patent 5,243,383, Sep. 7, 1993; Other US Patents 6,426,630; 6,771,912; 6,792,220.

7. T.A. Hayes, Jr., Charge control system, US Patent 4,456,370, Jun. 26, 1984.

8. B.T. Williams, High speed electrostatic voltmeter, US Patent 4,205,267, May 27, 1980; Other US Patents 3,870,968; 4,205,257; 4,330,749; 4,804,922; 4,853,639; 4,853,639; 4,868,907; 4,928,057; 5,212,451; 5,227,270; 5,270,660; 5,323,115; 6,166,550; 6,320,387.

9. T.A. Henderson, System for identifying areas in pre-existing image data as test patches for print quality, US Patent 5,450,165, Sep. 12, 1995; Other US Patents 4,684,243; 5,060,013; 5,253,018; 5,266,997; 5,315,352; 6,697,582.

10. L.K. Mestha, Method for measurement of tone reproduction curve using a single structured patch, US Patent 5,543,896, Sep. 13, 1995; Other US Patents 5,652,946; 6,167,217; 6,204,869.

11. L.K. Mestha, Method to model a xerographic system, US Patent 5,717,978, Feb. 10, 1998. Other US Patents 5,243,383; 5,481,337; 5,523,831; 5,606,395.

12. J. Kautzky and N.K. Nichols, Robust eigenstructure assignment in state feedback control, Numerical Analysis Report NA/2/83, School of Mathematical Sciences, Flinders U., Bedford Park, SA 5042, Australia. Algorithm coded in Control System Toolbox for use with MATLAB, The Math Works, MA, 1992.

13. K. Ogata, *Discrete-Time Control Systems*, 2nd edn., Prentice Hall, Upper Saddle River, NJ, 1995; T. Kailath, *Linear Systems*, Prentice Hall, Englewood Cliffs, NJ, 1980.

14. F.F. Hubble, III and J.P. Martin, Infrared reflectance densitometer, US Patent 4,553,033, Nov. 12, 1985; Other US Patents 4,054,391; 4,200,391; 4,553,033; 4,568,191; 4,796,065; 4,986,665; 4,989,985; 5,122,835; 5,173,750; 5,204,538; 5,519,497; 6,229,972; 6,331,832; 6,384,918; 6,633,382; 7,122,800; 7,262,853.

15. T.P. Sim and P.Y. Li, Optimal time-sequential sampling of tone reproduction function, in *Proceedings of the 2006 American Control Conference*, Minneapolis, MN, Jun. 14–16, 2006.

16. G.B. Raj, Adaptive process controller for electrophotographic printing, US Patent 5,436,705, Jul. 25, 1995; Other US Patents 4,113,371; 4,866,481; 5,749,023; 6,034,703; 6,741,816.

17. L.K. Mestha, Developed mass per unit area (DMA) controller to correct for development errors, US Patent 5,749,021, May 5, 1998; Other US Patents 4,348,099; 4,639,117; 4,989,043; 5,122,842; 5,383,005; 5,416,564; 5,436,705; 5,812,903; 5,873,010; 6,035,152; 611,556; 6,198,886; 6,285,840; 6,694,109; 6,741,816; 7,024,126; 7,162,187; 7,203,433.

18. L.J. Fantozzi, Development control of a reproduction machine, US patent 4,341,461, Jul. 27, 1982; Other US Patents 4,003,650; 4,113,371; 4,256,401; 4,866,481; 4,965,634; 4,967,211; 5,045,882; 5,200,783; 5,250,988; 5,544,258; 6,741,816.

19. S. Kobayashi, K. Sato, and K. Ono, Image recording apparatus for controlling image in high quality and image quality control method, US Patent 5,576,811, Mar. 16, 1995.

20. S. Kobayashi, K. Sato, and K. Ono, Image recording apparatus for controlling image in high quality and image quality control method thereof, US Patent 5,576,811, Nov. 19, 1996; Other US Patents 7,162,169; 7,221,477; 7,246,005; 7,239,819.

21. T. Nakai and Y. Maebashi, Image forming apparatus with transfer voltage control for transferring toner patterns, US Patent 6,564,021, May 13, 2003.

22. S.R. Schmidt and R.G. Launsby, *Understanding Industrial Designed Experiments*, AIR Academy Press & Associates, ISBN 1-880156-03-2, Colorado Springs, CO, 2005.

23. L.K. Mestha, Y.R. Wang, S. Dianat, E. Jackson, T. Thieret, P.P. Khargonekar, and D.E. Koditschek, Toward a control oriented method of xerographic marking engine, in *Proceedings of IEEE Conference on Decision and Control*, Kobe, Japan, Dec 11–13, 1996.

24. F.L. Lewis, *Applied Optimal Control & Estimation: Digital Design & Implementation*, Prentice Hall, Englewood Cliffs, NJ, ISBN 0-13-040361-X, 1992.

25. C.L. Phillips and H.T. Nagle, Jr., *Digital Control System Analysis and Design*, Prentice Hall, Englewood Cliffs, NJ, ISBN 0-13-212043-7, 1984.

26. P.A. Crean and M.S. Maltz, Marking engine and method to optimize tone levels in a digital output system, US Patent 6,643,032, Nov. 4, 2003.

27. L.K. Mestha and S. Dianat, TRC smoothing algorithm to improve image contours in 1-D color controls, US Patent 7,397,581, Jul. 8, 2008.

28. P.M. DeRusso, R.J. Roy, C.M. Close, and A.A. Desrochers, *State Variables for Engineers*, 2nd edn., John Wiley & Sons, New York, ISBN 0-471-57795-2, 1998.

29. J.H. Hubbard, G.W. Van Cleave, Toner concentration control apparatus, US Patent 4,032,227, Jun. 28, 1977.

30. D.R. Rathbun, M.D. Borton, and J. Buranicz. Toner concentration sensing apparatus, US Patent 5,166,729, Nov. 24, 1992; Other US Patents 3,572,551; 3,604,939; 3,830,401; 3,876,106; 3,970,036; 4,706,032; 4,962,407; 4,974,025; 4,980,727; 5,012,286; 5,111,247; 5,166,729; 5,353,103; 5,581,335; 5,895,141; 5,812,903.

31. P.J. Donaldson, Toner age calculation in print engine diagnostics, US Patent 6,047,142, Apr. 4, 2000; P. Ramesh, Quantification of toner aging in two component development systems, *NIP21: Proceedings of IS&T's International Conference on Digital Printing Technologies*, pp.544–547, Baltimore, MD, Sep. 18–23, 2005.

32. J.M. Pacer, B.C. Casey, G.F. Bergen, P.J. Weber, W.M. Ouyang, V. Lopez-Heroux, Automatic compensation for toner concentration drift due to developer aging, US Patent 5,410,388, Apr. 25, 1995.

33. R.E. Grace, Signature sensing for optimum toner control with donor roll, US Patent 5,887,221, Mar. 23, 1999.

34. M.A. Scheuer, J. Buranicz, P.J. Donaldson, P.A. Garsin, E.M. Gross, E.S. Hamby, D.W. MacDonald, P. Padmanabhan, E.W. Smith, Jr., and J.W. Ward, Toner concentration control for an imaging system, US Patent 6,175,698, Jan. 16, 2001; Other US Patents 4,369,733; 4,419,010; 4,514,480; 4,734,737; 4,829,336; 4,901,115; 5,710,958; 6,160,970; 6,160,971; 6,175,698.

35. Y.R. Wang and L.K. Mestha, Filter for reducing effect of noise in TC control, US Patent 5,839,022, Nov. 17, 1998.

36. P.J. Donaldson, J. Buranicz, P.A. Garsin, E.S. Hamby, D.W. MacDonald, M.A. Scheuer, E.W. Smith, Jr., and E.M. Gross, Feedback toner concentration control for an imaging system, US Patent 6,173,133, Jan. 9, 2001; Other US Patents 3,756,192; 4,326,646; 4,572,102; 4,980,727; 5,678,131; 6,167,213; 6,167,214; 6,169,861; 6,321,045; 6,498,909; 6,580,882; 6,718,147; 7,298,980.

37. K. Gu, V.L. Kharitnov, and J. Chen, *Stability and Robust Stability of Time-Delay Systems*, Birkhauser, Boston, MA, 2003.

38. R. Balakrishnan and G.M. Butler, Control system for sheetmaking, US Patent 5,121,332, Jun. 9, 1992; Other US Patents 5,144,549; 5,777,872; 5,892,679; 6,052,194.

39. A. Nortcliffe and J. Love, Varying time delay Smith predictor process controller, *ISA Transactions*, 43(1), 61–71, Jan. 2004.

40. S.H. Ahn, B.S. Sim, D.Y. Chi, K.N. Park, C.Y. Lee, and Y.J. Kim, Smith predictor control for water pressure control system with time delay, *Industrial Electronics Society, 2004. IECON 2004. 30th Annual Conference of IEEE*, 1(2–6), 663–666, Nov. 2004.

41. B. Friedland, *Control System Design: An Introduction to State-Space Method*, McGraw-Hill Primis Custom Publishing, New York, ISBN 0-07-286376-5, 2002.

42. D.P. DeBoer and D.W. Jiral, Anti-windup proportional plus integral controller, US Patent 5,298,845, Mar. 29, 1994.

43. P. Hippe, *Windup in Control: Its Effect and Their Prevention*, Springer, London, ISBN 1-84628-322-1, 2006.
44. P.Y. Tiwari, E.F. Mulder, and M.V. Kothare, Multivariable anti-windup controller synthesis incorporating multiple convex constraints, *ACC*, 9–13, 5212–5217, Jul. 2007.
45. S. Tarbouriech, I. Queinnec, and G. Garcia, *Advanced Strategies in Control Systems with Input and Output Constraints*, Springer, Berlin/Heidelberg, ISBN 978-3-540-37009-3, 2007.
46. J. Sofrony, M.C. Turner, and I. Postlethwaite, Anti-windup synthesis using Riccati equations, *International Journal of Control*, 80, 112–128, 2007.
47. L. Zaccarian, D. Nesic, and A.R. Teel, L-2 anti-windup for linear dead-time systems, *Systems & Control Letters*, 54, 1205–1217, 2005.
48. L. Zaccarian and A.R. Teel, Nonlinear scheduled anti-windup design for linear systems. *IEEE Transactions on Automatic Control*, 49, 2055–2061, 2004.
49. S.-F. Mo, J.R. Wagner, W.K. Apton, S.F. Randall, P.G. Medina, and P.J. Walker, Method and system for using toner concentration as an active control actuator for TRC control, US Patent 7,158,732, Jan. 2, 2007.
50. L.K. Mestha, S.B. Bolte, E.S. Saber, and S.P. Updegraff, Systems and methods for obtaining a spatial color profile, and calibrating a marking system, US Patent 7,295,340, Nov. 13, 2007.
51. H.A. Mizes, Systems and methods for compensating for streaks in images, US Patent 7,347,525, Mar. 25, 2008.
52. M. Sampath, H.A. Mizes, and S. Zoltner, Method and system for automatically compensating for diagnosed banding defects prior to the performance of remedial service, US Patent 7,400,339, Jul. 15, 2008.
53. L.K. Mestha, Y.R. Wang, S.A. Dianat, D.E. Koditschek, E. Jackson, and T.E. Thieret, Coordinitization of tone reproduction curve in terms of basis functions, US Patent 5,749,020, May 5, 1998.
54. Z. Fan, L.K. Mestha, Y.R. Wang, R. Loce, and Y. Zhang, Optimal test patch level selection for systems that are modeled using low rank eigen functions, with applications to feedback controls, US Patent Application, 20070140552, Jun. 21, 2007.

10 Printing System Models

10.1 INTRODUCTION

The xerographic printing process is a unique discipline that incorporates domain-specific ideas from physical sciences to engineering. Developing a complete model of the printer is very difficult and can limit its use for analysis purposes. We therefore present a nonlinear model of the printing system with reasonable abstractions of key elements of the digital color printing processes and yet retain the simplicity necessary to study the effects of control techniques on each of these processes. We first develop the underlying physics of the processes used for developing a colored dot that is fundamental to the creation of digital images for the color *CMYK* process. The process models are then cascaded in sequence where the output of one process model becomes the input to the next one. The dot spread is then modeled using a halftoning strategy and modulation transfer functions of the key segments of the electrophotographic process (EP). These models are used for designing feedback controllers for each of the major subsystems, such as controllers for generating multidimensional profiles, to understand their interactions, to manage the complexity of the system through careful design of control loops, and to achieve the overall system objective. We present the transformations that an electronic image goes through, before being reproduced on paper using analytical abstractions of the printing processes.

10.2 PROCESS MODELS

There are at least six fundamental steps to monochrome (B/W) digital EP printing. Color digital printing carried out in several different architectures [1,2] involves different combinations of these steps. The photoconductor is charged in the "charging" station, and the electrostatic latent image is formed on the photoconductor in the "exposure" station. Then, in the next step, the latent image is rendered into a real visible image on the photoconductor in the "development" station using electrostatically charged toner cloud. The developed latent image is transferred to the media in a "transfer" station. The transferred image is fused to the media by heat and mechanical pressure in the "fusing" station. The residual toner on the photoconductor is removed in the "cleaning" station. While each of the six stations of EP process is critical to the proper functioning of the monochrome printing, we ignore the cleaning station for our modeling purposes. The remaining five steps are modeled as nonlinear lumped localized transfer functions (LTF models) with actuators as key inputs, and sensed or measured parameters as key outputs. The LTF models capture only the local aspect of the color "dot printer." The spatial aspects (i.e., dot growth, dot spread, edge enhancement) of these processes are captured using modulation transfer function (MTF) models.

10.2.1 CHARGING MODEL

With the aid of a corotron or a scorotron device (Figures 10.1 and 10.2), charge is uniformly deposited on the photoconductor surface (in the dark). In the corotron setup, corona wires are placed above a photoconductive plate, whereas in the scorotron system, a metallic grid is inserted between the corotron wires and the photoconductor. When electric field strength in the vicinity of a thin corotron wire becomes sufficiently large to ionize the gas molecules in the air surrounding the photoconductor, corona discharge occurs. The corona discharge produces ionized gas molecules that flow to the surface of the photoconductor which absorbs them, generating a charge distribution on its surface. A uniform surface charge is created on the whole photoconductor when it moves with a constant velocity through a stationary corotron that is steadily emitting corona ions. Corona wires are generally maintained at a high potential ($\sim$5–10 kV) throughout the charging process. To improve the uniformity of the charge distribution in the scorotron, a metallic grid is

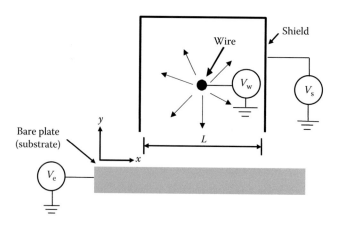

FIGURE 10.1 Corotron charging system.

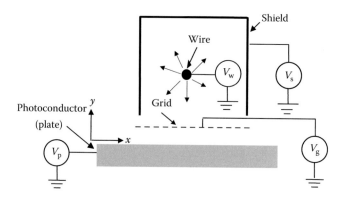

FIGURE 10.2 Scorotron charging system.

introduced between the corona wires and the dielectric surface. The grid is biased to a voltage that approximates the potential required for charging the photoconductor (typically -600 to -900 V) which provides better control over the amount of charge on the photoconductor surface.

As charge builds up on the surface of the photoconductor, its voltage increases and the current flowing to the photoconductor decreases. When this current reaches zero, charging stops. If V_p is the photoconductor surface voltage, then the charging rate in a scorotron charging system can be expressed as the plate current per unit length, which is given by

$$I_p = S(V_g - V_p) \qquad (10.1a)$$

where
I_p is the plate current per unit length [A/m]
V_p is the plate voltage
V_g is the grid voltage
S is the slope of the current–voltage response curve

See Figure 10.3 for an ideal current–voltage response curve [3] and Figure 10.4 for an actual response curve shown for a typical scorotron charging system when the wire potential is held at 6 kV. Photoconductor current density (the surface current per unit area) as a function of photoconductor position is shown in Figure 10.5.

Characterizing the PR by its capacitance per unit area C $(C/V\text{-}m^2)$ given by

$$\frac{\varepsilon_r \varepsilon_0}{d} \qquad (10.1b)$$

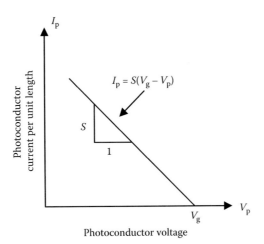

FIGURE 10.3   Idealized current vs. voltage response curve for the scorotron charging system constructed in the nominal operating regime.

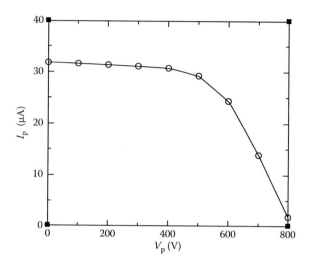

FIGURE 10.4 Actual current vs. voltage response curve for the scorotron charging system ($V_w = 6000$ V, $V_g = 800$ V, and $V_s = 0$).

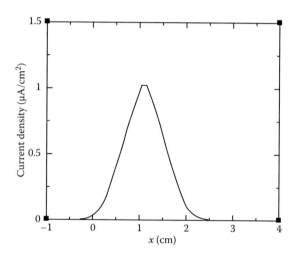

FIGURE 10.5 Current density with respect to photoconductor position.

where
    ε_r is the PR (photoreceptor) relative permittivity
    ε_0 is the free air permittivity
    d is the PR thickness

Equation 10.1a becomes

$$C\frac{dV_p}{dt} = J(x) = \frac{I_p}{L}j(x) = S(V_g - V_p)\frac{1}{L}j(x) \tag{10.2}$$

where

  L is the charging nip length

  $j(x)$ is the current profile of the bare plate current that satisfies

$$\frac{1}{L} \int_0^L j(x)dx = 1 \tag{10.3}$$

When the photoconductor is moved by a distance x at a velocity v m/s, Equation 10.2 becomes

$$Cv\frac{dV_p}{dx} = S(V_g - V_p)\frac{1}{L}j(x) \tag{10.4}$$

By rewriting Equation 10.4, we get

$$\frac{dV_p}{V_g - V_p} = \frac{S}{Cv} \cdot \frac{1}{L}j(x)dx \tag{10.5}$$

Integrating Equation 10.5 (the left-hand side from $V_p(0)$ to $V_p(L)$ and the right-hand side from 0 to L), we get

$$\int_{V_p(0)}^{V_p(L)} \frac{dV_p}{V_g - V_p} = \frac{S}{Cv}\frac{1}{L}\int_0^L j(x)dx = \frac{S}{Cv} \tag{10.6}$$

Therefore,

$$-\ln(V_g - V_p)\Big|_{V_p(0)}^{V_p(L)} = \frac{S}{Cv} \tag{10.7}$$

Applying the upper and lower limits, we have

$$\ln\frac{V_g - V_p(0)}{V_g - V_p(L)} = \frac{S}{Cv} \tag{10.8}$$

Solving Equation 10.8 for $V_p(L)$ results in

$$V_p(L) = V_g(1 - e^{-a}) + V_p(0)e^{-a} \tag{10.9}$$

where

  $V_p(0)$ is the initial plate voltage

  $a = \frac{S}{Cv}$

This model is linear with respect to the input voltage V_g with a constant gain of $(1 - e^{-a})$. Moreover, there is no dependence on the charging nip length because

TABLE 10.1
Charging Model

$$V_p(L) = V_g(1 - e^{-a}) + V_p(0)e^{-a} \qquad (10.9)$$

where

$a = \frac{S}{Cv}$

$S =$ slope of the current–voltage response curve

$C =$ capacitance per unit area

$v =$ photoconductor velocity

$L =$ charging nip length

TABLE 10.2
Parameters of Charging Model

Charging Model Parameters	Units
S	0.94×10^{-6} A/V-m
v	0.254 m/s
V_g	600–800 V
ε_r	3
ε_0	8.854×10^{-12} C/V-m
d	24 μm
C	9.486×10^{-7} C/V-m^2
a	3.9013
$V_p(0)$	0 V

of the integration. All uncertainties are ignored by the model; they can be handled as disturbances in the system dynamics when this equation is used in the controller design. More detailed charging models can be found in Refs. [4–7]. A summary of the charging model and its related parameters is shown in Tables 10.1 and 10.2.

10.2.2 EXPOSURE MODEL

A photoconductor acts like an insulator in the dark and a conductor when it is exposed to light. A photoconductor has at the minimum two layers, a charge-transport layer (CTL) and a charge-generation layer (CGL) or photosensitive layer. The cross section of a dual layer PR is shown in Figure 10.6 [7]. In the CGL, photogeneration of charge carriers takes place at the bottom of the layer when the PR is exposed with photons. It has photoconductive material that generates electron–hole pairs in response to the photon exposure. These charge carriers drift and migrate to the top surface, and neutralize the preexisting surface charges (deposited during the charging stage) in the illuminated areas to form the latent electrostatic image. The CTL contains materials that subsequently allow these charge carriers to be transported. The undercoat layer (UCL) is an uncharged layer used to block unwanted

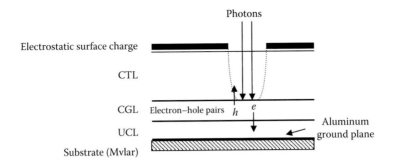

FIGURE 10.6   Cross section of a two-layer PR.

charge. For example, during imaging no change in charge occurs in the surface voltage whenever dark characters exist on the image. Wherever the white background existed, the photoconductor is discharged, thus producing an electrostatic image of the document on the photoconductor. The photoconductor could be a thin film, usually about 25 μ in thickness, or a drum that is used not only for electrophotography in copying machines, printers, and facsimile machines, but also in photoelectric transducers such as solar batteries and electrolytic luminescence elements, photo transducers, and in materials for optical disks. Detailed investigation of the physics of charge-generation and charge-transport properties in photoconductors can be found in Ref. [8]. Our main goal is to develop an exposure model for photoconductors (with CGL and CTL) used in imaging systems such as copying or printing systems that can be used in design of controller for the exposure process.

After the charging process, the charge on the photoconductor decays with time due to light leakage between the charging and exposure stations. The amount of decay depends on the amount of charge and the amount of thermally released holes in the photoconductor. The dark decay period can be different from charging to exposure stations and then from exposure to development stations. This decay rate should be comprehended by the model if the period is long. The time evolution of the surface charge and the photoinduced discharge (PID) is shown schematically in Figure 10.7 [4]. During exposure, the charge generated during PID can spread laterally within the CTL (shown schematically in Figure 10.6) due to electrostatic repulsion resulting in the loss of image resolution. This repulsion is a function of the photoconductor properties like mobility, thickness, etc. [7,9,10].

PID taking place in the CGL is a nonlinear function of exposure light intensity. This nonlinear relationship is commonly called photoinduced discharge curve (PIDC). Next we derive an expression for the PIDC based on the physics of the photoconductor [11–13].

The field-dependent injection of charge from the CGL to CTL is usually explained by the field-dependent collection efficiency (CE) term, η, which gives the probability that an absorbed photon will generate an electron–hole pair that will actually contribute to the discharge of the photoconductor [7]. In the Springett–Melnyk model [11], the field-dependent CE of a photogenerator is taken to be of the form

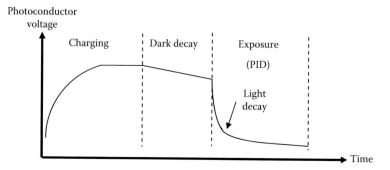

FIGURE 10.7   Photoconductor charge–discharge cycle.

$$\eta(E) = \frac{\eta_0}{1 + \left(\frac{E_c}{E - E_r}\right)^p} \tag{10.10}$$

where
   E is the electric field
   E_c is a scaling parameter that depends on the initial state of the photoconductor
   p is the shape of the field dependence
   η_0 is the high field limit of the CE
   E_r is the residual field

Figure 10.8 shows the Melnyk–Springett CE for different values of p. Given the differential number of photons per unit area (exposure energy per unit area in ergs/cm^2) absorbed by the CGL, dX, the differential charge density produced is

$$d\sigma = e\eta(E)dX \tag{10.11}$$

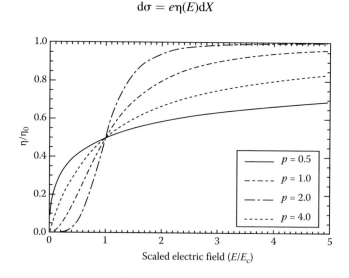

FIGURE 10.8   Melnyk–Springett collection efficiency curves.

where e is the charge of an electron. The positive and negative charges then move to the ground plane and to the surface of the photoconductor and discharge the photoconductor (the polarity depends on the surface charge polarity). By modeling the photoconductor as a simple capacitor with C as the capacitance per unit area, the differential discharge per unit area is given by

$$dV = -\frac{1}{C}d\sigma \tag{10.12}$$

The negative sign indicates that V decreases as X increases (assuming positive charging). If the thickness of the photoconductor is L and its dielectric permittivity is ε, then $C = \varepsilon/L$. Hence, the PIDC is given by

$$\frac{dV}{dX} = -\frac{e}{C}\eta(E) \tag{10.13}$$

Since the photoconductor can be treated as a simple capacitor with all the charge residing at the surfaces, the surface voltage V after exposure and the electric field E are related by $V = EL$. Using this representation in Equation 10.10, Equation 10.13 becomes

$$\left[1 + \left(\frac{V_c}{V - V_r}\right)^p\right]dV = -\frac{e\eta_0}{C}dX \tag{10.14}$$

where $V_c = E_c L = \alpha V_i$ with parameter α as a constant and voltage V_i as the initial voltage on the photoconductor. If the dark decay is zero, then this voltage is same as the photoconductor surface voltage obtained by Equation 10.6 during the charging process. The residual voltage V_r is represented by following expression:

$$V_r = E_r L = V_{r0}\left(\frac{L}{L_0}\right) \tag{10.15}$$

where
   L/L_0 represents the ratio of the photoconductor thickness to a reference photoconductor thickness
   V_{r0} is the residual surface potential under maximum exposure

The ratio L/L_0 can be considered equal to one for our purpose. Integrating Equation 10.14 from $X = 0$ to $X = X$, with corresponding values of left integral from V_i to $V(X)$, we have

$$\int_{V_i}^{V(X)}\left[1 + \left(\frac{V_c}{V - V_r}\right)^p\right]dV = -\frac{e\eta_0}{C}\int_0^X dX = -\frac{e\eta_0}{C}X \tag{10.16}$$

The above integral has a different closed-form solution for different values of the parameter p. The Springett–Melnyk PIDC equations for different values of p are given below.

If $p = 1$, then

$$V(X) - V_i + V_c \ln\left[\frac{V(X) - V_r}{V_i - V_r}\right] + \frac{e\eta_0}{C}X = 0 \qquad (10.17)$$

If $p \neq 1$, then

$$V(X) - V_i + \frac{V_c}{1-p}\left[\left(\frac{V_c}{V(X) - V_r}\right)^{p-1} - \left(\frac{V_c}{V_i - V_r}\right)^{p-1}\right] + \frac{e\eta_0}{C}X = 0 \qquad (10.18)$$

where V_i is the initial voltage. It is not possible to obtain an explicit general form for $V(X)$ in terms of p. However, for the special case of $p = 2$, it is possible to obtain an explicit form. This case is sometimes referred to as the quadratic Melnyk (QM) PIDC model and is given by

$$V(X) = V_r + \beta + \sqrt{\beta^2 + V_c^2} \qquad (10.19)$$

where $\beta = \frac{1}{2}[V_i - V_r - \frac{V_c^2}{V_i - V_r} - \frac{e\eta_0}{C}X]$

It is customary to define a coefficient of exposure as S. Hence,

$$S = \frac{e\eta_0}{C} = \frac{e\eta_0 L}{\varepsilon} \qquad (10.20)$$

The initial slope of the PIDC, that is, dV/dX at $X = 0$ is called the sensitivity of the photoconductor. It is calculated from Equation 10.14 by using $V = V_i$ at $X = 0$:

$$\left.\frac{dV}{dX}\right|_{X=0} = -\frac{S}{1 + \left(\frac{V_c}{V_i - V_c}\right)^p} \qquad (10.21)$$

Generally, in photoconductors of practical interest, $V_c \lll V_i$. This implies that the initial slope of the PIDC is practically independent on the initial voltage V_i, thus all PIDCs start with the same slope regardless of the initial voltage:

$$\left.\frac{dV}{dX}\right|_{X=0} \cong -\frac{S}{1 + \left(\frac{V_c}{V_i}\right)^p} \qquad (10.22)$$

This deficiency in the PIDC model is overcome by modifying the exposure parameter S, so that the model captures the functional form of a PIDC throughout the operating region. The new parameter introduced is $S = LS_0$, where $S_0 = e\eta_0/\varepsilon$ and the ratio L/L_0 can be considered equal to 1 for our purpose. It is important to note, however, that the ratio L/L_0 is significant when photoconductors of varying

thickness are considered. It is especially important while comparing the perform-
ances of various photoconductors to one another:

$$S = S_0\left[1 - \exp\left(-\frac{V_k}{V_i}\right)\right]$$

(10.23)

where $S_0 = S_{00}(\frac{L}{L_0})$ and $V_k = V_{k0}(\frac{L}{L_0})$, with S_{00} and V_{k0} as two additional parameters obtained from the physical properties of the photoconductor. The final PIDC
model is presented in Tables 10.3 and 10.4 that captures the functional form of the

---

TABLE 10.3

Exposure Model

General Springett–Melnyk Exposure Model:

$$V(X) - V_i + V_c \ln\left(\frac{V(X) - V_r}{V_i - V_r}\right) + \frac{e\eta_0}{C}X = 0, \text{ if } p = 1$$

(10.17)

$$V(X) - V_i + \frac{V_c}{1-p}\left[\left(\frac{V_c}{V(X) - V_r}\right)^{p-1} - \left(\frac{V_c}{V_i - V_r}\right)^{p-1}\right] + \frac{e\eta_0}{C}X = 0, \text{ if } p \neq 1$$

(10.18)

QM Model ($p = 2$):

$$V(X) = V_r + \beta + \sqrt{\beta^2 + V_c^2} \quad \text{where} \quad \beta = \frac{1}{2}\left[V_i - V_r - \frac{V_c^2}{V_i - V_r} - SX\right]$$

(10.19)

$V_r = V_{r0}\left(\frac{L}{L_0}\right)$ (approximately, $L/L_0 = 1$)

$S = S_0\left[1 - \exp\left(-\frac{V_k}{V_i}\right)\right], \quad S_0 = S_{00}\left(\frac{L}{L_0}\right), \quad V_k = V_{k0}\left(\frac{L}{L_0}\right)$

Dark Decay Model: Figure 10.7
Measured quantities: $\{V(X)\}$ at different exposure points
Actuators: $\{V_g, X\}$

TABLE 10.4

List of Parameters for a Photoconductor

QM PIDC Model Parameters	Photoconductor
X	9 ergs/cm^2
S_{00}	334 V/(ergs/cm^2)
V_{k0}	325 V
α	0.258
V_{r0}	20 V
$L = L_0$	24 μm
V_i	$\pm(400$–$800)$ V
p	2

PIDC for most of the practical photoconductors used in digital electrophoto-graphic printers based on electrophotography. Often, the quadratic model fails to capture the performance at the knee of the PIDC. To resolve this issue, it is necessary to use a more general model (Equation 10.18) to achieve a good fit. It is also interesting to cite an analytical solution we were able to find for the case $p = 1$. It is easy to show that the solution can be expressed in terms of the well-studied Lambert function $W(x)$, which is, by definition, the solution of Equation 10.24 with respect to W:

$$We^W = x \qquad (10.24)$$

Then one can show that the solution of Equation 10.17 is given by

$$V(X) = V_r + V_c W \left[F \exp\left(-\frac{S}{V_c} X \right) \right] \qquad (10.25)$$

where

$$F = \frac{1}{V_c} \exp\left[\frac{V_c \ln(V_i - V_r) + V_i - V_r}{V_c} \right]$$

$$S = \frac{e \eta_0}{C}$$

One can also find closed-form solutions for the cases when $p = 3$ and $p = 4$ by solving the polynomial equations (with respect to $V(X)$) that originate from Equation 10.18. For the rest of the cases, numerical solutions are recommended.

It is also interesting to observe that the exposure model developed in this section is based on the assumption that the photoconductor is positively charged and this charge decreases with exposure. This is true in the case of charged area development (CAD) but not true in the case of discharged area development (DAD) (for definitions, see Section 10.2.3). In that case, one can simply consider the following generalized relation that works for both cases:

$$V(X) = \text{sgn}(V_i) \, \text{expose}(|V_i|, X) \qquad (10.26)$$

where $\text{sgn}(x)$ is the signum function defined by

$$\text{sgn}(x) = \begin{cases} 1 & x \geq 0 \\ -1 & x < 0 \end{cases} \qquad (10.27)$$

expose(.,.) denotes the exposure model developed above
$|\cdot|$ stands for the absolute value

Figure 10.9 shows PIDC curves for a photoconductor for different values of initial voltage V_i whose parameters are shown in Tables 10.3 and 10.4.

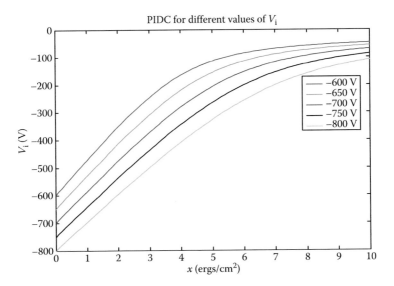

FIGURE 10.9 PIDC for photoconductor of Table 10.4 for five different values of V_i.

10.2.3 DEVELOPMENT MODEL

In this step, the electrostatic latent image is converted into a visible image by developing charged toner particles or liquid ink on the surface of the photoconductor. Various types of development systems that are employed today (Table 10.5) use a dry powder development [13,14]. They are classified based on the method with which the toner particles are transported to the photoconductor surface. These systems use various mechanisms to charge the toner and transport them to the

TABLE 10.5
Types of Electrophotographic Development Systems

Development Technology	Acronym	Product Example
Insulative magnetic brush	IMB	1025, 9200
Conductive magnetic brush	CMB	1075, 4850
Wrapped development	HAZE	1065, 5090
Magnetically agitated development	MAZE	Kodak ColorEdge
Single component development	SCD	1012, Canon GP55
Single component magnetic brush development	SCMB	5775, A-Color, Canon CLC series
Positive/negative development	Trilevel	4850
Hybrid scavengeless development	HSD	iGen3
Hybrid jumping development	HJD	Information not available
Recharge and development	REaD	Konica 5028

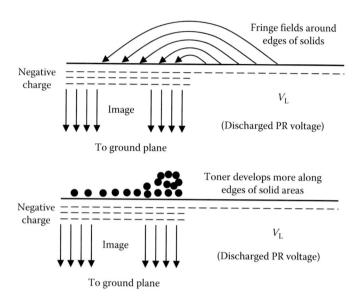

FIGURE 10.10   Development without electrode.

development nip. For a deeper review of these development systems, refer to Refs. [3,4,9,13] and various proceedings of IS&T conferences on nonimpact printing.

To understand the basics of image development, we first consider the electric field in the region of the latent image. If the toner is developed on the fully charged area of the photoconductor and not on the discharge area, the sharp electrostatic contrast between the charged and discharge areas of the photoconductor has fringe fields (Figure 10.10) [14,15]. Due to fringe fields, toner develops more along the edges of the solid areas. Small areas, for example, lines, are well developed since without a development electrode on top of the photoconductor only fringe fields associated with the edges of solids and lines exist in the development zone. These edge effects produce halos giving rise to poor image quality.

When a grounded development electrode is placed near the photoconductor, the field pattern in the solid area is improved. Hence, solid areas will be filled-in with more toner, giving improved solid area development. On the otherhand, the non-image areas (background) appear toned, showing a light-gray appearance due to the developed toner particles as shown schematically in Figure 10.11.

When the development electrode is biased with a control voltage, V_{bias}, background development can be significantly reduced. Figure 10.12 shows shift in development field pattern with a proper bias voltage, which is used to adjust the gray scale in many copiers and printers. Figures 10.13 and 10.14 show the actual development of an image without and with a development electrode biased with a control voltage, respectively. Halos can be seen in Figure 10.13 due to fringe fields around edges of solids.

For CAD, toner is developed only on the charged areas of the photoconductor since the toner is repelled due to a more positively biased discharged region. If V_h is the voltage on the charged area of the photoconductor and V_L is the voltage on the

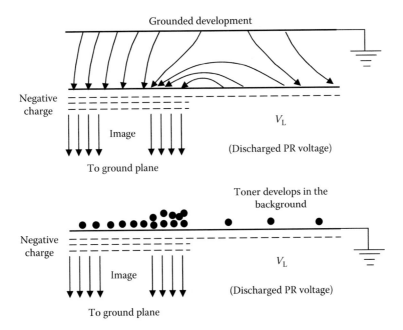

FIGURE 10.11   Development with grounded electrode.

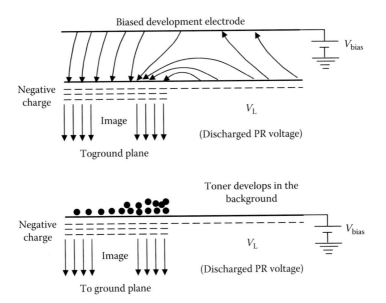

FIGURE 10.12   Development with biased electrode.

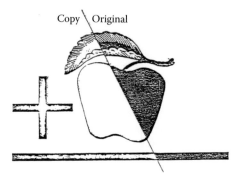

FIGURE 10.13   Example of the gray-scale image without a biased development electrode.

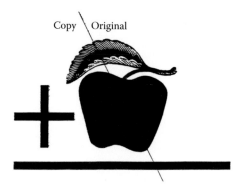

FIGURE 10.14   Example of the gray-scale image with a biased development electrode.

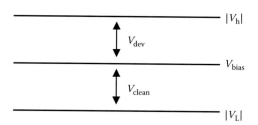

FIGURE 10.15   Development and cleaning voltages in a CAD system.

discharged region, then the development and cleaning voltages can be written as (Figure 10.15)

$$V_{dev} = V_h - V_{bias}$$
$$V_{clean} = V_{bias} - V_L$$
(10.28)

For DAD, which is more commonly used than CAD, only the discharged areas of the photoconductor will attract toner from the development zone. Other negatively

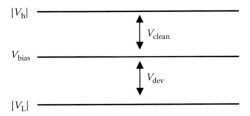

FIGURE 10.16 Development and cleaning voltages in a DAD system.

biased areas of the photoconductor will repel toner since the field over these areas is more negatively biased than the development electrode. The development and cleaning voltages are given by (Figure 10.16)

$$V_{\text{dev}} = V_{\text{bias}} - V_{\text{L}}$$
$$V_{\text{clean}} = V_{\text{h}} - V_{\text{bias}}$$

(10.29)

It is important to note that the quantities V_{h} and V_{L} satisfy $|V_{\text{h}}| \geq |V_{\text{L}}|$, so in most cases when we have negative charging, we have $V_{\text{h}} \leq V_{\text{L}}$. This is the reason we use absolute values in Figures 10.15 and 10.16. V_{bias} is typically positive (Figure 10.12).

A typical two component developer consists of (1) toner powder that consists of pigmented thermoplastic irregularly shaped particles between 5 and 25 μm in size, and (2) carrier beads usually circular in shape made out of magnetic core material (e.g., ferrite) with a typical dimension of around 120 μm (see Refs. [16–19] for more details on the physics of toner electrostatic charging). A thin dielectric coating gives the desired charging characteristics and surface properties required for the toner particles to adhere to the carrier beads. Charge control agents are added to the toner to adjust the magnitude of charge. Mixing toner and carrier beads will result in oppositely charged particles due to triboelectrification. For example, when active matrix (AMAT) photoconductors are used, during the mixing process, toner is given positive charge and carrier beads are given negative charge. An example of the toner particles attached to the carrier bead is shown in Figure 10.17.

Carrier—transport toner

Toner—provide color

Additives—control charging and flow

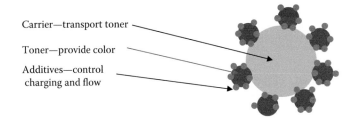

FIGURE 10.17 Example of toner particles attached to carrier beads.

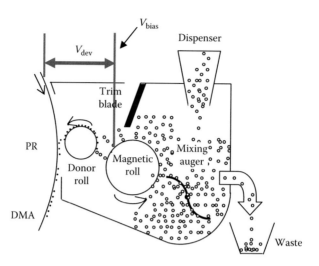

FIGURE 10.18 Example of a hybrid two-component development process.

Figure 10.18 shows an example of the two-component development system hardware. The charge-to-mass ratio, Q/M, also called "the tribo" of the toner particles, is one of the fundamental parameters in any EP printer. The blackness of the image depends heavily on the magnitude of the tribo. The ratio between the number of toner particles to the number of carrier particles in the developer mixture, called the toner concentration (TC), is also an important parameter that changes over time (unless efficiently controlled) due to a continuous toner development in high-speed printers. In solid area development, the developed mass per unit area (DMA) on the photoconductor depends on the development voltage V_{dev}, physical and electrical parameters of the development system, and the toner.

For a two-component conductive magnetic brush development system, several quantitative theories have been developed that are available in the published literature. A simple theory is based on the assumption that the toner charge per unit area completely neutralizes the photoconductor charge per unit area [13]. According to this theory, the DMA (M/A, measured in mg/cm^2) can be predicted by the following expression (see Problem 10.5):

$$D = \frac{M}{A} = \frac{V_{dev}\varepsilon_0}{(Q/M)(d_p/K_p)} \tag{10.30}$$

where
  d_p is the thickness of the photoconductor
  K_p is the dielectric constant of the photoconductor
  ε_0 is the dielectric permittivity of air
  Q/M is the toner tribo (charge-to-mass ratio)

The parameterized DMA Equation 10.30 suggests that a linear relationship between the DMA and the development voltage exists when the toner tribo is held constant throughout the printing process. Development physicists have shown that the Q/M of toner in a two-component mixture degrades with time [16–18] and toner aging [20]. Ignoring toner aging, tribo and TC, T_c, have the following relationship:

$$\frac{Q}{M} = \frac{A_t}{T_c + k_0} \qquad (10.31)$$

where
  A_t is proportional to the carrier surface area that varies with time due to toner attaching to the carrier
  k_0 is a constant unaffected by the developer aging

Figure 10.19 shows a plot of toner tribo vs. TC for a test system.
  A typical solid area curve is shown in Figure 10.20 for an experimental printer for various values of TC. This is called the developability curve. Clearly, DMA is a nonlinear function of development voltage at low development-voltage values. Accordingly, the functional behavior of DMA captured by the first principle model

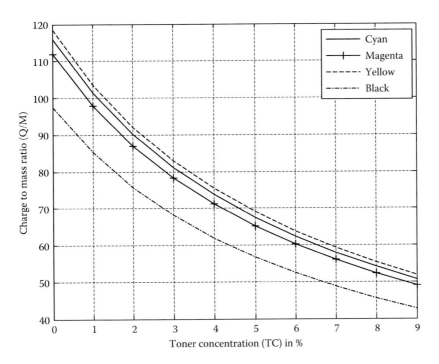

FIGURE 10.19 Toner tribo vs. TC relationships for cyan, magenta, yellow, and black toners (note slight variations in the slopes among different toners).

Solid area development

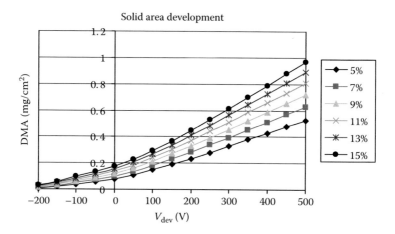

FIGURE 10.20 Example of solid area development curves with respect to development voltage for different TC.

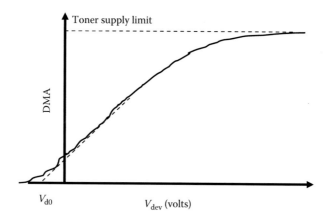

FIGURE 10.21 Example of the complete solid area development curve (a schematic view).

(Equation 10.30) does not predict the curve well at low development voltages. The nonlinear developability characteristic is also prevalent near the toner supply limits, as shown schematically in Figure 10.21. To improve the predictability, additional physical, electrical, and material parameters should be included in the DMA expression. Analytical forms of those expressions are very complicated and can lead to increasing difficulty in designing controllers. Hence, in such cases, one can resort to experimentally validated phenomenological expressions for characterizing the developability.

V_{d0} represents the voltage at which the DMA vanishes. One such useful phenomenological model of the DMA, D, is described by the following equation:

TABLE 10.6

Development Model

Developability

CAD: $V_{dev} = V_h - V_{bias}$; $V_{clean} = V_{bias} - V_L$ (10.28)

DAD: $V_{dev} = V_{bias} - V_L$; $V_{clean} = V_h - V_{bias}$ (10.29)

$$D = \frac{M}{A} = \frac{V_{dev}\varepsilon_0}{(Q/M)(d_p/K_p)}$$ (10.30)

$$\text{or} \quad D = \alpha\left\{1 - \exp\left[-\frac{\gamma}{\alpha}(V_{dev} - V_{d0})\right]\right\}$$ (10.32)

$$\frac{Q}{M} = \frac{A_t}{T_c + k_0}$$ (10.31)

$$\alpha = \text{MOR}\left(\frac{u_{br}}{u_{pr}}\right)\frac{T_c}{100 + T_c}$$ (10.33)

$$\gamma = \frac{\varepsilon_0}{\frac{Q}{M}\left[(t/K)_{pr} + \frac{u_{pr}}{u_{br}}(t/K)_{br}\right]}$$ (10.34)

Measurement quantities: $\{V_h, V_L, D, T_c\}$, D measured at different gray
levels

Actuators: $\{V_g, X, V_{bias}\}$ or $\{V_g, X, V_{dev}, V_{clean}\}$

$$D = \alpha\left\{1 - \exp\left[-\frac{\gamma}{\alpha}(V_{dev} - V_{d0})\right]\right\}$$ (10.32)

where the parameter α represents the supply limit and γ represents the development slope.

They have a dependency on TC and tribo as follows [21]:

$$\alpha = \text{MOR}\left(\frac{u_{br}}{u_{pr}}\right)\frac{T_c}{100 + T_c}$$ (10.33)

$$\gamma = \frac{\varepsilon_0}{\frac{Q}{M}\left[(t/K)_{pr} + \frac{u_{pr}}{u_{br}}(t/K)_{br}\right]}$$ (10.34)

The tribo is obtained from Equation 10.31. The parameters $(t/K)_{pr}$ and $(t/K)_{br}$ are fit to experimental data and are measured in μm. Parameters u_{pr} and u_{br} denote the speed (mm/s) of the PR and the brushing roll, respectively, and MOR is the mass on roll (in mg/cm^2). Table 10.6 summarizes the model and shows actuators and measured quantities. Table 10.7 below shows parameters for a two-component development system.

10.2.4 TRANSFER MODEL

In the transfer station, the developed image on the photoconductor is transferred to the media. In their simplest form, the transfer methods are similar to the methods

TABLE 10.7

Parameters of a Phenomenological Model for a Typical Two-Component Development System

	Parameters	Cyan	Magenta	Yellow	Black
Equation 10.26	k_0	7	7	7	7
	A_t [μC/g]	810.59	783.04	828.84	681.83
Equations	$(t/K)_{pr}$ [μm]	8	8	8	8
10.30, 10.34,	$(t/K)_{br}$ [μm]	7.28	7.28	7.28	8.45
and 10.35	V_{d0} [V]	−55	−55	−55	−49
	MOR [mg/cm^2]	35	35	35	35
	u_{pr} [mm/s]	165	165	165	165
	u_{br} [mm/s]	260	260	260	260
	ε_0 [C/V-m]	8.854×10^{-12}	8.854×10^{-12}	8.854×10^{-12}	8.854×10^{-12}

used for developing the electrostatic images formed on the photoconductor surfaces. Toner is attached to the photoconductor surface by an adhesion force.

Understanding the mechanism of transfer is important because the image density, the ultimate image quality, and the degree of cleaning required in the printing system are all directly related to the transfer efficiency. The transfer efficiency is the fraction of the toner transferred to the receiver layer (in this case media) from the donor layer. To capture the essential underlying physics in the transfer process, Yang–Hartmann transfer model [22] is considered. A physical model is reviewed below.

A simple transfer system configuration consists of two electrodes, as shown in Figure 10.22, and includes three regions of dielectric materials (of different thickness and properties) sandwiched between the two electrodes. The particle layer with charge density $\rho(z)$, thickness d_2, and an effective dielectric constant ε_2 exists between two dielectric substrates with thicknesses d_1 and d_3 and dielectric constants ε_1 and ε_3, respectively; z is a distance variable between 0 and d_2. An electric potential, V_a, is applied between the electrodes adjacent to the dielectric layers. As

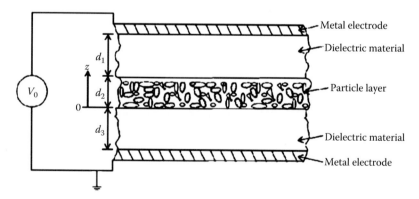

FIGURE 10.22 Configuration of particle and dielectric layers. (From Yang, C.C. and Hartmann, G.C., *IEEE Transactions on Electron Devices*, 23, 1976. With permission.)

the potential is increased, the particle layer ruptures and splits when the sum of the electrostatic forces and mechanical stress exceeds the cohesive strength. If z_s is the spatial distance at which the layer splits, the fraction F of the material attached to layer 1 is defined as the transfer efficiency, which is

$$F = 1 - \frac{z_s}{d_2} \tag{10.35}$$

The transfer efficiency depends on (1) the space charge density of particle layer, (2) the electric field between electrodes, and (3) the cohesive forces between particles. The stress caused by space charge and electric field can be modeled using Euler's equation:

$$P(z) = \int_0^z \rho(z')E_2(z')dz' + P_m - \varepsilon_3 E_3^2 \ [\mathrm{N/m^2}] \tag{10.36}$$

where
  $\rho(z)$ is the space charge density $[\mathrm{C/m^3}]$
  $E_2(z)$ is the electric field in the particle layer $[\mathrm{V/m}]$

The last two terms represent the constant of integration resulting from the electrostatic attraction between the two electrodes. At the boundary between layers 2 and 3 ($z = 0$), the total stress must equal the sum of the mechanical stress P_m (in $\mathrm{N/m^2}$) and the compression $\varepsilon_3 E_3^2$ due to electrostatic attraction between the two metal electrodes, where E_3 is the electric field in layer 3. Cohesive forces between particles are represented by the quantity $C(z)$ (in $\mathrm{N/m^2}$) resulting from the particle–particle interaction and particle–electrode adhesion due to electrostatic, dispersion, or chemical forces.

When the mechanical stress at the boundary, P_m, is increased, the layer separates where $P(z)$, which is the sum of the electrostatic forces and mechanical stress, exceeds the cohesive force, $C(z)$. At the particle layer separation (i.e., at $z = z_s$), the value of P_m can be found by determining the value of z that corresponds to a minimum in total stress $P(z) - C(z)$. Therefore,

$$\frac{d}{dz}[P(z) - C(z)] = 0 \quad \text{and} \quad P(z) - C(z) = 0 \tag{10.37}$$

The solution that gives the smallest value of P_m corresponds to the mechanical stress for which $P(z)$ first exceeds $C(z)$.

When the particle layer does not have a spatial variation, then Equations 10.36 and 10.37 reduce to the condition

$$\rho(z)E_2(z) = 0 \tag{10.38}$$

Equation 10.38 has the trivial solution $\rho(z_s) = 0$ or $E_2(z_s) = 0$. The field within the particle layer, $E_2(z)$, is derived by using Poisson's equation:

$$E_2(z) = \frac{1}{\varepsilon_2} \int_0^z \rho(z')dz' + E_0 \qquad (10.39)$$

The constant E_0 is determined using Gauss's law at the boundary condition. The electric fields E_1 and E_3 at the boundary are given by

$$E_1 = \frac{\varepsilon_2}{\varepsilon_1} E_2(d_2)$$

$$E_3 = \frac{\varepsilon_2}{\varepsilon_3} E_2(0) \qquad (10.40)$$

The applied potential, V_a, is equal to the sum of the potentials across each layer that is obtained from the fields and distances:

$$V_a = -E_1 d_1 - \int_0^{d_2} E_2(z)dz - E_3 d_3 \qquad (10.41)$$

Equations 10.39 through 10.41 are solved to find the unknown field, E_0:

$$E_0 = \frac{V_a}{\varepsilon_2 D} - \frac{D_1}{\varepsilon_2 D} \int_0^{d_2} \rho(z)dz - \frac{1}{\varepsilon_2^2 D} \int_0^{d_2} \int_0^z \rho(z')dz'dz \qquad (10.42)$$

where

$$D_i = \frac{d_i}{\varepsilon_i} \quad i = 1,2,3 \quad \text{and} \quad D = \sum_{i=1}^3 D_i \qquad (10.43)$$

From Equation 10.39, since $E_2(z_s)=0$, using this useful result and Equations 10.41 and 10.42, we obtain

$$\int_0^{z_s} \rho(z)dz = -\varepsilon_2 E_0 = \frac{V_a}{D} + \frac{D_1}{D} \int_0^{d_2} \rho(z)dz + \frac{1}{\varepsilon_2^2 D} \int_0^{d_2} \int_0^z \rho(z')dz'dz \qquad (10.44)$$

A solution for Equation 10.39 is obtained for three cases: (1) for constant charge density in the layer 3 containing insulating particles (i.e., $\rho = \rho_0$), (2) for exponential charge distribution in the toner particles (i.e., $\rho = \rho_0 e^{-\gamma z}$ where γ represents the charge-penetration parameter), and (3) for a more complex charge distribution with two exponentials that can be found in a photoactive particle electrophotography (i.e., $\rho = \rho_1 e^{-\gamma z} - \rho_2 e^{-\gamma(d-z)}$). The two exponentials in the expression can be thought of as the result of two processes: discharging the particle layer by exposure to light through one dielectric layer in the presence of an electric field, and corona charging of the other side of the particle layer. If $V_\rho = \rho_0 d_2^2/(2\varepsilon_2) =$ the space charge potential across layer 2, then for a constant density case, the transfer efficiency equation becomes

$$F = 1 \quad \text{for} \quad \frac{V_a}{V_\rho} \leq -\left(\frac{2D_1}{D_2} + 1\right) \tag{10.45}$$

$$F = 1 - \frac{2D_1 + D_2 + D_2\left(V_a/V_\rho\right)}{2D} \quad \text{for} \quad -\left(\frac{2D_1}{D_2} + 1\right) \leq \frac{V_a}{V_\rho} \leq \frac{2D_3}{D_2} + 1 \tag{10.46}$$

$$F = 0 \quad \text{for} \quad \frac{V_a}{V_\rho} \geq \frac{2D_3}{D_2} + 1 \tag{10.47}$$

Thus, from these equations, we can find the transfer limits. The transfer begins (i.e., at $F = 0$) when the applied voltage is equal to

$$V_a = V_\rho\left(\frac{2D_3}{D_2} + 1\right) \tag{10.48}$$

and complete transfer occurs (i.e., $F = 1.0$) when the applied voltage is equal to

$$V_a = -V_\rho\left(\frac{2D_1}{D_2} + 1\right) \tag{10.49}$$

These results were compared to experimental data in a laboratory setup. We show the use of this model for one of the EP transfer system below.

One of the dielectrics in the EP system is the plain paper and the other is the photoconductor. The particle layer is the layer of image surface carried by the toner that is crammed between the paper and the photoconductor. For complete transfer from the photoconductor surface to the paper, the transfer efficiency equation can be written with respect to key parameters of the xerographic transfer system and the transfer voltage.

The space charge density (ρ_0) of the toner layer is given by [9]

$$\rho_0 = \left(\frac{Q}{M}\right)_d p d_t = 3\sigma\frac{p}{R} \ (\mu C/cm^3) \tag{10.50}$$

where
p is the packing fraction of the toner ($p = \sim 0.5$ for most toners)
d_t is the toner mass [mg/cm^3]
$\left(\frac{Q}{M}\right)_d$ is the toner tribo on the photoconductor [$\mu C/gr$], which is related to the tribo shown in Equation 10.31 with a simple linear relation of the form

$$\left(\frac{Q}{M}\right)_d = c\frac{Q}{M} \tag{10.51}$$

due to time decay of the tribo

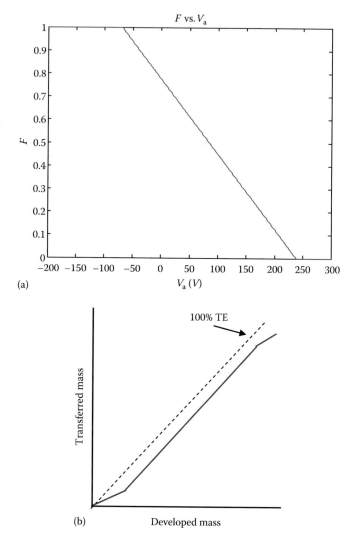

(a)

(b)

FIGURE 10.23 (a) Transfer efficiency F vs. V_a and (b) relationship between transferred mass to developed mass.

With σ as the surface charge density (in $\mu C/cm^2$), the space charge density can be expressed in terms of toner radius R (in cm). Dependence of F on ratio V_a is shown in Figure 10.23a for the parameters shown in Table 10.8.

The final transfer model is shown in Table 10.9, which captures the relationship between the transfer efficiency and the transfer voltage. Transferred mass per unit area (TMA) onto the paper from the DMA on the PR is obtained by the following equation:

$$\text{TMA} = \text{DMA} \times F \tag{10.52}$$

TABLE 10. 8
Yang and Hartmann Transfer Model

Transfer Efficiency:

$$F = 1 \quad \text{for} \quad \frac{V_a}{V_\rho} \leq -\left(\frac{2D_1}{D_2} + 1\right) \tag{10.45}$$

$$F = 1 - \frac{2D_1 + D_2 + D_2\left(V_a/V_\rho\right)}{2D} \quad \text{for} \quad -\left(\frac{2D_1}{D_2} + 1\right) \leq \frac{V_a}{V_\rho} \leq \frac{2D_3}{D_2} + 1 \tag{10.46}$$

$$F = 0 \quad \text{for} \quad \frac{V_a}{V_\rho} \geq \frac{2D_3}{D_2} + 1 \tag{10.47}$$

$$V_\rho = \rho_0 d_2^2/(2\varepsilon_2)$$

$$\rho_0 = c\frac{Q}{M}pd_t = 3\sigma\frac{P}{R} \tag{10.50}$$

$$D_i = \frac{d_i}{\varepsilon_i} \quad (i = 1, 2, 3) \quad \text{and} \quad D = \sum_{i=1}^{3} D_i$$

TMA:
Measured quantities: {TMA}
Actuators: $\{V_a\}$

TABLE 10.9
Transfer Model Parameter

Transfer Model Parameters	Units
d_1	24 μm
d_2	20 μm
d_3	120 μm
ε_1	$3\varepsilon_0$
ε_2	$2\varepsilon_0$
ε_3	$3\varepsilon_0$
ε_0	8.854×10^{-12} C/V-m
c	0.6
p	0.5
d_t	1.1 g/cm^3

This model is highly dependent on the transfer efficiency of the system. As the DMA increases, so does the TMA, but not infinitely. Figure 10.23b shows the relationship between the transferred mass and the developed mass. Note that the dotted line represents 100% transfer efficiency, which is impossible for a practical system to achieve. The TMA increases proportional to the DMA and linearly for most of the operating range of the applied voltage. Beyond this operating range, the TMA saturates. As fields build up and decay in the transfer nip, there is a constraint on the maximum field strength allowable across the air gap. If voltages rise above this value, known as the Paschen voltage, air molecules will undergo ionization to reduce the electric field below this constraint, which limits transfer [23]. There is also enough latitude needed in the applied voltage to ensure smooth transfer over a range of toner and paper states.

10.2.5 Fusing Model

In the fusing process, toner is fixed to the paper using heat and pressure permanently adhering it to the paper. Prior to the fusing, toner is only loosely bonded to the paper fibers due to electrostatic forces; it can easily be disturbed or rubbed off. Most high-speed fusing is done using hot roll pressure fusers, in which parameters like time, temperature, and pressure determine the quality of the prints. Other techniques include pressure only fixing and solvent fixing. Thermal fixing with pressure rolls has proven to be very efficient. The fusing assembly typically has two rollers: a heated upper roller and a lower rubber pressure roller. The fuser roller heats the toner sufficiently to cause it to melt, and the pressure roller presses the melted toner into the paper, causing it to bond to the paper. There is usually a long warm-up time (5–10 min) associated with the process of thermal fixing with pressure. Typical fusing temperatures are controlled around 140°C. When the heat rods are switched on, the entire fuser roll surface is heated through conduction. The nip between the fuser roll and the pressure roll induces the fused print to self-strip from the fuser roll. To ensure that the toner is released from the fuser roll surface reliably, a release agent (such as amino functional oil) management system is incorporated into the design. There are many important (printer-specific) practical considerations required for fuser design. Pressure and temperature change the physical form of toner particles from solid to viscoelastic then to rubber, and finally the toner coalesces to fluid spreads and penetrates into the paper fibers. The complex interactions between the toner, the substrate, and the fusing method give a glossy appearance to the bonded toner that has significant effect on perceived color of the image [24].

To determine its fusing latitude for variety of media thickness and toner materials, a comprehensive (physics-based) parametric study is required. For the purpose of controlling the overall image reproduction process, we require a simplified model of gloss as a function of the TMA and the fusing temperature. Generally, fuser roll pressures are held constant for a given media thickness. Hence, we present an experimental gloss model at a constant roll pressure below. Since, for now, we are interested only in the color quality of a solid area dot, effects of the fusing process on line width, blurriness and edge quality, etc. are ignored. More complex time-domain models for designing the fusing system are available in Refs. [25–27]. They can be easily integrated with the process models for a variety of media thickness and coatings.

Figure 10.24 shows the measured gloss, G, as a function of the TMA in mg/cm^2 and temperature. Data was created by fusing toner patches for cyan, magenta, yellow, black, red (magenta with yellow), green (cyan with yellow), and blue

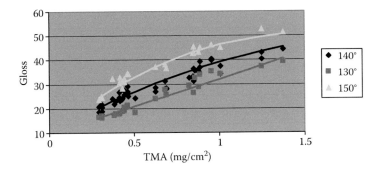

FIGURE 10.24 TMA vs. gloss for different temperatures.

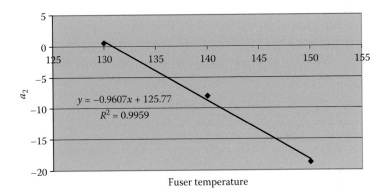

FIGURE 10.25 Coefficient a_2 as function of fuser temperature.

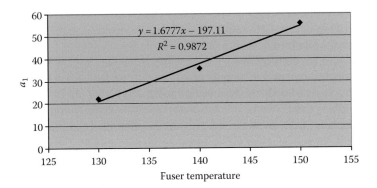

FIGURE 10.26 Coefficient a_1 as function of fuser temperature.

(cyan with magenta) for varying TMA and temperatures. The plot of the quadratic fit represented by Equation 10.53 is superimposed on the experimental data, which is shown by solid lines with the coefficients a_2 and a_1 as functions of fuser temperature in Figures 10.25 and 10.26, respectively. The coefficient $a_0 = 10.1$ is

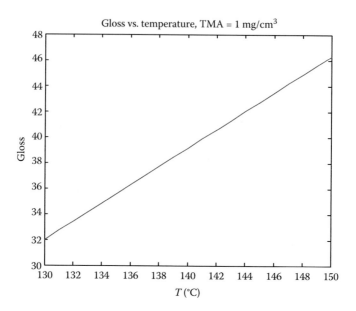

FIGURE 10.27   Gloss vs. temperature for fixed TMA.

independent of temperature for the experimental fusing system and is obtained from Figure 10.24.

$$G = a_2(T) \cdot (\text{TMA})^2 + a_1(T) \cdot \text{TMA} + a_0 \qquad (10.53)$$

At this point, it is constructive to note that, according to this model, the gloss (G) is a linear function of temperature (T) for a fixed TMA and a quadratic function of the TMA (with a negative leading coefficient a_2) for fixed temperature (T). This dependency is further visualized in Figures 10.27 and 10.28.

In Figure 10.28, the measured gloss decreases as the gray level or TMA increases. Since the samples were only minimally fused, we can conclude that at low gray levels, gloss is controlled primarily by the substrate and that at higher gray levels, the under-fused, non-glossy toner covering the substrate reduces the contribution of the substrate to the gloss. Apart from the toner, the gloss of the substrate is an important contributing factor to the overall gloss of the image.

10.2.6   COLOR MODEL

In this section, we describe a mathematical color model that can be found in Refs. [28–30]. The model predicts the output reflectance (R) at each wavelength, which is the most complete characterization of the output color. The output reflectance curve, that is, the reflectance with respect to wavelength (λ), can then be used to derive the tristimulus values in any of the device-independent color spaces, for example, L^*, a^*, b^* or X, Y, Z.

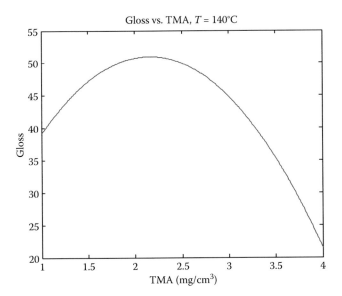

FIGURE 10.28 Gloss vs. TMA for fixed temperature.

This section develops a comprehensive color predictive model that determines the desired output based on scientific reason and principles. Physical parameters are used as input and new-dependent physical parameters are the resulting output. This color predictive model utilizes most of the main formulations of the well-known Kubelka–Munk model [31], but it also provides an analytical approach to evaluate the involved quantities based on an empirical set of equations whose parameters are derived by fitting models to experimental data. Hence, this color model considers the reflectance as a function of the toner masses (TMAs) for all four toner separations (C, M, Y, K), which are, obviously, the most important parameters that determine the output color. This model also follows the main assumption of Ref. [32], namely, that the reflectance is a function of gloss. Consequently, it contains a generalization of these models, and simulations show that it performs reasonably well in predicting the reflectance of fused toner layer on paper for a large range of paper types.

The main equation of the reflectance spectral model is

$$R(m_i, g; \lambda) = p_0(g) + [1 - p_1(n_1, n_2, i)]\tau(m_i; \lambda)R_s(\lambda)\tau'(m_i; \lambda) \qquad (10.54)$$

where

m_i are the masses of the four different toner layers used in printing, $i = 1, 2, 3, 4$, that is, m_1, m_2, m_3, m_4 for the masses of C, M, Y, K (m_C, m_M, m_Y, m_K, respectively)

g is the surface gloss that is dependent on the masses as described by the fusing model

$R(m_i, g; \lambda)$ is the output reflectance as a function of wavelength (λ), gloss (g), the masses of different toner separations m_i, and the measuring instrument geometry

$p_0(g)$ is the measurable front surface reflectance (FSR) (function of gloss as described in Ref. [32])

$p_1(n_1, n_2, i)$ is the total FSR (function of refractive indices n_1, n_2, and incident angle i)

$R_s(\lambda)$ is the substrate reflectance (function of paper type)

$\tau(m_i; \lambda)$ is the transmittance from air through toner to paper (function of TMA and toner spectral transmittance)

$\tau'(m_i; \lambda)$ is the return transmittance from paper through toner to air (again, function of TMA and toner spectral transmittance)

The physical process considered here is fairly simple (see also Figure 10.29 and Ref. [32]). Essentially, incident light hits the surface of the paper and a portion of it, namely, p_1, the so-called total FSR, gets reflected. This is the total amount of light reflected, which is assumed to be constant and independent of gloss. It depends only on the ratio of the refractive indices of the two surfaces and the angle of the incident light. This is captured by the well-known Fresnel equations (see also Ref. [32]) given by

$$p_{\parallel} = \left[\frac{\cos i - \sqrt{(n_2/n_1)^2 - \sin^2 i}}{\cos i + \sqrt{(n_2/n_1)^2 - \sin^2 i}}\right] \tag{10.55a}$$

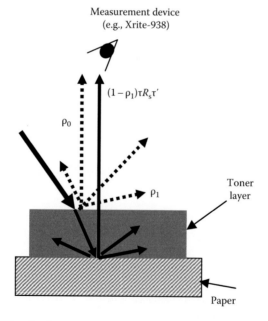

FIGURE 10.29   Visualization of the physical process considered by the color model.

$$p_\perp = \left[\frac{(n_2/n_1)^2 \cos i - \sqrt{(n_2/n_1)^2 - \sin^2 i}}{(n_2/n_1)^2 \cos i + \sqrt{(n_2/n_1)^2 - \sin^2 i}}\right] \qquad (10.55b)$$

$$p_1(n_1, n_2, i) = \frac{p_\parallel + p_\perp}{2} \qquad (10.55c)$$

where the parallel and perpendicular subscripts refer to the polarization of the incident beam. Hence, the total reflectance, $p_1(n_1, n_2, i)$, can be simply calculated by averaging $p_\parallel$ and $p_\perp$, as described by Equation 10.55c. It is known that the FSR dilutes the colored light reflected from within the object, thus lowering the perceived chroma [32]. According to the same model, gloss affects the direction rather than the magnitude of the reflected light. The perceived color depends on how much of the surface reflected light is detected. Some instruments, for example, integrating spheres, detect virtually all of this light, in which case the measured color is independent of the sample gloss. On the contrary, instruments with 0/45 or 45/0 (or generally, 0/D or D/0) geometry, reject almost all specularly reflected light, making the perceived color a strong function of gloss; the glossier samples have a larger specular component and, consequently, the perceived chroma is higher. The captured portion of the reflected light (which is a fraction of the FSR $p_1(n_1, n_2, i)$) is defined as $p_0(g)$. It is expected to be a decreasing function of gloss, which by fitting a logistic curve to data was found to be

$$p_0(g) = \frac{1}{1 + (g/18.55)^{2.22}} p_1(n_1, n_2, i) \qquad (10.56)$$

It is important to point out that in order to calculate the gloss using the fusing model described above, the total TMA has to be used; that is, we need to add the TMAs for the different toner layers:

$$\text{TMA} = \text{TMA}_C + \text{TMA}_M + \text{TMA}_Y + \text{TMA}_K \qquad (10.57)$$

This is based on the standard assumption that the different toners have similar thermal properties, hence when the fusing model is used to determine the surface gloss, only the total mass is important and not the individual masses.

The part of the light that is not reflected, that is, $[1 - p_1(n_1, n_2, i)]$, is transmitted through the toner layers, reflected by the substrate (paper), and then transmitted back through the toner (Figure 10.29). This gives rise to the factor $[1 - p_1(n_1, n_2, i)]$ $\tau(m_i; \lambda) R_s(\lambda) \tau'(m_i; \lambda)$ in Equation 10.54. For generality, the term $\tau'(m_i; \lambda)$ is not considered to be equal to $\tau(m_i; \lambda)$. Indeed, they are not always equal since in practice transmittance depends on the order of the different layers through which light passes. For example, transmittance is different when light passes through the yellow layer followed by the magenta layer than vice versa. Even if the transmittance is considered to be approximately the same regardless of layer order for the case described above, this will not be the case for halftoned images (see Section 10.3). This is

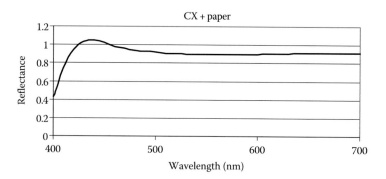

FIGURE 10.30   Substrate reflectance, $R_s(\lambda)$.

because for half-toned images, light may enter the toner, get reflected by the paper, and then exit through air instead of toner. The abstraction of the process using the three terms, that is, $\tau(m_i;\ \lambda)$, $R_s(\lambda)$, and $\tau'(m_i;\ \lambda)$, is very effective since the transmittance factors $\tau(m_i;\ \lambda)$ and $\tau'(m_i;\ \lambda)$, are independent of the characteristics of the paper type. In fact, the effect of changing the media type is captured only by $R_s(\lambda)$. This is somewhat similar to the idea considered in Ref. [31] but gives a more elegant and complete representation. The substrate reflectance can be found through measurements, and a typical curve is shown in Figure 10.30.

We still need to calculate the transmittance as a function of the masses (TMAs) of the four toners denoted as $m_C, m_M, m_Y,$ and m_K. This is done using an empirical model derived by fitting the model to experimental data and is presented in Equations 10.58 through 10.61. The transmittance through air, through toner, and finally to paper surface, $\tau(m_i;\ \lambda)$, is first written as

$$\tau(m_C, m_M, m_Y, m_K; \lambda) = \exp\left[-\sum_{i=1}^{3} M_i(\lambda)\sigma_i(m^*) - \sigma_4(m_K)\right] \qquad (10.58)$$

where

     $i = 1, 3$ for C, M, Y and $i = 4$ for K

     $M_i(\lambda)$ $(i = 1, 2, 3)$ is the toner absorption spectra (also called the toner master curve), which is assumed to be a function of the toner material

     σ_i denotes the mathematical mass for separation (i), which is a function of the TMAs (m_i) and is given by

$$\sigma_i(m^*) = \alpha_i\left\{1 - \exp\left[-\beta_i\left(\sum_{j=1}^{7} C_{ji}m_j^*\right)^{\gamma_i}\right]\right\} \quad \text{for } i = 1,2,3 \qquad (10.59)$$

$$\sigma_4(m_K) = \alpha_4\left(\frac{m_K}{\rho_t}\right) + \beta_4\left(\frac{m_K}{\rho_t}\right)^2 \qquad (10.60)$$

TABLE 10.10
General Form of Color Mixing Coefficients

	C	M	Y
C	1	0	0
M	0	1	0
Y	0	0	1
C + M	C_{41}	C_{42}	0
M + Y	0	C_{52}	C_{53}
C + Y	C_{61}	0	C_{63}
C + M + Y	C_{71}	C_{72}	C_{73}

TABLE 10.11
Typical Values of Color Mixing Coefficients

	C	M	Y
C	1.000	0.000	0.000
M	0.000	1.000	0.000
Y	0.000	0.000	1.000
C + M	−0.965	−1.027	0.000
M + Y	0.000	−0.263	−1.364
C + Y	−0.348	0.000	−0.629
C + M + Y	2.830	−0.087	3.240

where
α_i, β_i, γ_i, $M_i(\lambda)$ are from single separation (C, M, Y, and K) data
C_{ji} are the color mixing coefficients (Tables 10.10 and 10.11)
ρ_t is the toner mass density (typically, $\approx 1 \text{ gr/cm}^3$)
m_i^* are the entries of the vector defined as

$$m^* := \begin{bmatrix} m_1 \\ m_2 \\ m_3 \\ m_1 m_2 \\ m_2 m_3 \\ m_1 m_3 \\ m_1 m_2 m_3 \end{bmatrix} \tag{10.61}$$

The mathematical masses computed from Equation 10.59 are plotted as a function of the actual masses in Figure 10.31.

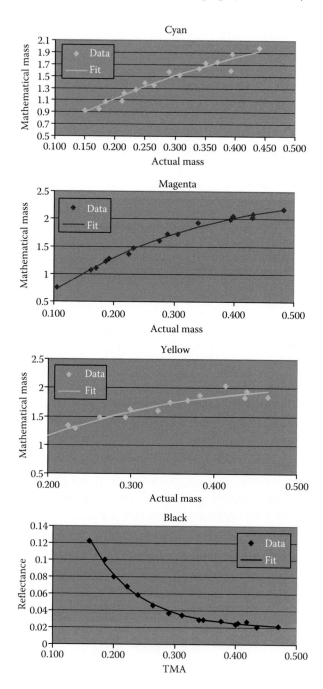

FIGURE 10.31 Curves showing actual mass to mathematical mass.

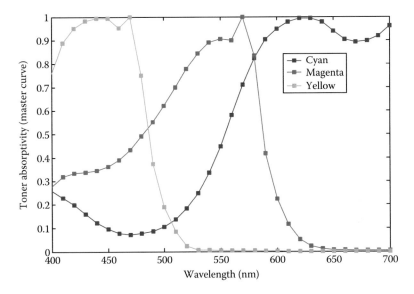

FIGURE 10.32 Toner master curves, $M_i(\lambda)$.

The described color model appears to be robust with respect to paper type. For a given paper type, the model can predict solid *CMY* patches with roughly 2.2 ΔE_{ab}^* prediction error on CX+ (color expression) paper. In addition, keeping the model fixed and switching between different paper types give a prediction error of roughly 3.0 ΔE_{ab}^*. Yet, the mathematical-mass model and master curves are very robust to paper type (Figure 10.32).

We conclude this section by sharing plots for the reflectance vs. TMA for *C, M, Y* separations (Figure 10.33) and the effects of gloss on chroma (*C**) and lightness (*L**) for different TMAs (Figures 10.34 and 10.35). For the sake of visualization and model validation, we cite plots of the variation of tristimulus values *L*, a*, b** vs. the masses for the three basic subtractive colors (*C, M, Y*) and for black (*K*) (Figure 10.35b). Note the decrease of lightness with the increase of mass, which becomes apparent in all cases.

10.2.6.1 Sensitivity Analysis of the Model

Next, we perform some model sensitivity analysis. Such analysis is fundamental for the application of linear state feedback methods described in Chapter 9 for the control of printers. This is because the models examined above are static nonlinear MIMO (multi-input and multi-output) systems with no direct time dependency, that is, the system is of the form

$$y = f(x) \tag{10.62}$$

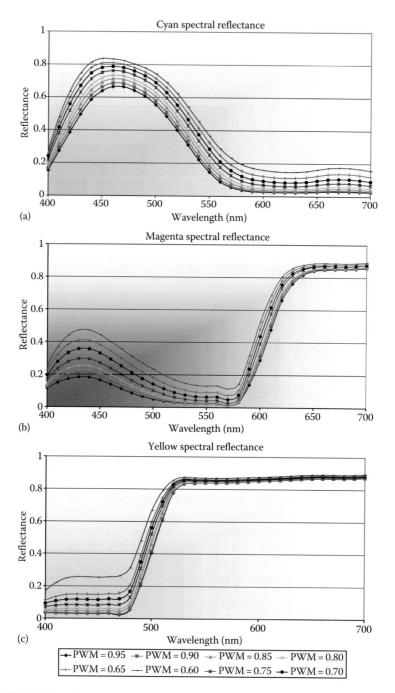

FIGURE 10.33 Reflectance vs. TMA for *C*, *M*, and *Y* separations: (a) cyan, (b) magenta, and (c) yellow.

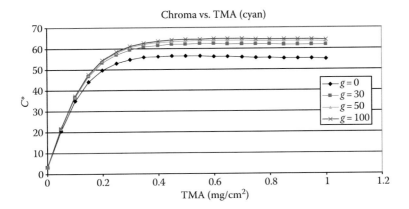

FIGURE 10.34   Effect of gloss on Chroma, C^* (for different values of TMA).

where
    x is the vector of the inputs to the system
    y is the one of the outputs

    It is interesting to observe the nice symmetry coming out of the models since the entire system, which can be viewed as a complete deterministic characterization of a dot printer, consists of a cascading of the process models (where the output of a model becomes an input to the next one), the fusing and the color model. Therefore, we can get the reflectance curve or tristimulus values as functions of the actuators, as illustrated in block diagram form in Figure 10.36.

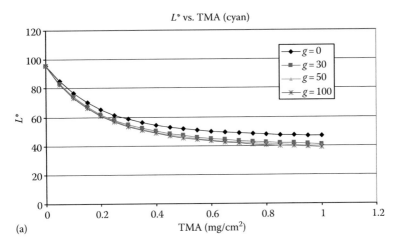

(a)

FIGURE 10.35   (a) Effect of gloss on lightness, L^* (for different values of TMA) and
(*continued*)

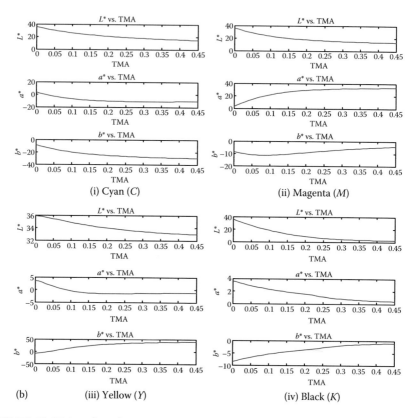

FIGURE 10.35 (continued)   (b) L^*, a^*, and b^* plotted as a function of TMA for different patches.

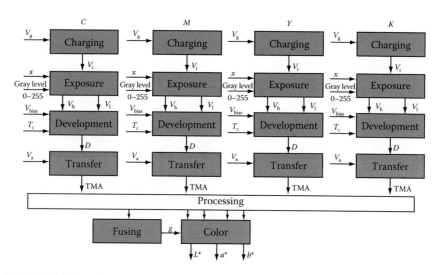

FIGURE 10.36   Block diagram representation of the complete model of the dot printer.

The compact characterization of a printer as an analytical MIMO system is very important for applications. However, this is not enough for control. The dot printer model shown above for electrophotographic printers is meant to represent how a single rendered dot results into color when produced with four separations. In an attempt to print a single dot, EP prints may print a light dot or no dot at all depending on the state of the ROS (raster output scanner) and development process. In a single dot system, not only do the neighboring dots interact, but they also spread as the laser beam cannot expose binary pixels perfectly at the micron level. Even if the laser exposes the dots well, the development process cannot make the toner perfectly adhere to the exposed dot, thereby injecting nonlinearities in the tone curves. At least the nonlinearities have to be modeled so that the analytical models presented above can become useful for developing system-level calibration, profiling, and control algorithms.

10.3 MODULATION TRANSFER FUNCTIONS

The MTF approach is used to characterize the spatial response of the printing system when dots are spread over an entire page to form a color image. This approach is based on the spatial frequency response of an imaging system. High spatial frequencies correspond to fine details in the image that are lost or enhanced due to dot spread caused by laser spreading or developability defects, light scattering, diffusion of light through the paper, etc. The dot spread and light spread functions can be measured and incorporated at each step in the EP process.

The exposure step involves generating a laser intensity profile for each separation. The input image consisting of *CMYK* separations is halftoned prior to exposure. The halftoned separations are exposed on the PR (to form a latent image) using a Gaussian or sinc^2 beam profile. This beam profile models the dot gain due to the laser spreading effect at the micron level. The resulting latent image contains the spatial frequencies that are not filtered out by the MTF for the exposure station since the dot spreading produces a low-pass filtering effect on the image. Thus, with this process, we create the halftoned image using the dot exposure model including the majority of spatial effects. The schematics in Figure 10.37a and b show the exposure process and Figure 10.38a and b show the functions used to model the effects of exposure MTF.

The rotation of the polygon mirror moves the laser beam across the width of the PR belt, and the transition to the next face on the polygon corresponds to the next line in the latent image. This direction of motion is termed as the scan direction. The perpendicular direction that is controlled by the angular rotation of the PR drum is called the cross scan or process direction.

The image is written on the PR as a series of overlapping envelopes. We consider the case when the laser exposes the PR in a raster scan format. We assume that the pixels where the halftone pattern takes a binary "1" need to be exposed by the laser and the spreading effect of the laser beam is modeled using a Gaussian beam or a sinc^2 function (Figure 10.38a and b) or an experimentally measured function.

The main parameter of the laser profile is the width of the beam at 50% of the maximum intensity. The beam profiles are obtained as a matrix containing the laser

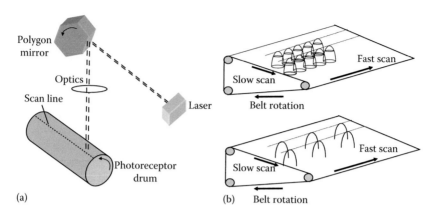

FIGURE 10.37   (a) ROS imager and (b) schematic for the raster scanning scheme.

intensity values. Let this matrix be denoted as $g(n)$. Note that the matrix is only one-dimemsional (1-D), that is, a vector. Due to the raster scanning nature of the laser, we need to differentiate between the fast-scan and slow-scan directions, which means obtaining the ROS intensity profile is not exactly the same as convolving halftoned image with a two-dimensional (2-D) Gaussian or sinc^2 filter. The step-by-step approach involved in modeling this process is captured in Figure 10.39 and described with a simple example below.

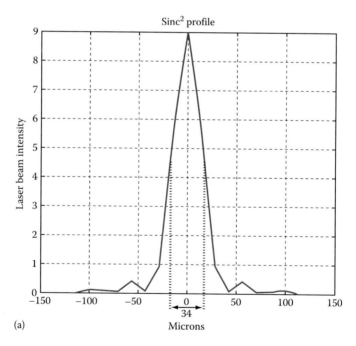

FIGURE 10.38   (a) Sinc^2 function beam profile.

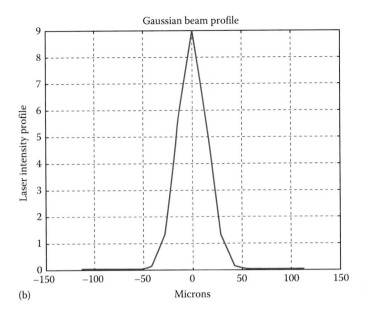

(b)

FIGURE 10.38 (continued)   (b) Gaussian beam.

Example 10.1

Consider a halftone dot matrix $Q(x)$ given by Equation 10.63 and as shown in Figure 10.40 where the dimensions of the x–y axes are given in microns. The matrix representing the halftone patch contains series of ones and zeros as shown below:

$$Q(x) = \begin{bmatrix} 0 & 1 & 0 & 0 & 0 & 0 & 0 & 0 \\ 1 & 1 & 1 & 0 & 0 & 0 & 0 & 0 \\ 1 & 1 & 1 & 0 & 0 & 0 & 0 & 0 \\ 0 & 1 & 0 & 0 & 0 & 0 & 0 & 0 \\ 0 & 0 & 0 & 0 & 0 & 0 & 1 & 0 \\ 0 & 0 & 0 & 0 & 0 & 1 & 1 & 1 \\ 0 & 0 & 0 & 0 & 0 & 1 & 1 & 1 \\ 0 & 0 & 0 & 0 & 0 & 0 & 1 & 0 \end{bmatrix} \qquad (10.63)$$

In this example, the resolution of the halftone dot is selected to be approximately 35 microns per pixel. However, to model the effect of the laser beam, the resolution has to be enhanced to around 7–15 microns. This is done by upsampling the laser profile (Figure 10.38) five times per pixel for a 7 micron resolution. We assume here that the laser beam is exposed at the center of each scan line (Figure 10.41).

The lighter (highlighted) regions of Figure 10.40 reflect the pixels that need to be exposed by the laser beam. This step is identical to converting the image to ideal video on a simulation grid. The upsampling factor is usually chosen to be an odd number. For an upsampling factor $p = 5$, we upsample each scan line five

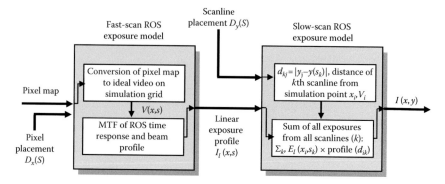

FIGURE 10.39   Steps involved in modeling the ROS profile.

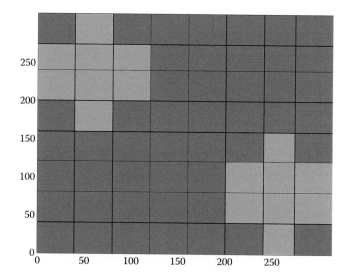

FIGURE 10.40   Halftoned patch.

times and expose the center third row leaving $(p-1)/2$ rows on either side unexposed since the laser beam is exposed at the center of each scan line. The upsampled image can be mathematically obtained by taking a Kronecker product with a suitably chosen matrix. Define the matrix S as follows to reflect the upsampling mentioned above:

$$S = \begin{bmatrix} 0 & 0 & 0 & 0 & 0 \\ 0 & 0 & 0 & 0 & 0 \\ 1 & 1 & 1 & 1 & 1 \\ 0 & 0 & 0 & 0 & 0 \\ 0 & 0 & 0 & 0 & 0 \end{bmatrix}_{p \times p} \tag{10.64}$$

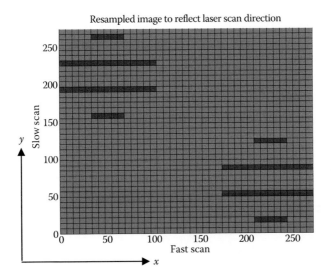

FIGURE 10.41   Upsampled halftone patch.

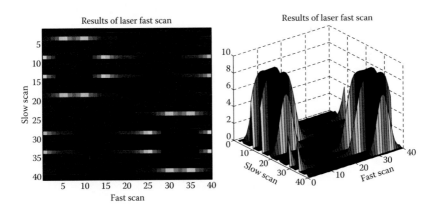

FIGURE 10.42   Effect of laser profile along fast-scan direction (a linear exposure profile).

Then the upsampled image is given by $R(x) = Q(x) \otimes S$, where the symbol $\otimes$ stands for Knocker product. The given S is obtained for $p = 5$, for arbitrary p the scaling matrix S had the same structure with ones along the appropriate row. Thus, if the halftoned image is of size $n \times n$, the rescaled matrix will be of size $np \times np$. In the next stage, the upsampled image is convolved along the fast-scan direction with a filter representing the laser beam. This is being shown in Figure 10.42. Here the trailing edges in the fast-scan direction show the effect of convolution. The z-axis in the surf plot shows the photon intensity after convolution. This step gives us the linear exposure profile (Table 10.12).

The final ROS profile is obtained by convolving the linear exposure profile with the Gaussian or $sinc^2$ filter along the slow-scan direction. This step models the

TABLE 10.12

Typical Values for Parameters α_i, β_i, and γ_i

	C	M	Y	K
α	2.798	2.490	2.064	6.907
β	2.706	4.897	8.447	−1.355
γ	1.033	1.180	1.448	
ρ	0.007	0.000	0.000	0.019

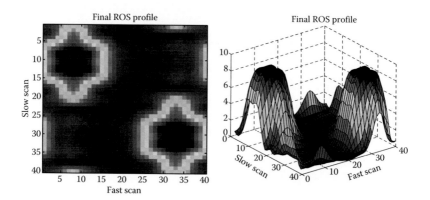

FIGURE 10.43   Effect of convolution along the slow-scan direction.

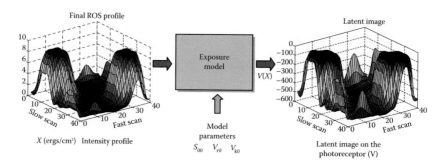

FIGURE 10.44   Exposure of latent image on the photoconductor.

effect of beam spreading along the slow-scan direction. The effect of convolution along the slow-scan direction is shown in Figure 10.43.

The next step is to obtain the latent image on the photoconductor by using the laser intensity profile generated in Figure 10.43 with the exposure model shown in Tables 10.3 and 10.4. Figure 10.44 shows the latent image after the dot printer model pixel by pixel on the photoconductor.

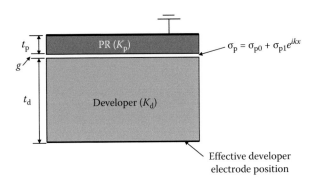

FIGURE 10.45   Schematic of a development nip.

Similarly, the development process can be modeled completely using the dot development models shown in Table 10.6 and the MTF equations shown in Equation 10.65.

In a development nip, toners are recruited from the top of the magnetic brush to the PR by electric fields due to the latent image. Figure 10.45 shows a schematic of a development nip. The MTF for the normal electric field at the surface of the magnetic brush may be written as

$$\text{MTF}(k) = \frac{K_d}{A_k + B_k} \cosh(kt_d) \tag{10.65}$$

where A_k and B_k are given by

$$A_k = t_p \coth(kt_p)\left\{\frac{K_d + 1}{2}\sinh[k(t_d + g)] - \frac{K_d - 1}{2}\sinh[k(t_d - g)]\right\}$$

$$B_k = \frac{t_p}{K_p}\left\{\frac{K_d + 1}{2}\cosh[k(t_d + g)] + \frac{K_d - 1}{2}\cosh[k(t_d - g)]\right\} \tag{10.66}$$

k is the angular spatial frequency
t_p is the PR thickness
K_p is the PR dielectric constant
g is the air gap
t_d is the effective developer layer thickness
K_d is the dielectric constant of the developer
sinh(.), cosh(.), and coth(.) are hyperbolic sine, hyperbolic cosine, and hyperbolic cotangent functions, respectively

Note that Equation 10.65 is similar to the MTF derived by Neugebauer [33] and others. The effective developer layer thickness may be expressed as

$$t_d = \frac{g t_{d,\infty} \exp\left(-\frac{T_d}{\tau}\right)}{g + \left(\frac{t_{d,\infty}}{K_d}\right)\left[1 - \exp\left(-\frac{T_d}{\tau}\right)\right]} \tag{10.67}$$

where

$t_{d,\infty}$ is the actual developer layer thickness

T_d is the nip dwell time $= \frac{L}{U_{mag}}$, where L is the nip width and U_{mag} is the speed of the magnetic roll

τ is the developer charge relaxation time constant, which is given by

$$\tau = \frac{K_d \varepsilon_0}{\gamma_d} \frac{g + \frac{t_{d,\infty}}{K_d} + \frac{t_p}{K_p}}{g + \frac{t_p}{K_p}} \tag{10.68}$$

Figure 10.46 show the image after applying the development process. Table 10.13 shows typical parameters used for modeling the development MTF. The input to the model is the latent image on the PR and the output is the DMA on the PR. Each dot is developed using the dot printer model.

Similarly, the transfer process can be modeled using the dot transfer models shown in Table 10.8 and the profile functions of the transfer process. The input to the

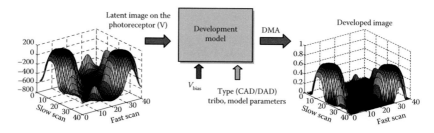

FIGURE 10.46 Developed image on the photoconductor.

TABLE 10.13
Parameters Used in the Development MTF

$t_d = 400 \times 10^{-3}$ mm	Thickness of the brush
$K_d = 4.6$	Dielectric constant of the developer
$g = 10 \times 10^{-3}$ mm	Air gap
$t_p = 25 \times 10^{-3}$ mm	PR thickness
$K_p = 3$	PR dielectric constant
$\varepsilon_0 = 8.854 \times 10^{-14}$ C/V-cm	Permittivity of free space
$\gamma_d = 1.2 \times 10^{-10}$ (ohm-cm)$^{-1}$	Developer conductivity
$T_d = 4 \times 10^{-3}$ s	Nip dwell time

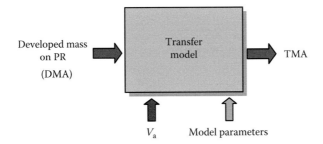

FIGURE 10.47   Transferred image on the paper.

transfer model is the output of the development process and the output is the TMA on the paper for all the *CMYK* separations (Figure 10.47).

Fusing and paper MTFs can also be incorporated based on the need to approximate the actual spatial response of the printed image to the model output. The paper MTFs are particularly useful for modeling the appearance of high-resolution images, since the images are strongly affected by the diffusion of light through the paper. Light scattering effects in xerographic images is modelled in Ref. [34]. Coleman and Li [35] present an approach to model the paper MTF using an experimentally determined paper light-spread function to describe the scattering phenomenon.

10.4   TONE REPRODUCTION CURVE

The tone reproduction curve (TRC) of a printer is the relationship between the input pixel value and the tone level that appears at various points in the imaging process. The TRC at the end of the development process refers to the function of the toner mass deposited with respect to the input pixel value represented in terms of the requested gray level or area coverage. Ideally, this graph should be linear for a linear development. Due to various printer nonlinearities like laser spreading in the exposure process, dot spreading in the development process, light diffusion and scattering on paper, etc., the TRC is a nonlinear function. The curve also depends on various process parameters. TRCs can be computed for various halftoning schemes using the printer model described in this chapter by creating a step change in gray levels (Figure 10.48). Figure 10.49 shows TRC functions for cyan separation for two different halftoning schemes.

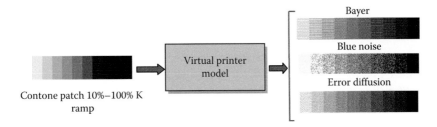

FIGURE 10.48   Schematic for computing the TRC.

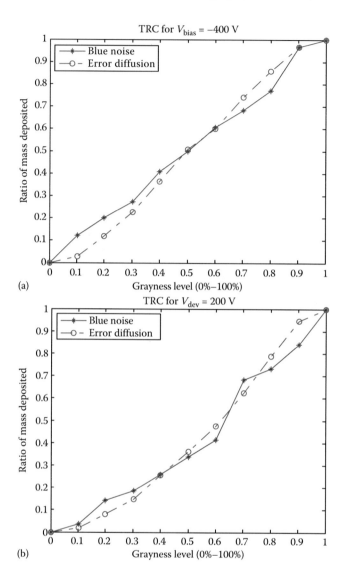

(a)

(b)

FIGURE 10.49 (a) TRC for $V_{bias} = -400$ V and (b) TRC for $V_{dev} = 200$ V.

10.5 IMAGE SIMULATION WITH FUSING AND COLOR MODELS

Once transferred to the paper, the *CMYK* separations are fused in the fusing process and then transformed into reflectance spectra using the equations described in Section 10.2.6 (Figure 10.50). This process is computationally intense for a full page image simulation on a 8.5″ × 11″ size paper is required.

Digital color printers are capable of reproducing images of different sorts such as text, lines, *RGB* color, etc. To describe the key steps in the full page image

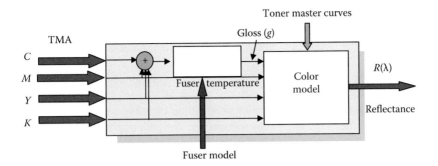

FIGURE 10.50 Effect of transferred image on the paper.

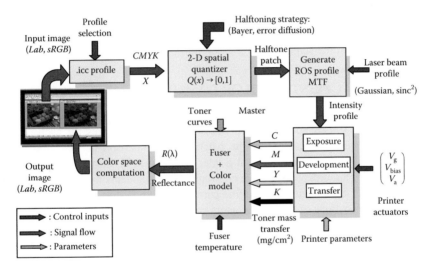

FIGURE 10.51 Image flow in the virtual printer model.

simulation, we assume that the input image is given in the $L^*a^*b^*$ or $sRGB$ format. This input image needs to be transformed to a $CMYK$ separated image before it can be processed by the model. For this purpose, it is necessary to develop a machine profile that is an inverse mapping between $L^*a^*b^*$ (or XYZ) and device $CMYK$ values using techniques described in Chapter 7.

Figure 10.51 shows an overview of the image flow in the printer model. The input to the model is an electronic image in $sRGB$ format and the output of the model is an electronic image in the CIELab format, which can be transformed to the $sRGB$ format to display on the screen for evaluation purposes. The different stages that the input image goes through are summarized as follows. We omit the computational details like scaling of images and the resolution concerns for a later section.

- The input image is transformed from its source space (e.g., *sRGB*) to *L*a*b**
  coordinates and then on to the *CMYK* space by using source and destination
  profiles. The destination profile is customized to the virtual printer model of
  Figure 10.51.
- The transformed image consisting of *CMYK* separations is halftoned using a
  desired halftoning strategy.
- The halftoned separations are exposed on the PR using a Gaussian or $sinc^2$
  beam profile. This takes into account the dot gain due to laser spreading
  effect at the microlevel. This has significant effect on the perceived color of
  the images.
- Next the image consisting of four separations are passed through the dot
  printer model. The output is the transferred mass density for each element
  of the grid for each *CMYK* separation.
- The mass density of four separations is used to generate *L*a*b** values for
  each pixel using standard color models.
- The output image is transformed to the image source space (e.g., *sRGB*) and
  displayed on the screen for evaluation purposes.

10.6   VIRTUAL PRINTER COLOR GAMUT

The color gamut for the virtual printer can be obtained using a series of color patches
spread over the *CMYK* space (Figure 10.52). The *L*a*b** values for each of the
patches are obtained by averaging the patches. A comparison between the color
gamut of the virtual printer and a real printer is shown in Figure 10.53. The actual
available color gamut depends on the colorants of the pigments that are captured in
the toner master curves and other engine parameters like DMA, TMA, paper gloss,
etc. Since combinations of these parameters are not tuned to the actual printer shown
in Figure 10.53, the gamut volume, shape, and limits come out different. However,
this is not a trivial task (see Section 10.7), and requires careful design of experiments
and must be done in a laboratory environment. A comparison of color gamuts created
using the virtual printer can be seen in Figures 10.54 and 10.55. In Figure 10.54,
the solid gamut is created by using 9% TC for all the four separations while the
mesh gamut is created by using 12% TC for all the four separations. As a result of

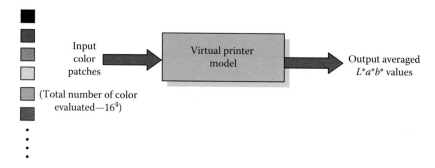

FIGURE 10.52   Schematic for obtaining the color gamut.

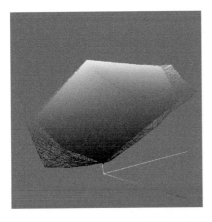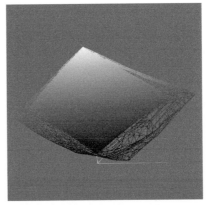

FIGURE 10.53 Comparison of virtual printer gamut (solid) with a representative color printer gamut (mesh).

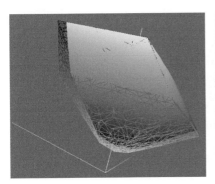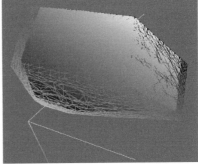

FIGURE 10.54 Comparison of virtual printer gamut from cell 1 (9% TC) (solid) with another gamut from cell 2 (12% TC) (mesh).

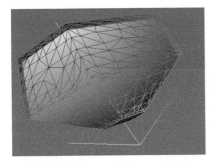

FIGURE 10.55 Comparison of virtual printer gamut from medium 1 (solid) with another gamut from medium 2 (mesh) on the same print engine.

the increase in TC, the volume of the gamut has increased by a significant amount. In Figure 10.55, the two gamuts are created by using different media. This is achieved by plugging in the reflectance spectrum of the substrate $(R_s(\lambda))$ in Equation 10.54.

10.7 VIRTUAL PRINTER MODEL TUNING TO AN EXPERIMENTAL PRINTER

High-volume digital color printers designed with image-on-image EP technology perform recharge and exposure (several times) on the developed toner layer. Accordingly, light exposure depends on the optical transmissivity of multiple toner layers. The voltage across a previously developed toner layer, development suppression due to image voltage, etc. are other contributing factors that play an important role in the actual response of the printer. In a real printer, these physical processes interact in a more complicated way as compared with the processes modeled by the virtual printer and are specific to a particular architecture of the print engine. To make the models more representative of the physical process, a few key parameters can be tuned to generate colors similar to the real printer. A few important steps used in tuning the parameters of the virtual printer models are as follows:

(1) Use realistic values for the process parameters/actuators like voltage levels, PR/toner/media optical properties, etc. for a nominal print engine.
(2) Tune the toner master curves using experimental data that represent the absorption spectra of each of the C, M, and Y toners.
(3) Tune the single separation coefficients in the color model for each of the C, M, Y, and K separations.
(4) Tune the color mixing coefficients for C, M, and Y separations.
(5) Match the channel-wise TRCs, assuming each separations of the real printer is linearized with a TRC.

10.7.1 TUNING TONER MASTER CURVES

In the main equation of the reflectance spectral model (Equation 10.54), assume the transmittance from air through toner to paper equals the return transmittance from paper through toner to air (i.e., $\tau(m_i; \lambda) = \tau'(m_i; \lambda)$). Equation 10.54 can then be simplified as

$$R(m_i, g; \lambda) = p_0(g) + [1 - p_1(n_1, n_2, i)]\tau^2(m_i; \lambda)R_s(\lambda) \qquad (10.69)$$

The total FSR term, $p_1(n_1, n_2, i)$, depends on the refractive indices n_1 and n_2 of the air and toner layer as well as the angle of incidence of light. The measurable FSR term, $p_0(g)$, depends on the total FSR and gloss, which (again) is a function of the fusing temperature and transferred mass on the substrate. For our tuning purpose, all the above terms are assumed constant for a single toner. So let $p_0(g) = C_1$ and $1 - p_1(n_1, n_2, i) = C_2$.

In Equation 10.69, the substrate reflectance is obtained from the reflectance measurement for the patch of $C=M=Y=K=0$ of the real printer $R(0; \lambda)$ using Equation 10.70 that is modeled based on Equation 10.69:

$$R_s(\lambda) = \frac{R(0, g; \lambda) - C_1}{C_2} \qquad (10.70)$$

In order to determine the toner master curve for a particular printer for a single separation, for example, the cyan channel, a single patch $C=255$, $M=0$, $Y=0$, and $K=0$ is printed and the spectral reflectance measurements are taken. Keep the other three channels fixed at 0 so that their toners will not be deposited and will not affect the transmittance term, $\tau(m_i; \lambda)$. The reason for choosing 100% area coverage for cyan is to avoid the effects of halftoning on the spectral measurements. Equation 10.69 can then be written as

$$R(m_C; \lambda) = C_1 + C_2 \tau^2(m_C; \lambda) R_s(\lambda) \qquad (10.71)$$

$$\tau^2(m_C; \lambda) = \frac{R(m_C; \lambda) - C_1}{C_2 R_s(\lambda)} \qquad (10.72)$$

For the patch $C=255$, $M=0$, $Y=0$, and $K=0$, $m_1=m_C$ and $m_2=m_3=0$ in Equations 10.58 through 10.61, Equations 10.59 and 10.60 can be simplified to

$$\begin{aligned}
\sigma_C(m^*) &= \alpha_C \left[1 - \exp\left(-\beta_C m_C^{\gamma_C}\right)\right] \\
\sigma_M(m^*) &= 0 \\
\sigma_Y(m^*) &= 0 \\
\sigma_K(m^*) &= 0
\end{aligned} \qquad (10.73)$$

From Equations 10.58 and 10.73, we have

$$\tau(m_C; \lambda) = \exp\left\{-M_C(\lambda)\alpha_C \left[1 - \exp\left(-\beta_C m_C^{\gamma_C}\right)\right]\right\} \qquad (10.74)$$

Taking the natural logarithm of both sides of Equation 10.74, we obtain

$$\ln \tau(m_C; \lambda) = -M_C(\lambda)\alpha_C \left[1 - \exp\left(-\beta_C m_C^{\gamma_C}\right)\right] \qquad (10.75)$$

Taking the natural logarithm of both sides of Equation 10.72 and simplifying, we obtain

$$\ln \tau(m_C; \lambda) = \frac{\ln\left[\frac{R(m_C;\lambda)-C_1}{C_2}\right] - \ln R_s(\lambda)}{2} \qquad (10.76)$$

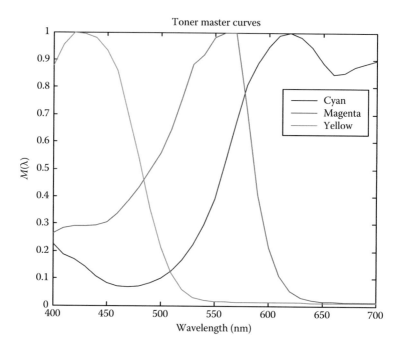

FIGURE 10.56   Toner master curves used for tuning the virtual printer model.

Solving Equations 10.75 and 10.76 for the toner master curve, $M_C(\lambda)$, we obtain

$$M_C(\lambda) = \frac{\ln R_s(\lambda) - \ln\left[\frac{R(m_C;\lambda)-C_1}{C_2}\right]}{2\alpha_C\left[1 - \exp\left(-\beta_C m_C^{\gamma_C}\right)\right]} \qquad (10.77)$$

Since it is required to normalize the master curves in such a way that the maximum value is 1, we can ignore the denominator. Hence,

$$M_C(\lambda) \cong \ln R_s(\lambda) - \ln\left[\frac{R(m_C;\lambda) - C_1}{C_2}\right]\Bigg|_{\text{MaxVal}=1} \qquad (10.78)$$

The master curves for the other two separations can be determined in a similar fashion. Equation 10.78 gives the best approximation of master curves used in the virtual printer model while trying to emulate a real printer. The toner master curves derived from a real printer using Equation 10.78 are shown in Figure 10.56 for C, M, and Y toners.

10.7.2   Tuning of Single Separation Coefficients

Equation 10.78 shows how to obtain the toner master curves from real experimental data with paper white and 100% color patches. Below we show how to optimize the single separation coefficients based on the measurements made at various area coverages along each of the primaries. For cyan or magenta or yellow channels, if the primaries are not mixed, the transmittance (Equation 10.58) reduces to Equation 10.79. Similarly, the black channel is reduced to Equation 10.80:

$$\tau(m_i; \lambda) = \exp\{-M_i(\lambda)\alpha_i[1 - \exp(-\beta_i m_i^{\gamma_i})]\}, \quad i = C/M/Y \quad (10.79)$$

$$\tau(m_i; \lambda) = \exp\left\{-\left[\alpha_K\left(\frac{m_K}{\rho_t}\right) + \beta_K\left(\frac{m_K}{\rho_t}\right)^2\right]\right\} \quad (10.80)$$

This transmittance term is used in Equation 10.69 to obtain the spectral reflectance values for any color patches that are then converted to tristimulus values and $L^*a^*b^*$.

Let us consider the cyan channel for illustration. Consider 10 uniformly spaced points along the cyan primary while keeping $M = Y = K = 0$. Print these patches on the real printer. The $L^*a^*b^*$ values are measured from the printed patches using a spectral measurement device. The measured $L^*a^*b^*$ values represent the reference values. The cost function shown in Equation 10.81 is then minimized to find the optimal values of α, β, and γ:

$$F(\alpha_i, \beta_i, \gamma_i) = \sum_{i=1}^{10} \left\| (L^*a^*b^*)_i^T - (L^*a^*b^*)_{iref}^T \right\|_2^2 \quad (10.81)$$

α and β do not have any constraints, but γ has a lower bound as it cannot go negative since it is in the exponent. So a bound-constrained nonlinear least-squares optimization algorithm is used to minimize Equation 10.81. Such an algorithm was developed by Coleman and Li [35]. The optimized single separation coefficients are shown in Table 10.14.

The color gamuts of the real printer and the partially tuned virtual printer model are shown in Figure 10.57. One can observe that the color gamuts are well matched near the primaries C, M, and Y. Matching at the bottom of the gamut is not good because of the differences in the physics of the virtual printer model and the real printer in that region. The other regions are also not well matched because we have not yet optimized the color mixing coefficients.

10.7.3 DETERMINATION OF COLOR MIXING COEFFICIENTS $\{C_{ji}\}$

The color mixing coefficients shown in Equation 10.59 determine the way the transferred masses on paper are mixed in the color model to generate the mathematical masses, which in turn affect the reflectance spectra of each color. There are nine color mixing coefficients that are required to determine the mathematical masses,

TABLE 10.14
Single Separation Coefficients

	C	M	Y	K
α	5.1886	4.3987	3.8420	26.5352
β	7.4488	6.4264	9.2957	−62.7119
γ	1.2416	1.1029	1.1302	—

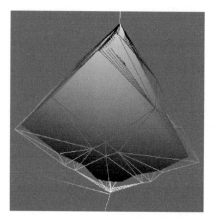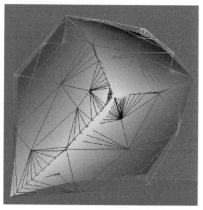

FIGURE 10.57 Two views of the color gamuts after single separation coefficient optimization (solid: real printer and wire mesh: virtual printer model).

which are shown in Tables 10.10 and 10.11. The combination of the transferred masses that these color mixing coefficients affect are shown in Equation 10.61. The first three masses are just primaries and hence the color mixing coefficients are all 1's. The fourth element is a combination of C and M channels and there are two coefficients that affect them, namely C_{41} and C_{42}. Similarly, the combination of M and Y channels has coefficients of C_{52} and C_{53}, the combination of C and Y channels has coefficients of C_{61} and C_{63}, the combination of C, M, and Y channels has coefficients of C_{71}, C_{72}, and C_{73}. To determine C_{41} and C_{42}, we use 25 uniformly spaced points with combinations of C and M primaries while keeping $Y = K = 0$. To determine C_{52} and C_{53}, we use 25 uniformly spaced points with combinations of M and Y primaries while keeping $C = K = 0$. To determine C_{61} and C_{63}, we use 25 uniformly spaced points with combinations of C and Y primaries while keeping $M = K = 0$. To determine C_{71}, C_{72}, and C_{73}, we use 125 uniformly spaced points with combinations of C, M, and Y primaries while keeping $K = 0$.

The $L^*a^*b^*$ values for a real printer patch set can be regarded as the reference values. The cost function shown in Equation 10.82 is minimized to find the optimal values of the corresponding color mixing coefficients:

$$F(C_{ji}) = \sum_{i=1}^{N} \left\| (L^*a^*b^*)_i^{\mathrm{T}} - (L^*a^*b^*)_{\mathrm{ref}}^{\mathrm{T}} \right\|_2^2 \qquad (10.82)$$

All the color mixing coefficients are unbounded and unconstrained. Therefore, we use an unconstrained nonlinear least-squares optimization algorithm to minimize Equation 10.82. The Levenberg–Marquardt algorithm provides a good tool for this purpose. The complete description of the method can be found in Ref. [36]. The optimized color mixing coefficients are shown in Table 10.15.

The color gamuts of the real printer and the tuned virtual printer model are shown in Figure 10.58. One can observe a major issue at the bottom of the gamut

TABLE 10.15

Optimized Color Mixing Coefficients for Virtual Printer Model

	C	M	Y
C	1	0	0
M	0	1	0
Y	0	0	1
$C+M$	−1.6308	−1.4360	0
$M+Y$	0	−1.9811	−2.2318
$C+Y$	−2.08	0	−3.0025
$C+M+Y$	14.3163	16.0697	29.0905

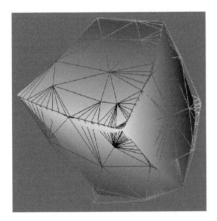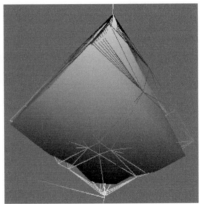

FIGURE 10.58 Two views of the color gamuts after color mixing coefficient optimization (solid: real printer and wire mesh: virtual printer model).

because the virtual printer model does not capture the physics of the real printer in that region. The other regions are also not well matched, which indicates an opportunity for further investigation into optimizing the color mixing coefficients.

10.7.4 ONE-DIMENSIONAL CHANNEL-WISE TRC MATCHING

As a final step, to match the virtual printer model to a real printer, we perform the matching of single separation TRCs on paper (ΔE_{ab}^* from paper) of the virtual printer model to the real printer. A complete methodology for TRC matching can be found in Chapter 8. The original TRC of the virtual printer model, TRC of a real printer, and TRC of the virtual printer model after performing 1-D channel-wise matching for cyan channel can be seen in Figure 10.59. The TRC correction lookup table (LUT) for the cyan channel is shown in Figure 10.60.

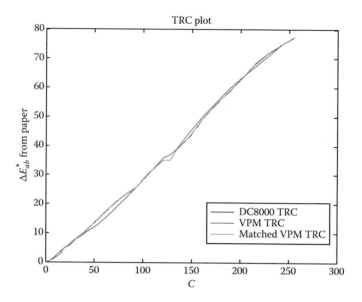

FIGURE 10.59   TRC (on paper) for cyan channel.

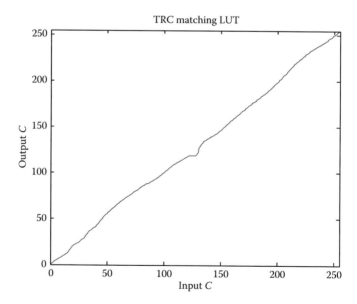

FIGURE 10.60   TRC correction LUT used for cyan channel.

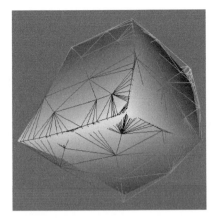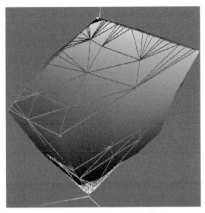

FIGURE 10.61   Two views of the color gamuts after 1-D channel-wise TRC matching (solid: real printer and wire mesh: virtual printer model).

The color gamuts of the real printer and the tuned virtual printer model after 1-D channel-wise TRC matching are shown in Figure 10.61. There is a significant improvement in the internal color matching after performing this step though it is not visible in Figure 10.61.

10.7.5   TUNING RESULTS

The ΔE_{ab}^* statistics between the tuned virtual printer model and a real printer for the 1617 GRAYCol color set are shown in Table 10.16. There is still a significant error between the tuned model and the real printer. There are two main reasons for this are (1) the physical color model does not accurately represent the dark part of the gamut and (2) there is a need for better optimization of color mixing coefficients.

TABLE 10.16
ΔE_{ab}^* Statistics between Tuned VPM and DC8000 Printer Model

Condition	Number of Test Patches	ΔE_{ab}^* mean	ΔE_{ab}^* 95%	ΔE_{ab}^* mean
—	1617	3.6996	7.92	11.5891
$K = 0$	818	3.29	6.637	11.0417
$L^* \geq 30$	1308	3.1481	6.2816	11.0417
$L^* \geq 50$	870	2.8764	5.6197	11.0417
$L^* \geq 70$	429	2.3722	4.9918	7.846

10.7.6 SUMMARY

The virtual printer model parameters can be further matched to a real printer using the above techniques if the process parameters (e.g., process voltages, TC, tribo, etc.) in the steps before the color model (like development station and transfer station) are updated with more realistic values of a nominal state of the real printer. There is also a need for improving the structure of the model to capture the physical process in the darker part of the gamut. However, these partially tuned models based on numerous process parameters accurately predict the general behavior of the printer and are, therefore, considered adequate for developing advanced feedback systems and evaluating their effects regarding quality. Various process control architectures, calibration approaches (1-D, 2-D, etc.), multidimensional profiling, and image-processing techniques can be matured with these models prior to print testing.

PROBLEMS

10.1 Let the parameters of a charging system be

$$S = 0.9 \times 10^{-6} \frac{A}{V\text{-m}}, \quad C = 9.486 \times 10^{-7} \frac{C}{V\text{-m}^2}, \quad \text{and} \quad v = 0.254 \frac{m}{s}$$

(a) Plot $V_P(L)$ as a function of grid voltage V_g. Assume $V_P(0) = 0$.
(b) Find sensitivity of the photoconductor surface voltage with respect to the grid voltage V_g.
(c) Repeat part b if v is decreased by 10%. Assume all the other parameters remain the same.

10.2 Consider a charging system given by differential equation

$$C \frac{dV_p}{dt} = S(V_g - V_p)$$

where the grid voltage $V_g = 800$ V, $S = 0.9 \times 10^{-6} \frac{A}{V\text{-m}}$, and $C = 9.486 \times 10^{-7} \frac{C}{V\text{-m}^2}$.
(a) Find photoconductor surface voltage $V_p(t)$ if $V_p(0) = 0$.
(b) Plot $V_p(t)$ as a function of time.

10.3 Consider the Springett–Melnyk exposure model given by

$$V(X) - V_i + V_c \ln\left(\frac{V(X) - V_r}{V_i - V_r}\right) + \frac{e\eta_0}{C} X = 0$$

(a) Solve the above equation numerically and plot $V(X)$ as a function of $\frac{e\eta_0}{C} X$. Assume the following parameters for the model: $V_i = 500$ V, $V_c = 4$ V, and $V_r = 20$ V.
(b) Solve the Springett–Melnyk exposure model for $p = 3$.

10.4 A square sheet 2 m on a side situated in the x–y plane and centered on the origin has a charge of 1 nC uniformly distributed over its surface.
   (a) Find the electric potential V at a distance of 1 m perpendicular from center of the square.
   (b) Find the electric field $\vec{E}$ at the same point.

10.5 For a two-component conductive magnetic-brush development system, assume that the toner charge per unit area completely neutralizes the photoconductor charge per unit area. Show that with this assumption, the DMA can be predicted by the following expression:

$$D = \frac{M}{A} = \frac{V_{dev}\varepsilon_0}{(Q/M)(d_p/K_p)}$$

where
   d_p is the thickness of the photoconductor
   K_p is the dielectric constant
   V_{dev} is the development voltage
   ε_0 is the dielectric permittivity of air
   Q/M is the toner tribo (charge to mass ratio)

10.6 For a four-color printing system, using the techniques described in Section 10.7, obtain the printer gamut for five different paper types. Use measured paper reflectance spectra for each type.

REFERENCES

1. D.A. Hays and K.R. Ossman, Electrophotographic copying and printing (xerography), in *The Optics Encyclopedia*, Berlin, Germany: Wiley-VCH, 2003, pp. 587–607.
2. R. Lux and H.-J. Yuh, Is image-on-image color printing a privileged printing architecture for production digital printing applications? *NIP20: Proceeding of the IS&T's International Conference on Digital Printing Technologies*, pp. 323–327, Salt Lake City, UT, Oct. 31–Nov. 5, 2004.
3. D.M. Pai and B.E. Springett, Physics of electrophotography, *Reviews of Modern Physics*, 65(1), January 1993.
4. R.M Schaffert, *Electrophotography*, London and New York: The Focal Press, 1975.
5. J.A. Giacometti, S. Fedosov, and M.M. Costa, Corona charging charging of polymers: Recent advances on constant current charging, *Brazilian Journal of Physics*, 29(2), Jun. 1999.
6. J.Q. Feng, P.W. Morehouse, and J.S. Facci, Steady-state corona charging behavior of a corotron over a moving dielectric substrate, *IEEE Transactions on Industry Applications*, 37(6), November/December 2001.
7. S. Maitra, P. Ramesh, and S. Jeyadev, Impact of photoreceptor design on digital electrophotography, in *Color Imaging: Device-Independent Color, Color Hard Copy, and Graphic Arts*, San Jose, CA, Jan. 29–Feb. 1, 1996; S. Maitra, P. Ramesh, and S. Jayadev, Impact of photoreceptor design on digital electrophotography, *Proceedings of the SPIE*, 2658, 112–122, 1996.
8. A.C. Tam and R.D. Balanson, Lasers in electrophotography, *IBM Journal of Research Development*, 26(2), Mar. 1982.

9. E.M. Williams, *The Physics and Technology of Xerographic Processes*, New York: John Wiley, 1984.

10. I. Chen, Optimization of photoreceptors for digital electrophotography, *Journal of Imaging Science*, 34, 15–20, 1990.

11. B.E. Springett, Maintaining photoreceptor charge potential constant; vol. 4, no. 5, Sep.–Oct. 1979; K.W. Pietrowski, S.W. Erig, R.L. Bullock, T. Fluk, C.H. Tabb, Zhao-Zhi Yu, J.J. Folkins, D.M. Bray, C.G. Edmunds, Method and apparatus for reducing residual toner voltage, US Patent, 5,581,330, Dec. 3, 1996; Springett–Melnyk Model (Internal Xeror Communications).

12. S. Jeyadev, Derivation of Springett–Melnyk PIDC model, Internal Memorandum, Xerox Corporation, January 2000; Other US Patents 6,771,912.

13. L.B. Schein, *Electrophotography and Development Physics*, 2nd edition, New York: Springer-Verlag, 1992.

14. B. Nash, Toner developer physics, Xerox Lecture Handouts, Aug. 25, 1993.

15. E. Gutman and M. Scheuer, Development in the xerographic process, Xerox Internal Notes, 1995.

16. L.B. Schein, Theory of toner charging, *Journal of Imaging Science and Technology*, 37(1), Jan./Feb. 1993.

17. L.B. Schein, W.S. Czarnecki, B. Christensen, T. Mu, and G. Galliford, Theory of toner adhesion, *NIP20: Proceedings of the IS&T's International Conference on Digital Printing Technology*, pp.163–168, Salt Lake City, Utah, Oct.–Nov. 2004.

18. P.S. Ramesh, Simulating digital exposure of xerographic photoreceptors using the domain-decomposition method, *IEEE Transactions on Industrial Applications*, 42, 392–398, 2006.

19. E.J. Gutman and G.C. Hartmann, Triboelectric properties of two component developers for xerography, *Journal of Imaging Science and Technology*, 36(4), Jul./Aug. 1992.

20. P. Ramesh, Quantification of toner aging in two component development systems, *NIP21: Proceedings of the IS&T's Non-Impact Printing Conference*, pp. 544–547, Baltimore, MD, Sep. 2005.

21. Jeff Folkin and P.S. Ramesh (Private Communication).

22. C.C. Yang and G.C. Hartmann, Electrostatic separation of a charged-particle layer between electrodes, *IEEE Transactions on Electron Devices*, vol. 23, no. 3, Mar. 1976.

23. M.C. Zaretsky, Performance of an electrically biased transfer roller in a Kodak ColorEdge CD copier, *Journal of Imaging Science and Technology*, 37(2), Mar./Apr. 1993.

24. J.C. Briggs, J. Cavanaugh, M.-K. Tse, and D.A. Telep, The effect of fusing on gloss in electrophotography, *NIP14: Final Proceedings of the IS&T's International Conference on Digital Printing Technologies*, pp. 456–461, Toronto, Canada, Oct. 18–23, 1998.

25. M.K. Tse, F.Y. Wong, and D.J. Forrest, A fusing apparatus for toner development and quality control, *NIP13: Final Proceedings of the IS&T's International Conference on Digital Printing Technologies*, pp. 310–313, Seattle, WA, Nov. 2–7, 1997.

26. M.K. Tse, J.C. Briggs, and D.J. Forrest, Optimization of toner fusing using a computer-controlled hot roll fusing test system, *ICISH'98, 3rd International Conference on Imaging Science and Hardcopy*, Chongqing, China, May 26–29, 1998.

27. J.J. Folkins, S. Haque, A.M. Loeb, H.R. Till, Control system for regulating the dispensing of marking particles in an electrophotography printing machine, US Patent 4,492,179, Jan. 8, 1985.

28. P.S. Ramesh and Y. Gartstein, Modeling toner transport in single-component non-contact development, *NIP13: International Conference on Digital Printing Technologies*, Seattle, WA, pp. 67–70, Nov. 2–7, 1997.

29. J.G. Shaw, T. Retzlaff, and P.S. Ramesh, Particle based simulations of image quality defects, *Proceedings of the IS&T's 50th Annual Conference*, Webster, NY, pp. 269–272, 1997.

30. Contributions of P.S. Ramesh of Xerox Innovation Group, Xerox Corporation.

31. P. Kubelka, New contributions to the optics of intensely light-scattering materials: Part I, *Journal of the Optical Society of America*, 38, 448–457, 1947.

32. E.N. Dalal and K.M. Natale-Hoffman, The effect of gloss on color, *Color Research & Application*, vol. 24, Issue 5, pp. 369–376, Aug. 1999.

33. H.E.J. Neugebauer, A describing function of modulation transfer of xerography, *Applied Optics*, 4(4), 453, 1995.

34. M. Maltz, Light scattering in xerographic images, *Journal of Applied Photographic Engineering*, 9(3), Jun. 1983.

35. T.F. Coleman and Y. Li, An interior, trust region approach for nonlinear minimization subject to bounds, *SIAM Journal on Optimization*, 6, 418–445, 1996.

36. D.P. Bertsekas, *Nonlinear Programming*, 2nd edition, Belmont, MA: Athena Scientific, 1995.

Appendix A

A.1 INTRODUCTION

Color imaging systems for color reproduction, storage, and manipulations are widely available for consumer and industrial applications. Examples of such devices are scanners, digital cameras, LCD monitors, and digital printers. Color engineering is a discipline that applies color science to the design of color imaging systems.

A.2 HUMAN VISUAL SYSTEM

When incident light is reflected from an object, it is focused by the cornea and the lens to form image of the object on the retina. The retina contains the rods and cones. They act as sensors for the human visual system (HVS). The rods are responsible for sensing the monochromic vision at low light intensity and the cones are sensing colors. There are three different kinds of cones in a HVS for a normal observer: short, medium, and long. The short cone (S) is sensitive to the short wavelength of light, and the medium (M) and long (L) cones are sensitive to the medium and long wavelengths, respectively. The three cone sensitivity functions can be designated by $S_i(\lambda)$, $i = 1, 2, 3$.

A.3 COLOR MATCHING FUNCTIONS

Let $R(\lambda)$ be the reflectance spectra of the object and $L(\lambda)$ be the light source incident on the object, then the spectral power distribution of the incident light on the retina is $f(\lambda) = R(\lambda)L(\lambda)$ and the response of the cones to the incident light is given by

$$c_i = \int_{\lambda_{min}}^{\lambda_{max}} f(\lambda)S_i(\lambda)d\lambda = \int_{\lambda_{min}}^{\lambda_{max}} R(\lambda)L(\lambda)S_i(\lambda)d\lambda \tag{A.1}$$

where $S_i(\lambda)$, $i = 1, 2, 3$, are the short, medium, and long cone sensitivity functions of the observers. The above integral can be approximated by sampling the spectra uniformly from λ_{min} to λ_{max} using N samples. The resulting equation is

$$c_i \approx \sum_{k=0}^{N-1} R(\lambda_k)L(\lambda_k)S_i(\lambda_k)\Delta\lambda = S_i^T f \tag{A.2}$$

where
$$f = [R(\lambda_0)L(\lambda_0)\Delta\lambda \quad R(\lambda_1)L(\lambda_1)\Delta\lambda \quad \cdots \quad R(\lambda_{N-1})L(\lambda_{N-1})\Delta\lambda]^{\mathrm{T}}$$
$$\Delta\lambda = \frac{\lambda_{max} - \lambda_{min}}{N - 1}$$
$$\lambda_k = \lambda_{min} + k\Delta\lambda$$
$$S_i = [S_i(\lambda_0) \quad S_i(\lambda_1) \quad \cdots \quad S_i(\lambda_{N-1})]^{\mathrm{T}}$$

Equation A.2 can be written in matrix form as

$$c = Sf \tag{A.3}$$

where
   $c = [c_1 \quad c_2 \quad c_3]^{\mathrm{T}}$ is the 3×1 tristimulus vector
   f is the $N \times 1$ column vector
   $S = [S_1 \quad S_2 \quad S_3]^{\mathrm{T}}$ is the $3 \times N$ matrix of cone sensitivity vectors known as
      the color matching function

If f and g are two different spectra, their corresponding tristimulus vectors are Sf and Sg. If $f = g$, then $Sf = Sg$ and the two identical spectra have same color appearance to the observer. It is important to point out that two different spectra $f \neq g$ may have $Sf = Sg$. This means that different spectra may have the same color appearance. This is known as metamerism and the two spectra f and g are called metamers. Metamerism is kind of color aliasing. This is due to a reduction of dimensionality from the N-dimensional spectral space to the 3-D tristimulus vector space. The tristimulus vector defined by Equation A.3 is observer dependent and since it is impossible to measure the color matching function S for different observers, we need to have one color matching function for a standard observer. The Commission Internationale de l'Eclairage (CIE) that is an international body in charge of color standardization has defined a standard set of color matching functions that are used by the color community [1]. The first one is CIE *RGB* color matching functions shown in Figure A.1 and the other one is CIE *XYZ* color matching functions shown in Figure A.2. The CIE *RGB* color matching functions are not realizable since for portion of the spectra from roughly 450–550 nm one component becomes negative.

   The CIE *XYZ* color matching functions are all positive and, therefore, physically realizable using optical filters. The tristimulus values are defined using the *XYZ* color matching functions and are given by

$$t = \alpha Af \tag{A.4}$$

where
   $t = [X \quad Y \quad Z]^{\mathrm{T}}$
   A is the $3 \times N$ color matching functions
   $f = [R(\lambda_0)L(\lambda_0)\Delta\lambda \quad R(\lambda_1)L(\lambda_1)\Delta\lambda \quad \cdots \quad R(\lambda_{N-1})L(\lambda_{N-1})\Delta\lambda]^{\mathrm{T}}$

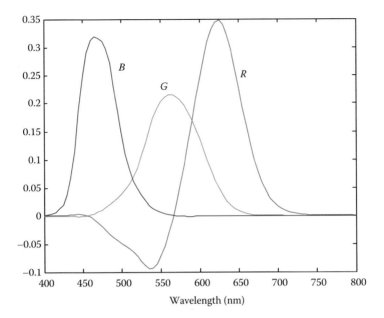

FIGURE A.1   CIE *RGB* color matching functions.

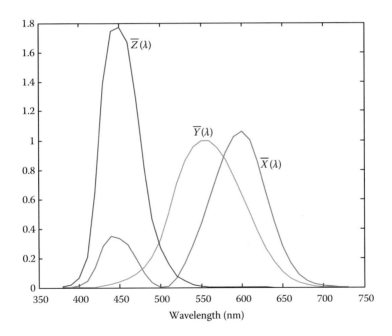

FIGURE A.2   CIE *XYZ* color matching functions.

The scale factor α is used to normalize the tristimulus values such that $Y = 100$ for perfect white and is given by

$$\alpha = \frac{100}{\sum_{k=0}^{N-1} \bar{Y}(\lambda_k)l(\lambda_k)} \tag{A.5}$$

Applying the scale factor of Equation A.5, Equation A.4 can be expanded as

$$X = 100 \frac{\sum_{k=0}^{N-1} \bar{X}(\lambda_k)R(\lambda_k)l(\lambda_k)}{\sum_{k=0}^{N-1} \bar{Y}(\lambda_k)l(\lambda_k)} \tag{A.6}$$

$$Y = 100 \frac{\sum_{k=0}^{N-1} \bar{Y}(\lambda_k)R(\lambda_k)l(\lambda_k)}{\sum_{k=0}^{N-1} \bar{Y}(\lambda_k)l(\lambda_k)} \tag{A.7}$$

$$Z = 100 \frac{\sum_{k=0}^{N-1} \bar{Z}(\lambda_k)R(\lambda_k)l(\lambda_k)}{\sum_{k=0}^{N-1} \bar{Y}(\lambda_k)l(\lambda_k)} \tag{A.8}$$

where

$\bar{X}(\lambda_k)$, $\bar{Y}(\lambda_k)$, and $\bar{Z}(\lambda_k)$ are the unscaled color matching functions
$R(\lambda_k)$ is the reflectance spectra
$l(\lambda_k)$ is the incident light source

Example A.1

Assuming a D50 light source, compute the tristimulus values of perfect white.

SOLUTION

Using the numerical data provided in Appendix B for the CIE XYZ and D50 light source and using $R(\lambda_k) = 1$ for perfect white, we have

$$X = 100 \frac{\sum_{k=0}^{N-1} \bar{X}(\lambda_k)D50(\lambda_k)}{\sum_{k=0}^{N-1} \bar{Y}(\lambda_k)D50(\lambda_k)} = 96.42$$

$$Y = 100 \frac{\sum_{k=0}^{N-1} \bar{Y}(\lambda_k)D50(\lambda_k)}{\sum_{k=0}^{N-1} \bar{Y}(\lambda_k)D50(\lambda_k)} = 100$$

$$Z = 100 \frac{\sum_{k=0}^{N-1} \bar{Z}(\lambda_k)D50(\lambda_k)}{\sum_{k=0}^{N-1} \bar{Y}(\lambda_k)D50(\lambda_k)} = 82.49$$

Example A.2

Assuming a D50 light source, compute the XYZ values of a color patch with a reflectance spectra of

$$R(\lambda) = -6.25 \times 10^{-6}\lambda^2 + 0.0063\lambda - 1.225 \quad 380 \le \lambda \le 730$$

SOLUTION

Using the numerical data provided in Appendix B for the CIE XYZ and D50 light source, we have

$$X = 100 \frac{\sum_{k=0}^{N-1} \bar{X}(\lambda_k)R(\lambda_k)D50(\lambda_k)}{\sum_{k=0}^{N-1} \bar{Y}(\lambda_k)D50(\lambda_k)} = 27.59$$

$$Y = 100 \frac{\sum_{k=0}^{N-1} \bar{Y}(\lambda_k)R(\lambda_k)D50(\lambda_k)}{\sum_{k=0}^{N-1} \bar{Y}(\lambda_k)D50(\lambda_k)} = 33.2$$

$$Z = 100 \frac{\sum_{k=0}^{N-1} \bar{Z}(\lambda_k)R(\lambda_k)D50(\lambda_k)}{\sum_{k=0}^{N-1} \bar{Y}(\lambda_k)D50(\lambda_k)} = 19.39$$

A.4 CHROMATICITY VALUES AND CHROMATICITY DIAGRAM

The tristimulus vector is a 3-D vector. Because it is useful for visualization, the XYZ tristimulus values are normalized to yield the $x\ y\ z$ chromaticity values, which are defined as

$$x = \frac{X}{X + Y + Z} \qquad (A.9)$$

$$y = \frac{Y}{X + Y + Z} \qquad (A.10)$$

$$z = \frac{Z}{X + Y + Z} \qquad (A.11)$$

Note that $x + y + z = 1$ and, as a result of this constraint, the dimensionality of the space is two. The chromaticity diagram is the loci of all realizable colors in x–y plane and is shown in Figure A.3. Each point on the boundary of the chromaticity diagram is obtained using constant illumination and a spectrum that is zero except for one single wavelength. Region inside the closed curve is the physically realizable chromaticites. Any output color device such as a digital printer has a gamut that is a subset of the chromaticity diagram.

A.5 COLOR SYSTEMS REPRODUCTION

There are two common color reproduction systems: additive and subtractive. In additive color system reproduction, colors are generated by combining light sources with different wavelengths. These different sources are called primaries. For example, red, green, and blue primaries are combined with different weight to produce colors seen on screen of a monitor or TV set. In subtractive color system reproduction, colors are produced by removing portions of spectra of the light source. For example, color reproduction in an output device such as a digital color printer is subtractive. The primaries are cyan, magenta, and yellow (CMY printers) or cyan, magenta, yellow, and black for CMYK printers, and these primaries remove portion of the spectra from paper white.

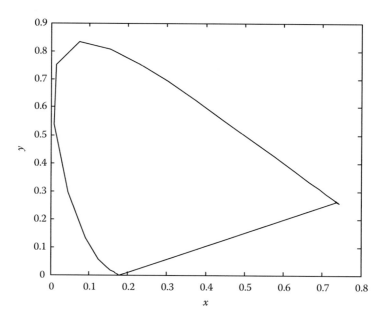

FIGURE A.3   Chromaticity diagram.

A.6   DEVICE-INDEPENDENT AND DEVICE-DEPENDENT COLOR SPACES

The CIE *XYZ* tristimulus color space is an example of a device-independent color space. Any nonsingular transformation from *XYZ* color space to another color space is a device-independent color space. If the color space is not related to CIE *XYZ* through a nonsingular transformation, then it is device-dependent color space. An example of a device-dependent color space is the *RGB* color space. Since *RGB* is device dependent, any given scanner *RGB* is different from any given monitor *RGB*. In the following sections, we discuss device-independent and device-dependent color spaces.

A.6.1   DEVICE-INDEPENDENT COLOR SPACES

The most common device-independent color spaces used in practice are the CIE *L*u*v**, CIE *L*a*b**, and *LCH* color spaces.

(a) CIE *L*u*v** Color Space
    The CIE *L*u*v** color space is obtained from *XYZ* using the following nonlinear transformations:

$$L^* = 116 f\left(\frac{Y}{Y_n}\right) - 16 \tag{A.12}$$

$$u^* = 13L^*\left(u' - u'_n\right) \quad \text{and} \quad v^* = 13L^*\left(v' - v'_n\right) \tag{A.13}$$

where

$$f(x) = \begin{cases} x^{\frac{1}{3}} & x \geq 0.008856 \\ 7.787x + \frac{16}{116} & x \leq 0.008856 \end{cases}$$

$$u' = \frac{4X}{X + 15Y + 3Z}$$

$$u'_n = \frac{4X_n}{X_n + 15Y_n + 3Z_n}$$

$$v' = \frac{9X}{X + 15Y + 3Z}$$

$$v'_n = \frac{9X_n}{X_n + 15Y_n + 3Z_n}$$

X_n, Y_n, and Z_n are the tristimulus values of reference white, which may be particular to the light source and paper white.

The numerical values of X_n, Y_n, and Z_n for a D50 light source are $X_n = 96.42$, $Y_n = 100$, and $Z_n = 82.49$.

(b) CIE $L*a*b*$ Color Space

The CIE $L*a*b*$ color space is obtained from XYZ using the following nonlinear transformations:

$$L* = 116f\left(\frac{Y}{Y_n}\right) - 16 \tag{A.14}$$

$$a* = 500\left[f\left(\frac{X}{X_n}\right) - f\left(\frac{Y}{Y_n}\right)\right] \tag{A.15}$$

$$b* = 200\left[f\left(\frac{Y}{Y_n}\right) - f\left(\frac{Z}{Z_n}\right)\right] \tag{A.16}$$

where

$$f(x) = \begin{cases} x^{\frac{1}{3}} & x \geq 0.008856 \\ 7.787x + \frac{16}{116} & x \leq 0.008856 \end{cases}$$

X_n, Y_n, and Z_n are tristimulus values of reference white

(c) CIE LCH Color Space

The LCH (brightness, chroma, and hue angle) color space is obtained from $L*a*b*$ by using the following equations:

$$L = L*, \quad C = \sqrt{a*^2 + b*^2}, \quad \text{and} \quad H = \tan^{-1}\left(\frac{b*}{a*}\right) \tag{A.17}$$

Note that the hue angle is in units of degrees between $0°$ and $360°$. The MATLAB function $atan2(b^*, a^*)$ can be used to compute the hue angle.

Example A.3

Assuming a D50 light source, compute the $L^*a^*b^*$ values of a color patch with a reflectance spectra of

$$R(\lambda) = -6.25 \times 10^{-6}\lambda^2 + 0.0063\lambda - 1.225 \quad 380 \le \lambda \le 730$$

SOLUTION

For a D50 light source, the CIE XYZ values (Example A.2) are

$$X = 27.59, \quad Y = 33.2, \quad \text{and} \quad Z = 19.39$$

The $L^*a^*b^*$ values are computed using Equations A.12 through A.14:

$$L^* = 116f\left(\frac{Y}{Y_n}\right) - 16 = 116f\left(\frac{33.2}{100}\right) - 16 = 116 \times \left(\frac{33.2}{100}\right)^{\frac{1}{3}} - 16 = 53.24$$

$$a^* = 500\left[f\left(\frac{X}{X_n}\right) - f\left(\frac{Y}{Y_n}\right)\right] = 500\left[\left(\frac{27.59}{96.42}\right)^{\frac{1}{3}} - \left(\frac{33.2}{100}\right)^{\frac{1}{3}}\right] = -16.73$$

$$b^* = 200\left[f\left(\frac{Y}{Y_n}\right) - f\left(\frac{Z}{Z_n}\right)\right] = 200\left[\left(\frac{33.2}{100}\right)^{\frac{1}{3}} - \left(\frac{19.39}{82.49}\right)^{\frac{1}{3}}\right] = 15.05$$

The above transformation has a unique inverse that maps $L^*a^*b^*$ back to the XYZ color space. The inverse is given by the following equations:
First compute $f_y = \frac{L^*+16}{116}$, $f_x = \frac{a^*}{500} + f_y$, and $f_z = f_y - \frac{b^*}{200}$. Then find X, Y, and Z using the following equations:

$$X = X_n f^{-1}(f_x), \quad Y = Y_n f^{-1}(f_y), \quad \text{and} \quad Z = Z_n f^{-1}(f_z) \quad \text{(A.18)}$$

where

$$f^{-1}(u) = \begin{cases} u^3 & u \ge 0.206893 \\ \dfrac{1}{7.787}\left(u - \dfrac{16}{116}\right) & u \le 0.206893 \end{cases}$$

(d) $L^*a^*b^*$ Relative to Paper

The above computation assumes that the $L^*a^*b^*$ of paper white is $L^*a^*b^* = [100\ 0\ 0]$. This means that we are computing $L^*a^*b^*$ with respect to an ideal paper white. If the $L^*a^*b^*$ of paper is different from the ideal case, we can still compute the $L^*a^*b^*$ with respect to paper. The transformation from absolute $L^*a^*b^*$ to relative to paper $L^*a^*b^*$ are given as follows:

Step 1: Convert the paper $L^*a^*b^*$ ($L^*a^*b^*_{paper}$ obtained without toner/ink) to paper XYZ using Equation A.18.

Step 2: Convert the absolute $L^*a^*b^*$ of the color patch (with toner ink) to XYZ.

Step 3: Using the paper XYZ values as X_n, Y_n, and Z_n, the $L^*a^*b^*$ relative to paper is computed using Equations A.14 through A.16.

Example A.4

Find the $L^*a^*b^*$ with respect to paper of a color patch with absolute $L^*a^*b^* = [53 \quad -16 \quad 15]$. Assume that the paper $L^*a^*b^* = [90 \quad 0 \quad 0]$.

SOLUTION

Using Equation A.18, the tristimulus X_n, Y_n, and Z_n of the reference paper white with $L^*a^*b^*_{paper} = [90 \quad 0 \quad 0]$ are $[X_n \quad Y_n \quad Z_n] = [73.56 \quad 76.30 \quad 62.94]$. The XYZ corresponding to $L^*a^*b^* = [53 \quad -16 \quad 15]$ are $[X \quad Y \quad Z] = [17.19 \quad 21.05 \quad 11.60]$. The $L^*a^*b^*$ with respect to paper are computed using Equations A.14 through A.16.

$$L^* = 116f\left(\frac{Y}{Y_n}\right) - 16 = 116f\left(\frac{21.05}{76.30}\right) - 16 = 116 \times \left(\frac{21.05}{76.30}\right)^{\frac{1}{3}} - 16 = 59.51$$

$$a^* = 500\left[f\left(\frac{X}{X_n}\right) - f\left(\frac{Y}{Y_n}\right)\right] = 500\left[\left(\frac{17.19}{73.56}\right)^{\frac{1}{3}} - \left(\frac{21.05}{76.30}\right)^{\frac{1}{3}}\right] = -17.50$$

$$b^* = 200\left[f\left(\frac{Y}{Y_n}\right) - f\left(\frac{Z}{Z_n}\right)\right] = 200\left[\left(\frac{21.05}{76.30}\right)^{\frac{1}{3}} - \left(\frac{11.60}{62.94}\right)^{\frac{1}{3}}\right] = 16.41$$

A.6.2 DEVICE-DEPENDENT COLOR SPACES

Of the many different device-dependent color spaces, the most common one is RGB. Since RGB is device-dependent, each device has a different RGB. For example, Apple RGB, Photoshop RGB, Xerox RGB, etc. are different types of RGB color spaces. We now discuss a few important RGB color spaces.

(a) Reference Output Medium Metric RGB ($ROMMRGB$)

The $ROMMRGB$ color space is used for representing rendered output images in a device-independent fashion. To convert from $L^*a^*b^*$ color space to $ROMMRGB$ color space, the following computational steps are taken [2]:

Step 1: Convert $L^*a^*b^*$ to XYZ using Equation A.18.

Step 2: Normalize XYZ:

$$XYZ_{nor} = \frac{XYZ}{100} \tag{A.19}$$

Step 3: Convert *XYZ* to linear *ROMMRGB* using the transformation given as follows:

$$
\begin{bmatrix} R'_{\text{ROMM}} \\ G'_{\text{ROMM}} \\ B'_{\text{ROMM}} \end{bmatrix} = \begin{bmatrix} 1.3460 & -0.2556 & -0.0511 \\ -0.5446 & 1.5082 & 0.0205 \\ 0.0000 & 0.0000 & 1.2123 \end{bmatrix} \begin{bmatrix} X_{\text{nor}} \\ Y_{\text{nor}} \\ Z_{\text{nor}} \end{bmatrix} \tag{A.20}
$$

Step 4: Convert linear *ROMMRGB* to nonlinear *ROMMRGB* using the following 1-D nonlinearity:

$$
R_{\text{ROMM}} = f(R'_{\text{ROMM}}), \quad G_{\text{ROMM}} = f(G'_{\text{ROMM}}), \quad \text{and}
$$
$$
B_{\text{ROMM}} = f(B'_{\text{ROMM}}) \tag{A.21}
$$

The function $f(x)$ is given by

$$
f(x) = \begin{cases} 0 & x < 0 \\ 16 I_{\text{max}} x & 0 \le x < E_{\text{t}} \\ I_{\text{max}} x^{\frac{1}{1.8}} & E_{\text{t}} \le x < 1 \\ I_{\text{max}} & x \ge I_{\text{max}} \end{cases} \tag{A.22}
$$

where $E_{\text{t}} = 0.001953$ and I_{max} is the integer corresponding to the maximum number of bits used to represent *ROMMRGB*. For example, for an 8-bit representation, $I_{\text{max}} = 2^8 = 256$. It is also possible to convert *ROMMRGB* to *XYZ*. The first step is to convert *ROMMRGB* to linear *ROMMRGB* by using the following transformation:

$$
R'_{\text{ROMM}} = g(R_{\text{ROMM}}), \quad G'_{\text{ROMM}} = g(G_{\text{ROMM}}), \quad \text{and}
$$
$$
B'_{\text{ROMM}} = g(B_{\text{ROMM}}) \tag{A.23}
$$

where $g(u) = f^{-1}(u)$ and is given by

$$
g(u) = \begin{cases} \frac{u}{16 I_{\text{max}}} & 0 \le u < 16 E_{\text{t}} I_{\text{max}} \\ \left(\frac{u}{I_{\text{max}}}\right)^{1.8} & 16 E_{\text{t}} I_{\text{max}} \le u < I_{\text{max}} \end{cases} \tag{A.24}
$$

Once the linear *ROMMRGB* are computed, the following transformation is used to convert linear *ROMMRGB* to *XYZ*:

$$
\begin{bmatrix} X \\ Y \\ Z \end{bmatrix} = \begin{bmatrix} 1.3460 & -0.2556 & -0.0511 \\ -0.5446 & 1.5082 & 0.0205 \\ 0.0000 & 0.0000 & 1.2123 \end{bmatrix}^{-1} \begin{bmatrix} R'_{\text{ROMM}} \\ G'_{\text{ROMM}} \\ B'_{\text{ROMM}} \end{bmatrix} \tag{A.25}
$$

Example A.5

Convert the following $L^*a^*b^*$ values to ROMMRGB:

$$L^*a^*b_1^* = [60 \quad 0 \quad 0] \quad \text{and} \quad L^*a^*b_2^* = [40 \quad -16 \quad 50]$$

SOLUTION

Using the transformations given by Equations A.19 through A.24, we have

$$ROMMRGB_1 = [125 \quad 126 \quad 126] \quad \text{and}$$
$$ROMMRGB_2 = [66.74 \quad 79.17 \quad 22.46]$$

(b) Video RGB (sRGB)

The video RGB is used to display an image on display devices such as CRT. Therefore, we need to convert ROMMRGB to sRGB for visualization. The following algorithm is used to convert ROMMRGB to sRGB:

Step 1: Convert ROMMRGB to linear ROMMRGB using Equation A.23.

Step 2: Apply the linear transformation T to linear ROMMRGB to obtain linear sRGB:

$$\begin{bmatrix} R'_s \\ G'_s \\ B'_s \end{bmatrix} = T \begin{bmatrix} R'_{ROMM} \\ G'_{ROMM} \\ B'_{ROMM} \end{bmatrix} = \begin{bmatrix} 2.0564 & 0.7932 & 0.2632 \\ 0.2118 & 1.249 & 0.0372 \\ 0.0152 & 0.1405 & 1.1556 \end{bmatrix} \begin{bmatrix} R'_{ROMM} \\ G'_{ROMM} \\ B'_{ROMM} \end{bmatrix}$$

(A.26)

Step 3: Apply the nonlinear transformation $h(.)$ to each component of linear sRGB:

$$R_s = h(R'_s), \quad G_s = h(G'_s), \quad \text{and} \quad B_s = h(B'_s)$$ (A.27)

where

$$h(u) = \begin{cases} 255(12.92u) & u \le 0.0031308 \\ 255\left(1.055u^{\frac{1}{2.4}} + 0.055\right) & u > 0.0031308 \end{cases}$$ (A.28)

Other common RGB color spaces are Adobe RGB, Apple RGB, NTSC RGB, and ProPhoto RGB. The reference white and the XYZ tristimulus values of the red, green, and blue primaries of different RGB color systems are shown in Table A.1. Their corresponding chromaticity coordinates are shown in Table A.2.

TABLE A.1

XYZ Tristimulus Values of the Red, Green, and Blue Primaries

Name	Reference White	Red Primary			Green Primary			Blue Primary		
		X	Y	Z	X	Y	Z	X	Y	Z
Adobe *RGB*	D65	57.67	29.73	2.70	18.55	62.73	7.06	26.34	7.52	99
Apple *RGB*	D65	44.97	24.46	2.51	31.62	67.20	14.11	33.32	8.33	92
NTSC *RGB*	C	60.67	29.88	0	17.35	58.68	6.61	30.01	11.43	111
ProPhoto *RGB*	D50	79.76	28.80	0	13.52	71.18	0	13.72	0.00	82.84
sRGB	D65	41.24	21.26	1.93	35.76	71.51	11.91	36.09	7.21	95.04
ROMMRGB	D50	71.63	25.81	0	10.09	72.49	5.17	10.78	1.68	77.34

TABLE A.2

Chromaticity Coordinates of the Red, Green, and Blue Primaries

Name	Reference White	Red Primary		Green Primary		Blue Primary	
		x	y	x	y	x	y
Adobe *RGB*	D65	0.648431	0.330856	0.230154	0.701572	0.155886	0.066044
Apple *RGB*	D65	0.634756	0.340596	0.301775	0.597511	0.162897	0.079001
NTSC *RGB*	C	0.671910	0.329340	0.22591	0.710647	0.142783	0.096145
ProPhoto *RGB*	D50	0.734700	0.265300	0.159600	0.840400	0.036600	0.000100
sRGB	D65	0.648431	0.330856	0.321152	0.597871	0.155886	0.066044
ROMMRGB	D50	0.735000	0.265000	0.115000	0.826000	0.157000	0.018000

A.7 COLOR DIFFERENCE FORMULA

A distance metric is used to measure color difference between two colors in device-independent color space. The Euclidian CIE color difference is denoted by ΔE_{ab}^* and is given by

$$\Delta E_{ab}^* = \sqrt{\left(L_2^* - L_1^*\right)^2 + \left(a_2^* - a_1^*\right)^2 + \left(b_2^* - b_1^*\right)^2} \qquad (A.29)$$

Figures A.4 through A.6 show color patches near the gray axis with ΔE_{ab}^* color difference of 1, 2, and 5, respectively.

Figures A.7 through A.9 show color patches near the gray axis with ΔE_{ab}^* color difference of 1, 2, and 5, respectively.

Color values and respective color difference numbers in ΔE_{2000} and ΔE_{ab}^* space with respect to the reference patch is summarized in Table A.3.

The Euclidian CIE color difference weights the brightness, hue, and chroma components equally. As a result, the Euclidian CIE color difference will be different from the perceived color difference for a certain part of device color gamut. The new CIE color difference known as ΔE_{2000} is a better measure of the color difference and is more consistent with the perceived color difference. It is a major revision to the

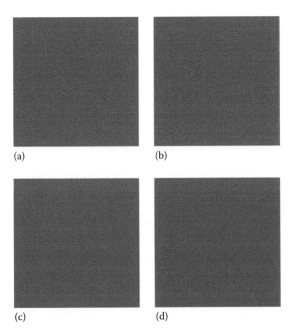

(a)   (b)

(c)   (d)

FIGURE A.4 (See color insert following page 428.)   Patches near gray with ΔE_{ab}^* color difference of 1 (a is the reference patch).

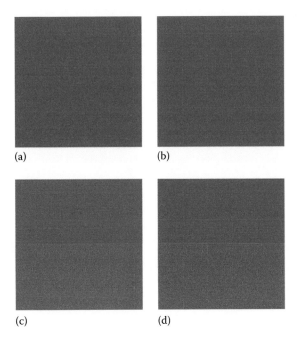

(a)   (b)

(c)   (d)

FIGURE A.5 (See color insert following page 428.)   Patches near gray with ΔE_{ab}^* color difference of 2 (a is the reference patch).

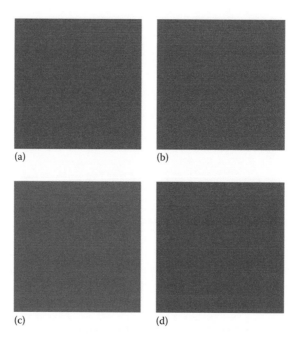

(a)   (b)

(c)   (d)

FIGURE A.6 (See color insert following page 428.)   Patches near gray with ΔE_{ab}^* color difference of 5 (a is the reference patch).

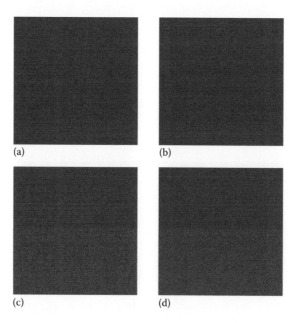

(a)   (b)

(c)   (d)

FIGURE A.7 (See color insert following page 428.)   Patches off gray with ΔE_{ab}^* color difference of 1 (a is the reference patch).

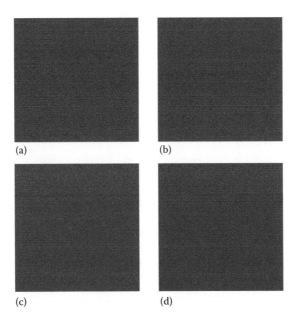

FIGURE A.8 (See color insert following page 428.) Patches off gray with ΔE_{ab}^* color difference of 2 (a is the reference patch).

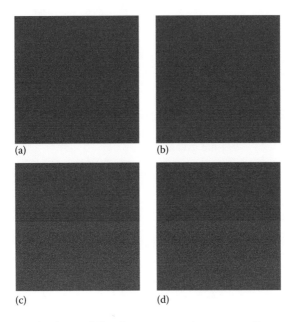

FIGURE A.9 (See color insert following page 428.) Patches off gray with ΔE_{ab}^* color difference of 5 (a is the reference patch).

TABLE A.3

Color Patches with Respective Color Differences Used in Figures A.4 through A.9

Figures	Patches	$L*$	$a*$	$b*$	ΔE_{2000}	ΔE_{ab}^*
A.4	a	60.53	0.00	0.00	0.00	1
	b	60.53	1.00	0.00	1.45	1
	c	61.53	0.00	0.00	0.87	1
	d	60.53	0.00	1.00	0.98	1
A.5	a	60.53	0.00	0.00	0.00	2
	b	60.53	2.00	0.00	2.81	2
	c	62.53	0.00	0.00	1.72	2
	d	60.53	0.00	2.00	1.91	2
A.6	a	60.53	0.00	0.00	0.00	5
	b	60.53	5.00	0.00	6.42	5
	c	65.53	0.00	0.00	4.22	5
	d	60.53	0.00	5.00	4.49	5
A.7	a	40.38	102.43	69.63	0.00	1
	b	40.38	103.43	69.63	0.26	1
	c	41.38	102.43	69.63	0.89	1
	d	40.38	102.43	70.63	0.35	1
A.8	a	40.38	102.43	69.63	0.00	2
	b	40.38	104.43	69.63	0.52	2
	c	42.38	102.43	69.63	1.79	2
	d	40.38	102.43	71.63	0.70	2
A.9	a	40.38	102.43	69.63	0.00	5
	b	40.38	107.43	69.63	1.29	5
	c	45.38	102.43	69.63	4.59	5
	d	40.38	102.43	74.63	1.75	5

CIE94 color difference formula. The steps to calculate the ΔE_{2000} color difference between two color patches with $L*a*b*$ values of $(L*a*b*)_1 = \lfloor L_1^* \quad a_1^* \quad b_1^* \rfloor$ and $(L*a*b*)_2 = \lfloor L_2^* \quad a_2^* \quad b_2^* \rfloor$ are given below:

Step 1: Find $\begin{bmatrix} L_1' & a_1' & b_1' \end{bmatrix}$ and $\begin{bmatrix} L_2' & a_2' & b_2' \end{bmatrix}$

$$C_1^* = \sqrt{(a_1^*)^2 + (b_1^*)^2}, \quad C_2^* = \sqrt{(a_2^*)^2 + (b_2^*)^2}, \quad C^* = \frac{C_1^* + C_2^*}{2}, \quad \text{and}$$

$$G = 0.5 - 0.5\sqrt{\frac{(C^*)^7}{(C^*)^7 + 25^7}}$$

$$\begin{bmatrix} L_1' \\ a_1' \\ b_1' \end{bmatrix} = \begin{bmatrix} L_1^* \\ (1+G)a_1^* \\ b_1^* \end{bmatrix} \quad \text{and} \quad \begin{bmatrix} L_2' \\ a_2' \\ b_2' \end{bmatrix} = \begin{bmatrix} L_2^* \\ (1+G)a_2^* \\ b_2^* \end{bmatrix} \tag{A.30}$$

Step 2: Compute

$$C_1'^* = \sqrt{(a_1')^2 + (b_1')^2}, \quad C_2'^* = \sqrt{(a_2')^2 + (b_2')^2}, \quad \text{and} \quad C' = \frac{C_1'^* + C_2'^*}{2} \quad \text{(A.31)}$$

$$h_1' = \tan^{-1}\left(\frac{b_1'}{a_1'}\right), \quad h_2' = \tan^{-1}\left(\frac{b_2'}{a_2'}\right), \quad \text{and} \quad h' = \frac{h_1' + h_2'}{2} \quad \text{(A.32)}$$

Step 3: Find T

$$T = 1 - 0.17\cos(h'' - 30) + 0.24\cos(2h'') + 0.32\cos(3h'' + 6) - 0.2\cos(4h'' - 63) \quad \text{(A.33)}$$

where

$$h'' = \begin{cases} h' & h' \le 180 \\ h' - 180 & h' > 180 \end{cases}$$

Step 4: Compute S_L, S_C, and S_H

$$L_{av} = 0.5(L_1' + L_2'), \quad S_L = 1 + \frac{0.015(L_{av} - 50)^2}{\sqrt{20 + (L_{av} - 50)^2}}, \quad \text{(A.34)}$$

$$S_C = 1 + 0.045C', \quad \text{and} \quad S_H = 1 + 0.015C'T \quad \text{(A.35)}$$

Step 5: Compute $\Delta L'$, $\Delta C'$, and $\Delta H'$

$$\Delta L' = L_2' - L_1', \quad \Delta C' = C_2'^* - C_1'^*, \quad \text{and} \quad \Delta H' = 2\sqrt{C_2'^* C_1'^*}\,\sin\frac{h_2' - h_1'}{2} \quad \text{(A.36)}$$

Step 6: Compute R_C and R_T

$$R_C = 2\sqrt{\frac{C'^7}{C'^7 + 25^7}}, \quad \Delta\theta = 30\exp\left[-\left(\frac{h'' - 275}{25}\right)^2\right], \quad \text{and}$$

$$R_T = -R_C \sin 2\Delta\theta \quad \text{(A.37)}$$

Step 7: Select the weights k_L, k_C, and k_H and compute ΔE_{2000}

$$\Delta E_{2000} = \sqrt{\left(\frac{\Delta L'}{k_L S_L}\right)^2 + \left(\frac{\Delta C'}{k_C S_C}\right)^2 + \left(\frac{\Delta H'}{k_H S_H}\right)^2 + R_T\frac{\Delta C'}{k_C S_C}\frac{\Delta H'}{k_H S_H}} \quad \text{(A.38)}$$

FIGURE A.10 (See color insert following page 428.) Two color patches of Example A.5.

Example A.6

Find ΔE_{ab}^* and ΔE_{2000} color difference between the following two color patches $L^*a^*b^*_2 = [65 \ -120 \ -70]$ and $L^*a^*b^*_1 = [68 \ -117 \ -68]$. These color patches are shown in Figure A.10

SOLUTION

$$\Delta E_{ab}^* = \sqrt{\left(L_2^* - L_1^*\right)^2 + \left(a_2^* - a_1^*\right)^2 + \left(b_2^* - b_1^*\right)^2}$$

$$= \sqrt{(65 - 68)^2 + (-117 + 120)^2 + (-68 + 70)^2} = 4.69$$

ΔE_{2000} is computed by coding the above equations into MATLAB. The resulting color difference for $k_L = k_C = k_H = 1$ is $\Delta E_{2000} = 2.47$.

REFERENCES

1. International Color Consortium, File Format for Color Profiles, Specification ICC.1A:1999–04.
2. Kelvin E.S., Geoffrey J.W., and Edward J.G., Reference input/output medium metric RGB color encodings (RIM/ROMM RGB), *PICS 2000 Conference*, Portland, OR, March 26–29, 2000.

Appendix B

TABLE B.1
CIE *XYZ* Color Matching Functions

Wavelength (nm)	$\bar{x}$	$\bar{y}$	$\bar{z}$
380	0.0041	0	0.0065
390	0.0042	0.0001	0.0201
400	0.0143	0.0004	0.0679
410	0.0435	0.0012	0.2074
420	0.1344	0.004	0.6456
430	0.2839	0.0116	1.3856
440	0.3483	0.023	1.7471
450	0.3362	0.038	1.7721
460	0.2908	0.06	1.6692
470	0.1954	0.091	1.2876
480	0.0956	0.139	0.813
490	0.032	0.208	0.4652
500	0.0049	0.323	0.272
510	0.0093	0.503	0.1582
520	0.0633	0.71	0.0782
530	0.1655	0.862	0.0422
540	0.2904	0.954	0.0203
550	0.4334	0.995	0.0087
560	0.5945	0.995	0.0039
570	0.7621	0.952	0.0021
580	0.9163	0.87	0.0017
590	1.0263	0.757	0.0011
600	1.0622	0.631	0.0008
610	1.0026	0.503	0.0003
620	0.8544	0.381	0.0002
630	0.6424	0.265	0.0001
640	0.4479	0.175	0
650	0.2835	0.107	0
660	0.1649	0.061	0
670	0.0874	0.032	0
680	0.0468	0.017	0
690	0.0227	0.0082	0
700	0.0114	0.0041	0
710	0.0058	0.0021	0
720	0.0029	0.001	0
730	0.0014	0.0005	0

TABLE B.2
D50 and D65 Light Sources

Wavelength (nm)	*D*50	*D*65
380	24.49	49.98
390	29.87	54.65
400	49.31	82.75
410	56.51	91.49
420	60.03	93.43
430	57.82	86.68
440	74.82	104.86
450	87.25	117.01
460	90.61	117.81
470	91.37	114.86
480	95.11	115.92
490	91.96	108.81
500	95.72	109.35
510	96.61	107.8
520	97.13	104.79
530	102.1	107.69
540	100.75	104.41
550	102.32	104.05
560	100	100
570	97.74	96.33
580	98.92	95.79
590	93.5	88.69
600	97.69	90.01
610	99.27	89.6
620	99.04	87.7
630	95.72	83.29
640	98.86	83.49
650	95.67	80.03
660	98.19	80.21
670	103	82.28
680	99.13	78.28
690	87.38	69.72
700	91.6	71.61
710	92.89	74.35
720	76.85	61.6
730	86.51	69.89

TABLE B.3
CIE RGB Color Matching Functions

Wavelength (nm)	$\bar{r}$	$\bar{g}$	$\bar{b}$
380	0.0000	0.0000	0.0012
390	0.0001	0.0000	0.0036
400	0.0003	−0.0001	0.0121
410	0.0008	−0.0004	0.0371
420	0.0021	−0.0011	0.1154
430	0.0022	−0.0012	0.2477
440	−0.0026	0.0015	0.3123
450	−0.0121	0.0068	0.3167
460	−0.0261	0.0149	0.2982
470	−0.0393	0.0254	0.2299
480	−0.0494	0.0391	0.1449
490	−0.0581	0.0569	0.0826
500	−0.0717	0.0854	0.0478
510	−0.0890	0.1286	0.0270
520	−0.0926	0.1747	0.0122
530	−0.0710	0.2032	0.0055
540	−0.0315	0.2147	0.0015
550	0.0228	0.2118	−0.0006
560	0.0906	0.1970	−0.0013
570	0.1677	0.1709	−0.0014
580	0.2453	0.1361	−0.0011
590	0.3093	0.0975	−0.0008
600	0.3443	0.0625	−0.0005
610	0.3397	0.0356	−0.0003
620.	0.2971	0.0183	−0.0001
630	0.2268	0.0083	−0.0001
640	0.1597	0.0033	0.0000
650	0.1017	0.0012	0.0000
660	0.0593	0.0004	0.0000
670	0.0315	0.0001	0.0000
680	0.0169	0.0000	0.0000
690	0.0082	0.0000	0.0000
700	0.0041	0.0000	0.0000
710	0.0021	0.0000	0.0000
720	0.0010	0.0000	0.0000
730	0.0005	0.0000	0.0000

Appendix C

This appendix contains definitions for some useful tensor operations.

1. *n-Mode multiplication*

   Let A be a tensor of order N of size $I_1 \times I_2 \times \cdots \times I_N$ and U a matrix of size $J_n \times I_n$. Their product using n-mode multiplication $A \otimes_n U$ has size $I_1 \times I_2 \times \cdots \times I_{n-1} \times J_n \times I_{n+1} \times \cdots \times I_N$ and is defined by

$$(A \otimes_n U)_{i_1,\ldots,i_{n-1},j_n,i_{n+1},\ldots,i_N} = \sum_{i_n=1}^{I_n} A_{i_1,i_2,\ldots,i_N} U_{j_n i_n} \tag{C.1}$$

   The following identities hold for n-mode multiplication

   a) $A \otimes_n (UV) = A \otimes_n V \otimes_n U$

   b) $B = A \otimes_n U \Rightarrow A = B \otimes_n Z, \quad \text{where } ZU = 1 \tag{C.2}$

   c) $A \otimes_n U \otimes_m V = A \otimes_m V \otimes_n U$

2. *Tensor multiplication*

   If tensors A and B have the same sizes, then their tensor product scalar and is defined as

$$\langle A,B \rangle = \sum_{i_1=1}^{I_1} \sum_{i_2}^{I_2} \cdots \sum_{i_N}^{I_N} A_{i_1,i_2,\ldots,i_N} B_{i_1,i_2,\ldots,i_N} \tag{C.3}$$

Index